Selective Catalytic Reduction of NO$_x$

Selective Catalytic Reduction of NO$_x$

Special Issue Editor

Oliver Kröcher

MDPI • Basel • Beijing • Wuhan • Barcelona • Belgrade

MDPI

Special Issue Editor
Oliver Kröcher
Paul Scherrer Institut
Switzerland

Editorial Office
MDPI
St. Alban-Anlage 66
4052 Basel, Switzerland

This is a reprint of articles from the Special Issue published online in the open access journal *Catalysts* (ISSN 2073-4344) from 2017 to 2018 (available at: https://www.mdpi.com/journal/catalysts/special_ issues/selective_catalytic_reduction)

For citation purposes, cite each article independently as indicated on the article page online and as indicated below:

LastName, A.A.; LastName, B.B.; LastName, C.C. Article Title. *Journal Name* **Year**, *Article Number*, Page Range.

ISBN 978-3-03897-364-5 (Pbk)
ISBN 978-3-03897-365-2 (PDF)

Cover image courtesy of Oliver Kröcher.

Contents

About the Special Issue Editor

Oliver Kröcher earned a diploma in chemistry from the University of Wurzburg in 1993 and a PhD in the field of heterogeneous catalysis from ETH Zurich in 1997. After that, he worked for a couple of years for Degussa in Germany on catalytic processes for chemical production in different positions, before returning to academia in 2001. He took over the Exhaust Gas Aftertreatment Group at the Paul Scherrer Institute, working on catalysis for cleaning exhaust gases. In 2010 he became the head of the Bioenergy and Catalysis Laboratory with more than 40 coworkers, in which he integrated his own group with a broadened scope on catalysis for energy conversion. Since 2013, he has been teaching at the post-graduate level at École Polytechnique Fédérale de Lausanne (EPFL) as an adjunct professor. He runs another small research group at EPFL on catalysis for biofuels. In 2014 he became the head of the Swiss Competence Centre for Energy Research in the field of biomass (SCCER BIOSWEET), comprising 13 research groups. Besides his more recent research activities in catalysis for bioenergy and biofuels, his main research interest since 2001 has been exhaust gas catalysis, working on the selective catalytic reduction for stationary and mobile applications in almost all facets, including diesel oxidation catalysts, soot oxidation, SOx removal, three-way-catalysis, and methane abatement. In his career, Oliver Kröcher has published more than 100 peer-reviewed publications and submitted eight patent applications, resulting in the transfer of some of the technologies in industrial applications in the field of exhaust gas catalysis. He has supervised 20 PhD students in addition to numerous Post-Doctorate, Master, and Bachelor students in their projects. Oliver Kröcher was the recipient of the Swiss Technology Award in 2005 and received the Special Prize of the ABB Schweiz AG for Saving Resources in 2005. He is member of the steering committee of the European Energy Research Alliance (EERA) Bioenergy; a member of the board of directors of the Hydromethan AG, a spin-off company of the Paul Scherrer Institute; a member of the scientific advisory board of the International Congress on Catalysis and Automotive Pollution Control (CAPoC); and a member of the energy commission of the Swiss Academies of Arts and Sciences (a+).

Preface to "Selective Catalytic Reduction of NO$_x$"

The recent diesel scandal made the public aware of the fact that NOx emissions from combustion processes are a major threat to human health and by no means easy to avoid. The most efficient process to reduce NOx emissions from lean exhaust gases is selective catalytic reduction (SCR) with ammonia, which has undergone tremendous development over the past decades. Originally only applied in stationary power plants and industrial installations, SCR systems are now installed in millions of mobile diesel engines, ranging from off-road machineries, to heavy-duty and light-duty trucks and passenger cars, to locomotives and ships. All of these applications involve specific challenges due to tighter emission limits, new internal combustion engine technologies, or alternative fuels.

The review articles and original research papers in this edited book contribute to the solution of these challenges with a broad range of innovative ideas, covering many aspects of SCR technology. Two papers deal with the proper dosage of the reducing agents, ammonia and urea, which is a pre-requisite for high NOx reduction efficiencies, in addition to the selected type of SCR catalyst. Vanadia-based SCR has been established for stationary applications for decades, and a series of contributions describe ways to improve its activity, selectivity to nitrogen, or resistance to biomass-related poisons, comprising the addition of heteropolyacids, iron oxide, magnesia, sepiolite, manganesia-ceria, or the application of other methods. In another contribution, post-treatment with SCR-active components is shown to be suitable to regenerate deactivated vanadia-based catalysts. When iron vanadate was used as a vanadium component, the obtained catalyst was not only active for the SCR reaction but also for the oxidation of soot, opening doors for the design of combined exhaust gas cleaning systems. In the last few years, alternative mixed metal oxides have started to emerge as SCR catalysts. In this book, a number of contributions show the potential of niobia, ceria, titania, manganesia, and iron oxide for the further development and optimization of SCR catalysts, particularly for low-temperature applications. A problem for many of these catalysts is limited sulfur or water resistance, which is particularly severe for manganese-based catalysts. The state of development of these catalysts is reviewed in one contribution, which also offers potential solutions to overcome or at least mitigate their sulfur and water sensitivity. When material costs are a special issue, natural minerals are also of interest for the development of active SCR catalysts, as shown by the contributions using Bayer red mud and vermiculite as base materials.

In recent years, copper-zeolite based catalysts have been developed for mobile applications with the most demanding requirements concerning low-temperature and high-temperature activity and long-term stability. However, this class of catalysts also suffers from poisoning, hydrothermal deactivation, and other issues, which require further studies. According to this fact, a deeper understanding of this catalyst type is required, as addressed in two contributions with a focus on the active sites in copper-zeolites and their characterization. Another paper addresses the deactivation of copper-zeolites by sodium from biofuels.

Despite the success of the established SCR process with ammonia or urea, research is ongoing to develop alternatives for future applications using other reducing agents for NOx. Two papers lead in this direction, addressing the mechanistic aspects of hydrogen-assisted SCR and hydrocarbon-SCR with plasma activation. The diversity of the contributions in this book reflects the plethora of research questions in the SCR process, whose development potential is still not fully exploited. Considering all the important work on SCR done in the world, which is partly reported in this book, it is predicted that SCR will remain the most important processes employed to clean lean exhaust gases from NOx emissions in the future.

I would like to express my sincerest gratitude to all authors for the valuable contributions, without which this book would not have been possible.

Oliver Kröcher
Special Issue Editor

Editorial

Selective Catalytic Reduction of NO$_x$

Oliver Kröcher

Bioenergy and Catalysis Laboratory, Paul Scherrer Institut (PSI), 5232 Villigen, Switzerland;
oliver.kroecher@psi.ch

Received: 12 October 2018; Accepted: 15 October 2018; Published: 17 October 2018

The recent diesel scandal made the public aware of the fact that NO$_x$ emissions from combustion processes are a major threat to human health and by no means easy to avoid. The most efficient process to reduce NO$_x$ emissions from lean exhaust gases is selective catalytic reduction (SCR) with ammonia, which has undergone tremendous development over the past few decades. Originally only applied in stationary power plants and industrial installations, SCR systems are now installed also in millions of mobile diesel engines, ranging from off-road machineries, heavy-duty and light-duty trucks, and passenger cars, to locomotives and ships. All these applications involve specific challenges due to tighter emission limits, new internal combustion engine technologies, or alternative fuels. The review articles and original research papers in this edited book contribute to the solution of these challenges with a broad range of innovative ideas, covering many aspects of SCR technology.

One of the bottlenecks of the application of SCR technology in vehicles, and particularly passenger cars, is the threshold temperature of about 200 °C for injection of the urea solution in hot exhaust gas, which is above the exhaust gas temperature for some modes of engine operation. Sala, Bielaczyc, and Brzezanski have addressed this problem by preheating and evaporation of the urea solution before injection into the engine's exhaust gas [1]. The concept was checked at a diesel test rig and significantly improved NO$_x$ reduction efficiencies were observed. A critical factor for the efficiency of SCR systems is the uniformity of the reducing agent distribution across the frontal area of the catalytic converter. To this end, Sala et al. have developed an automated sampling and analysis system to probe the NO$_x$ and NH$_3$ concentration profiles of a passenger car SCR catalytic converter indirectly from the downstream side of the converter [2].

Regarding stationary SCR applications, firing of biomass is a special challenge due to the high concentration of impurities and particularly potassium contained in this type of fuel, deactivating the used vanadia–titania-based catalysts. Schill and Fehrmann reviewed different strategies for SCR systems to cope with the high potassium loading from biomass with a focus on intrinsically potassium-resistant SCR catalysts [3]. Such catalysts can be prepared by coating vanadia–titania systems with thin protective layers of, for example, magnesia or sepiolite, using zeolites as support, replacing tungsta with heteropoly acids, and preparation methods to achieve unusual high surface areas.

Zhao, Mao, and Dong have worked on ways to improve vanadia–titania systems and achieved good low-temperature activity combined with water and sulfur tolerance, when adding manganesia and ceria [4]. The low-temperature SCR activity of this catalyst type can also be enhanced by promotion with heteropolyacids of the Keggin structure, as shown by Wu et al. in their study [5]. Diffuse Reflectance Infrared Fourier Transform Spectroscopy (DRIFTS) revealed that addition of the heteropolyacids increased the number of Brønsted and Lewis acid sites, in addition to other beneficial effects. Selectivity to nitrogen is another important property of SCR catalysts. Kim and Young added iron oxide to a vanadia–titania system and observed suppression of N$_2$O formation at high temperatures, which was assigned to the formation of strong V–O–Fe interactions and tetrahedrally coordinated polymeric vanadates [6].

When deactivation cannot be prevented, the regeneration of SCR catalysts is an interesting option. Ye et al. successfully demonstrated that treatment of a deactivated vanadia–titania catalyst in a stationary SCR application with solutions containing a vanadia precursor and oxalic acid as well as a tungsta or molybdena precursor completely reactivated the used catalyst [7].

The combination of NO_x reduction and soot oxidation activity in a single device is particularly attractive for mobile applications, where weight and construction space is limited. Casanova, Colussi, and Trovarelli addressed this topic in their study and showed that when iron vanadates were combined with $CeZrO_2$ as active components on titania support, good SCR activity and soot oxidation activity could be reached with one single material [8].

In the last few years, rare-earth metal oxides have started to emerge as SCR catalysts. Mosrati et al. found remarkable low-temperature activity at excellent nitrogen selectivity over a niobium-modified ceria–titania system [9]. On one side, the introduction of niobium decreased the surface area, but on the other side the surface acidity was strengthened, which was explained by strong ceria–titania interactions and high concentration of highly dispersed niobia.

Manganese-based oxides are in the focus of interest since a few years due to their excellent low-temperature SCR activity at moderate costs. However, this type of catalysts suffers from insufficient nitrogen selectivities and moderate water and sulfur resistance. Gao et al. provided a comprehensive review of the state of knowledge about water and sulfur resistance of manganese-based SCR catalysts and potential solutions to overcome, or at least mitigate this problem [10]. They concluded that, at the current state of development, much research is still required to reach a level where commercial application of manganese-based SCR catalysts comes into sight. The research article of Zhang et al. deals with an economically interesting manganese-based catalyst system, combining manganesia and iron oxide as active components with natural vermiculite ore as support. Despite the low costs of the components, the catalyst material excels by considerable low-temperature activity and sulfur resistance [11].

When the costs for the raw materials are the main development target, even waste materials such as Bayer red mud from the aluminum industry can be used for the manufacture of SCR catalysts, as demonstrated by Wu et al. in their research paper [12]. Iron oxide is the main component in this type of catalyst, whose activity can be enhanced by cerium addition.

At the other end of the scale, Cu–SSZ-13 catalysts are used in automotive SCR applications, where the highest volumetric activity and high hydrothermal stability are required, but costs are less of an issue compared to stationary applications. Gao and Peden reviewed the recent progress in the understanding of the mechanistic functionality of Cu–SSZ-13 in the SCR reaction with a focus on the work of their own group [13]. In addition to the coherent explanation of NO–SCR, the review also addresses catalyst deactivation and the description of possible fast-SCR reaction mechanisms over Cu–SSZ-13. Rizzotto, Chen, and Simon add an interesting facet to the understanding of the unique activity of Cu–SSZ-13 for the SCR reaction by applying in situ impedance spectroscopy to explore the mobility of NH_3-solvated Cu^{II} ions [14]. The found out that the unique cage structure of Cu–SSZ-13 favors the local motion of $Cu^{II}(NH_3)_n$ species, which has been described before as an important factor for the SCR activity of this type of material.

In the years to come, an increasing fraction of biofuels is expected in the market. Since biofuels are produced from biomass containing significant amounts of inorganic impurities, such as alkaline metals, there is a higher risk of poisoning the SCR catalysts. This motivated Tarot et al. to investigate the influence of sodium poisoning on the activity of copper-exchanged zeolites at the example of Cu–FER [15]. They found a decrease in NO_x reduction efficiency, which is governed by a loss of ammonia adsorption capacity at low temperatures and a change of the ratio of exchanged copper due to copper oxide formation at high temperatures, respectively.

Despite the success of the established SCR process with ammonia or urea as reducing agents for NO_x, research is ongoing to develop alternatives for future applications. Ström et al. have studied alumina-supported silver and indium catalysts, which are interesting catalysts for SCR with hydrocarbons as a reducing agent [16]. The activity of these catalysts can be promoted by hydrogen addition. Since ammonia has been identified as intermediate in hydrocarbon–SCR, the researchers have used ammonia as a reducing agent in this study and added hydrogen to the model exhaust gas. Their fundamental study comprising detailed characterization of the catalysts revealed interesting structure–function relationships which may help to improve SCR activity and the development of systems which work without hydrogen addition. The activity of alumina-supported silver catalysts in hydrocarbon–SCR can be improved not only by hydrogen addition, but also by application of cold plasma. The study of Lee, Kang, Jo, and Mok traced back the effect of the plasma on the decomposition of the long-chain hydrocarbon fuel and formation of partially oxidized hydrocarbons, particularly aldehydes [17].

The diversity of contributions in this book reflects the plethora of research questions in the SCR process, whose development potential is still not exploited. Considering all the important work on SCR done in the world, which is partly reported in this book, it is probably not too daring to predict that SCR will also, in the future, remain the most important process to clean lean exhaust gases from NO_x emission.

I would like to express my sincerest thanks to all authors for the valuable contributions, without which this book would not have been possible.

Funding: This research received no external funding.

Conflicts of Interest: The authors declare no conflicts of interest.

References

1. Sala, R.; Bielaczyc, P.; Brzezanski, M. Concept of Vaporized Urea Dosing in Selective Catalytic Reduction. *Catalysts* **2017**, *7*, 307. [CrossRef]
2. Sala, R.; Dzida, J.; Krasowski, J. Ammonia Concentration Distribution Measurements on Selective Catalytic Reduction Catalysts. *Catalysts* **2018**, *8*, 231. [CrossRef]
3. Schill, L.; Fehrmann, R. Strategies of Coping with Deactivation of NH_3-SCR Catalysts Due to Biomass Firing. *Catalysts* **2018**, *8*, 135. [CrossRef]
4. Zhao, X.; Mao, L.; Dong, G. Mn-Ce-V-WO_x/TiO_2 SCR Catalysts: Catalytic Activity, Stability and Interaction among Catalytic Oxides. *Catalysts* **2018**, *8*, 76. [CrossRef]
5. Wu, R.; Zhang, N.; Li, L.; He, H.; Song, L.; Qiu, W. DRIFT Study on Promotion Effect of the Keggin Structure over V_2O_5-MoO_3/TiO_2 Catalysts for Low Temperature NH_3-SCR Reaction. *Catalysts* **2018**, *8*, 143. [CrossRef]
6. Kim, M.H.; Yang, K.H. The Role of Fe_2O_3 Species in Depressing the Formation of N_2O in the Selective Reduction of NO by NH_3 over V_2O_5/TiO_2-Based Catalysts. *Catalysts* **2018**, *8*, 134.
7. Ye, T.; Chen, D.; Yin, Y.; Liu, J.; Zeng, X. Experimental Research of an Active Solution for Modeling *In Situ* Activating Selective Catalytic Reduction Catalyst. *Catalysts* **2017**, *7*, 258. [CrossRef]
8. Casanova, M.; Colussi, S.; Trovarelli, A. Investigation of Iron Vanadates for Simultaneous Carbon Soot Abatement and NH_3-SCR. *Catalysts* **2018**, *8*, 130. [CrossRef]
9. Mosrati, J.; Atia, H.; Eckelt, R.; Lund, H.; Agostini, G.; Bentrup, U.; Rockstroh, N.; Keller, S.; Armbruster, U.; Mhamdi, M. Nb-Modified Ce/Ti Oxide Catalyst for the Selective Catalytic Reduction of NO with NH_3 at Low Temperature. *Catalysts* **2018**, *8*, 175. [CrossRef]
10. Gao, C.; Shi, J.-W.; Fan, Z.; Gao, G.; Niu, C. Sulfur and Water Resistance of Mn-Based Catalysts for Low-Temperature Selective Catalytic Reduction of NO_x: A Review. *Catalysts* **2018**, *8*, 11. [CrossRef]
11. Zhang, K.; Yu, F.; Zhu, M.; Dan, J.; Wang, X.; Zhang, J.; Dai, B. Enhanced Low Temperature NO Reduction Performance via MnO_x-Fe_2O_3/Vermiculite Monolithic Honeycomb Catalysts. *Catalysts* **2018**, *8*, 100. [CrossRef]
12. Wu, J.; Gong, Z.; Lu, C.; Niu, S.; Ding, K.; Xu, L.; Zhang, K. Preparation and Performance of Modified Red Mud-Based Catalysts for Selective Catalytic Reduction of NO_x with NH_3. *Catalysts* **2018**, *8*, 35. [CrossRef]
13. Gao, F.; Peden, C.H.F. Recent Progress in Atomic-Level Understanding of Cu/SSZ-13 Selective Catalytic Reduction Catalysts. *Catalysts* **2018**, *8*, 140.

14. Rizzotto, V.; Chen, P.; Simon, U. Mobility of NH_3-Solvated Cu^{II} Ions in Cu-SSZ-13 and Cu-ZSM-5 NH_3-SCR Catalysts: A Comparative Impedance Spectroscopy Study. *Catalysts* **2018**, *8*, 162. [CrossRef]

15. Tarot, M.-L.; Barreau, M.; Duprez, D.; Lauga, V.; Iojoiu, E.E.; Courtois, X.; Can, F. Influence of the Sodium Impregnation Solvent on the Deactivation of Cu/FER-Exchanged Zeolites Dedicated to the SCR of NO_x with NH_3. *Catalysts* **2018**, *8*, 3. [CrossRef]

16. Ström, L.; Carlsson, P.-A.; Skoglundh, M.; Härelind, H. Surface Species and Metal Oxidation State during H_2-Assisted NH_3-SCR of NO_x over Alumina-Supported Silver and Indium. *Catalysts* **2018**, *8*, 38. [CrossRef]

17. Lee, B.J.; Kang, H.-C.; Jo, J.O.; Mok, Y.S. Consideration of the Role of Plasma in a Plasma-Coupled Selective Catalytic Reduction of Nitrogen Oxides with a Hydrocarbon Reducing Agent. *Catalysts* **2017**, *7*, 325. [CrossRef]

catalysts

MDPI

Article

Concept of Vaporized Urea Dosing in Selective Catalytic Reduction

Rafal Sala [1],*, Piotr Bielaczyc [1] and Marek Brzezanski [2]

[1] BOSMAL Automotive Research & Development Institute Ltd., 43-300 Bielsko-Biala, Poland; piotr.bielaczyc@bosmal.com.pl

[2] Faculty of Mechanical Engineering, Cracow University of Technology, 31-864 Cracow, Poland; mbrzez@usk.pk.edu.pl

* Correspondence: rafal.sala@bosmal.com.pl; Tel.: +48-33-8130-417

Received: 7 September 2017; Accepted: 13 October 2017; Published: 19 October 2017

Abstract: This work tried to identify the influence of dosing vaporized urea solution in a selective catalytic reduction (SCR) system. In the SCR method, optimising the urea evaporation and mixing properties can significantly improve the NO_x conversion efficiency in the catalyst. It can also exert a positive effect on the uniformity of NH_3 concentration distribution across the catalyst face. The concept of an electrically evaporated urea-dosing system was investigated and it was found that urea pre-heating prior to introduction into the exhaust gas is favourable for enhancing NO_x removal under steady-state and transient engine operation. In the urea evaporating system the heating chamber was of a cylindrical tube shape and the urea vapour was introduced into the exhaust by means of a Venturi orifice. The concept urea dosing was only a custom-made solution, but proved to be superior to the regular dosing system operating in the liquid phase.

Keywords: urea; vapor; ammonia; nitrogen oxides; catalytic reduction

1. Introduction

Controlling nitrogen oxide (NO_x) emissions from Euro 6 vehicles with compression ignition (CI) engines is one of the biggest technical challenges facing vehicle manufacturers today. There are three main technologies currently available for this purpose: in-engine modifications combined with high- and low-pressure exhaust gas recirculation (EGR), lean-burn NO_x adsorbers (also called lean NO_x traps, or LNTs) and selective catalytic reduction (SCR) [1,2]. Although diesel engine car manufacturers have managed to meet tight legislation figures for NO_x emission during regulatory laboratory tests, it is widely known that the "real-world" NO_x emissions of diesel passenger vehicles are substantially higher (roughly 3–8 times) than the certified limit. One of the major issues is cold-start emission in the time period before light-off temperature is reached and the catalyst becomes activated. For an SCR system, the urea injection is released at 180–200 °C of exhaust gas temperature, mainly in order to prevent formation of solid urea deposits [3,4].

The divergence in laboratory and real-world NO_x emission values was the core reason for introducing new amendments to the Euro 6 standard, which requires Original Equipment Manufacturer (OEM) to perform real-driving emissions (RDE) tests using portable emission measurement systems (PEMSs) for the approval of passenger cars in the Europe [5–8]. With RDE testing legally enforced in 2017, the OEM approach to exhaust system layouts of passenger cars will have to be thoroughly modified by implementing the SCR technology as a baseline solution for the CI engines. Moreover, due to the cold-start emission issue, the NO_x adsorbers (DeNO$_x$ traps) will likely be used in series with SCR systems, having an important role in NO_x adsorbing before the SCR is activated [9,10].

One of constraints for urea dosing in SCR is excessive ammonia emissions (slip), which may necessitate deploying a clean-up catalyst (CUC) to oxidise the ammonia. The CUC is fitted downstream of the SCR catalyst, and ammonia slip occurs especially at elevated exhaust gas temperatures.

There is an effort and an ongoing development process on shortening the SCR activation time. Research has focused on enhancing the urea evaporation process and on elaborating alternative methods of volatile ammonia introduction upstream of the SCR catalyst [7,11].

2. SCR Chemistry

In the SCR applications, urea is the preferred reducing agent due to toxicological and safety reasons. An aqueous urea solution (32.5% urea) is injected into the hot exhaust gas upstream of the SCR catalyst. The decomposition of urea into ammonia and carbon dioxide precedes the SCR reaction [12].

The first step is the evaporation of water from the droplets, thus leading to solid or molten urea:

$$NH_2\text{-}CO\text{-}NH_2 \text{ (aqueous)} \rightarrow NH_2\text{-}CO\text{-}NH_2 \text{ (molten)} + x\,H_2O \text{ (gas)}. \tag{1}$$

Molten urea will then heat up and decompose thermally according to:

$$NH_2\text{-}CO\text{-}NH_2 \text{ (molten)} \rightarrow NH_3 \text{ (gas)} + HNCO \text{ (gas)} \; \Delta H_{298} = +186 \text{ kJ}. \tag{2}$$

Equimolar amounts of ammonia and isocyanic acid are thus formed. Isocyanic acid is very stable in the gas phase, but hydrolyzes easily on many solid oxides, reacting with water vapor originating from the combustion process:

$$HNCO \text{ (gas)} + H_2O \text{ (gas)} \rightarrow NH_3 \text{ (gas)} + CO_2 \text{ (gas)} \; \Delta H_{298} = -96 \text{ kJ}. \tag{3}$$

The thermo-hydrolysis of urea is globally an endothermic process. The Reactions (1) and (2) may also occur in the gas phase upstream of the catalyst, whereas the hydrolysis of the isocyanic acid (Reaction (3)) proceeds mainly on the SCR catalyst itself.

3. Results and Discussion

3.1. Steady-State Engine Operation

The Figure 1 presents comparisons of test results obtained under steady-state engine operation according to test steps as determined in Table 1. The resulting evaluations aim to compare the NO_x conversion efficiency as a function of the urea-dosing method applied. It was found that the gaseous urea system brought a significant benefit in terms of NO_x reduction in nearly the entire cycle range. That was further confirmed by test cycles run at a different α value.

One of the critical parameters related to the gas-phase concept was the urea–water solution (UWS) temperature level obtained at the outlet of the heating chamber. The UWS temperature rapidly rose at the beginning of step 1, reaching its maximum of 160 °C, and later stabilised at around 110 °C. Across all the remaining test steps, the UWS temperature steadily decreased down to a value of 103 °C, measured at the end of the test run. The UWS temperature consistently declined, as the heating chamber was more and more intensively cooled down by urea flow, which increased its rate with each test step. The heating power delivered to the chamber was, on the other hand, maintained at a constant level.

As the evaporation of water was known to be independent of the urea evaporation [13,14], a point of interest was the composition of the feed-gas dosed upstream of SCR catalyst.

Before chemical reactions occur, the dosed UWS aerosol is heated up in an evaporation chamber and later by the surrounding exhaust gas that contains evaporated water. The exact state of aggregation of urea during decomposition is still uncertain [15,16]. Two recent theoretical studies [17,18] relying on experimental data point toward urea evaporation from liquid aerosols and decomposition in the gas phase. However, another recent study supposed that the mentioned chemical reactions take place in solid aerosols [19].

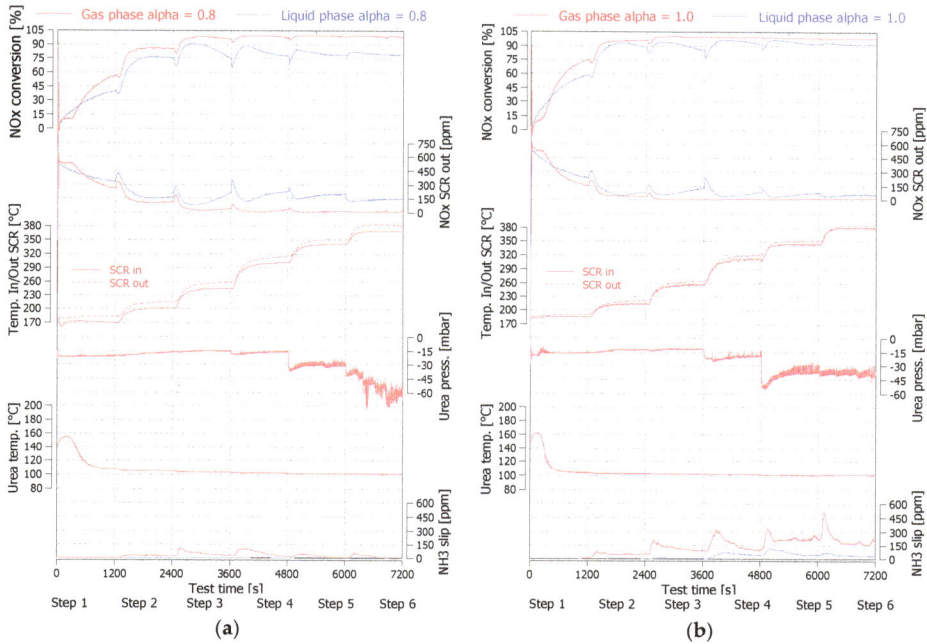

Figure 1. Comparison of results of standard and gaseous urea dosing at $\alpha = 0.8$ (**a**) and $\alpha = 1.0$ (**b**).

Table 1. Engine operation points during steady-state test cycle.

Test Step Number	SCR Inlet Gas Temp. (°C)	Engine Load (Nm)
Step 1	180	60
Step 2	214	82
Step 3	255	111
Step 4	308	180
Step 5	344	316
Step 6	380	400

At the UWS temperature ranging from 160 °C to 103 °C, the water content vaporizes as well as the urea vaporisation and decomposition processes taking place. Figure 2 presents the results of an experiment of heating up the UWS in an open vessel. It was found that urea decomposition to NH_3 and CO_2 occurred also at temperatures below 103 °C [12].

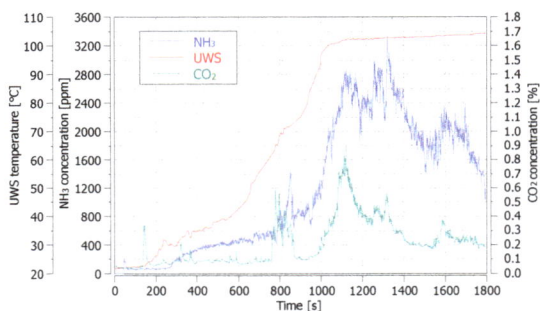

Figure 2. Results of NH_3 and CO_2 concentration measured over urea–water solution heat-up.

The biggest impact of the gaseous urea system on NO_x reduction was obtained at α ratio 0.8, and in that case, the decrease in total value of NO_x emission [g/kWh] reached 60%. In the initial phase of the test, corresponding to the lowest gas temperature of 170 °C, the NO_x conversion value for the gaseous urea system was significantly lower than that for the liquid counterpart. The decrease in performance was due to the time delay of gaseous urea reaching the exhaust gas upstream of the SCR. There was the necessity of heating the chamber saturated with urea. The NO_x conversion for gaseous urea subsequently increased and remained greater than that for the liquid-urea dosing until the end of the test. The difference in NO_x traces for both dosing methods was greatest in the middle and upper range of exhaust-gas temperature (steps 3 to 6). The ammonia slip downstream of the SCR catalyst was also measured. It was found to be greater for the gaseous urea system for both $\alpha = 0.8$ and $\alpha = 1.0$, and that phenomenon needs to be further investigated. The exhaust line was not equipped with a clean-up catalyst (CUC) to oxidise excessive ammonia, and therefore, sudden release of NH_3 was noted with the gas-temperature increase. The NH_3 storage capacity of an SCR catalyst is inversely proportional to its temperature, and therefore, the greater the temperature the less NH_3 can be stored inside the catalyst [20–22]. The NH_3 concentration measured downstream of the SCR was greater for the gaseous urea system for both $\alpha = 0.8$ and $\alpha = 1.0$. The reason behind that was presumably greater urea-to-NH_3 conversion for the gaseous urea system, that is, the key point of the concept. In order to explain this phenomenon, ammonia-mass calculations based upon measured NO_x and NH_3 concentrations downstream of the SCR, and the amount of urea injected, were carried out. The results are presented in Table 2.

Table 2. Ammonia-mass calculations for standard and gaseous urea dosing at $\alpha = 1.0$.

	Amount of NH_3 Reacted (g)	Amount of NH_3 Dosed (g)	Percentage of NH_3 Reacted (%)
Gas phase	252	426	59
Liquid phase	185	374	49

It was found that for the gas-phase system, the amount of NH_3 obtained from urea was 10% greater than that from the liquid-phase solution. The excessive (compared to liquid solution) ammonia was accumulated in the SCR catalyst, especially at the initial phase of the test run at low gas temperature, and later released, causing a greater slip.

Figure 3 presents a comparison of results of NO_x emissions for individual test steps and the total value. Improvement in NO_x reduction was achieved for all test steps and both α values. Total NO_x emission was lower by 60% at $\alpha = 0.8$ and by 45% at $\alpha = 1.0$. Other emission compounds, expressed in [g/kWh], are shown in Figure 4. The measured CO_2 emission value was greater in the case of the gaseous urea system for both α values, explained by increased exhaust-gas backpressure caused by the mixing element introducing gaseous urea upstream of the SCR catalyst. The mixing element was a Venturi-type orifice narrowed in a section where the heating chamber was connected and causing exhaust gas to increase in speed, creating a pressure drop. It was also noticed that the CO emission was slightly higher for the gaseous urea solution, while the total hydrocarbons (THC) values remained at a similar level for both urea-dosing methods.

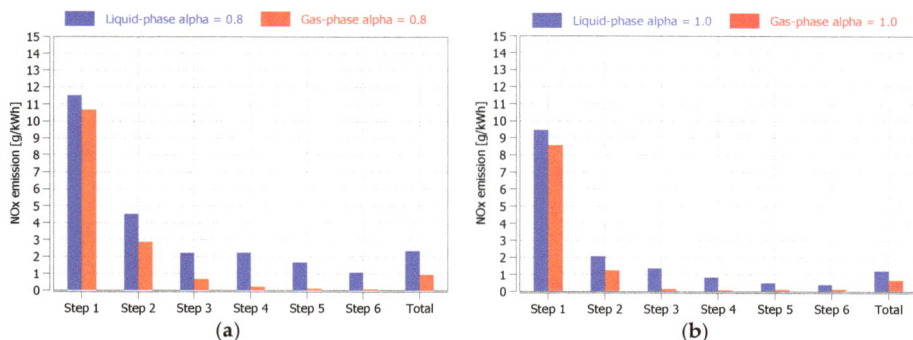

Figure 3. NO_x emission results for standard and gaseous urea dosing at $\alpha = 0.8$ (**a**) and $\alpha = 1.0$ (**b**).

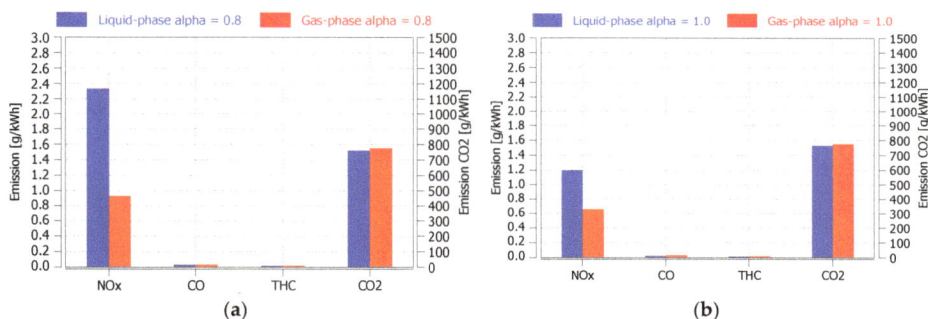

Figure 4. Emission results of gas and standard urea dosing at $\alpha = 0.8$ (**a**) and $\alpha = 1.0$ (**b**).

3.2. Transient Engine Operation

Part of the study included concept-system verification under transient conditions. Figure 5b presents calculated emission results of the WHTC cycle with hot-engine start. The NO_x emission was found to be 18% lower for the gaseous urea system ($\alpha = 1.2$), while the CO_2 and NH_3 emissions were consequently higher. Traces of NO_x conversion, gas temperature and ammonia are shown in Figure 5a. The NH_3 slip was measured in the last part of the test cycle at high gas-temperature range. It occurred at about 1300 s into the test and lasted until test completion. The NH_3 concentration was higher for the gaseous urea system.

Based on those test results, it was found that application of the gaseous urea system had a positive impact on decreasing NO_x, but it was disadvantageous for CO_2 and NH_3 emission.

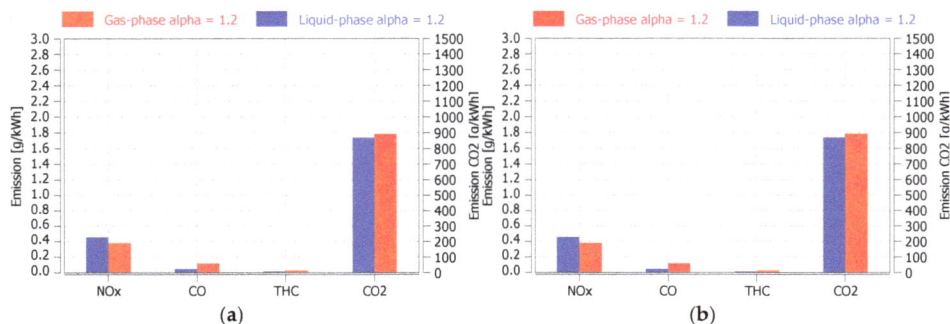

Figure 5. Comparison test results of liquid and gaseous urea-dosing on World Harmonized Transient Cycle (WHTC), hot-cycle at α = 1.2, (**a**) modal analysis, (**b**) emission results.

3.3. Practical Implementation

The analysis of test results raises a question of possible application of the gaseous urea concept. The idea of dosing NO_x reductant in the gas phase is currently the subject of research work and it is seen as a solution for decreasing or eliminating issues related to poor urea evaporation, inefficient ammonia and urea conversion, and deposit formation, when SCR is operated at low exhaust-gas temperature [14].

The key observations made in the course of the study were the following.

Firstly, for urea heat-up and evaporation, electric energy was applied at a power of 200 W. In that case, this level of power was suitable to evaporate mainly the water in urea solution, although it needs to be correlated with urea flow-rate for obtaining best results.

Secondly, urea introduction into the gas stream via a mixing element unit has become an issue due to the accompanying increase in backpressure. An optimization and redesign of the mixing element is required. The properties of material that the mixing element is made of might influence the rate of thermal urea decomposition. Optimized selection of material composition could be advantageous for enhancing urea-to-ammonia conversion.

Thirdly, the start-up conditions of the gaseous urea system remain to be improved. At certain engine operating points, solid urea-deposit formation started to occur inside the heating chamber.

4. Concept Description

The studies on concept gaseous urea dosing systems were performed in an engine-testing laboratory on a 3.0 L CI Euro V medium-duty engine. The exhaust line consisted of a diesel oxidation catalyst (DOC), diesel particle filter (DPF), both belonging to tested engine type and an Fe–zeolite SCR catalyst combined with a development urea-dosing system. The dosing system was based on Siemens NO_x sensors, Bosch urea injector, while the automation software was prepared by Bosmal. The solution allowed for fast changeover between concept (gas) and standard (liquid) urea-dosing methods. The core of the concept method was to heat up and evaporate the urea water solution (UWS) in a special heating chamber prior to its introduction into the exhaust gas stream at the SCR inlet. For this purpose, an external energy source was utilized. Figure 6a depicts the exhaust line with the inbuilt gaseous urea concept implemented on an engine test bed. Figure 7 presents physical realisation of the system. The urea heating chamber was supplied with electrical energy with a heating power of 200 W [12].

The gaseous urea dosing system utilized a standard OEM urea Bosch injector, which dosed liquid urea into a thermally insulated heating chamber of tubular shape with a diameter of 10 mm. Urea vapours were then subsequently drawn down by the pressure drop induced by the mixing element mounted upstream of the SCR catalyst. The mixing element was made of aluminium alloy, while the heating chamber was equipped with a thermostatic temperature control set to 220 °C. During test runs, urea vapour pressure and temperature were measured at the outlet of the heating chamber.

The standard urea dosing system is depicted in Figure 6b. The NO_x reductant was injected in the liquid phase directly into the exhaust gas stream. The urea mixer is typically used for providing good ammonia uniformity in the SCR catalyst. The mixer is placed between the urea injector and the SCR brick. It is made of fixed-geometry metal blades to help urea droplets break up [23–25].

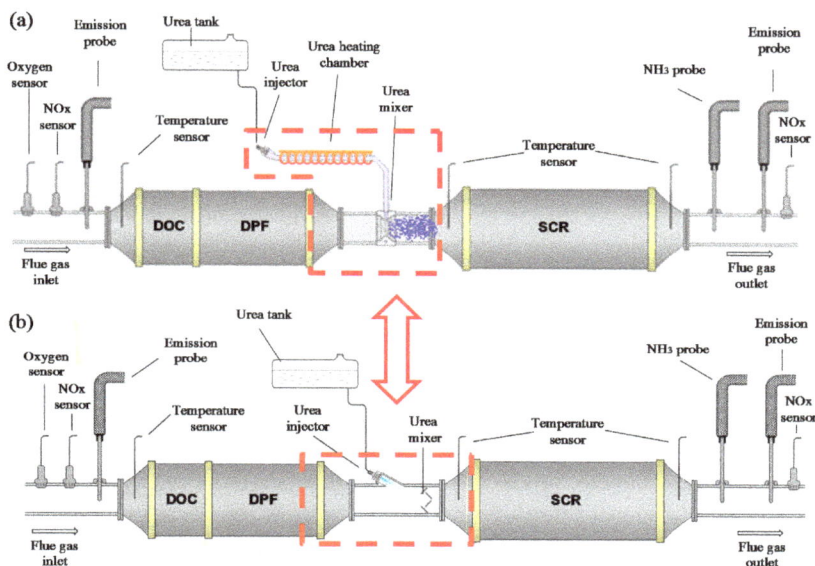

Figure 6. Comparison of SCR systems: (**a**) gaseous urea dosing, (**b**) liquid urea dosing.

The demanded value of urea-mass injected was realized in either open or closed-loop transfer functions. In the open loop, the urea injection took fixed values, while in the closed loop it was expressed by the α ratio. The α value was calculated by the control unit as the function of the actual exhaust-mass flow rate and the NO_x concentration value measured by the portable NO_x sensor. The sensor was mounted at the engine outlet. The urea was dispensed at a pressure of approximately 5 bar, depending on the dosing-system configuration: directly into the exhaust gas stream (liquid) or into the heating chamber (gas).

The main part of the study was carried out at steady-state engine operation. A dedicated test procedure was prepared that covered as wide a range of exhaust-gas temperatures as possible. The procedure was applied to the concept and standard urea-dosing methods in order to compare their respective performance. The test consisted of six 20-min test steps lasting two hours in total. The gas temperature measured at the SCR inlet ranged from 180 °C to 380 °C and was increased by approximately 40 °C at each subsequent step, starting from 180 °C. The engine speed was set to a constant value of 1900 rpm and the gas temperature was controlled by varying engine load in the range form 60 to 400 Nm. The EGR valve remained shut-off during testing, and therefore, the NO_x concentration in the raw gas was high, varying from 580 ppm to 1100 ppm. Exhaust mass flow varied within the range 230 to 400 kg/h.

The quantity of urea dosed was expressed by the α ratio, defined as the quotient of the amount of ammonia molecules derived from the reducing agent (NH_3) and the amount of nitrogen oxides molecules (NO_x) present in the elementary exhaust gas mass flow.

$$\alpha = \frac{NH_3}{NO_x} \tag{4}$$

Steady-state testing was carried out at α values of 0.8 and 1.0. The exhaust emission and gas temperature were measured continuously (at 1 Hz) upstream of the DOC and downstream of the SCR, where the ammonia concentration was also measured. The exhaust emission bench was an AVL AMA i60 measurement system with standalone ammonia analyser with an accuracy of $\pm1\%$. The injected urea flow-rate was expressed in mg/s.

Before commencing testing, the exhaust line underwent a pre-conditioning cycle to ensure stability and repeatability of initial conditions and to eliminate ammonia saturation of the SCR catalyst. The cycle was run at a high gas-temperature generated by engine load and with the urea dosing system remaining switched off.

The work was extended to evaluate the gaseous urea-dosing systems also in engine transient conditions. To do that, the World Harmonized Transient Cycle (WHTC) was chosen and run in a simplified version on eddy-current dynamometer. The hot-start WHTC test was only taken into consideration.

Figure 7. Gaseous urea dosing system: 1—heating chamber, 2—urea pressure port, 3—urea temperature port, 4—insulated tube, 5—mixing element, 6—DPF filter, 7—SCR catalyst.

5. Conclusions

The SCR method still requires further development work, especially when operated under engine warm-up and idle conditions. Urea injection at low exhaust-gas temperature is critical due to the limited ability of evaporation and chemical decomposition into ammonia.

The purpose of this study was to evaluate an effect of a water–urea solution preparation on NO_x conversion in an SCR catalyst. The preparation process relies on urea heating up and evaporating prior to its introduction upstream of the SCR catalyst. An expected effect was to accelerate urea

Catalysts **2017**, *7*, 307

thermal decomposition and ultimately to increase NO_x conversion efficiency and possibly reduce urea consumption.

A custom-made gas-urea system was designed and constructed and put under test. Although it was a very simple solution that needed further development, it gave an insight into the benefits that can be achieved by urea vaporization.

It was found that the gaseous urea-dosing method had significantly increased the NO_x conversion in the SCR catalyst compared to standard (liquid) urea-dosing. The benefits were achieved under both steady-state and dynamic engine conditions.

The work carried out demonstrated the potential of a gaseous urea concept for an SCR application, and the results obtained serve as a great incentive for undertaking further research in that field.

Acknowledgments: This work was financed and carried out in Engine Testing Laboratory at BOSMAL R&D Institute based in Bielsko-Biala, Poland and was a part of a PhD dissertation.

Author Contributions: For research articles with several authors, a short paragraph specifying their individual contributions must be provided. R.S. and M.B. conceived and designed the experiments; R.S. performed the experiments; R.S. and M.B. analyzed the data; P.B. contributed reagents and materials; R.S. wrote the paper.

Conflicts of Interest: The authors declare no conflict of interest.

References

1. Yang, L.; Franco, V.; Campestrini, A.; German, J.; Mock, P. *NO$_x$ Control Technologies for Euro 6 Diesel Passenger Cars*; ICCT, White Paper: Washington, DC, USA, 2015.
2. Brzezanski, M.; Sala, R. In-service problems of selective catalytic reduction systems for reduction of nitrogen oxides. *Combus. Engines* **2013**, *154*, 969–976.
3. Brzezanski, M.; Sala, R. A study on the indirect urea dosing method in the Selective Catalytic Reduction system. In Proceedings of the Scientific Conference on Automotive Vehicles and Combustion Engines, Materials Science and Engineering, (KONMOT 2016), Krakow, Poland, 22–26 September 2016; Volume 148.
4. Bielaczyc, P.; Woodburn, J. Current directions in LD powertrain technology in response to stringent exhaust emissions and fuel efficiency requirements. *Combus. Engines* **2016**, *166*, 62–75. [CrossRef]
5. Merkisz, J.; Pielecha, J.; Bielaczyc, P.; Woodburn, J. *Analysis of Emission Factors in RDE Tests as Well as in NEDC and WLTC Chassis Dynamometer Tests*; No. 2016-01-0980; SAE Technical Paper: Washington, DC, USA, 2016.
6. European Environment Agency (EEA). *Exceedance of Air Quality Limit Values in Urban Areas*; European Environment Agency: Copenhagen, Denmark, 2015.
7. Yang, L.; Franco, V.; Mock, P.; Kolke, R.; Zhang, S.; Wu, Y.; German, J. Experimental assessment of NO$_x$ emissions from 73 Euro 6 diesel passenger cars. *Environ. Sci. Technol.* **2015**, *49*, e14409–e14415. [CrossRef] [PubMed]
8. Johnson, T. Vehicular Emissions in Review. *SAE Int. J. Engines* **2016**, *9*, 1258–1275. [CrossRef]
9. Maunula, T. *NO$_x$ Reduction with the Combinations on LNT and SCR in Diesel Applications*; No. 2013-24-0161; SAE Technical Paper: Washington, DC, USA, 2013.
10. Chi, J. *Control Challenges for Optimal NO$_x$ Conversion Efficiency from SCR Aftertreatment Systems*; No. 2009-01-0905; SAE Technical Paper: Washington, DC, USA, 2009.
11. Marotta, A.; Pavlovic, J.; Ciuffo, B.; Serra, S.; Fontaras, G. Gaseous emissions from Light-Duty Vehicles: Moving from NEDC to the new WLTP test procedure. *Environ. Sci. Technol.* **2015**, *49*, 8315–8322. [CrossRef] [PubMed]
12. Sala, R. Wpływ Sposobu Dozowania Roztworu Mocznika na Sprawność Selektywnej Redukcji Katalitycznej (The Influence of the Urea Dosing Method on the Efficiency of the Selective Catalytic Reduction Process). Ph.D. Thesis, Cracow University of Technology, Cracow, Poland, 2014.
13. Bernhard, A.M.; Czekaj, I.; Elsener, M.; Wokaun, A.; Kröcher, O. Evaporation of Urea at Atmospheric Pressure. *J. Phys. Chem. A* **2011**, *115*, 2581–2589. [CrossRef] [PubMed]
14. Kröcher, O.; Elsener, M.; Jacob, E. A model gas study of ammonium formate, methanamide and guanidinium formate as alternative ammonia precursor compounds for the selective catalytic reduction of nitrogen oxides in diesel exhaust gas. *Appl. Catal. B* **2009**, *88*, 66–82. [CrossRef]

15. Birkhold, F.; Meingast, U.; Wassermann, P.; Deutschmann, O. Modeling and simulation of the injection of urea-water-solution for automotive SCR DeNO$_x$-systems. *Appl. Catal. B* **2007**, *70*, 119–127. [CrossRef]

16. Ström, H.; Lundström, A.; Andersson, B. Choice of urea-spray models in CFD simulations of urea-SCR systems. *Chem. Eng. J.* **2009**, *150*, 69–82. [CrossRef]

17. Abu-Ramadan, E.; Saha, K.; Li, X. Modeling the depleting mechanism of urea-watersolution droplet for automotive selective catalytic reduction systems. *AIChE J.* **2011**, *57*, 3210–3225. [CrossRef]

18. Lundström, A.; Waldheim, B.; Ström, H. Modelling of urea gas phase thermolysis and theoretical details on urea evaporation. *Proc. Inst. Mech. Eng. Part D J. Autom. Eng.* **2011**, *255*, 1392–1398. [CrossRef]

19. Ebrahimian, V.; Nicolle, A.; Habchi, C. Detailed modeling of the evaporation and thermal decomposition of urea-water solution in SCR systems. *AIChE J.* **2012**, *58*, 1998–2009. [CrossRef]

20. Gong, J.; Narayanaswamy, K.; Rutland, C.J. Heterogeneous Ammonia Storage Model for NH$_3$–SCR Modeling. *Ind. Eng. Chem. Res.* **2016**, *55*, 5874–5884. [CrossRef]

21. Seneque, M.; Can, F.; Duprez, D.; Courtois, X. NO$_x$ Selective Catalytic Reduction (NO$_x$-SCR) by Urea: Evidence of the Reactivity of HNCO, Including a Specific Reaction Pathway for NO$_x$ Reduction Involving NO + NO$_2$. *ACS Catal.* **2016**, *6*, 4064–4067. [CrossRef]

22. Karjalainen, P.; Ronkko, T.; Lahde, T.; Rostedt, A.; Keskinen, J.; Saarikoski, S.; Aurela, M.; Hillamo, R.; Malinen, A.; Pirjola, L.; et al. Reduction of heavy-duty diesel exhaust particle number and mass at low exhaust temperature driving by the DOC and the SCR. *SAE Int. J. Fuels Lubr.* **2012**, *5*, e1114–e1122. [CrossRef]

23. Koebel, M.; Strutz, E.O. Thermal and Hydrolytic Decomposition of Urea for Automotive Selective Catalytic Reduction Systems: Thermochemical and Practical Aspects. *Ind. Eng. Chem. Res.* **2003**, *42*, 2093–2100. [CrossRef]

24. Sung, D.Y.; Soo, J.K.; Joon, H.B.; In-Sik, N.; Young, S.M.; Jong-Hwan, L.; Byong, K.C.; Se, H.O. Decomposition of Urea into NH$_3$ for the SCR Process. *Ind. Eng. Chem. Res.* **2004**, *43*, 4856–4863. [CrossRef]

25. Kleemann, M.; Elsener, M.; Koebel, M.; Wokaun, A. Hydrolysis of Isocyanic Acid on SCR Catalysts. *Ind. Eng. Chem. Res.* **2000**, *39*, 4120–4126. [CrossRef]

catalysts

MDPI

Article

Ammonia Concentration Distribution Measurements on Selective Catalytic Reduction Catalysts

Rafal Sala *, Jakub Dzida and Jaroslaw Krasowski

BOSMAL Automotive Research & Development Institute Ltd., 43-300 Bielsko-Biala, Poland;
jakub.dzida@bosmal.com.pl (J.D.); jaroslaw.krasowski@bosmal.com.pl (J.K.)
* Correspondence: rafal.sala@bosmal.com.pl; Tel.: +48-33-8130-417

Received: 18 April 2018; Accepted: 22 May 2018; Published: 1 June 2018

Abstract: This work presents the methodology and accurate evaluation of ammonia concentration distribution measurements at the selective catalytic reduction (SCR) catalyst outlet cross-section. The uniformity of ammonia concentration is a crucial factor influencing overall SCR effectiveness, and it contributes to the necessity of employing a reliable test method. The aftertreatment system design (mainly its geometrical features) can be evaluated in detail. The ammonia concentration is measured at the SCR catalyst outlet at grid points covering from the center to the outer edges of the catalyst. Its execution requires the introduction of a probe hovering over the back face of the SCR. To obtain the expected accuracy, it is necessary to measure a sufficient number of points in a reasonable timeframe. In order to achieve that, a fully automatic sampling device was developed. Sample results are presented showing the capabilities of the created test stand and its importance for the design development and validation stages of SCR-based engine aftertreatment.

Keywords: ammonia distribution; ammonia uniformity index; selective catalytic reduction; urea; automated measurement

---

1. Introduction

Selective catalytic reduction (SCR) has become the principle method of nitrogen oxides removal (NO_x) from exhaust gases of diesel engine powertrains in the automotive industry. The SCR system utilizes water–urea solution (UWS) as a reducing agent, which is injected directly into the exhaust upstream of the SCR catalyst. The SCR system working at the rated operating conditions allows the NO_x emission to be decreased by more than 98%. The excellent NO_x removal efficiency of the SCR systems allows calibration of the engine for more efficient combustion strategies, thus increasing the fuel efficiency and reducing the CO_2 emissions. Considering a typical NO_x-particulate matter (PM) trade-off, shifting towards higher engine-out NO_x emission leads to an improvement of fuel efficiency and decreases the soot emission. This in turn lowers the diesel particulate filter (DPF) soot loading and decreases the frequency of the active regeneration events.

In the automotive sector, the SCR method allows the elimination of the divergence in NO_x emission values between laboratory certification driving cycles and the real-driving emissions test (RDE) that were legally enforced in 2017 for passenger vehicles [1–4]. However, it is complex to implement the SCR system onboard a vehicle, as it requires the deployment of a urea-dosing infrastructure with a urea tank, NO_x/NH_3 catalysts, and sensors. To decrease the aftertreatment system dimensions under the vehicle's compartment, the SCR can be coated onto a DPF, and this type of solution is referred as an SCR-Catalyzed Diesel Particulate Filter (SDPF) [5,6].

To the achieve high NO_x reduction efficiency in an SCR system, the urea dosed needs to be mixed evenly with the exhaust gas stream in order to obtain as uniform an ammonia distribution at the SCR inlet face as possible. The mixing uniformity challenge has to be met throughout the wide and continuously varying range of exhaust flow rate and temperature, which produces variability in the injected urea spray dispersion and interaction with mixing elements. This correlates the catalyst efficiency to the degree of uniformity of the ammonia. Consequently, with poor ammonia distribution, the SCR system will be prone to increased urea consumption, decreased NO_x reduction, and increased NH_3 slip [7–9]. The uniformity of ammonia entering the SCR reactor has become even more critical, since SCR technology is widely applied to light duty vehicles. This type of vehicle usually incorporates a close-coupled aftertreatment layout, leaving minimal volume and time to mix in the ammonia. As a result, a dependable methodology of urea uniformity evaluation is required. An investigation into the distribution of ammonia in SCR systems has already been investigated and characterized, as published by Song, Naber and Johnson in [10].

In the methodology presented below, in order to maximize the accuracy of the ammonia distribution evaluation at the SCR front face, a high-density grid of measuring locations at the SCR outlet was applied. A dedicated test stand was developed featuring an automatically operated gas sampling probe, allowing the efficient and flexible testing of various SCR system designs. The test method was elaborated to support Euro 6 emission-compliant SCR applications, from which NO_x conversion efficiency above 95% is demanded, maintaining at the same time low ammonia slip and UWS consumption. The objective is to accurately estimate the ammonia distribution at the SCR front face. The actual measurement is impossible at this location, since it is unworkable to insert a sampling probe into such a tight environment. Assuming a linear flow of the exhaust gas throughout the SCR brick, a uniformity level of NH_3 and NO_x is the same on both sides of the reactor. Consequently, it was decided to measure gaseous compounds' concentrations at the SCR back face. This location is accessible to the moving probe, and the obtained results are post-processed to calculate ammonia distribution at SCR reactor front face.

A uniformity index (UI) is commonly used to describe the ammonia distribution across the exhaust pipe or the catalyst cross-section [11]. It is based on the formula with local and mean NH_3 concentrations related to the catalyst surface. Well-optimized Euro 6/VI mixing systems for light-duty and heavy-duty applications are able to achieve UI values greater than 0.98. Improper integration of a mixing element can hinder system performance and introduce urea deposit formation concerns [12].

There are three key SCR catalyst chemical reactions [13], and their precedence depends on NO_2/NO_x ratio. The standard reaction, which reduces only NO, is the following:

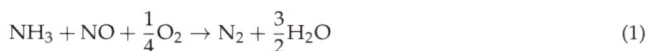

$$NH_3 + NO + \frac{1}{4}O_2 \rightarrow N_2 + \frac{3}{2}H_2O \tag{1}$$

Since there is always NO_2 present to some extent in the exhaust e.g., 10%, the reaction treating both NO and NO_2 is (2). This reaction is the fastest and the most preferred. For this reason, the diesel oxidation catalyst (DOC) positioning upstream of SCR benefits in an enhanced NO_2 share in the exhaust by platinum coating.

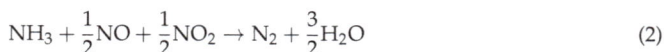

$$NH_3 + \frac{1}{2}NO + \frac{1}{2}NO_2 \rightarrow N_2 + \frac{3}{2}H_2O \tag{2}$$

When the NO_2/NO_x ratio exceeds 50%, then reaction (3) becomes operative:

$$NH_3 + \frac{3}{4}NO_2 \rightarrow \frac{7}{8}N_2 + \frac{3}{2}H_2O \tag{3}$$

This reaction is undesirable, as the excess of NO_2 can yield N_2O, which is a strong greenhouse gas:

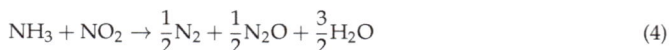

$$NH_3 + NO_2 \rightarrow \frac{1}{2}N_2 + \frac{1}{2}N_2O + \frac{3}{2}H_2O \tag{4}$$

16

The quantity of urea that was dosed is expressed by a stoichiometric ratio, α, which was calculated as a quotient of the amount of ammonia molecules from the urea $NH_{3_{in}}$ and the amount of nitrogen oxides molecules $NO_{x_{in}}$ in the elementary exhaust gas mass flow:

$$\alpha = \frac{NH_{3_{in}}}{NO_{x_{in}}} \tag{5}$$

Consequently, $\alpha = 1$ is defined as the theoretical flow of UWS required for converting 100% of the incoming NO_x flow.

2. Results and Discussion

2.1. Test Methodology

The test methodology is based on the measurements of NO_x and NH_3 concentration distribution at SCR back face and the post-process of collected results in order to calculate the ammonia uniformity index at the SCR inlet. The input data that was necessary for the NH_3 uniformity index calculation was obtained by two independent test runs.

The first run was aimed at the measurement of NO_x distribution, without urea water solution (UWS) injection to the exhaust. The values were further used in formula (1) for the results calculation that was acquired during the second test run.

The second run was executed with an active UWS injection, and NO_x and NH_3 concentrations were measured. Both test runs consisted of the same number of probe locations evenly distributed over the SCR outlet face (Figure 1).

This paper presents the example test on an SCR reactor of 184-mm diameter and with 231 probe measuring positions. The grid of measurement points covered 82% of total SCR cross-section area. The sticky characteristics of NH_3 demanded a relatively long stabilization time, which was needed to secure the exhaust gas exchange in the sampling lines. Trial measurements at different stabilization periods were performed to establish the minimum time that was needed to reach a stable reading of NH_3. As a result, a single point location measurement lasted 40 s, and consisted of 30 s of stabilization time and 10 s of measurement phase. Consequently, a single test-run lasted 2.57 h, which was required to ensure a stable engine operating condition over the entire test time.

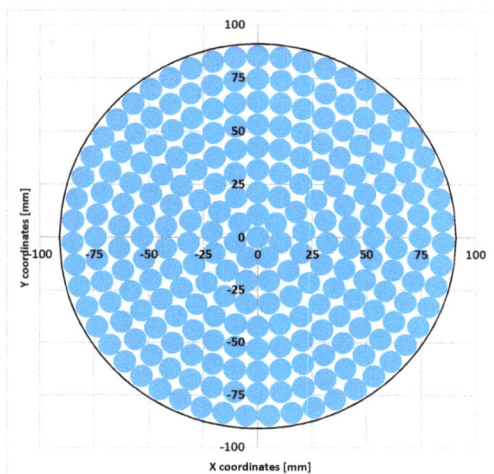

Figure 1. Measuring probe positions over the SCR outlet face.

Before the first test run (without UWS injection), the engine and aftertreatment system were thermally stabilized for 1.5 h at test conditions, before the actual NO_x measurement started. The extended preconditioning time was aimed at eliminating the urea saturation inside the SCR catalyst.

During the second test run, the UWS was injected into the exhaust gas upstream of SCR at an α coefficient equal to 1. The chosen α value corresponded to an UWS flow rate of 12.4 mg/s for the set engine operation conditions. The UWS injector was calibrated in order to accurately follow requested flow rates. Prior to NH_3 measurements beginning, the engine and exhaust aftertreatment system (ATS) were also stabilized by running the engine at test parameters. The preconditioning phase was important to achieve steady-state test conditions, as the UWS that was introduced into the exhaust gas lowers its temperature due to evaporation. Moreover, the SCR catalyst needs a certain time to achieve an equilibrium in ammonia saturation level that corresponds to the particular exhaust gas conditions. The engine with ATS was considered to be preconditioned once stable readings of NO_x and NH_3 concentrations at a central location, as well as the gas temperature downstream of the SCR catalyst, were both measured. Nevertheless, the preconditioning was continued to reach the fixed time of 1.5 h to ensure that equilibrium conditions at regions reach with ammonia. The engine operating condition is presented in Table 1.

Table 1. Engine operation conditions during test run. SCR: selective catalytic reduction.

Engine Parameter	Setpoint Value
Engine speed	1300 rpm
Engine torque	50 Nm
NO_x concentration at engine outlet	200 ppm
Exhaust mass flow rate	70 kg/h
Exhaust gas temperature at SCR inlet	230 °C

2.2. Acquired Results

The acquired NO_x and NH_3 concentration results at each sampling point of both test runs were plotted against X and Y coordinates. Figure 2 presents NO_x distribution measured without UWS dosing. The average NO_x concentration is 3% lower than the raw engine out emission. The possible explanations of such a phenomenon is DOC lean NO_x performance and reactions with diesel PM and HCs (hydrocarbons). The Figure 3 plots correspond to the test run with an active UWS dose. The SCR substrate cross-section maps that were generated clearly indicate regions of high and low concentration of both measured compounds (Figure 3).

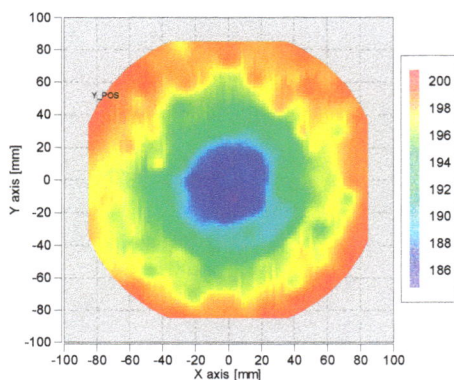

Figure 2. NO_x concentration [ppm] measured at the SCR outlet face without an active water–urea solution (UWS) dose.

The inhomogeneity of ammonia distribution affects the SCR's overall effectiveness. The areas of high NO_x emissions are the results of insufficient ammonia concentration upstream of the SCR to react with NO_x. Since the target average α coefficient is close to 1, the regions that are lean with ammonia are compensated by other regions of NH_3 enrichment across SCR outlet face. At the locations with excess NH_3, the NO_x emissions downstream of the SCR are 10 ppm or lower, and these areas are sources of unwanted ammonia slip. For the given test conditions, the NO_2/NO_x ratio at the SCR inlet was 67%.

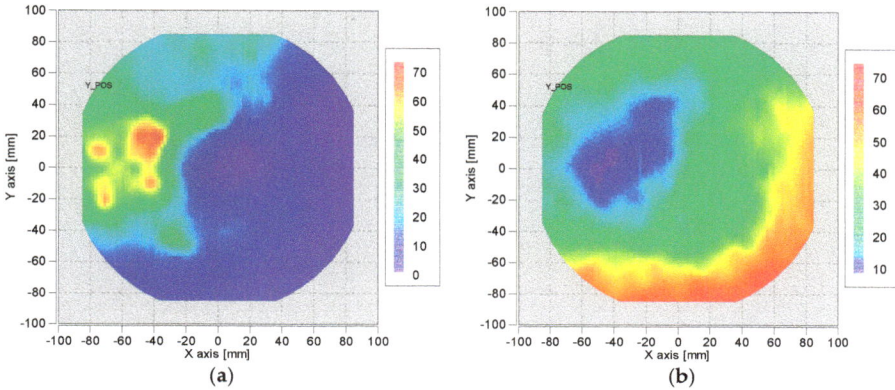

Figure 3. Results of test run with an active UWS dose; (**a**) NH_3 concentration [ppm] measured at the SCR outlet face; (**b**) NO_x concentration [ppm] measured at the SCR outlet face.

2.3. Data Post-Processing

Assuming no NO_x conversion in the SCR, the NO_x concentrations measured downstream of the SCR are reflective of the NO_x distribution at the SCR inlet face. Based on the obtained measurement results, the ammonia concentration at SCR inlet face NH_{3in_i} is calculated according to the formula:

$$NH_{3in_i}[ppm] = NO_{xin_i} - NO_{xout_i} + NH_{3out_i} \qquad (6)$$

where:

NO_{xin_i}—local NO_x concentration at selective catalytic reduction (SCR) inlet at the i-th location
NO_{xout_i}—local NO_x concentration measured at SCR outlet at the i-th location
NH_{3out_i}—local NH_3 concentration measured at SCR outlet at the i-th location

The results of the calculations are presented on Figure 4.
Following the obtained test results, the ammonia distribution can be uniformly evaluated. For this purpose, the uniformity index (UI) is calculated according to the formula:

$$UI = 1 - \frac{\sum_i |NH_{3in_i} - \overline{NH_{3in}}| A_i}{2 \sum_i NH_{3in_i} A_i} \qquad (7)$$

where:

$\overline{NH_{3in}}$—average ammonia concentration at SCR inlet
A_i—the area of the i-th measuring point

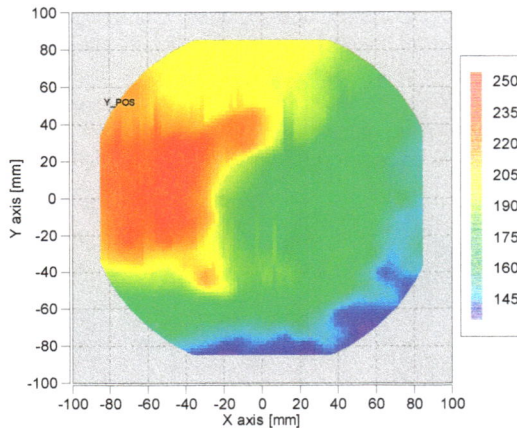

Figure 4. Calculated NH_3 concentration [ppm] at the SCR inlet face.

The uniformity index value for the SCR system measured was equal to 0.934. The measurements were repeated in order to establish its repeatability. The uniformity index obtained from the second run was equal to 0.926, giving a UI mean value of 0.929 ± 0.05. Formula (7) is based on the equation presented by Nova and Tronconi in [13]. The original equation takes into account the actual mass fraction of ammonia rather than its concentration. Assuming a uniform exhaust mass flow over the SCR cross-section, the concentration figures reflect the trend of the mass fraction. Consequently, the calculation can be simplified solely to the concentration readings. The measurement correctness was subsequently verified by calculation of a mean α value (α_{mean}) of local α values at each probe position. The α_{mean} was calculated by the formula:

$$\alpha_{mean} = \frac{\sum_i \frac{NH_{3ini}}{NO_{xini}}}{i} \tag{8}$$

In this particular case, the mean calculated α value was 0.937. It was expected that the mean α value was lower than the setpoint value. This phenomenon can be explained by incomplete urea thermolysis and hydrolysis processes, and therefore, the total mass of ammonia in SCR catalyst is always lower than the theoretical mass derived from the urea quantity introduced into the exhaust stream. Moreover, the slow SCR reaction (3) with a low reaction rate is active at the $NO2/NO_x$ shares over 50%. The NO_x reduction of the slow SCR reaction mechanism is limited by kinetic factors other than the inlet NH3 maldistribution [10]. Furthermore, the possible formation of urea solid deposits inside the exhaust system decreases the amount of the reducing agent available for NO_x conversion [14,15]. Cases of α_{mean} values higher than the setpoint α should be investigated in detail for measurement errors. There are several factors affecting the correctness of the measurement. The drift of the UWS injector's calibration curve may lead to a faulty urea injection rate, and consequently a wrong α value. Additionally, the drift of gas analyzers' measurement, such as for instance understating the NO_x and amplifying the NH_3 readings, leads to an overstated α coefficient value.

3. Materials and Methods

The NH_3 and NO_x concentration measurements were performed in an engine testing laboratory on a 2.3 L CI Euro 5 light-duty engine with a customized exhaust aftertreatment system. The close-coupled exhaust line consisted of a diesel oxidation catalyst (DOC), urea injector, urea mixer and an open-ended SCR catalyst. With an SCR substrate diameter of 184 mm, the sampling probe active diameter was 10 mm. The total area of locations covered by the probe represented 82% of the total cross-section area of the SCR brick. The SCR housing was modified by removing the outlet cone, in order to allow the moving probe access to the entire outlet face. The open-ended exhaust line was further extended by 20 cm with a metal pipe of the same diameter as the SCR to stabilize the gas flow as presented on Figure 5. Since the setup featured an open outlet, an exhaust extraction fan was placed right after the end of extension pipe, leaving a minimal gap for the moving probe.

The sampling probe was conveyed across the back face with a clearance of 3–4 mm. It consisted of an L-shaped steel pipe fixed to a ball screw linear guide driven by a stepper motor, with an accuracy of 0.05 mm, allowing for a transition in the Y-direction. The X-direction probe movement was realized by a pair of linear guides based perpendicularly to the Y-direction with a second stepper motor (Figure 5). The sampling probe trace was defined by the X,Y coordinates implemented to the automation system, together with the number of target sampling points.

The sample gas was transferred to gas analyzers, where NH_3 and NO_x concentrations were measured. The exhaust emission benches were an AVL AMA i60 CLD gas analyzer and an AVL LDD standalone ammonia analyzer with an accuracy of $\pm 1\%$. Both emission benches were produced by AVL, Graz, Austria.

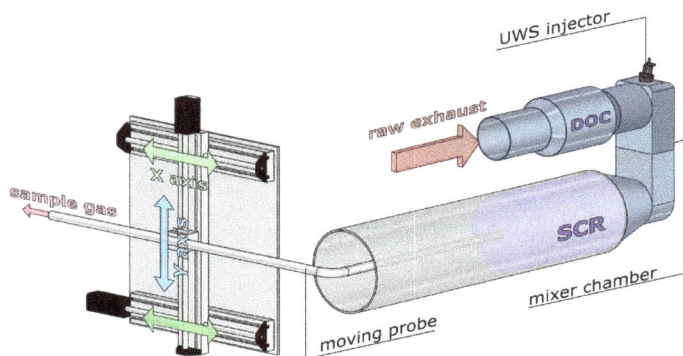

Figure 5. View of a modified SCR reactor and exhaust gas probing system for NH_3 and NO_x uniformity measurement.

4. Conclusions

There is an ongoing effort and continuous development process on SCR system optimization for automotive applications. For light-duty vehicles equipped with diesel engines, the main constraints for SCR systems are the warm-up phase, the external dimensions of entire system, and cost.

A close-coupled aftertreatment system layout is widely chosen as a reasonable compromise for light-duty vehicles, providing a compact system package that fits into the vehicle's engine compartment. Furthermore, its proximity to the engine ensures a short warm-up phase. However, high component density dramatically reduces the distance between the UWS injector and the SCR reactor. This, in return, is detrimental for the mixing capabilities of UWS that is introduced to the exhaust. Consequently, the development of highly effective mixers is required to ensure an acceptable uniformity level of the ammonia entering the SCR reactor. Due to the variety of the layout geometry and flow characteristics, every application design is developed separately. The preliminary design stages are based upon simulation results, which are sufficient input for the initial shape of the canning and mixers. Further development incorporates prototype validation. At this point, a reliable measurement of ammonia distribution at the SCR inlet is recommended. The results examine the correctness of the design and provide input to simulation improvements.

The experimental uniformity evaluation features challenges that are uncommon for typical automotive emission measurements. The elaboration of a new methodology was needed to meet objective requirements. In contrast to the most common transient emission measurements of light-duty applications, the distribution evaluation is executed under firmly stable conditions. In order to maximize the efficiency of the measurement campaign, a fully automatic sampling device was developed. Coupled with the test cell automation system, it is capable of performing gaseous compound distribution measurements at a rate of 90 positions per hour. Since its elaboration, the procedure has been applied to testing various SCR reactor designs at various engine-operating points. The early test runs were executed twice in order to evaluate their repeatability. Since the reliability of the method was proved, the test runs are executed once.

This research focuses on enhancing the urea evaporation process, shortening the activation time, and elaborating alternative methods of volatile ammonia introduction upstream of the SCR catalyst [16]. The methodology and test results presented here give an insight into the urea uniformity challenge in SCR catalysts and an indication of technical issues related to uneven NH_3 distribution.

Author Contributions: R.S. and J.D. conceived and designed the experiments; J.K. performed the experiments; J.K. and J.D. analyzed the data and contributed reagents and materials; J.D. and R.S. wrote the paper.

Acknowledgments: This work was carried out in Engine Testing Laboratory at BOSMAL Automotive Research and Development Institute Ltd., based in Bielsko-Biala, Poland. The funding to conduct the study and the costs for open access publication were covered by BOSMAL.

Conflicts of Interest: The authors declare no conflict of interest.

References

1. Yang, L.; Franco, V.; Campestrini, A.; German, J.; Mock, P. *NOx Control Technologies for Euro 6 Diesel Passenger Cars*; International Council on Clean Transportation: Washington, DC, USA, 2015. Available online: www.theicct.org/nox-tech-euro6-diesel-cars (accessed on 8 March 2018).

2. Merkisz, J.; Pielecha, J.; Radzimirski, S. *New Trends in Emission Control in the European Union*; Springer Tracts on Transportation and Traffic; Springer: Berlin, Germany, 2014; Volume 4. [CrossRef]

3. Yang, L.; Franco, V.; Mock, P.; Kolke, R.; Zhang, S.; Wu, Y.; German, J. Experimental assessment of NO_x emissions from 73 Euro 6 diesel passenger cars. *Environ. Sci. Technol.* **2015**, *49*, e14409–e14415. [CrossRef] [PubMed]

4. Marotta, A.; Pavlovic, J.; Ciuffo, B.; Serra, S.; Fontaras, G. Gaseous emissions from Light-Duty Vehicles: Moving from NEDC to the new WLTP test procedure. *Environ. Sci. Technol.* **2015**, *49*, 8315–8322. [CrossRef] [PubMed]

5. Merkisz, J.; Pielecha, J.; Bielaczyc, P.; Woodburn, J. *Analysis of Emission Factors in RDE Tests as Well as in NEDC and WLTC Chassis Dynamometer Tests*; SAE Technical Paper 2016-01-0980; SAE: Washington, DC, USA, 2016.

6. Johnson, T. Vehicular Emissions in Review. *SAE Int. J. Engines* **2016**, *9*, 1258–1275. [CrossRef]

7. McKinley, T.L.; Alleyne, A.G.; Lee, C.F. Mixture non-uniformity in SCR systems: Modeling and uniformity index requirements for steady-state and transient operation. *SAE Int. J. Fuels Lubr.* **2010**, *3*, 486–499. [CrossRef]

8. Brzezanski, M.; Sala, R. *A Study on the Indirect Urea Dosing Method in the Selective Catalytic Reduction System*; Materials Science and Engineering Volume 148; IOP Publishing: Bristol, UK, 2016. [CrossRef]

9. Sala, R.; Bielaczyc, P.; Brzezanski, M. Concept of Vaporized Urea Dosing in Selective Catalytic Reduction. *Catalysts* **2017**, *7*, 307. [CrossRef]

10. Song, X.; Naber, J.D.; Johnson, J.H. A study of the effects of NH3 maldistribution on a urea-selective catalytic reduction system. *Int. J. Engine Res.* **2015**, *16*, 213–222. [CrossRef]

11. Weltens, H.; Bressler, H.; Terres, F.; Neumaier, H.; Neumaier, H.; Rammoser, D. *Optimisation of Catalytic Converter Gas Flow Distribution by CFD Prediction*; SAE Technical Paper 930780; SAE: Warrendale, PA, USA, 1993. [CrossRef]

12. Nishad, K.; Sadiki, A.; Janicka, J. Numerical Investigation of AdBlue Droplet Evaporation and Thermal Decomposition in the Context of NOx-SCR Using a Multi-Component Evaporation Model. *Energies* **2018**, *11*, 222. [CrossRef]

13. Nova, I.; Tronconi, E. (Eds.) *Urea-SCR Technology for DeNOx after Treatment of Diesel Exhaust*; Fundamental and Applied Catalysis; Springer: Berlin, Germany, 2014; pp. 467–471. [CrossRef]

14. Zheng, G.; Fila, A.; Kotrba, A.; Floyd, R. *Investigation of Urea Deposits in Urea SCR Systems for Medium and Heavy Duty Trucks*; SAE Technical Paper 2010-01-1941; SAE: Warrendale, PA, USA, 2010. [CrossRef]
15. Koebel, M.; Strutz, E.O. Thermal and hydrolytic decomposition of urea for automative decomposition selective catalytic reduction systems: Thermochemical and practical aspects. *Ind. Eng. Chem. Res.* **2003**, *10*, 2093–2100. [CrossRef]
16. Baleta, J.; Martinjak, M.; Vujanović, M.; Pachler, K.; Wang, J.; Duić, N. Numerical analysis of ammonia homogenization for selective catalytic reduction application. *J. Environ. Manag.* **2017**, *203 Pt 3*, 1047–1061. [CrossRef] [PubMed]

catalysts

MDPI

Review

Strategies of Coping with Deactivation of NH$_3$-SCR Catalysts Due to Biomass Firing

Leonhard Schill and Rasmus Fehrmann *

Center for Catalysis and Sustainable Chemistry, Department of Chemistry, Building 207,
Technical University of Denmark, DK-2800 Kongens Lyngby, Denmark; leos@kemi.dtu.dk
* Correspondence: rf@kemi.dtu.dk; Tel.: +45-25-23-89

Received: 28 February 2018; Accepted: 22 March 2018; Published: 30 March 2018

Abstract: Firing of biomass can lead to rapid deactivation of the vanadia-based NH$_3$-SCR catalyst, which reduces NO$_x$ to harmless N$_2$. The deactivation is mostly due to the high potassium content in biomasses, which results in submicron aerosols containing mostly KCl and K$_2$SO$_4$. The main mode of deactivation is neutralization of the catalyst's acid sites. Four ways of dealing with high potassium contents were identified: (1) potassium removal by adsorption, (2) tail-end placement of the SCR unit, (3) coating SCR monoliths with a protective layer, and (4) intrinsically potassium tolerant catalysts. Addition of alumino silicates, often in the form of coal fly ash, is an industrially proven method of removing K aerosols from flue gases. Tail-end placement of the SCR unit was also reported to result in acceptable catalyst stability; however, flue-gas reheating after the flue gas desulfurization is, at present, unavoidable due to the lack of sulfur and water tolerant low temperature catalysts. Coating the shaped catalysts with thin layers of, e.g., MgO or sepiolite reduces the K uptake by hindering the diffusion of K$^+$ into the catalyst pore system. Intrinsically potassium tolerant catalysts typically contain a high number of acid sites. This can be achieved by, e.g., using zeolites as support, replacing WO$_3$ with heteropoly acids, and by preparing highly loaded, high surface area, very active V$_2$O$_5$/TiO$_2$ catalyst using a special sol-gel method.

Keywords: biomass firing; NH$_3$ SCR; potassium resistant catalysts; alumino silicate addition; coal ash; tail end placement; basic coating; KCl; aerosol

---

1. Introduction

The amount of electricity generated from firing solid biomass has been rising steeply in Europe over the last decades and is expected to continue to do so [1]. Similar trends are seen in other regions of the world [2,3]. Replacing fossil fuels, especially coal, by biomass aims at reducing the CO$_2$ emissions associated with thermal power plants [2,4–7]. Even though renewable energy sources like solar and wind power are more and more cost competitive [8] and make up an increasing share of power generation in most regions [9], some thermal power plant capacity is still needed due to the renewables' fluctuating nature and the current lack of sufficient storage capacity [10]. Firing and co-firing of biomass can cause several problems in the power plant like slagging and fouling problems in boilers [11], ash deposition on heat exchangers, and increased catalyst deactivation in the NO$_x$ removing unit [12–18]. This review deals with the last-mentioned problem.

NO$_x$ gases cause formation of photochemical smog, acid rain (HNO$_3$), and ground level ozone formation. These conditions in turn have adverse consequences on human life and ecosystems. NO$_x$ emissions from power plants can be reduced by modifications to the combustion process (primary measures) or post-combustion techniques (secondary measures). Secondary measures are typically more expensive but also afford a higher degree of NO$_x$ removal. Due to ever stricter environmental regulations, secondary measures are increasingly needed for power plants to be compliant. The highest degree of NO_x removal is achieved with selective catalytic reduction (SCR) using ammonia as the reductant [18,19]. The most widespread kind of catalyst is V_2O_5-WO_3/TiO_2 (VWT) [20,21]. The loading of the active species vanadia is typically between 1 and 5 wt.%, depending on the temperature of operation and the SO_2 content in the flue gas. Tungsta adds acid sites, reduces SO_2 oxidation, and reduces rutilization of anatase. The typical loading is between 5 and 10 wt.%.

The increased rate of catalyst deactivation experienced in biomass-fired plants is mostly caused by the relatively high alkali- and alkali-earth metal contents in most biomasses [11,17,20–24]. Alkaline metals cause deactivation by neutralizing the catalyst's acid sites, hence reducing the adsorption of NH_3 [13,25–30]. Potassium, in the form of submicron aerosols of mainly KCl and K_2SO_4 [31–33], is the most important poison due to both its relative abundance and high basicity [24,34]. Equation (1) gives a simplified neutralization reaction with M being any metal.

$$MCl + V - OH \rightarrow HCl + V - M \tag{1}$$

Other modes of deactivation like change in redox properties [35,36] and pore plugging [31] were reported to be of minor importance.

We have identified four kinds of strategies to deal with the high potassium content in biomasses: (1) potassium removal by adsorption; (2) tail-end placement of the SCR unit; (3) alkali barrier materials on the catalyst surface; and (4) intrinsically potassium resistant catalysts.

2. Strategies Coping with Potassium Rich Fuels

2.1. Potassium Removal by Adsorption

One way of reducing the impact of potassium salts is to minimize the amount taken up by the catalyst bed(s). An obvious strategy is to use an acidic guard bed in front of the catalyst modules. However, due to the high space velocities (5000–10,000 h^{-1}) in SCR units and the high KCl content of about 0.2^{-1} g Nm^{-3} of the flue gas [37,38], such a guard bed would probably be saturated too rapidly and require substantial space. Assuming a KCl concentration of 0.2 g Nm^{-3} in the flue gas, a "guard bed space velocity" of 20,000 h^{-1}, and a monolith density of 300 kg m^{-3}, 1 h of exposure translates into about 180 µmol K per gram. Even highly acidic substances like H-type zeolites with low Si/Al ratios only possess around 5000 µmol of acid sites per gram [39]. To the best of our knowledge, no guard beds have been implemented so far.

Wang et al. [24] have published a critical review on additives mitigating ash related problems. They have grouped the additives by the following four capture mechanisms: (1) chemical absorption and reaction; (2) physical absorption; (3) dilution and inert elements enrichment and (4) restraining and powdering effects. The first mentioned mechanism was singled out to be the most effective and is based on converting troublesome ash elements into high temperature stable compounds. Additives causing chemical binding can be based on alumino silicates such as, e.g., kaolin, coal fly ash, cat litter, clay minerals, and detergent zeolites. Alumino silicates bind potassium according to the simplified Equation (2).

$$Al_2O_3 \cdot xSiO_2 + 2KCl + H_2O \rightarrow K_2O \cdot Al_2O_3 \cdot SiO_2 + 2HCl_{(g)} \tag{2}$$

Addition of fly ash obtained from coal-fired plants is an industrially used strategy [40] to bind potassium. Coal fly-ash contains high levels of alumino silicates, which can bind potassium [14,40,41]. Coal fly-ash has the advantage of being abundant and low-cost. Diarmaid et al. [11] have very recently studied the efficacy of coal-fly ash in reducing the release of potassium from various biomass (white wood pellets, straw, and olive cake) pellets suspended in a methane flame. Additive loadings of 5, 15, and 25 wt.% were used. Olive cake requires larger amounts of alumino silicates to minimize potassium release, probably because it contains more potassium than the other two biomasses. In the presence of additive, up to 100% of K is retained, and in the wood and olive cake ash up to 80% is retained, demonstrating the effectiveness of alumino silicates even when burning pure biomasses with high potassium contents.

Firing coal with up to 10% [42,43] or even 20% [44] of biomass has also been reported to result in acceptable catalyst stability, probably because the resulting coal fly ash adsorbs released potassium compounds.

Sulfates of, e.g., ammonia, iron, aluminum, and phosphates of ammonia and calcium, as well as phosphoric acid, have also been listed by Wang et al. A possible issue with using sulfates is an increased formation of SO_3. Injection of phosphorous-based "K-getter" compounds leads to the formation of, e.g., K_3PO_4 and $K_4P_2O_7$. Dahlin et al. [27] performed a multivariate analysis of six catalyst poisons (Na, K, Mg, P, S, and Zn) by impregnating monolithic VWT catalysts with corresponding metal precursor solutions. The obtained model showed that P dampens the deactivating effect of K and was explained by the formation of phosphates, preventing the interaction of potassium with vanadia. The effect of K_3PO_4 on the stability of a vanadia-based catalyst was investigated by Castellino et al. [45] by exposing full length monoliths to a flue gas containing between 100 mg of K_3PO_4 per Nm^3. 720 h of exposure caused almost 40% deactivation, which was mainly ascribed to potassium neutralizing the catalyst's acid sites and thereby resembling the deactivation by KCl. The authors concluded that binding K by P is not advantageous to the SCR unit.

2.2. Tail-End Placement of the SCR Unit

Wieck-Hansen et al. [15] studied the catalyst stability using a slip stream from a 150 MW coal-straw (80%/20%) fired power plant. The catalyst was exposed to the flue gas at 350 °C without prior de-dusting, simulating high-dust placement, and at 280 °C downstream of a baghouse filter, which reduced the particulate concentration from 100 to a few mg Nm^{-3}, simulating low-dust placement of the SCR unit. 2860 h of high-dust exposure caused about 35% activity loss, while 2350 h of low-dust exposure only caused 15% activity loss. The difference in stability can probably be explained by the removal of, e.g., KCl particles by the dust-filter. Tail-end placement would probably lead to an even higher stability because of the desulphurization unit further reducing the potassium content in the flue gas. Tail-end operation at the biomass co-fired Amager plant in Denmark indeed showed promising results between 2010 and 2012 [44]. Laboratory studies by Putluru et al. [46] have furthermore shown that heteropoly acid (instead of WO_3)-promoted catalysts with a high (3 and 5 wt.%) vanadia loading can retain more than 90% of their activity at 225 °C when poisoned with 100 µmol K $g_{catalyst}^{-1}$. A corresponding WO_3 promoted catalyst lost almost 50% of its activity. At 400 °C, the loss was reported to be around 70% [47]. Generally, potassium poisoning has a stronger relative effect at high temperatures [23,48], which is reflected by a lower apparent activation energy upon potassium poisoning [23], which is consistent with acid neutralization being the main mode of deactivation.

Kristensen et al. [49] reported excellent potassium tolerance and activity of sol-gel prepared 20 wt.% V_2O_5/TiO_2 at temperatures below 250 °C. The potassium loading introduced by KNO_3 impregnation was 280 µmol K $g_{catalyst}^{-1}$. A commercial reference catalyst got completely deactivated.

The major drawback with tail-end placement is that wet and dry SO_2 scrubbers typically reduce the flue gas temperature to about 50 and 150 °C, respectively. The VWT catalyst is not active enough at these temperatures, making costly reheating to 180–280 °C necessary. Over the last 10 to 15 years, a high number of reports on low-temperature SCR catalysts have appeared [50]. The aim of these studies is to make re-heating redundant. However, most of the reported catalysts are based on manganese, making them extremely sulfur and water sensitive. In 2014, we summarized literature findings on the effects of SO_2 and H_2O and could not find any convincing reports on sulfur and water-resistant manganese-based catalysts [51]. Here we only give some examples of reports on catalysts being severely affected by SO_2 and H_2O. Casapu et al. [52] studied $MnCeO_x$ and reported a 79% activity reduction at 150 °C by adding 5 vol.% of water to the simulated flue gas. Flue gases typically contain at least 5 vol.% of water. Exposing the same catalyst to 50 ppm of SO_2 for 30 min at 250 °C reduced the NO conversion from about 70 to 25%. Our group has experienced rapid and severe deactivation of MnFe/TiO$_2$ and MnFeCe/TiO$_2$ at 150 °C by SO$_2$ levels as low as 5 ppm [51,53]. The modes of deactivation were formation of (NH$_4$)$_2$Mn$_2$(SO$_4$)$_3$ and ammonium sulfates. Regeneration by heating to 400 °C was only effective with prior washing with base. 20 vol.% of water in the flue gas reduced the NO conversion over a MnFe/TiO$_2$ from over 90% to 30.6%. Doping with ceria did not improve the water tolerance. In 2018, Gao et al. [54] reviewed the sulfur and water tolerance of Mn-based catalysts at low temperature and concluded, among other things, that more long term studies are needed to validate the viability of this kind of catalyst under realistic conditions.

2.3. Coating Monoliths with Basic Substances

In order to reach the catalyst's acid sites, potassium, typically originating from submicron aerosols of KCl and K$_2$SO$_4$, first needs to be deposited on the external catalyst (monolith) surface [48]. From there, potassium needs to separate from its counter-ion and diffuse into the catalyst pores, most likely through a surface transport mechanisms involving acid sites [31,55]. In other words, potassium mobility becomes a determining factor in the poisoning mechanism of monolithic samples. A pilot plant study performed by Jensen et al. [48] investigated the potassium uptake and the resulting deactivation of plate type samples with various WO$_3$ (0, 7 wt.%) and V$_2$O$_5$ (1, 3, 6 wt.%) contents. According to ammonia chemisorptions measurements, both tungsta and vanadia add acid sites to the fresh samples, thereby favoring the potassium uptake. This, in turn, leads to an increased rate of deactivation, e.g., 600 h of KCl aerosol (0.12 μm) at 350 °C leads to 76, 81, 89, and 98% relative deactivation for 1%V$_2$O$_5$–0% WO$_3$, 3%V$_2$O$_5$–0% WO$_3$, 1%V$_2$O$_5$–7% WO$_3$, and 3%V$_2$O$_5$–7% WO$_3$, respectively. Based on these results, it is highly questionable if the commonly used strategy of simply increasing the number of surface acid sites is realistic under real life conditions. Despite the just quoted deactivation data, tungsta-free catalysts are not an option for biomass fired plants, because they start from a significantly lower base activity and probably suffer from rutilization over time.

Since the potassium uptake relies on acid sites on the outer monolith surface, it can be reduced by coating this surface with a basic material, thus reducing the relative rate of deactivation [23,56,57]. MgO and Sepiolite (Mg$_4$Si$_6$O$_{15}$(OH)$_2$·6H$_2$O) have been reported as effective barrier materials. These substances are, on the one hand, basic enough to hinder potassium from penetrating the catalyst wall, and, on the other hand, they do not cause deactivation on their own. Olsen et al. [56] coated a plate type catalyst with composition of 3 wt.% V$_2$O$_5$–7 wt.% WO$_3$/TiO$_2$ with 8.06 wt.% MgO resulting in a roughly 200 μm thick layer and performed a pilot plant exposure campaign with KCl aerosols for several hundred hours at 350 °C. The coating layer reduced the rate of deactivation from 0.91% to 0.24% per day. These percentages refer to the initial activity of the uncoated sample. However, the decreased rate of deactivation comes at the cost of an initial activity reduction of about 42%. This activity reduction was ascribed to increased gas phase diffusion limitations introduced by the MgO layer, slight poisoning by MgO on the outer layer of the catalyst, or a combination thereof. SEM-EDS measurements confirmed that the outer MgO layer very effectively prevented potassium from diffusing into the catalyst and that magnesium did not diffuse into the catalyst. Kristensen [23] very successfully used sepiolite as a binder material for making plate type catalysts from 20 wt.% V$_2$O$_5$/TiO$_2$ powder, reinforced silica sheets, and 20 wt.% sepiolite as binder. The resulting catalyst was exposed to a KCl aerosol for 632 h at 380 °C and thereafter crushed to a powder for lab scale activity measurements. A commercial of 3 wt.% V$_2$O$_5$–7 wt.% WO$_3$/TiO$_2$ plate type catalyst was used as reference. When tested at 400 °C after KCl exposure, the 20 wt.% V$_2$O$_5$/TiO$_2$-Sepiolite composite retained 68% of its activity, translating into a first order rate constant of about 1650 cm^3·g^{-1}·s^{-1}. The activity loss of the reference catalyst was 84%, and the resulting first order rate constant was reported to be only about 200 cm^3·g^{-1}·s^{-1}. These activity losses were compared with data from a corresponding incipient wetness (KNO$_3$) poisoning study. The losses experienced by the 20 wt.% V$_2$O$_5$/TiO$_2$-Sepiolite composite and the reference translate into impregnated K loadings of 75 and 172 μmol K g$^{-1}_{catalyst}$, strongly suggesting that sepiolite acts as a barrier material. This was confirmed by SEM-EDS measurements, showing that potassium mainly accumulated on the outer surface of the plate.

2.4. Intrinsically Potassium Resistant Catalysts

In this review "intrinsically potassium resistant" refers to catalysts that retain a high share of their activity, even when potassium is taken up from the flue gas and diffuses into the catalysts pore system. To the best of our knowledge, there is up to now no review on potassium tolerant catalysts.

The majority of studies mimic potassium poisoning by impregnation with potassium salts like e.g., KNO_3, K_2CO_3, and KCl followed by calcination. The resulting K-loaded catalysts are typically tested in powder form in lab scale reactors. Studies performed by different laboratories are often difficult to compare due to vastly different experimental conditions and benchmark catalysts. For example, using different potassium loadings and activity testing in different temperature regimes might lead to different conclusions. Benchmarking against catalyst of different potassium tolerance might also lead to different conclusions. Because of these shortcomings in comparability, we start this section with results from our laboratory, which tested a high number of alternative catalysts using identical or very similar experimental conditions.

Figures 1 and 2 present the potassium tolerance for an assortment of catalyst with various active metals (Fe, Cu, and V) and support materials (TiO_2, tungsto phosphoric acid (TPA) promoted TiO_2, mordernite (MOR), and sulfated ZrO_2). The retained activity clearly depends on the number of acid sites of the fresh catalysts, which in turn is very much a function of the support material.

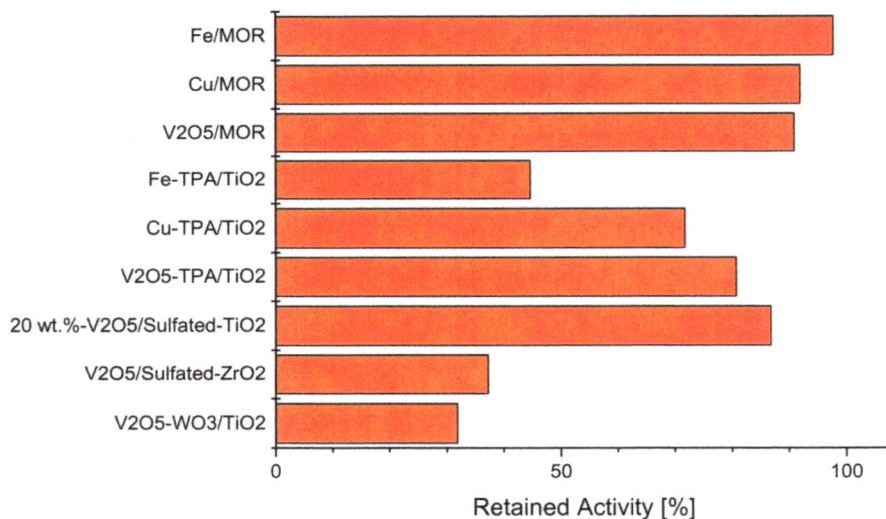

Figure 1. Retained activity at 400 °C upon impregnation with 100 µmol K $g_{catalyst}^{-1}$. (130 µmol K $g_{catalyst}^{-1}$ for V_2O_5/sulfated-ZrO_2). Reproduced from [47].

In this study, the highest alkali tolerance was obtained with MOR (Si/Al = 10)-based catalysts. Putluru et al. [58] optimized the Cu loading and tested the effect of 0, 250, and 500 µmol K/$g_{catalyst}$. 4 wt.% Cu/MOR retains about 60% of its initial activity after poisoning with 500 µmol K $g_{catalyst}^{-1}$, while only half that potassium loading causes more than 80% deactivation on a reference catalyst containing 3 wt.% vanadia and 7 wt.% tungsta. Cu/BEA (Si/Al = 25) and Cu/ZSM5 (Si/Al = 15) exhibit only slightly lower potassium tolerance than Cu/MOR does. Cu/Zeolite catalysts are not only very potassium resistant but also very active at 400 °C with first order rate constants of up to 1800 $cm^3g^{-1}s^{-1}$, while this value is only about 1000 $cm^3g^{-1}s^{-1}$ for the VWT reference catalyst [49]. Since the high potassium tolerance is at least in part due to the high number of acid sites on the zeolites, these materials will probably have to be protected by a thin layer of, e.g., MgO in order to avoid increased uptake of potassium containing particles. Another issue with Cu-based catalyst is their sulfur intolerance [59,60]. Vanadia supported on zeolites are very potassium tolerant but suffer from relative low activities. Likewise, iron-zeolite catalysts show comparatively low activities below 400 °C.

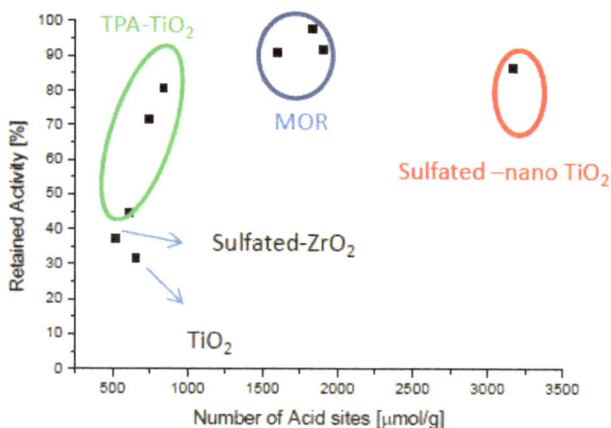

Figure 2. Retained activity at 400 °C upon impregnation with 100 μmol K $g_{catalyst}^{-1}$ (130 μmol K $g_{catalyst}^{-1}$ for V_2O_5/sulfated-ZrO_2) as a function of the number of acid sites of fresh catalysts. Generated with data from [47].

Putluru et al. [61] also demonstrated that the WO_3 component of the VWT catalyst can be replaced by heteropoly acids such as $H_3PW_{12}O_{40}$, $H_4SiW_{12}O_{40}$, $H_3PMo_{12}O_{40}$, and $H_4SiMo_{12}O_{40}$. Heteropoly acids contain more acid sites than WO_3, and these can probably serve as sacrificial sites, which is reflected by a higher potassium tolerance. Tungsto phosphoric acid (TPA, $H_3PW_{12}O_{40}$) resulted in the highest activity and the highest number of acid sites and is thermally more stable than the other heteropoly acids. Note that preparation of HPA-promoted catalyst is entirely based on impregnation and could therefore relatively easily be upscaled. A corresponding study on HPA-promoted Cu/TiO_2 and Fe/TiO_2 delivered similar results regarding activity and potassium tolerance [62]. The best HPAs were reported to be $H_3PW_{12}O_{40}$ and $H_3PMo_{12}O_{40}$. Another study by Putluru et al. [46] showed the effect of vanadia loading (3–6 wt.%) on the activity, and potassium tolerance of HPA promoted V_2O_5/TiO_2 catalysts at temperatures below 300 °C. The optimum vandia loading was 5 wt.%, and the resulting catalysts were almost unaffected by 100 μmol K $g_{catalyst}^{-1}$ when tested at 225 °C.

The most active and potassium-tolerant catalyst published by our laboratory is a 20 wt.% V_2O_5/TiO_2 prepared by a sol-gel route [23,47,49]. This catalyst contains about 5 times as many acid sites as the VWT reference and is at least twice as active. The conversion of SO_2 to SO_3 at 380 and 420 °C was reported to be less pronounced than over the VWT reference. This is probably due to the amorphous nature of vanadia, which is a result of the special sol-gel method of preparation. Impregnation with 500 μmol K $g_{catalyst}^{-1}$ resulted in the catalyst being about as active as the VWT reference loaded with only 150 μmol K $g_{catalyst}^{-1}$. Pilot scale exposure to KCl aerosols has demonstrated that a 20 wt.% V_2O_5/TiO_2—sepiolite composite catalyst suffers relatively little deactivation under more realistic conditions because of sepiolite impeding the surface diffusion of potassium.

Other research groups have also made many contributions over the last 10 years. Peng et al. [63] reported on the effect of doping V_2O_5-WO_3/TiO_2 with Ce. $V_{0.4}Ce_5W_5$/Ti and $V_{0.4}W_{10}$/Ti loaded with 1% K convert 30 and 18% NO, respectively, when tested at 400 °C. Du et al. [64] investigated the effect of Sb and Nb additives to V_2O_5/TiO_2. Both Sb and Nb have promotional effects on their own and can act synergistically. At 300 °C, potassium loaded VTi and $VSb_{0.5}NbTi$ show NO conversions of 22 and 43%, respectively. Gao et al. [65] reported on CeV mixed oxides supported on sulfated zirconia showing resistance to both SO_2 and potassium. The formation of $CeVO_4$ hinders the formation of $Ce_2(SO_4)_2$, and vanadia suppresses the absorption of SO_2, thus inhibiting NH_4HSO_4 formation. The potassium-loaded CeV mixed oxide catalyst maintains more than 95% NO conversion over 400 min of exposure to 600 ppm SO_2, while the conversion over the V free catalyst drops to about 65%.

To the best of our knowledge, very few reports exist on the potassium tolerance of hydrocarbon-SCR. Ethanol-SCR using Ag/Al_2O_3 is comparable in activity to NH_3-SCR over a 3 wt.% V_2O_5–7wt.% WO_3/TiO_2, however, is almost equally affected by potassium [66]. The mechanism of poisoning is not well understood but involves oxidation of ethanol to CO_2. Another problem with using ethanol instead of NH_3 as reductant is its much higher price. Furthermore, Ag/Al_2O_3 suffers from poor sulfur tolerance.

3. Conclusions

Different strategies of dealing with high concentrations of potassium in flue gases, typically present in biomass fired plants, were discussed. Addition of coal fly ash or other substances rich in alumino silicates like, e.g., kaolin is already an industrial practice and can very effectively bind potassium-containing aerosols. Lab scale experiments have demonstrated that this approach can be applied to various biomasses. The drawback of these additives is an increased concentration of particulates that need to be filtered off the flue gas. Tail-end placement of the SCR unit has also been demonstrated to work industrially. The major disadvantage of the tail-end placement, the expensive flue gas reheating to at least 180 °C, can, at present, not be avoided due to lack of catalysts that are sufficiently active, as well as due to sulfur and water tolerant at the outlet temperature of the desulfurization unit. Coating of shaped (monolith, plates) catalysts with thin layers of MgO or sepiolite was demonstrated to strongly reduce the rate of deactivation in pilot plant studies. The mildly basic nature of the protective layer impedes the diffusion of potassium ions into catalyst pores. Some of the studies report that the protective layer reduces the base activity by almost 50%, whereas others report a much lower penalty. Also, catalysts designed to tolerate higher loadings of potassium have been developed on a lab scale and include V, Cu, and Fe as active metals and heteropoly acid-promoted TiO_2, sulfated ZrO_2, and zeolites with a low Si/Al ratio as support materials. Most of the alternative catalysts gain their increased potassium tolerance from the addition of sacrificial acid sites. Since an increased number of acid sites was demonstrated to increase the potassium uptake from the flue gas, the addition of sacrificial sites probably only makes sense in conjunction with a protective layer of, e.g., MgO. The most promising results in this regard were obtained with a sol-gel prepared 20 wt.% V_2O_5/TiO_2 in combination with sepiolite. This composite material is about twice as active as the commercial, takes up less potassium from the flue gas, and experiences less deactivation per amount of adsorbed potassium. Avoiding the issue of reduced ammonia adsorption due to potassium uptake by using hydrocarbons as reductants has so far not been promising. We believe that mitigating the effect of potassium in biomass-fired units requires a multidimensional approach. For example, researchers should, if possible, demonstrate, using pilot plant studies, that promising catalyst formulations are also combinable with effective barrier materials that can minimize potassium uptake. Cost benefit analyses should also compare the use of alumino silicate addition with the use of potentially more expensive catalysts and tail-end placement of the SCR unit.

Acknowledgments: Energinet.dk, Denmark, LAB S.A., Lyon, France and DONG Energy, Denmark are acknowledged for financial contribution to the ForskEL project 12096 "Low temperature deNOx technologies in waste incineration and power plants".

Author Contributions: Leonhard Schill carried out the experimental work, design of figures and drafted the manuscript with further contribution from Rasmus Fehrmann who also managed the project.

Conflicts of Interest: The authors declare no conflict of interest.

References

1. Bjerg, J.; Aden, R.; Ogand, J.A.; Arrieta, J.A.; Hahlbrock, A.; Holmquist, L.; Kellberg, C.; KIp, W.N.; Koch, J.; Langnickel, U.; et al. Biomass 2020: Opportunities, Challenges and Solutions. *Eurelectric* **2001**, *72*, 18.

2. Agbor, E.; Zhang, X.; Kumar, A. A review of biomass co-firing in North America. *Renew. Sustain. Energy Rev.* **2014**, *40*, 930–943. [CrossRef]

3. Livingston, W.R. *The Status of Large Scale Biomass Firing: The Milling and Combustion of Biomass Materials in Large Pulverised Coal Boilers*; IEA Bioenergy Report 2016; IEA Bioenergy: Paris, France, 2006; ISBN 9781910154267. Available online: http://www.ieabcc.nl/publications/IEA_Bioenergy_T32_cofiring_2016.pdf (accessed on 27 February 2018).

4. Khorshidi, Z.; Ho, M.T.; Wiley, D.E. Techno-economic study of biomass co-firing with and without CO_2 capture in an Australian black coal-fired power plant. *Energy Procedia* **2013**, *37*, 6035–6042. [CrossRef]

5. Nuamah, A.; Malmgren, A.; Riley, G.; Lester, E. Biomass co-firing. *Compr. Renew. Energy* **2012**, *5*, 55–73. [CrossRef]

6. Mann, M.; Spath, P. A life cycle assessment of biomass cofiring in a coal-fired power plant. *Clean Prod. Process.* **2001**, *3*, 81–91. [CrossRef]

7. Kadiyala, A.; Kommalapati, R.; Huque, Z. Evaluation of the Life Cycle Greenhouse Gas Emissions from Different Biomass Feedstock Electricity Generation Systems. *Sustainability* **2016**, *8*, 1181. [CrossRef]

8. Renewable Energy Agency. *Renewable Power Generation Costs in 2017*; International Renewable Energy Agency: Abu Dhabi, UAE, 2018.

9. British Petroleum. *BP Statistical Review of World Energy 2017*; British Petroleum: London, UK, 2017; pp. 1–52. Available online: http://www.bp.com/content/dam/bp/en/corporate/pdf/energy-economics/statistical-review-2017/bp-statistical-review-of-world-energy-2017-full-report.pdf (accessed on 26 February 2018).

10. Lüdge, S. The value of flexibility for fossil-fired power plants under the conditions of the Strommark 2.0. *VGB PowerTech J.* **2017**, *3*, 212–214.

11. Clery, D.S.; Mason, P.E.; Rayner, C.M.; Jones, J.M. The effects of an additive on the release of potassium in biomass combustion. *Fuel* **2018**, *214*, 647–655. [CrossRef]

12. Klimczak, M.; Kern, P.; Heinzelmann, T.; Lucas, M.; Claus, P. High-throughput study of the effects of inorganic additives and poisons on NH3-SCR catalysts-Part I: V_2O_5-WO_3/TiO_2 catalysts. *Appl. Catal. B Environ.* **2010**, *95*, 39–47. [CrossRef]

13. Kern, P.; Klimczak, M.; Heinzelmann, T.; Lucas, M.; Claus, P. High-throughput study of the effects of inorganic additives and poisons on NH_3-SCR catalysts. Part II: Fe-zeolite catalysts. *Appl. Catal. B Environ.* **2010**, *95*, 48–56. [CrossRef]

14. Jensen-holm, H.; Lindenhoff, P.; Safronov, S. *SCR Design Issues in Reduction of NOx Emissions from Thermal Power Plants*; Haldor Topsøe A/S: Kongens Lyngby, Denmark, 2007.

15. Wieck-Hansen, K.; Overgaard, P.; Larsen, O.H. Cofiring coal and straw in a 150 MWe power boiler experiences. *Biomass Bioenergy* **2000**, *19*, 395–409. [CrossRef]

16. Baxter, L. *Biomass Impacts on SCR Catalyst Performance*; IEA Task 32 Bioenergy Report; IEA Bioenergy: Paris, France, 2005; Available online: http://task32.ieabioenergy.com/wp-content/uploads/2017/03/Combined_Final_Report_SCR.pdf (accessed on 25 February 2018).

17. Argyle, M.; Bartholomew, C. Heterogeneous Catalyst Deactivation and Regeneration: A Review. *Catalysts* **2015**, *5*, 145–269. [CrossRef]

18. Kiełtyka, M.A. *Influence of Biomass Co-Firing on SCR Catalyst Deactivation 2*; Techniques of NOx Emission Reduction; Instituto Superior Técnico, Universidade de Lisboa: Lisbon, Portugal, 2010.

19. Mladenović, M.; Paprika, M.; Marinković, A. Denitrification techniques for biomass combustion. *Renew. Sustain. Energy Rev.* **2018**, *82*, 3350–3364. [CrossRef]

20. Forzatti, P. Present status and perspectives in de-NOx SCR catalysis. *Appl. Catal. A Gen.* **2003**, *222*, 221–236. [CrossRef]

21. Busca, G.; Lietti, L.; Ramis, G.; Berti, F. Chemical and mechanistic aspects of the selective catalytic reduction of NOx by ammonia over oxide catalysts: A review. *Appl. Catal. B Environ.* **1998**, *18*. [CrossRef]

22. Knudsen, J.N.; Jensen, P.A.; Dam-Johansen, K. Transformation and release to the gas phase of Cl, K, and S during combustion of annual biomass. *Energy Fuels* **2004**, *18*, 1385–1399. [CrossRef]

31

23. Kristensen, S.B. deNOx Catalysts for Biomass Combustion. Ph.D. Thesis, Technical University of Denmark, Kongens Lyngby, Denmark, April 2013.

24. Wang, L.; Hustad, J.E.; Skreiberg, Ø.; Skjevrak, G.; Grønli, M. A critical review on additives to reduce ash related operation problems in biomass combustion applications. *Energy Procedia* **2012**, *20*, 20–29. [CrossRef]

25. Shao, Y.; Wang, J.; Preto, F.; Zhu, J.; Xu, C. Ash deposition in biomass combustion or co-firing for power/heat generation. *Energies* **2012**, *5*, 5171–5189. [CrossRef]

26. Davidsson, K.O.; Steenari, B.M.; Eskilsson, D. Kaolin addition during biomass combustion in a 35 MW circulating fluidized-bed boiler. *Energy Fuels* **2007**, *21*, 1959–1966. [CrossRef]

27. Dahlin, S.; Nilsson, M.; Bäckström, D.; Bergman, S.L.; Bengtsson, E.; Bernasek, S.L.; Pettersson, L.J. Multivariate analysis of the effect of biodiesel-derived contaminants on V_2O_5-WO_3/TiO_2 SCR catalysts. *Appl. Catal. B Environ.* **2016**, *183*, 377–385. [CrossRef]

28. Khodayari, R.; Odenbrand, C.U.I. Regeneration of commercial SCR catalysts by washing and sulphation: Effect of sulphate groups on the activity. *Appl. Catal. B Environ.* **2001**, *33*, 277–291. [CrossRef]

29. Forzatti, P.; Nova, I.; Tronconi, E.; Kustov, A.; Thøgersen, J.R. Effect of operating variables on the enhanced SCR reaction over a commercial V_2O_5-WO_3/TiO_2 catalyst for stationary applications. *Catal. Today* **2012**, *184*, 153–159. [CrossRef]

30. Lisi, L.; Lasorella, G.; Malloggi, S.; Russo, G. Single and combined deactivating effect of alkali metals and HCl on commercial SCR catalysts. *Appl. Catal. B Environ.* **2004**, *50*, 251–258. [CrossRef]

31. Zheng, Y.; Jensen, A.D.; Johnsson, J.E.; Thøgersen, J.R. Deactivation of V_2O_5-WO_3-TiO_2 SCR catalyst at biomass fired power plants: Elucidation of mechanisms by lab- and pilot-scale experiments. *Appl. Catal. B Environ.* **2008**, *83*, 186–194. [CrossRef]

32. Zeuthen, J.H.; Jensen, P.A.; Jensen, J.P.; Livbjerg, H. Aerosol Formation during the Combustion of Straw with Addition of Sorbents. *Energy Fuels* **2007**, 860–870. [CrossRef]

33. Sippula, O.; Lind, T.; Jokiniemi, J. Effects of chlorine and sulphur on particle formation in wood combustion performed in a laboratory scale reactor. *Fuel* **2008**, *87*, 2425–2436. [CrossRef]

34. Chen, L.; Li, J.; Ge, M. The poisoning effect of alkali metals doping over nano V2O5-WO3/TiO2 catalysts on selective catalytic reduction of NOx by NH_3. *Chem. Eng. J.* **2011**, *170*, 531–537. [CrossRef]

35. Wu, X.; Yu, W.; Si, Z.; Weng, D. Chemical deactivation of V_2O_5-WO_3/TiO_2 SCR catalyst by combined effect of potassium and chloride. *Front. Environ. Sci. Eng.* **2013**, *7*, 420–427. [CrossRef]

36. Peng, Y.; Li, J.; Huang, X.; Li, X.; Su, W.; Sun, X.; Wang, D.; Hao, J. Deactivation mechanism of potassium on the V2O 5/CeO2 catalysts for SCR reaction: Acidity, reducibility and adsorbed-NOx. *Environ. Sci. Technol.* **2014**, *48*, 4515–4520. [CrossRef] [PubMed]

37. Putluru, S.S.R.; Jensen, A.D. *Alternative Alkali Resistant deNOx Technologies*; Appendix-I: PSO Project 7318; Department of Chemical and Biochemical Engineering, Technical University of Denmark: Kongens Lyngby, Denmark, 2011; Available online: http://orbit.dtu.dk/files/6454472/NEI-DK-5569.pdf (accessed on 25 February 2018).

38. Sorvajävi, T.; Maunula, J.; Silvennoinen, J.; Toivone, J. *Optical Monitoring of KCl Vapor in 4 MW CFB Boiler during Straw Comubstion and Ferric Sulfate Injection*; Optics Laboratory, Department of Physics, Tampere University of Technology: Tampere, Finnland, 2014.

39. Niwa, M.; Katada, N.; Okumura, K. *Characterization and Design of Zeolite Catalysts (Solid Acidity, Shape Selectivity and Loading Properties)*; Springer: Berlin/Heidelberg, Germany, 2010; p. 25, ISBN 9783642126192.

40. Wu, H.; Shafique, M.; Arendt, P.; Sander, B.; Glarborg, P. Impact of coal fly ash addition on ash transformation and deposition in a full-scale wood suspension-firing boiler. *Fuel* **2013**, *113*, 632–643. [CrossRef]

41. Steenari, B.M.; Lindqvist, O. High-temperature reactions of straw ash and the anti-sintering additives kaolin and dolomite. *Biomass Bioenergy* **1998**, *14*, 67–76. [CrossRef]

42. Overgaard, P.; Sander, B.; Junker, H.; Friborg, K.; Larsen, O.H. Small-scale CHP Plant based on a 75 kWel Hermetic Eight Cylinder Stirling Engine for Biomass Fuels—Development, Technology and Operating Experiences. In Proceedings of the 2nd World Conference on Biomass for Energy, Industry and Climate Protection, Rome, Italy, 10–14 May 2004; pp. 1261–1264.

43. Henderson, C. Cofiring of biomass in coal-fired power plants—European experience The role of biomass in Europe. Presented at the FCO/IEA CCC Workshops on Policy and Investment Frameworks to Introduce CCT in Hebei and Shandong Provinces, China, 8–9 and 13–14 January 2015.

44. Jensen-holm, H.; Castellino, F.; White, T.N. SCR DeNOx catalyst considerations when using biomass in power generation. In Proceedings of the Power Plant Air Pollutant Control "MEGA" Symposium, Baltimore, MD, USA, 20–23 August 2012; Available online: https://www.topsoe.com/sites/default/files/scr_denox_catalyst_considerations_when_using_biomass_in_power_generation_2012.ashx__0.pdf (accessed on 23 February 2018).

45. Castellino, F.; Jensen, A.D.; Johnsson, J.E.; Fehrmann, R. Influence of reaction products of K-getter fuel additives on commercial vanadia-based SCR catalysts. Part I. Potassium phosphate. *Appl. Catal. B Environ.* **2009**, *86*, 196–205. [CrossRef]

46. Putluru, S.S.R.; Schill, L.; Godiksen, A.; Poreddy, R.; Mossin, S.; Jensen, A.D.; Fehrmann, R. Promoted V2O5/TiO2 catalysts for selective catalytic reduction of NO with NH_3 at low temperatures. *Appl. Catal. B Environ.* **2016**, *183*, 282–290. [CrossRef]

47. Putluru, S.S.R.; Kristensen, S.B.; Due-Hansen, J.; Riisager, A.; Fehrmann, R. Alternative alkali resistant deNOx catalysts. *Catal. Today* **2012**, *184*, 192–196. [CrossRef]

48. Olsen, B.K.; Kügler, F.; Castellino, F.; Jensen, A.D. Poisoning of vanadia based SCR catalysts by potassium: Influence of catalyst composition and potassium mobility. *Catal. Sci. Technol.* **2016**, *6*, 2249–2260. [CrossRef]

49. Kristensen, S.B.; Kunov-Kruse, A.J.; Riisager, A.; Rasmussen, S.B.; Fehrmann, R. High performance vanadia-anatase nanoparticle catalysts for the Selective Catalytic Reduction of NO by ammonia. *J. Catal.* **2011**, *284*, 60–67. [CrossRef]

50. Fu, M.; Li, C.; Lu, P.; Qu, L.; Zhang, M.; Zhou, Y.; Yu, M.; Fang, Y. A review on selective catalytic reduction of NOx by supported catalysts at 100–300 °C—Catalysts, mechanism, kinetics. *Catal. Sci. Technol.* **2014**, *4*, 14–25. [CrossRef]

51. Schill, L. Alternative Catalysts and Technologies for NOx Removal from Biomass- and Waste-Fired Plants. Ph.D. Thesis, Technical University of Denmark, Kongens Lyngby, Denmark, April 2014.

52. Casapu, M.; Kröcher, O.; Elsener, M. Screening of doped MnOx-CeO2 catalysts for low-temperature NO-SCR. *Appl. Catal. B Environ.* **2009**, *88*, 413–419. [CrossRef]

53. Fehrmann, R.; Jensen, A.D. *Low Temperature deNOx Technologies for Biomass and Waste Fired Power Plants*; Forskel Energinet.dk Project No. 12096; Department of Chemistry, Technical University of Denmark: Kongens Lyngby, Denmark, 2017; Available online: https://energiforskning.dk/sites/energiteknologi.dk/files/slutrapporter/12096_slutrapport.pdf (accessed on 20 February 2018).

54. Gao, C.; Shi, J.-W.; Fan, Z.; Gao, G.; Niu, C. Sulfur and Water Resistance of Mn-Based Catalysts for Low-Temperature Selective Catalytic Reduction of NOx: A Review. *Catalysts* **2018**, *8*, 11. [CrossRef]

55. Zheng, Y.; Jensen, A.D.; Johnsson, J.E. Deactivation of V2O5-WO3-TiO2SCR catalyst at a biomass-fired combined heat and power plant. *Appl. Catal. B Environ.* **2005**, *60*, 253–264. [CrossRef]

56. Olsen, B.J.; Kügler, F.; Castellino, F.; Schill, L.; Fehrmann, R.; Jensen, A.D. Deactivation of SCR Catalysts by Potassium: A Study of Potential Alkali Barrier Materials. *VGB PowerTech J.* **2017**, *3*, 56–64.

57. Jensen, A.D.; Castellino, F.; Rams, P.D.; Pedersen, J.B.; Putluru, S.S.R. Deactivation Resistant Catalyst for Selective Catalytic Reduction of NOx. U.S. Patent 2012/0315206A1, 13 December 2012.

58. Putluru, S.S.R.; Riisager, A.; Fehrmann, R. Alkali resistant Cu/zeolite deNOxcatalysts for flue gas cleaning in biomass fired applications. *Appl. Catal. B Environ.* **2011**, *101*, 183–188. [CrossRef]

59. Cheng, Y.; Lambert, C.; Kim, D.H.; Kwak, J.H.; Cho, S.J.; Peden, C.H.F. The different impacts of SO2 and SO3 on Cu/zeolite SCR catalysts. *Catal. Today* **2010**, *151*, 266–270. [CrossRef]

60. Kumar, A.; Smith, M.A.; Kamasamudram, K.; Currier, N.W.; Yezerets, A. Chemical deSOx: An effective way to recover Cu-zeolite SCR catalysts from sulfur poisoning. *Catal. Today* **2016**, *267*, 10–16. [CrossRef]

61. Putluru, S.S.R.; Jensen, A.D.; Riisager, A.; Fehrmann, R. Heteropoly acid promoted V2O5/TiO2 catalysts for NO abatement with ammonia in alkali containing flue gases. *Catal. Sci. Technol.* **2011**, *1*, 631. [CrossRef]

62. Putluru, S.S.R.; Mossin, S.; Riisager, A.; Fehrmann, R. Heteropoly acid promoted Cu and Fe catalysts for the selective catalytic reduction of NO with ammonia. *Catal. Today* **2011**, *176*, 292–297. [CrossRef]

63. Peng, Y.; Li, J.; Shi, W.; Xu, J.; Hao, J. Design strategies for development of SCR catalyst: Improvement of alkali poisoning resistance and novel regeneration method. *Environ. Sci. Technol.* **2012**, *46*, 12623–12629. [CrossRef] [PubMed]

64. Du, X.; Gao, X.; Fu, Y.; Gao, F.; Luo, Z.; Cen, K. The co-effect of Sb and Nb on the SCR performance of the V2O5/TiO2catalyst. *J. Colloid Interface Sci.* **2012**, *368*, 406–412. [CrossRef] [PubMed]

65. Gao, S.; Wang, P.; Yu, F.; Wang, H.; Wu, Z. Dual resistance to alkali metals and SO$_2$: Vanadium and cerium supported on sulfated zirconia as an efficient catalyst for NH$_3$-SCR. *Catal. Sci. Technol.* **2016**, *6*, 8148–8156. [CrossRef]
66. Schill, L.; Sankar, S.; Putluru, R.; Funk, C.; Houmann, C.; Fehrmann, R.; Degn, A. Applied Catalysis B: Environmental Ethanol-selective catalytic reduction of NO by Ag/Al$_2$O$_3$ catalysts: Activity and deactivation by alkali salts. *Appl. Catal. B Environ.* **2012**, *127*, 323–329. [CrossRef]

catalysts

MDPI

Article

Mn-Ce-V-WO$_x$/TiO$_2$ SCR Catalysts: Catalytic Activity, Stability and Interaction among Catalytic Oxides

Xuteng Zhao [1], Lei Mao [1] and Guojun Dong [1,2,*]

[1] School of Material Science and Chemical Engineering, Harbin Engineering University, Harbin 150001, China; zhaoxuteng@hrbeu.edu.cn (X.Z.); maolei0119@163.com (L.M.)
[2] Key Laboratory of Superlight Material and Surface Technology of Ministry of Education, College of Material Science and Chemical Engineering, Harbin Engineering University, Harbin 150001, China
* Correspondence: dgj1129@163.com; Tel.: +86-180-4508-1027

Received: 28 January 2018; Accepted: 6 February 2018; Published: 12 February 2018

Abstract: A series of Mn-Ce-V-WO$_x$/TiO$_2$ composite oxide catalysts with different molar ratios (active components/TiO$_2$ = 0.1, 0.2, 0.3, 0.6) have been prepared by wet impregnation method and tested in selective catalytic reduction (SCR) of NO by NH$_3$ in a wide temperature range. These catalysts were also characterized by X-ray diffraction (XRD), Transmission Electron Microscope (TEM), in situ Fourier Transform infrared spectroscopy (in situ FTIR), H$_2$-Temperature programmed reduction (H$_2$-TPR) and X-ray photoelectron spectroscopy (XPS). The results show the catalyst with a molar ratio of active components/TiO$_2$ = 0.2 exhibits highest NO conversion value between 150 °C to 400 °C and good resistance to H$_2$O and SO$_2$ at 250 °C with a gas hourly space velocity (GHSV) value of 40,000 h^{-1}. Different oxides are well dispersed and interact with each other. NH$_3$ and NO are strongly adsorbed on the catalyst surface and the adsorption of the reactant gas leads to a redox cycle with the valence state change among the surface oxides. The adsorption of SO$_2$ on Mn^{4+} and Ce^{4+} results in good H$_2$O and SO$_2$ resistance of the catalyst, but the effect of Mn and Ce are more than superior water and sulfur resistance. The diversity of valence states of the four active components and their high oxidation-reduction performance are the main reasons for the high NO conversion in this system.

Keywords: composite oxide catalyst; NH$_3$-SCR of NO; lifetime; H$_2$O and SO$_2$ resistance

1. Introduction

The selective catalytic reduction (SCR) of NO$_x$ with NH$_3$ in the presence of O$_2$ has been widely used to control the emissions of NO$_x$ from mobile and stationary sources, such as coal-fired power plants and automobiles [1–4]. V$_2$O$_5$-WO$_3$/TiO$_2$ is the most used commercial catalyst for SCR of NO$_x$ at a relatively high temperature platform of 300 °C to 400 °C [5–7]. However, the drawbacks of this catalyst system cannot be ignored. The operating temperature window is narrow and it has low catalytic activity for deNO$_x$ at low temperature, for instance. Moreover, the catalytic performance can be seriously deactivated by H$_2$O and SO$_2$ in the emission [8–10]. Therefore, a low temperature catalyst with good resistance to SO$_2$ and H$_2$O is urgently needed for the SCR system to improve the situation.

So far, a lot of research has been done. Manganese-containing catalysts have attracted much attention due to their relatively high catalytic activity for the conversion of NO$_x$ at low temperature, including Ni-MnO$_x$/TiO$_2$ [6], MnO$_x$-TiO$_2$ [11,12], Cr-MnO$_x$ [13], Ce-MnO$_x$ [14,15] and Ca-MnO$_x$/TiO$_2$ [10]. Fang et al. [16] have compared the conversion of NO on MnO$_x$/TiO$_2$, MnO$_x$/CNT and nano-flaky MnO$_x$/CNT and found that nano-flaky MnO$_x$/CNTs presents favourable stability, H$_2$O resistance and better NO-SCR activity at a more extensive operating temperature window between 150 °C to 300 °C. However, SO$_2$ leads to the irreversible deactivation of MnO$_x$/TiO$_2$ [9]. In recent years, it has been reported that Ce-based catalysts reveal excellent SCR activity in the presence of SO_2 at 300–400 °C [17–19]. Yang and co-workers [18] have investigated the effect of SO_2 on the SCR reaction over CeO_2. The results have indicated that the adsorption of NH_3 over CeO_2 is obviously promoted with the sulfation of CeO_2, resulting in an obvious promotion of the Eley–Rideal mechanism.

In many cases, one or two active components can hardly handle all the situations of the SCR system. In this work, four elements of manganese, cerium, vanadium and tungsten have been used as the active components, with a reduction in the amount of vanadium. Then, a series of Mn-Ce-V-WO_x/TiO_2 composite oxide catalysts with different molar ratios of active components/TiO_2 have been prepared by an impregnation method. In addition, X-ray diffraction (XRD), Transmission electron microscopy (TEM), in situ Fourier transform infrared (FT-IR) spectroscopy, H_2 temperature-programmed reduction (H_2-TPR) and X-ray photoelectron spectroscopy (XPS) have been measured. The catalyst with a molar ratio of 0.2 exhibits the highest conversion of NO between 150 °C to 400 °C at a gas hourly space velocity (GHSV) of 40,000 h^{-1}. This catalyst shows long lifetime and superior resistance to H_2O and SO_2 at 250 °C. It is remarkable to note that the catalyst is not deactivated at all. Moreover, the analysis results also prove that the oxides on the catalyst surface interact with each other and enhance the redox properties of the oxides and the adsorption of reaction gas, resulting in a redox cycle with the valence state change of the surface oxides. This is probably the main reason for its unique performance.

2. Results and Discussion

2.1. Catalytic Behavior

All SCR activity of catalysts have been tested several times and error analysis have been performed, shown in the form of error bars in the figures. Figure 1A shows the SCR performances of Mn-Ce-V-WO_x/TiO_2 composite oxide catalysts with different molar ratios of active components/TiO_2 = 0.1, 0.2, 0.3, 0.6. For comparison, the SCR performances of V_2O_5/TiO_2, WO_3/TiO_2, MnO_2/TiO_2, CeO_2/TiO_2 and TiO_2 are given in Figure 1B. Compared with the single-component catalysts, composite catalysts show better catalytic activity. The NO conversion of all single component catalysts is less than 40% before 250 °C, while all Mn-Ce-V-WO_x/TiO_2 composite catalysts are more than 70% from 150 °C to 350 °C. Among them, the NO conversion of the catalyst with active components/TiO_2 = 0.2 molar ratio is even above 90% from 150 °C to 400 °C. Obviously, the addition of various active components can significantly improve the catalytic activity. The similar condition has also been observed through evaluating the SCR performances of two/three-components catalysts (Figure S1 in the Supplementary Materials). The catalytic activity of the three-components of MnO_2-V_2O_5-WO_3/TiO_2 and CeO_2-V_2O_5-WO_3/TiO_2 is significantly higher than that of reference catalysts. Furthermore, remarkably, the performance of CeO_2-V_2O_5-WO_3/TiO_2 is better than MnO_2-V_2O_5-WO_3/TiO_2, especially between 200 to 300 °C. It can be inferred that the effect of MnO_2 and CeO_2 on the V_2O_5-WO_3/TiO_2 catalyst is different. Moreover, with temperature increasing, the SCR activity of the Mn-Ce-V-WO_x/TiO_2 catalyst with the highest amount of loading decreases obviously, but others remain (shown in Figure 1A). It should also be mentioned that the catalyst with molar ratio of active components/TiO_2 = 0.2 has the highest catalytic performance at 150 °C. Thus, the optimal molar ratio (Mn-Ce-V-WO_x/TiO_2 = 0.2) is obtained, and the catalytic performance will drop when the ratio is low or high, especially when it is high. These results may be due to the different dispersion of the active components on the support surface and the strong interaction among them. Further explanation will be elaborated in the following analyses.

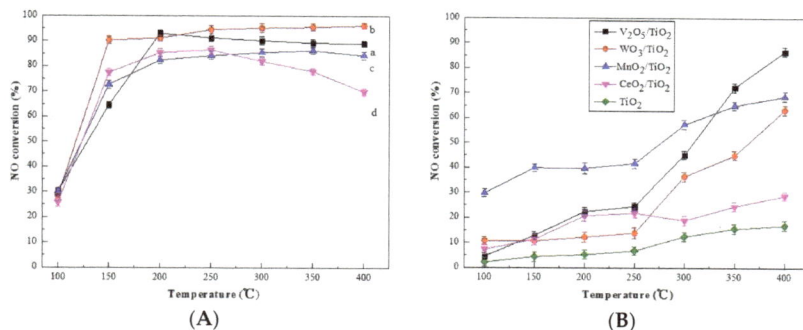

Figure 1. (A) Selective catalytic reduction (SCR) activity of Mn-Ce-V-WO$_x$/TiO$_2$ composite catalysts with the molar ratio of active components/TiO$_2$ at different values: (a) 0.1; (b) 0.2; (c) 0.3; (d) 0.6. **(B)** SCR activity of V$_2$O$_5$/TiO$_2$, WO$_3$/TiO$_2$, MnO$_2$/TiO$_2$, CeO$_2$/TiO$_2$ and TiO$_2$. Reaction conditions: [NO] = [NH$_3$] = 1500 ppm, [O$_2$] = 3%, gas hourly space velocity (GHSV) = 40,000 h^{-1}.

2.2. XRD Analysis

The XRD patterns of the Mn-Ce-V-WO$_x$/TiO$_2$ catalysts are shown in Figure 2. The XRD diffractions of the catalyst with the molar ratio of Mn-Ce-V-WO$_x$/TiO$_2$ = 0.1 are well indexed to anatase TiO$_2$ (JCPDS 21-1272) [20,21], indicating that active components are well dispersed on the surface of support. When the molar ratio reaches 0.2, the active components on the surface of the support begins to aggregate, but no strong diffraction peaks of crystal phase appear. If the molar ratio continues to be increased to 0.6, crystalline phases of all active components appear and TiO$_2$ peaks are diluted. The results are consistent with the SCR performances. Catalytic activity could be improved with the transition of active components from highly dispersed state to slightly aggregated state, yet a large amount of components' aggregation will also decrease the catalytic activity [22].

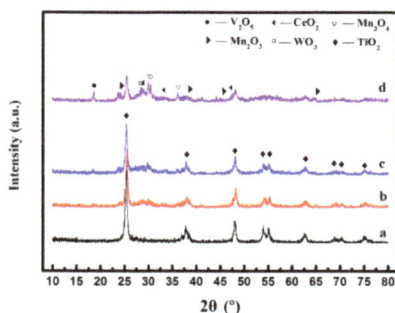

Figure 2. X-ray diffraction (XRD) patterns of Mn-Ce-V-WO$_x$/TiO$_2$ composite oxide catalysts with different molar ratio of active components/TiO$_2$: (a) 0.1; (b) 0.2; (c) 0.3; (d) 0.6.

XRD patterns of the reference catalysts are shown in Figure S2 in the Supplementary Materials. It can be seen that the reference catalysts with a single component only have the diffraction peak of anatase TiO$_2$. Interestingly, weak diffraction peaks of V$_2$O$_5$ and WO$_3$ are observed at approximately 23.2° and 34.3° in the spectrum of V$_2$O$_5$-WO$_3$/TiO$_2$ catalyst and 17.9°, 29.1°, 29.3° and 36.1° in the spectrum of MnO$_2$-V$_2$O$_5$-WO$_3$/TiO$_2$ catalyst, but the peaks of these angles disappear in the cerium containing systems; even the peaks of CeO$_2$ and MnO$_2$ have not appeared with the same molar ratio components added. The addition of Ce can be well dispersed, meanwhile, it can make other active components disperse more uniformly and avoid the large aggregation of active components on the support surface, which can make it easier for all active components to contact the reaction gas and improve the catalytic activity of this catalyst system.

2.3. TEM and HRTEM Analysis

TEM images of the catalyst with a molar ratio of 0.2 are presented in Figure 3. The diameter of the catalyst particle is about 20 nm (Figure 3a,b). It is similar to TiO_2 support. The HRTEM micrographs (Figure 3c,d) display better defined contours, exhibiting higher crystalline order (Figure 3e–h). The d-spacing is measured at ca. 0.352 nm, ascribable to the (101) crystal planes of anatase TiO_2. The detected spacing from different positions is the same, matching well with the XRD pattern. The result further indicates that active components are well dispersed on the surface. The addition of four active components occupy more positions on the surface of support, resulting in an increase of acidic sites exposed on the surface, which improves the ability of NH_3 adsorption; this indirectly corresponds to the in situ FTIR test results of NH_3 adsorption characteristic peak occurring at high temperature in the following.

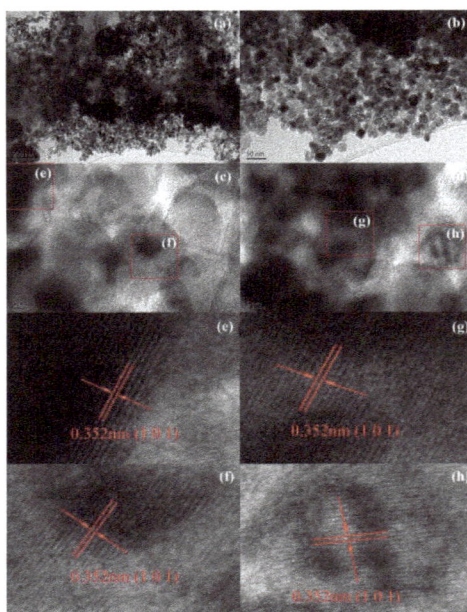

Figure 3. Transmission Electron Microscopy (TEM) and High Resolution Transmission Electron Microscopy (HRTEM) images of Mn-Ce-V-WO$_x$/TiO$_2$ catalyst with molar ratio of 0.2: (**a,b**) TEM images; (**c,d**) HRTEM images; (**e,f**) amplified images of (**c**); (**g,h**) amplified images of (**d**).

High Resolution Transmission Electron Microscopy (HRTEM) images of the reference catalysts with a molar ratio of 0.6 are presented in Figure S3 in the Supplementary Materials. Compared with the former, HRTEM images of the 0.6-ratio catalyst show different diffraction fringes with different distances. Only the spacing of the representative lattice fringes and the corresponding crystal planes are shown and corresponding crystals of active components are detected, indicating that mass aggregation of active components have appeared on the support surface. The results are similar to XRD analysis, and specific information is presented in Supplementary Materials. However, there is no characteristic fringes of TiO_2, indicating that TiO_2 may be completely covered by the active components.

2.4. Catalyst Stability and H_2O/SO_2 Resistance

The catalyst with the best SCR performance has been selected for the stability and H_2O/SO_2 resistance tests. From the above information of activity tests, catalysts begin to remain stable at 250 °C, and the reaction may reach balance at this temperature. So, H_2O/SO_2 resistance and the 100 h stability test of the catalyst have been performed at 250 °C, as Figure 4 and Figure S4 in the Supplementary Materials present, respectively. It can be seen that the activity of the catalyst remains at about 95.3% during 100 h with small fluctuations, and no decline is observed, which indicates that the catalyst has a longer service life. Furthermore, it is obvious that the SCR activity maintains a very high NO conversion with small fluctuations when 5 vol % H_2O, 100 ppm SO_2 or both are introduced into the typical reactant gas. The same results have been obtained after repeating the test several times. It has been reported that MnO_2, V_2O_5 and WO_3 are easily irreversible deactivated by SO_2; only CeO_2 shows high activity in the presence of SO_2 [6,7,9]. In this research, although MnO_2 may still be attacked by SO_2, the adulteration of CeO_2 is significant to the catalyst for resisting the effect of SO_2. The addition of these two elements makes V and W less affected by SO_2 in the system. The result also indicates that the Mn-Ce-V-WO$_x$/TiO$_2$ catalyst with molar ratio of active components/TiO$_2$ = 0.2 has strong resistance to H_2O and SO_2. However, it cannot be assumed that the addition of Mn and Ce is simply attacked instead of V and W because the high NO conversion of the catalyst can still last a long time rather than being deactivated quickly in the presence of SO_2 and H_2O. There is a stronger interaction among the four active ingredients and the role of each element is crucial. Further investigations on the adsorption of reactive molecules and the effects of SO_2 and H_2O on the catalyst have been performed by in situ FTIR below.

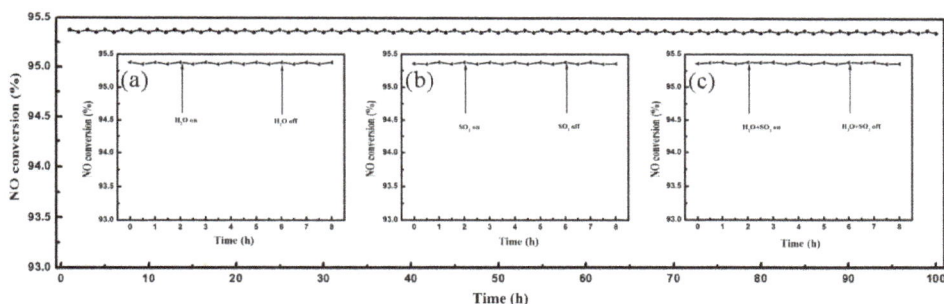

Figure 4. H_2O and SO_2 resistance of Mn-Ce-V-WO$_x$/TiO$_2$ catalyst with molar ratio of 0.2 at 250 °C: insert (a–c) H_2O and SO_2 resistance. Reaction conditions: [NO] = [NH$_3$] = 1500 ppm, [O$_2$] = 3%, [H$_2$O] = 5%, [SO$_2$] = 100 ppm, GHSV = 40,000 h^{-1}.

2.5. In Situ FTIR Analysis

The adsorption, transformation and desorption of the reactant gas on the catalyst's surface play an important role in the SCR reaction and they influence the reaction process. Under the condition of continuous exposure of typical reactant gas with or without SO_2, the in situ FTIR spectra of the catalyst with a molar ratio of 0.2 are shown in Figure 5. Several bands at 975, 1208, 1437, 1596, 1670, 3164, 3260, 3353 and 3395 cm^{-1} are observed under the condition of continuous exposure of typical reactant gas without SO_2 (shown in Figure 5a). The bands at 975 and 1670 cm^{-1} disappear over 150 °C, pointing to NH$_4^+$ species on Brønsted acid sites [23,24]. Then, ν(N$-$H) and δ(N$-$H) bands of NH$_3$ adsorbed on Lewis sites at 1208 cm^{-1} and between 3100 and 3400 cm^{-1} are observed [25,26]. Additionally, the band at 1437 cm^{-1} weakens gradually with increasing temperature, but exists even at 400 °C. So, it should be assigned to the adsorption of the nitro specie on the catalyst surface [27]. Moreover, the band at 1596 cm^{-1} is attributed to the characteristic band of bridged nitrate species and it is not detected when the temperature is higher than 150 °C due to its poor stability. It can be concluded that NH$_3$

and NO are strongly adsorbed on the catalyst surface even when the temperature reaches 400 °C, and the adsorption of NH_3 on Lewis acid sites and NO have strong interaction with catalyst surface oxides. The strong interaction on the support surface promotes the strong adsorption of NH_3 and the formation of nitrate species. At 400 °C, the activity of the catalyst remains at 90%, indicating that the catalyst still has a SCR reaction at this temperature. The results correspond well to the previous SCR activity test.

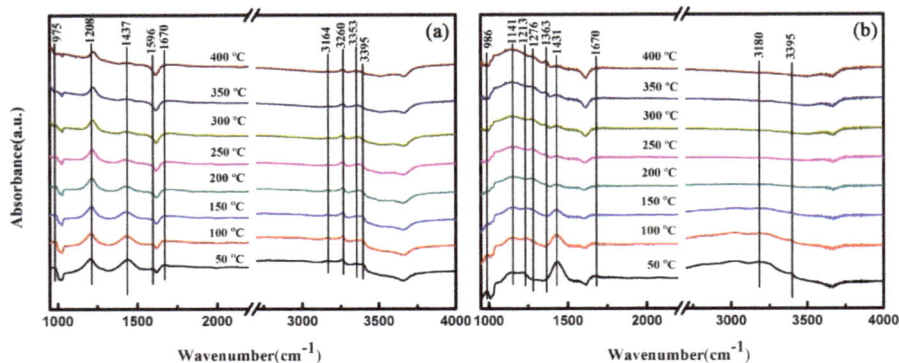

Figure 5. In situ Fourier Transform infrared spectroscopy (in situ FTIR) spectra of the catalyst with molar ratio of 0.2 under the condition of continuous exposure of typical reactant gas with or without SO_2: (a) without SO_2; (b) with SO_2.

When SO_2 is introduced into the typical reactant gas, several new bands at 986, 1141, 1213, 1276 and 1363 cm^{-1} have appeared (Figure 5b). The band at 1141 cm^{-1} is attributed to the stretching motion of adsorbed sulfate on the surface of the catalyst [1,28]. The bands at 986 and 1213 cm^{-1} are the shift of 975 and 1208 cm^{-1}, due to their disappearance above 150 °C. However, the band at 1276 cm^{-1} turns up at 200 °C, which can be assigned to the follow-up shift of 1213 cm^{-1} [25,29,30]. At the same time, the band at 1363 cm^{-1} may stem from asymmetric vibration of S–O bands of ammonium sulfate ((NH_4)$_2SO_4$) and grows in intensity over temperature [9,29]. The results indicate that there is no competitive adsorption between SO_2 and NO, and the adsorption of NH_3 on the catalyst can be promoted by adsorbed sulfate. Nevertheless, though SO_2 is surely adsorbed on the catalyst and sulfates are formed, the catalytic still maintains high activity, which is the same as before. It can be proposed that SO_2 is only adsorbed on specific sites on the catalyst and Ce^{4+} may be one of the specific sites according to a previous report [18].

2.6. H_2-TPR Analyses

To investigate the redox performance of the catalysts, H_2-TPR and H_2-TPR peak-differentiation-imitating of Mn-Ce-V-WO$_x$/TiO$_2$ catalyst profiles have been obtained, as Figure 6 shows. Figure 6A are the H_2-TPR curves of the catalysts. V_2O_5/TiO$_2$ shows a strong reduction peak at 428 °C, corresponding to the vanadia species reduction of $V^{5+}\rightarrow V^{3+}$ [31]. WO$_3$/TiO$_2$ shows two reduction peaks at 521 °C and 690 °C, and both of them are assigned to the reduction of $W^{6+}\rightarrow W^{4+}$ [32]. MnO$_2$/TiO$_2$ displays a wide reduction peak at 367 °C ascribed to the reduction of MnO$_2$ to Mn_2O_3 and Mn_2O_3 to Mn_3O_4 and a sharp reduction peak at 504 °C corresponds to the reduction of Mn_3O_4 to MnO [32–34]. The weak reduction peak for CeO$_2$/TiO$_2$ at 520 °C is caused by the reduction of low amount surface $Ce^{4+}\rightarrow Ce^{3+}$ [7,32,35]. Though the reference catalysts show multiple peaks in the temperature range of 300 °C to 800 °C, only one wide peak appears in the H_2-TPR profile of the catalyst with the molar ratio of 0.2 at 521 °C and peak shape is very wide, which is due to the overlapping of multiple peaks, even the possible interaction among oxides on the support surface.

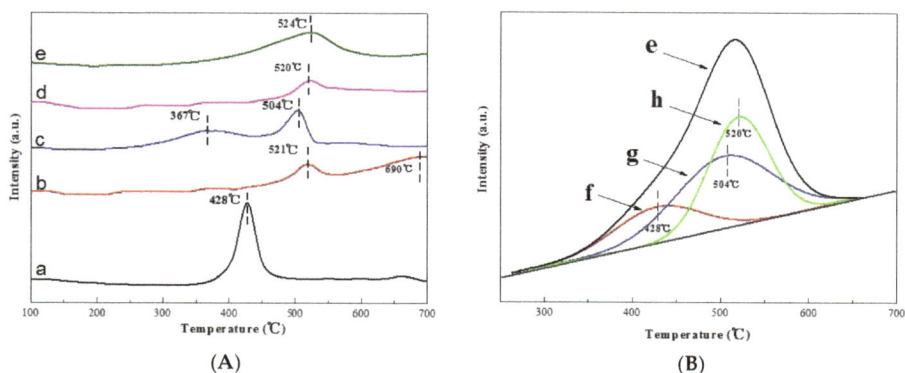

Figure 6. (A) H_2-Temperature programmed reduction (H_2-TPR) profiles of the catalysts: (a) V_2O_5/TiO_2; (b) WO_3/TiO_2; (c) MnO_2/TiO_2; (d) CeO_2/TiO_2; (e) Mn-Ce-V-WO_x/TiO_2 catalyst with molar ratio of 0.2. (B) H_2-TPR peak-differentiation-imitating of Mn-Ce-V-WO_x/TiO_2 catalyst: (e) Mn-Ce-V-WO_x/TiO_2 catalyst with molar ratio of 0.2; (f) the oxides of V; (g) the oxides of Mn; (h) the oxides of W and Ce.

To further investigate the important role of active components in the redox performance and the interaction among them, we have presented the H_2-TPR peak-differentiation-imitating of Mn-Ce-V-WO_x/TiO_2 catalyst in Figure 6B. Hydrogen consumption of single component catalysts and corresponding percent in composite catalysts have been listed in Table 1. Considering that the peak temperature of the WO_3 and CeO_2 is very close, peaks of the two oxides are unified into one peak here. The reduction peaks of one-component catalysts of V_2O_5 and MnO_2 are larger than those of WO_3 and CeO_2. However, in the composite catalyst, the single peak of MnO_2 and the overlap peak of CeO_2 and WO_3 are larger than the peak of V_2O_5, which is different from the redox properties of the four active components in the corresponding single-component catalysts. It can be inferred that the effect of Mn and Ce are more than superior water and sulfur resistance in the catalytic system. The interaction among the four active components enhances the redox properties of the oxides of the two elements. This can also be evidenced by the valence change of elements on the surface of the composite catalyst after H_2 reduction in the XPS below. It is well known that WO_3 acts as a promoter in traditional V_2O_5/WO_3-TiO_2 catalysts and contributes to the electron transfer among different valences of V element, which is also one of the main reasons for the high SCR activity of traditional catalysts; the oxide of Ce has a better ability to store and release oxygen. However, in the composite catalyst system, both the redox properties of Ce and W have been promoted, which shared part of the role of V_2O_5 with better redox ability in the catalytic activity. Therefore, both Ce and W contribute to the high NO conversion of the composite catalyst. The areas of reduction peaks of all oxides are different from the corresponding peaks of catalysts with a single component, which appeared in the case of the composite catalysts. The absolute amount of the four active components in the composite catalyst is less than that in the corresponding single-component catalysts, thus the hydrogen consumption of the four active components in the composite catalyst decreases. Among them, the reduction peaks of MnO_2 and CeO_2+WO_3 are strong but the reduction peak of V is relatively weak. The reduction of hydrogen consumption of V_2O_5 should not be attributed to the strong interaction among active components causing inhibition on V_2O_5 redox properties, but rather to the fact that in the composite catalyst, the amount of V_2O_5 is less than that in the other three active components. According to XPS test results, there is still a large amount of V_2O_5 that has been reduced in the composite catalyst. It can be inferred that they all play major roles in the redox reaction.

Table 1. Hydrogen consumption of single component catalysts and corresponding percent in composite catalyst.

	Catalysts						
	V2O5/TiO2	WO3/TiO2	MnO2/TiO2	CeO2/TiO2	Composite Catalyst	Peak of V2O5	Peak of MnO2	Peak of WO3 + CeO2
Hydrogen consumption (μmol/g)	90.27	33.01	126.24	22.89	99.0	19.44	43.09	40.02

The results of H_2-TPR suggest that active components of the composite catalyst reveal characteristic redox properties, which is different from that of being alone, and the results indicate that the existence of a strong interaction among the active components on the surface of support is conducive to the promotion of reduction performance.

2.7. XPS Analysis

In the case of NO-SCR with NH_3 over the composite oxide catalyst, the catalyst will take part in the reaction, which has been reported extensively [13,36,37]. In order to obtain the information about the oxidation states of the active components on the catalyst, V 2p, W 4f, Mn 2p, Ce 3d and O 1s XPS spectra of the catalysts are recorded as shown in Figure 7. According to the literature, two main peaks are attributed to V $2p_{3/2}$ and V $2p_{1/2}$ in the V 2p XPS spectra [38,39]. However, in this work, only the V $2p_{3/2}$ level can be used to distinguish vanadium oxide species in different chemical states, and the peak of V $2p_{1/2}$ is very weak and hindered by O 1s satellites. The peaks of the catalysts are separated into three peaks at the binding energies of 515.6, 516.3 and 516.8 eV, assigned to V^{3+}, V^{4+} and V^{5+}, respectively [40–42]. Two main peaks in the W 4f XPS spectra are due to W $4f_{7/2}$ and W $4f_{5/2}$, and the W $4f_{7/2}$ peak was divided into 34.6 (W^{5+}) and 35.2 eV (W^{6+}) and the W $4f_{5/2}$ peak is divided into 36.9 (W^{5+}) and 37.7 eV (W^{6+}), respectively [43,44]. Two main peaks assigned to Mn $2p_{3/2}$ and Mn $2p_{1/2}$ are observed, and the Mn $2p_{3/2}$ peaks are separated into three peaks at the binding energies of 640.3 ± 0.2, 641.3 ± 0.2 and 642.6 ± 0.2 eV, corresponding to Mn^{2+}, Mn^{3+} and Mn^{4+}, respectively [1,36,45]. The Ce 3d XPS spectra can be fitted into ten peaks: 880.3, 885.9, 898.8 and 903.9 eV assigned to Ce^{3+} and 882.5, 888.8, 898.4, 901.0, 907.5 and 916.7 eV associated with Ce^{4+}. The O 1s peak is fitted into two sub-bands, one at 529.6 eV and the other at 531.1 eV, which can be attributed to the lattice oxygen O^{2-} and the surface adsorbed oxygen such as O_2^{2-}, O^-, O_2^- or OH^-, respectively [13,46]. Compared with the XPS spectra of V 2p, W 4f and O 1s, Mn 2p and Ce 3d XPS spectra with the low intensity occur in 8 h using the composite catalyst with the molar ratio of 0.2 (shown in Figure 7C,D). The smaller nano-size and relative amount can lower the intensity of XPS spectra [45,47–49]. In this case, the low intensity can be attributed to sulfates covered on Mn and Ce species of the catalyst. Thus, the above conjecture that SO_2 adsorbed only on specific sites on the catalyst is proved to some extent.

The quantitative analyses of V, W, Mn, Ce and O species on the catalysts from XPS spectra have been listed in Table 2. For comparison, the quantitative analysis of each element after H_2 reduction has also been listed in Table 2. H_2 for the reduction of the catalyst is complete. The XPS analysis after reduction of H_2 shows that in the single-component catalysts, valence of V on the support surface is all converted from +5 to +3, and valence of Mn is also greatly converted from +4 to +2 and +3. Conversely, valence of W and Ce are relatively hard to be reduced, which proves the high redox properties of V and Mn compared to W and Ce as we know. However, the condition of valence change becomes different in the composite catalyst. It is easily determined that the proportion of different valence states of each element on a single active substance is quite different from those on the composite catalysts, which further indicates that the interaction exists among the oxides in the composite catalysts. The reduction of Mn increases, and the reduction of vanadium is only slightly reduced. It is worth noting that higher valence W and Ce are reduced, which is consistent with the result of H_2-TPR above. Interestingly, compared to the single-component catalysts, the amount of oxygen adsorbed on the composite catalyst surface decrease. Through the TEM and HRTEM analysis, we know that the addition of four active components occupy more positions on the surface of support, resulting in an increase of acidic sites exposed on the surface, which can explain why the surface adsorption of oxygen decreases. Although there are references supporting that the increase of adsorbed oxygen can promote the SCR catalytic reaction, more importantly, the catalytic reaction involves a large number of electron transfer processes and it is clear that the diversity of valence states of the four active components and their high oxidation-reduction performance are the main reason for the high SCR performance in this system.

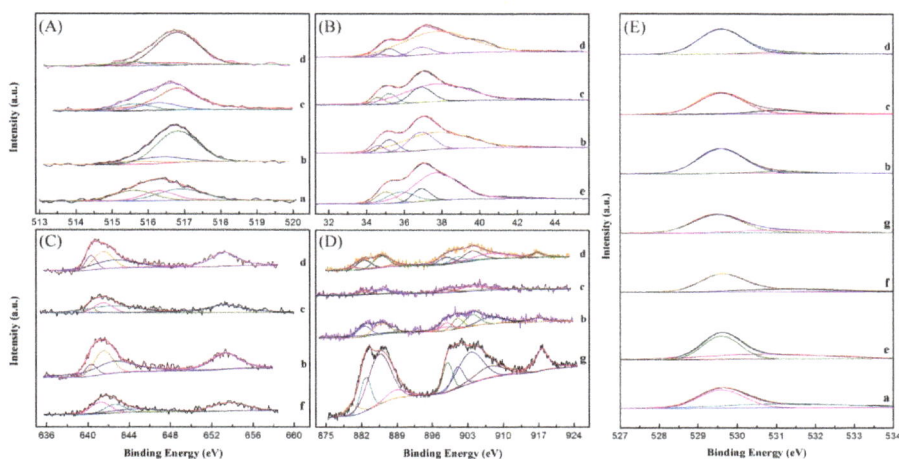

Figure 7. V 2p (**A**); W 4f (**B**); Mn 2p (**C**); Ce 3d (**D**) and O 1s (**E**) XPS spectra of the catalysts: (a) V_2O_5/TiO_2; (b) Fresh-catalyst with molar ratio of 0.2; (c) 8 h used-catalyst with molar ratio of 0.2; (d) 100 h used-catalyst with molar ratio of 0.2; (e) WO_3/TiO_2; (f) MnO_2/TiO_2; (g) CeO_2/TiO_2.

Table 2. The percent of different valence states for V 2p, W 4f, Mn 2p, Ce 3d and O 1s.

Catalysts	V 2p			W 4f		Mn 2p			Ce 3d		O 1s	
	V^{3+}	V^{4+}	V^{5+}	W^{5+}	W^{6+}	Mn^{2+}	Mn^{3+}	Mn^{4+}	Ce^{3+}	Ce^{4+}	Lattice Oxygen	Adsorbed Oxygen
V_2O_5/TiO_2	32.4	27.1	40.5	-	-	-	-	-	-	-	60.0	40.0
V_2O_5/TiO_2 **	100	0	0	-	-	-	-	-	-	-	53.5	46.5
WO_3/TiO_2	-	-	-	20.7	79.3	-	-	-	-	-	58.8	41.2
WO_3/TiO_2 **	-	-	-	46.8	53.2	-	-	-	-	-	58.4	41.6
MnO_2/TiO_2	-	-	-	-	-	3.9	52.7	43.5	-	-	68.7	31.3
MnO_2/TiO_2 **	-	-	-	-	-	22.3	61	16.7	-	-	28.1	71.9
CeO_2/TiO_2	-	-	-	-	-	-	-	-	40.3	59.7	80.4	19.6
CeO_2/TiO_2 **	-	-	-	-	-	-	-	-	46.1	53.9	63.5	36.5
Composite catalyst *	7.5	18.3	74.2	28.8	71.2	10.8	34.0	55.2	45.3	54.7	92.7	7.3
Composite catalyst **	79.9	3.6	16.5	54	46	35.4	50.2	14.4	60.6	39.4	78.9	21.1
Composite catalyst ***	14.1	15.9	70.0	24.2	75.8	6.9	30.8	62.3	57.3	42.7	80.2	19.8
Composite catalyst ****	7.9	9.0	83.1	10.8	89.2	18.3	38.5	43.2	47.3	52.7	92.5	7.5

* Fresh; ** After hydrogen reduction; *** 8 h used for resistance to H_2O and SO_2; ***** 100 h used for stability test.

After the 100 h test, the concentration of O specie almost remains the same. Low valence state V and W species partly transform into the high valence state while high valence state Mn and Ce species partly convert into the low valence state. However, the change is faint, which corresponds with its superior stability. However, the concentrations of V, Mn, Ce and O species display a novel change compared with the fresh-catalyst after the 8 h H_2O and SO_2 resistance test. The concentration of Ce^{4+} decreases, which can be caused by SO_2. As a reducing agent, SO_2 induces a transformation from Ce^{4+} to Ce^{3+} on the surface, resulting in the formation of $Ce_2(SO_4)_3$ [20]. The decline of Mn^{2+} concentration may be owing to the sulfation on Mn^{2+} bound by $SO_4{}^{2-}$ [50]. Furthermore, adsorption of H_2O and SO_2 on the sample can also affect the concentration of adsorbed oxygen, which makes it increase. Moreover, the increasing of V^{3+} concentration can be due to the enhanced adsorption of NH_3 by the sulfation on the catalyst surface. It can be inferred that the addition of manganese and cerium reduced the adverse effects of SO_2 on the catalytic activity owing to the result that SO_2 molecules are only adsorbed on Ce^{4+} and Mn^{2+} species on the composite catalyst.

3. Materials and Methods

3.1. Catalysts Preparation

All the catalysts were prepared by wet impregnation method. As active components, $Mn(CH_3COO)_2$, NH_4VO_3, $(NH_4)_{10}W_{12}O_{41} \cdot xH_2O$ (50 wt %) and $Ce(NO_3)_3 \cdot 6H_2O$ with a molar ratio of 1:0.2:1:1 were completely dissolved in 60 mL citric acid solution (10 wt % $C_6H_8O_7 \cdot H_2O$) and TiO_2 support powder was subsequently suspended in the obtained solution with the molar ratio of active components/TiO_2 = 0.1, 0.2, 0.3, 0.6, respectively. The obtained mixture was stirred at room temperature for 1 h, then heated to 100 °C with stirring until the excess water was evaporated completely. The obtained solid was further dried at 120 °C for 12 h, and then calcined in air at 500 °C for 3 h. For comparison, V_2O_5/TiO_2, WO_3/TiO_2, MnO_2/TiO_2, CeO_2/TiO_2, V_2O_5-WO_3/TiO_2, MnO_2-CeO_2/TiO_2, MnO_2-V_2O_5-WO_3/TiO_2 and CeO_2-V_2O_5-WO_3/TiO_2 catalysts with the same percentage content of corresponding active components were also prepared through the similar process.

3.2. NH$_3$-SCR Activity Test

The SCR activity tests were carried out in a fixed-bed quartz reactor (0.3 mL catalyst; 40–60 mesh). The typical reactant gas composition contained: 1500 ppm NO, 1500 ppm NH_3, 3% O_2, 5 vol % H_2O (when added), 100 ppm SO_2 (when added) and balance Ar. The total gas flow rate was 200 mL/min and regulated by mass flow controllers (Sevenstar D08 series Flow Readout Boxed, Beijing Sevenstar Electronics Co., Ltd., Beijing, China), corresponding to the gas hourly space velocity (GHSV) of about 40,000 h^{-1}. The activity tests were examined at the temperature range of 100–400 °C. The NO outlet concentration was continuously monitored by the ThermoStar Gas Analysis System GSD320 analyzer (Pfeiffer Vacuum GmbH, Berlin, Germany).

3.3. Catalyst Characterization

XRD patterns were obtained using a D/MAX-3A Auto X-ray diffractometer (Rigaku Corporation, Tokyo, Japan) with Cu Kα radiation. The X-ray source was operated at 40 kV and 40 mA. The diffraction patterns were taken in the 2θ range of 10–90° at a scan speed of 15° min^{-1} and a resolution of 0.02°. TEM and HRTEM were performed on a FEI Teccai G2S-Twin electron microscope (PHILIPS, Amsterdam, The Netherlands).

X-ray photoelectron spectra were obtained with K-Alpha spectrometer (Thermo Fisher Scientific, Waltham, MA, America) using Al Kα (1486.7 eV) radiation as the excitation source with a precision of ± 0.3 eV. All binding energies were referenced to the C 1s line at 284.6 eV.

H_2-TPR were performed on a sp-6801 gas chromatograph analyzer (Shandong Lunan Ruihong Chemical Instrument Co., Ltd., Tengzhou, China) using 0.1 g catalyst. The sample was first pretreated in Ar (30 mL·min^{-1}) at 50 °C for 1h and then heated up to 800 °C at a rate of 10 °C·min^{-1} under 5 vol % H_2/Ar. The consumption of H_2 was measured by a thermal conductivity detector (TCD, BEIJING BUILDER ELECTRONIC TECHNOLOGY CO., LTD., Beijing, China).

In situ FTIR spectra were recorded by a Fourier transform infrared spectrometer (Nicolet 6700, Thermo Fisher Scientific, Waltham, MA, America) equipped with a smart collector and MCT detector cooled by liquid N_2, collecting 32 scans with a resolution of 4 cm^{-1}. The catalysts were firstly treated at 500 °C in Ar for 1 h, then cooled down to 50 °C. Subsequently, the SCR reactant gas were introduced to the catalyst for 30 min, and then flushed with Ar for 10 min. The spectra were normally collected at temperatures ranging from 50 °C to 400 °C in a continuous NH_3 and NO flow. The background spectrum was recorded with the flowing of NH_3 and NO and subtracted from the sample spectrum.

4. Conclusions

Mn-Ce-V-WO_x/TiO_2 composite oxide catalysts with the molar ratio of active components/TiO_2 = 0.2, prepared by wet impregnation method, exhibit high NO conversion between 150 °C to 400 °C and good resistance to H_2O and SO_2 at 250 °C with a GHSV value of 40,000 h^{-1}. Four active components are well dispersed on TiO_2 surface and no crystalline phase is formed, but they can aggregate slightly, which is beneficial to the promotion of catalytic activity. NH_3 and NO are strongly adsorbed on the catalyst surface even at 400 °C, indicating that the catalyst still has a SCR reaction at this temperature. SO_2 is only adsorbed on Mn^{4+} and Ce^{4+} in this catalyst system, resulting in the formation of sulfates. However, the effect of Mn and Ce are more than superior water and sulfur resistance in the catalytic system. In addition, the characteristic redox properties of the catalyst are due to the existence of interaction among the active components on the support surface, and the interaction among them also enhances the redox properties of Ce and W oxides. Thus, all active components play major roles in the redox reaction, and the diversity of valence states of the four active components and their high oxidation-reduction performance are the main reason for the high SCR performance in this system.

Supplementary Materials: The following are available online at http://www.mdpi.com/2073-4344/8/2/76/s1, Figure S1: SCR activity test of two/three-component catalysts, Figure S2: XRD analysis of reference catalysts, Figure S3: HRTEM images of the composite oxide catalysts, Figure S4: The lifetime of Mn-Ce-V-WO_x/TiO_2 catalyst with 0.2 molar ratio.

Acknowledgments: The authors gratefully acknowledge financial support from the Fundamental Research Funds for the central universities (HEUCF20136910012), Internationl Cooperation Special of The State Ministry of Science and Technology (2015DFR60380) and Advanced Technique Project Funds of The Manufacture and Information Ministry. And we thank Zhijuan Zhao and Xiaoyu Zhang from Analysis and Test Center of Chinese Sciences Academy Institute of Chemistry for their help in XPS.

Author Contributions: Guojun Dong conceived and designed the experiments; Xuteng Zhao performed the experiments; Lei Mao contributed reagents/materials/analysis tools; Xuteng Zhao and Lei Mao analyzed the data and wrote the paper.

Conflicts of Interest: The authors declare no conflict of interest.

References

1. Park, E.; Kim, M.; Jung, H.; Chin, S.; Jurng, J. Effect of Sulfur on Mn/Ti Catalysts Prepared Using Chemical Vapor Condensation (CVC) for Low-Temperature NO Reduction. *ACS Catal.* **2013**, *3*, 1518–1525. [CrossRef]
2. Boudali, L.K.; Ghorbel, A.; Grange, P. SCR of NO by NH_3 over V_2O_5 supported sulfated Ti-pillared clay: Reactivity and reducibility of catalysts. *Appl. Catal. A* **2006**, *305*, 7–14. [CrossRef]
3. Lónyi, F.; Solt, H.E.; Pászti, Z.; Valyon, J. Mechanism of NO-SCR by methane over Co,H-ZSM-5 and Co,H-mordenite catalysts. *Appl. Catal. B* **2014**, *150–151*, 218–229. [CrossRef]
4. Nicosia, D.; Czekaj, I.; Kröcher, O. Chemical deactivation of V_2O_5/WO_3-TiO_2 SCR catalysts by additives and impurities from fuels, lubrication oils and urea solution. *Appl. Catal. B* **2008**, *77*, 228–236. [CrossRef]
5. Si, Z.; Weng, D.; Wu, X.; Li, J.; Li, G. Structure, acidity and activity of CuO_x/WO_x-ZrO_2 catalyst for selective catalytic reduction of NO by NH_3. *J. Catal.* **2010**, *271*, 43–51. [CrossRef]
6. Thirupathi, B.; Smirniotis, P.G. Nickel-doped Mn/TiO_2 as an efficient catalyst for the low-temperature SCR of NO with NH_3: Catalytic evaluation and characterizations. *J. Catal.* **2012**, *288*, 74–83. [CrossRef]
7. Chen, L.; Li, J.; Ge, M. Promotional Effect of Ce-doped V_2O_5-WO_3/TiO_2 with Low Vanadium Loadings for Selective Catalytic Reduction of NOx by NH_3. *J. Phys. Chem. C* **2009**, *113*, 21177–21184. [CrossRef]

8. Xie, G.; Liu, Z.; Zhu, Z.; Liu, Q.; Ge, J.; Huang, Z. Simultaneous removal of SO_2 and NOx from flue gas using a CuO/Al_2O_3 catalyst sorbentII. Promotion of SCR activity by SO_2 at high temperatures. *J. Catal.* **2014**, *224*, 42–49. [CrossRef]

9. Pan, S.; Luo, H.; Li, L.; Wei, Z.; Huang, B. H_2O and SO_2 deactivation mechanism of MnOx/MWCNTs for low-temperature SCR of NOx with NH_3. *J. Mol. Catal. A* **2013**, *377*, 154–161. [CrossRef]

10. Tian, Q.; Liu, H.; Yao, W.; Wang, Y.; Liu, Y.; Wu, Z.; Wang, H.; Weng, X. SO_2 Poisoning Behaviors of Ca-Mn/TiO_2 Catalysts for Selective Catalytic Reduction of NO with NH_3 at Low Temperature. *J. Nanomater.* **2014**, *2014*, 1–6. [CrossRef]

11. Xie, J.; Fang, D.; He, F.; Chen, J.; Fu, Z.; Chen, X. Performance and mechanism about MnOx species included in MnOx/TiO_2 catalysts for SCR at low temperature. *Catal. Commun.* **2012**, *28*, 77–81. [CrossRef]

12. Ettireddy, P.R.; Ettireddy, N.; Boningari, T.; Pardemann, R.; Smirniotis, P.G. Investigation of the selective catalytic reduction of nitric oxide with ammonia over Mn/TiO_2 catalysts through transient isotopic labeling and in situ FT-IR studies. *J. Catal.* **2012**, *292*, 53–63. [CrossRef]

13. Chen, Z.; Yang, Q.; Li, H.; Li, X.; Wang, L.; Tsang, S.C. Cr–MnOx mixed-oxide catalysts for selective catalytic reduction of NOx with NH_3 at low temperature. *J. Catal.* **2010**, *276*, 56–65. [CrossRef]

14. Zhang, D.; Zhang, L.; Shi, L.; Fang, C.; Li, H.; Gao, R.; Huang, L.; Zhang, J. In situ supported MnO(x)-CeO(x) on carbon nanotubes for the low-temperature selective catalytic reduction of NO with NH_3. *Nanoscale* **2013**, *5*, 1127–1136. [CrossRef] [PubMed]

15. Zhang, L.; Zhang, D.; Zhang, J.; Cai, S.; Fang, C.; Huang, L.; Li, H.; Gao, R.; Shi, L. Design of meso-TiO_2@MnO(x)-CeO(x)/CNTs with a core-shell structure as DeNO(x) catalysts: Promotion of activity, stability and SO_2-tolerance. *Nanoscale* **2013**, *5*, 9821–9829. [CrossRef] [PubMed]

16. Fang, C.; Zhang, D.; Cai, S.; Zhang, L.; Huang, L.; Li, H.; Maitarad, P.; Shi, L.; Gao, R.; Zhang, J. Low-temperature selective catalytic reduction of NO with NH_3 over nanoflaky MnOx on carbon nanotubes in situ prepared via a chemical bath deposition route. *Nanoscale* **2013**, *5*, 9199–9207. [CrossRef] [PubMed]

17. Jin, R.; Liu, Y.; Wang, Y.; Cen, W.; Wu, Z.; Wang, H.; Weng, X. The role of cerium in the improved SO_2 tolerance for NO reduction with NH_3 over Mn-Ce/TiO_2 catalyst at low temperature. *Appl. Catal. B* **2014**, *148–149*, 582–588. [CrossRef]

18. Yang, S.; Guo, Y.; Chang, H.; Ma, L.; Peng, Y.; Qu, Z.; Yan, N.; Wang, C.; Li, J. Novel effect of SO_2 on the SCR reaction over CeO_2: Mechanism and significance. *Appl. Catal. B* **2013**, *136–137*, 19–28. [CrossRef]

19. Zhang, D.; Zhang, L.; Fang, C.; Gao, R.; Qian, Y.; Shi, L.; Zhang, J. MnOx–CeOx/CNTs pyridine-thermally prepared via a novel in situ deposition strategy for selective catalytic reduction of NO with NH_3. *RSC Adv.* **2013**, *3*, 8811–8819. [CrossRef]

20. Shu, Y.; Aikebaier, T.; Quan, X.; Chen, S.; Yu, H. Selective catalytic reaction of NOx with NH_3 over Ce-Fe/TiO_2-loaded wire-mesh honeycomb: Resistance to SO_2 poisoning. *Appl. Catal. B* **2014**, *150–151*, 630–635. [CrossRef]

21. Thirupathi, B.; Smirniotis, P.G. Co-doping a metal (Cr, Fe, Co, Ni, Cu, Zn, Ce, and Zr) on Mn/TiO_2 catalyst and its effect on the selective reduction of NO with NH_3 at low-temperatures. *Appl. Catal. B* **2011**, *110*, 195–206. [CrossRef]

22. Tang, F.; Zhuang, K.; Yang, F.; Yang, L.; Xu, B.; Qiu, J.; Fan, Y. Effect of Dispersion State and Surface Properties of Supported Vanadia on the Activity of V_2O_5/TiO_2, Catalysts for the Selective Catalytic Reduction of NO by NH_3. *Chin. J. Catal.* **2012**, *33*, 933–940. [CrossRef]

23. Li, H.; Zhang, S.; Zhong, Q. Effect of nitrogen doping on oxygen vacancies of titanium dioxide supported vanadium pentoxide for ammonia-SCR reaction at low temperature. *J. Colloid Interface Sci.* **2013**, *402*, 190–195. [CrossRef] [PubMed]

24. Wu, Z.; Jiang, B.; Liu, Y.; Wang, H.; Jin, R. DRIFT Study of Manganese/Titania-Based Catalysts for Low-Temperature Selective Catalytic Reduction of NO with NH_3. *Environ. Sci. Technol.* **2007**, *41*, 5812–5817. [CrossRef] [PubMed]

25. Larrubia, M.A.; Ramis, G.; Busca, G. An FT-IR study of the adsorption of urea and ammonia over V_2O_5-MoO_3-TiO_2 SCR catalysts. *Appl. Catal. B* **2000**, *27*, L145–L151. [CrossRef]

26. Chen, L.; Li, J.; Ge, M. DRIFT Study on Cerium-Tungsten/Titania Catalyst for Selective Catalytic Reduction of NOx with NH_3. *Environ. Sci. Technol.* **2010**, *44*, 9590–9596. [CrossRef] [PubMed]

27. Mihaylov, M.; Chakarova, K.; Hadjiivanov, K. Formation of carbonyl and nitrosyl complexes on titania-and zirconia-supported nickel: FTIR spectroscopy study. *J. Catal.* **2004**, *228*, 273–281. [CrossRef]
28. Zhao, L.; Li, X.; Hao, C.; Raston, C.L. SO$_2$ adsorption and transformation on calcined NiAl hydrotalcite-like compounds surfaces: An in situ FTIR and DFT study. *Appl. Catal. B* **2012**, *117–118*, 339–345. [CrossRef]
29. Liu, F.; Asakura, K.; He, H.; Shan, W.; Shi, X.; Zhang, C. Influence of sulfation on iron titanate catalyst for the selective catalytic reduction of NOx with NH$_3$. *Appl. Catal. B* **2011**, *103*, 369–377. [CrossRef]
30. Vargas, M.A.L.; Casanova, M.; Trovarelli, A.; Busca, G. An IR study of thermally stable V$_2$O$_5$-WO$_3$-TiO$_2$ SCR catalysts modified with silica and rare-earths (Ce, Tb, Er). *Appl. Catal. B* **2007**, *75*, 303–311. [CrossRef]
31. Yu, W.; Wu, X.; Si, Z.; Weng, D. Influences of impregnation procedure on the SCR activity and alkali resistance of V$_2$O$_5$-WO$_3$/TiO$_2$ catalyst. *Appl. Surf. Sci.* **2013**, *283*, 209–214. [CrossRef]
32. Chang, H.; Li, J.; Yuan, J.; Chen, L.; Dai, Y.; Arandiyan, H.; Xu, J.; Hao, J. Ge, Mn-doped CeO$_2$-WO$_3$ catalysts for NH$_3$-SCR of NOx: Effects of SO$_2$ and H$_2$ regeneration. *Catal. Today* **2013**, *201*, 139–144. [CrossRef]
33. Li, Q.; Yang, H.; Nie, A.; Fan, X.; Zhang, X. Catalytic Reduction of NO with NH$_3$ over V$_2$O$_5$-MnO$_X$/TiO$_2$-Carbon Nanotube Composites. *Catal. Lett.* **2011**, *141*, 1237–1242. [CrossRef]
34. Wang, Z.; Li, X.; Song, W.; Chen, J.; Li, T.; Feng, Z. Synergetic Promotional Effects Between Cerium Oxides and Manganese Oxides for NH$_3$-Selective Catalyst Reduction Over Ce-Mn/TiO$_2$. *Mater. Express* **2011**, *1*, 167–175. [CrossRef]
35. Cen, W.; Liu, Y.; Wu, Z.; Wang, H.; Weng, X. A theoretic insight into the catalytic activity promotion of CeO$_2$ surfaces by Mn doping. *Phys. Chem. Chem. Phys.* **2012**, *14*, 5769–5777. [CrossRef] [PubMed]
36. Wan, Y.; Zhao, W.; Tang, Y.; Li, L.; Wang, H.; Cui, Y.; Gu, J.; Li, Y.; Shi, J. Ni-Mn bi-metal oxide catalysts for the low temperature SCR removal of NO with NH$_3$. *Appl. Catal. B* **2014**, *148–149*, 114–122. [CrossRef]
37. Yang, S.; Wang, C.; Chen, J.; Peng, Y.; Ma, L.; Chang, H.; Chen, L.; Liu, C.; Xu, J.; Li, J.; et al. A novel magnetic Fe-Ti-V spinel catalyst for the selective catalytic reduction of NO with NH$_3$ in a broad temperature range. *Catal. Sci. Technol.* **2012**, *2*, 915–917. [CrossRef]
38. Gao, R.; Zhang, D.; Liu, X.; Shi, L.; Maitarad, P.; Li, H.; Zhang, J.; Cao, W. Enhanced catalytic performance of V$_2$O$_5$-WO$_3$/Fe$_2$O$_3$/TiO$_2$ microspheres for selective catalytic reduction of NO by NH$_3$. *Catal. Sci. Technol.* **2013**, *3*, 191–199. [CrossRef]
39. Nefzi, H.; Sediri, F. Vanadium oxide nanotubes VOx-NTs: Hydrothermal synthesis, characterization, electrical study and dielectric properties. *J. Solid State Chem.* **2013**, *201*, 237–243. [CrossRef]
40. Guo, X.; Bartholomew, C.; Hecker, W.; Baxter, L.L. Effects of sulfate species on V$_2$O$_5$/TiO$_2$ SCR catalysts in coal and biomass-fired systems. *Appl. Catal. B* **2009**, *92*, 30–40. [CrossRef]
41. Boningari, T.; Koirala, R.; Smirniotis, P.G. Low-temperature catalytic reduction of NO by NH$_3$ over vanadia-based nanoparticles prepared by flame-assisted spray pyrolysis: Influence of various supports. *Appl. Catal. B* **2013**, *140–141*, 289–298. [CrossRef]
42. Yang, S.; Wang, C.; Ma, L.; Peng, Y.; Qu, Z.; Yan, N.; Chen, J.; Chang, H.; Li, J. Substitution of WO$_3$ in V$_2$O$_5$/WO$_3$-TiO$_2$ by Fe$_2$O$_3$ for selective catalytic reduction of NO with NH$_3$. *Catal. Sci. Technol.* **2013**, *3*, 161–168. [CrossRef]
43. Digregorio, F. Activation and isomerization of hydrocarbons over WO$_3$/ZrO$_2$ catalysts: I. Preparation, characterization, and X-ray photoelectron spectroscopy studies. *J. Catal.* **2004**, *225*, 45–55. [CrossRef]
44. Nishiguchi, T.; Oka, K.; Matsumoto, T.; Kanai, H.; Utani, K.; Imamura, S. Durability of WO$_3$/ZrO$_2$-CuO/CeO$_2$ catalysts for steam reforming of dimethyl ether. *Appl. Catal. A* **2006**, *301*, 66–74. [CrossRef]
45. Wang, L.; Huang, B.; Su, Y.; Zhou, G.; Wang, K.; Luo, H.; Ye, D. Manganese oxides supported on multi-walled carbon nanotubes for selective catalytic reduction of NO with NH$_3$: Catalytic activity and characterization. *Chem. Eng. J.* **2012**, *192*, 232–241. [CrossRef]
46. Shan, W.; Liu, F.; He, H.; Shi, X.; Zhang, C. A superior Ce-W-Ti mixed oxide catalyst for the selective catalytic reduction of NOx with NH$_3$. *Appl. Catal. B* **2012**, *115–116*, 100–106. [CrossRef]
47. Gao, X.; Jiang, Y.; Zhong, Y.; Luo, Z.; Cen, K. The activity and characterization of CeO$_2$-TiO$_2$ catalysts prepared by the sol-gel method for selective catalytic reduction of NO with NH$_3$. *J. Hazard. Mater.* **2010**, *174*, 734–739. [CrossRef] [PubMed]

48. Wang, X.; Shi, A.; Duan, Y.; Wang, J.; Shen, M. Catalytic performance and hydrothermal durability of CeO$_2$-V$_2$O$_5$-ZrO$_2$/WO$_3$-TiO$_2$ based NH$_3$-SCR catalysts. *Catal. Sci. Technol.* **2012**, *2*, 1386–1395. [CrossRef]
49. Li, Q.; Hou, X.; Yang, H.; Ma, Z.; Zheng, J.; Liu, F.; Zhang, X.; Yuan, Z. Promotional effect of CeO$_X$ for NO reduction over V$_2$O$_5$/TiO$_2$-carbon nanotube composites. *J. Mol. Catal. A* **2012**, *356*, 121–127. [CrossRef]
50. Yang, S.; Qi, F.; Liao, Y.; Xiong, S.; Lan, Y.; Fu, Y.; Shan, W.; Li, J. Dual Effect of Sulfation on the Selective Catalytic Reduction of NO with NH$_3$ over MnOx/TiO$_2$: Key Factor of NH$_3$ Distribution. *Ind. Eng. Chem. Res.* **2014**, *53*, 5810–5819. [CrossRef]

catalysts

MDPI

Article

DRIFT Study on Promotion Effect of the Keggin Structure over V$_2$O$_5$-MoO$_3$/TiO$_2$ Catalysts for Low Temperature NH$_3$-SCR Reaction

Rui Wu [1,†], Ningqiang Zhang [1,†], Lingcong Li [1], Hong He [1,2,*], Liyun Song [1,*] and Wenge Qiu [1]

[1] Laboratory of Catalysis Chemistry and Nanoscience, Department of Chemistry and Chemical Engineering, College of Environmental and Energy Engineering, Beijing University of Technology, Beijing 100124, China; wuruicoco@sina.com (R.W.); zhangningqiang08@163.com (N.Z.); lilingcong1988@163.com (L.L.); qiuwenge@bjut.edu.cn (W.Q.)

[2] Collaborative Innovation Center of Electric Vehicles in Beijing, Beijing 100081, China

* Correspondence: hehong@bjut.edu.cn (H.H.), songly@bjut.edu.cn (L.S.)

† These authors contributed equally.

Received: 10 March 2018; Accepted: 30 March 2018; Published: 3 April 2018

Abstract: Heteropoly acids (HPAs) with the Keggin structure have been widely used in NO$_x$ removal. Two kinds of catalysts (those with and without the Keggin structure) are prepared for studying the effect of the Keggin structure on the NH$_3$-SCR reaction. A series of in situ diffuse reflectance infrared Fourier transform spectroscopy (DRIFT) analyses are conducted to investigate the surface-adsorbed species on the catalysts during the SCR reaction. The mechanism for enhancing low-temperature activity of the catalysts is proposed. Furthermore, the effect of NH$_4^+$ in the Keggin structure is also investigated. Results indicate that both the Langmuir–Hinshelwood (L-H) and Eley–Rideal (E-R) mechanisms occurred in the NH$_3$-SCR reaction over the catalyst with the Keggin structure (Cat-A); in addition, when more acid sites are provided, NO$_x$ species activity is improved and more NH$_4^+$ ions participate in reaction over Cat-A, thus promoting SCR activity.

Keywords: NH$_3$-SCR; in situ DRIFT; Keggin structure; surface adsorption

1. Introduction

NO$_x$ is an air pollutant that can result in a variet of pollution, such acid rain, optical chemical smog, and the greenhouse effect. NO$_x$ is emitted when using fossil fuels in industrial facilities and automobiles [1,2]. Selective catalytic reduction (SCR) has been extensively used to remove NO$_x$ emissions from stationary sources such as coal-fired boilers [3]. The commercialized catalyst used in SCR has mainly been V$_2$O$_5$-WO$_3$/TiO$_2$ (or V$_2$O$_5$-MoO$_3$/TiO$_2$), which has high activity and good SO$_2$ tolerance. However, it has a very narrow operating temperature (300–400 °C), high conversion of SO$_2$ to SO$_3$, and low N$_2$ selectivity at high temperatures [4,5]. Therefore, developing an NH$_3$-SCR catalyst with good catalytic performance at low temperature is desirable.

Heteropoly acids (HPAs) have unique structure and extraordinarily strong acidity, and, as such, HPAs have attracted the attention of many researchers [6,7]. The main kind of HPAs used in catalytic applications is Keggin HPAs, such as H$_3$PW$_{12}$O$_{40}$, H$_3$PMo$_{12}$O$_{40}$, and H$_3$SiW$_{12}$O$_{40}$ [8]. It has been found that NO can be effectively absorbed on 12-tungstophosphoric acid at 150 °C and decomposed at about 450 °C upon rapid heating. It has been reported that small polar molecules (such as NO, NH$_3$, and H$_2$O) can access the secondary structure of Keggin HPAs and that SO$_2$ cannot influence adsorption of NO. Supplementary experiments showed that NO can replace the structural water present between Keggin units of HPAs [9,10]. Later, some researchers used aromatic hydrocarbons loaded with Pt and Pd to improve the activity of NO reduction [11,12]. Putluru et al. reported that HPAs can promote SCR activity and shows excellent alkali deactivation resistance [13]. Weng et al. used 12-tungstaphosphoric acid loaded on CeO_2 and found that the catalyst had significantly improved performance in the SCR reaction and improved SO_2 poisoning resistance [14]. Ammonium salts are more available than HPAs, and using 12-tungstophosphoric acid ammonium salt for NO_x removal has been reported. Moffat et al. found an interesting phenomenon; specifically, that the ammonium salt of 12-tungstophosphoric acid exhibited better NO_x removal capacity in the case without NH_3 because bound NH_4^+ on the solid can react with absorbed NO_2 to produce N_2, and ammonium salt can be regenerated from gaseous NH_3 [9]. However, there are few reports about using ammonium salts of HPAs for SCR, and further research on the effect of the Keggin structure and that of the NH_4^+ of ammonium salts on SCR is lacking.

In this research, V_2O_5-$(NH_4)_3PMo_{12}O_{40}/TiO_2$ and a conventional oxide catalyst are prepared for studying the effect of the Keggin structure on the NH_3-SCR reaction. A series of in situ diffuse reflectance infrared Fourier transform spectroscopy (DRIFT) analyses are conducted to investigate surface-adsorbed species. Fourier Transform infrared spectroscopy (FT-IR) is used to evaluate the effect of NH_4^+ of ammonium salts on the catalysts.

2. Results and Discussion

2.1. Catalytic Activity

Catalytic performances of the catalysts with and without the Keggin structure at elevated temperature are shown in Figure 1. Structures of the two samples were determined using X-ray diffraction (XRD) and Raman spectroscopy (Figures S1 and S2, respectively). X-ray Fluorescence Spectrometer (XRF) results of elemental analysis and the BET specific surface area data are shown in Table S1. Results show that Cat-A, which has the Keggin structure, had the same composition and BET surface area as Cat-B, which has an oxide phase. As shown in Figure 1, Cat-A had much higher SCR activity than Cat-B. NO conversion over Cat-A reached 93% at 220 °C and showed no obvious decline until 350 °C. In contrast, NO conversion over Cat-B was only 67% at 220 °C. Clearly, the catalyst with the Keggin structure had improved NO conversion in low temperature. Both catalysts had excellent N_2 selectivity of 99% below 250 °C. The time-on-stream stability was investigated (Figure S5). The NO_x conversion over Cat-A decreased from 100% to 95% in 20 h, while, for Cat-B, it decreased from 99% to 92%. Therefore, Cat-A exhibited better time-on-stream stability than Cat-B.

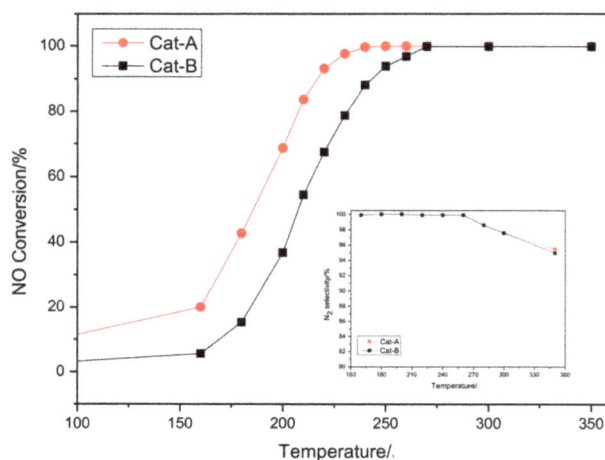

Figure 1. NO conversion of the two catalysts. (Reaction conditions: NO 1000 ppm, $NH_3/NO = 1$, O_2 5%, H_2O 5%, SO_2 350 ppm, N_2 balance, and GHSV = 40,000 h^{-1}.).

2.2. Adsorption Behaviors of Reactants on the Surface of the Catalysts (In Situ DRIFT)

2.2.1. NH$_3$ Adsorption on the Surface of the Catalysts

In situ DRIFT spectra of NH$_3$ adsorption on Cat-A and Cat-B at 250 °C are shown in Figure 2. As seen in Figure 2A, there were six bands: 1211, 1391, 1454, 1596, 1685, and 1737 cm^{-1}. The peaks centered at 1454 cm^{-1} and in the range of 1750–1650 cm^{-1} can be, respectively, attributed to asymmetric and symmetric bending (δ_{as} and δ_s) vibrations of NH$_4^+$ species on Brönsted acid sites [15–17]. The bands at 1211 and 1391 cm^{-1} can be assigned to N-H bonds in NH$_3$ that is chemisorbed on Lewis acid sites [18].

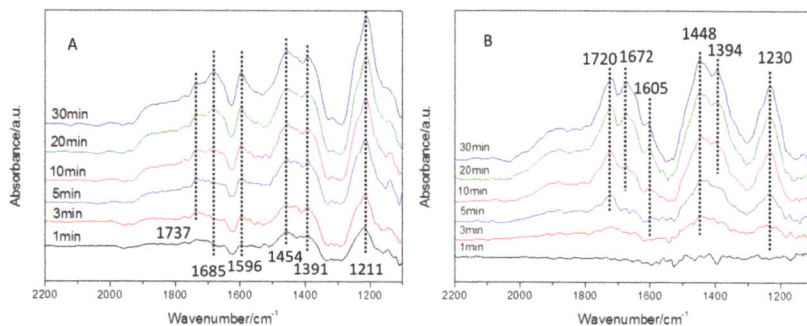

Figure 2. DRIFT spectra for NH$_3$ adsorption at 250 °C on: (**A**) Cat-A; and (**B**) Cat-B.

As illustrated in Figure 2B, DRIFT spectra for NH$_3$ adsorption over Cat-B were similar to those for NH$_3$ adsorption over Cat-A. A broad absorption band at 1448 cm^{-1} was ascribed to δ_{as} NH$_4^+$ vibration and the bands at 1750–1650 cm^{-1} were due to δ_s NH$_4^+$ vibration [19]. Coordinated δ_s NH$_3$ (1230 and 1394 cm^{-1}) and δ_{as} NH$_3$ species (1605 cm^{-1}) on Lewis acid sites were also observed [20,21]. It is evident that the intensity of the band assigned to Lewis acid sites (1230 cm^{-1}) is much weaker than that over Cat-A. It is known that HPAs have stronger (Brönsted) acidity, but it was reported that $(NH_4)_3PMo_{12}O_{40}$ provides Lewis acid sites that were strong at high temperature because of the oxygen-deficient Keggin structure [22]. The NH$_3$-desorption peaks in the temperature programmed desorption of ammonia (NH$_3$-TPD) was interconnected to the concentration of acid sites [23]. Combined with the results for NH$_3$-TPD (Figure S3), the peak area of Cat-A was more than Cat-B, which suggested that the concentration of Lewis acid sites was more than Cat-B. As reported, the high amount of Lewis acid site could improve the low temperature activity [24]. Therefore, the Keggin structure not only provides more Brönsted acid sites, but also offered Lewis acid sites that are beneficial for low-temperature SCR catalytic activity.

2.2.2. Co-Adsorption of NO + O$_2$ on the Surface of the Catalysts

Figure 3 shows DRIFT results of NO + O$_2$ co-adsorption on the surface of Cat-A and Cat-B. Adsorbed species on the two catalysts are clearly very different. As seen in Figure 3A, there are several distinct bands at 1347, 1513, 1640, 1668, 1786, 1751, 1860, and 1976 cm^{-1}. A broad band at 1347 cm^{-1} appeared first and can be ascribed to free nitrate ions [25]. Later, smaller bands at 1513, 1640, and 1668 cm^{-1} were observed, and these can be assigned to bidentate nitrate, adsorbed NO$_2$, and adsorbed N$_2$O$_4$, respectively [20,26]. Bands at 1751 and 1786 cm^{-1} can be ascribed to adsorption of trans-(NO)$_2$ [25,27]. A weaker band at 1860 cm^{-1} can be assigned to surface bound NO [28]. The band at 1976 cm^{-1} increased with time and can be ascribed to NO in the secondary structure of Keggin anions [28]. For Cat-B (Figure 3B), a series of peaks can be ascribed to monodentate nitrates (1317 and 1490 cm^{-1}), free nitrate ions (1340 cm^{-1}), and adsorbed NO$_2$ (1647 cm^{-1}) [29,30]. For Cat-B, there were fewer kinds of adsorbed NO_x species than for Cat-A. Over Cat-B, there were more monodentate nitrates than free nitrate ions. It is reported that, for NSR, NO_x was primarily trapped by the metal ions (such as K or Li) in free nitrate ions, thus leading to increased trapping capacity of NO_x [31,32]. It is likely that NH_4^+ in Cat-A plays the same role in trapping NO_x, and this indicates that the NO_x storage behavior of Cat-A is better than that of Cat-B.

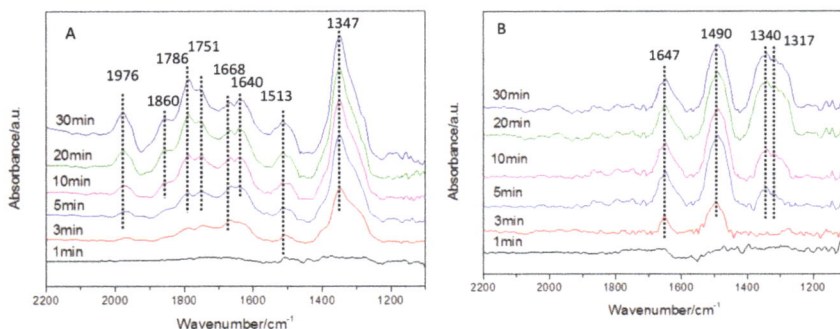

Figure 3. DRIFT spectra for NO + O_2 adsorption at 250 °C on: (**A**) Cat-A; and (**B**) Cat-B.

2.2.3. NO + O_2 Adsorption on the Surface of the Catalysts after NH_3 Pre-Adsorption

The reaction of NO + O_2 with pre-adsorbed NH_3 at 250 °C on the surface of both catalysts was investigated using DRIFT to understand the reactivity of NH_3 adsorbed on the catalysts. As seen in Figure 4A, several peaks that correspond to coordinated NH_3 (L) and NH_4^+ (B) species on the surface of Cat-A are observed. When NO + O_2 was added, Lewis and Brönsted acidity decreased gradually within 10 min, and, after 20 min, several bands (1349, 1514, 1638, 1786, and 1944 cm^{-1}) that correspond to adsorbed NO_x species appeared and grew with time. The adsorption behaviors of Cat-B (Figure 4B) were similar to those of Cat-A. Over time, all of the peaks gradually disappeared, accompanied by the appearance of nitrate species. These results imply that coordinated NH_3 and NH_4^+ species participate in the NH_3-SCR reaction and that both were active species.

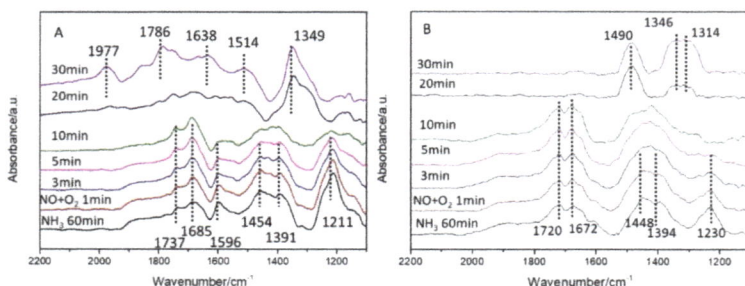

Figure 4. DRIFT spectra of NO + O_2 reacted with pre-adsorbed NH_3 species at 250 °C on: (**A**) Cat-A; and (**B**) Cat-B.

2.2.4. NH_3 Adsorption on the Surface of the Catalysts after NO + O_2 Pre-Adsorption

The catalysts were pre-treated with NO + O_2 at 250 °C for 60 min, and then NH_3 was introduced to investigate the reaction of NH_3 with nitrate species over the catalysts. As seen in Figure 5A, several peaks that correspond to nitrate species were found in the spectra after treatment with NO + O_2. With the introduction of NH_3, the amounts of free nitrate ions (1347 cm^{-1}), bidentate nitrates

(1513 cm^{-1}), NO$_2$ (1640 cm^{-1}), and N$_2$O$_4$ (1668 cm^{-1}) decreased quickly within 1 min. Surface NO (1840 cm^{-1}) and NOH$^+$ (1976 cm^{-1}) in the Keggin structure gradually decreased in 5 min, implying that NO in the secondary structure of Keggin anions participates in the reaction. However, the peak at 1786 cm^{-1}, which corresponds to trans-(NO)$_2$, is still present after purging with NH$_3$ for 30 min, and this suggests that the Eley–Rideal (E-R) mechanism occurred [26]. For Cat-B (Figure 5B), a series of peaks that correspond to nitrate species were observed in the spectra. After NH$_3$ was introduced for 1 min, NO$_2$ (1647 cm^{-1}) disappeared, and this can enhance the process of the "fast-SCR" reaction. Compared with Cat-A, free nitrate ions of Cat-B decreased more slowly and monodentate nitrates disappeared gradually within 5 min. It has been reported that monodentate nitrates and NO$_2$ molecules had higher activity in the SCR reaction than free nitrate ions. In our results, free nitrate ions for Cat-A, which had the Keggin structure, also easily participated in the reaction, and this was possibly because NO$_x$ was trapped by the NH$_4$$^+$ of the Keggin structure in free nitrate ions, such as in the case of NO$_x$ storage/reduction (NSR) catalysts where NO$_x$ can easily be trapped.

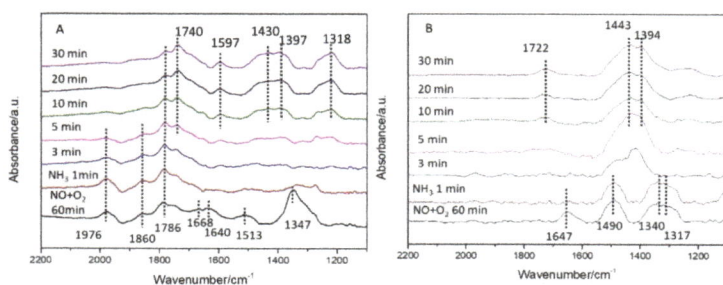

Figure 5. DRIFT spectra of NH$_3$ reacted with pre-adsorbed NO + O$_2$ species at 250 °C on: (**A**) Cat-A; and (**B**) Cat-B.

2.3. Effect of NH$_4$$^+$ in (NH$_3$)$_4$PMo$_{12}$O$_{40}$ for NH$_3$-SCR

It is known that NH$_3$ adsorption and storage is important for NH$_3$-SCR [33,34]. Compared with oxide SCR catalysts, Cat-A, which has the Keggin structure, has many NH$_4$$^+$ ions. Moffat et al. reported that NH$_4$$^+$ of ammonium 12-tungstophosphate can interact with NO$_2$ and NO and can be regenerated from ammonia gas at 150 °C [9]. Hence, it is essential to know whether NH$_4$$^+$ of the Keggin structure participates in NH$_3$-SCR and if NH$_4$$^+$ can be regenerated for Cat-A.

If NH$_4$$^+$ of the Keggin structure participates in the reaction, the SCR reaction on Cat-A would continue longer than that on Cat-B without NH$_3$. To verify this assumption, NO conversion of Cat-A and Cat-B after cutting off NH$_3$ was investigated, and the results are shown in Figure 6. As seen in Figure 6, NO conversions on the two catalysts decreased after cutting off NH$_3$. For Cat-A, NO conversion remained 100% for 14 min, whereas NO conversion over Cat-B kept remained 100% for only 4 min. Thus, it is obvious that Cat-A, which has the Keggin structure, had higher NH$_3$ storage than Cat-B, and this suggests that NH$_4$$^+$ of the Keggin structure participates in the reaction.

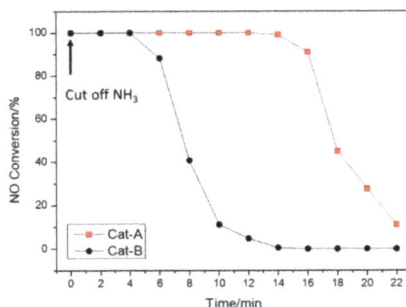

Figure 6. NO conversion of Cat-A and Cat-B over time after cutting off NH$_3$. Reaction conditions: T = 250 °C, [NO] = 1000 ppm, [O$_2$] = 5%, GHSV = 40,000 h^{-1}, and N$_2$ balance gas.

Change in the Keggin structure during successive adsorption of NO + O$_2$ and NH$_3$ was investigated using FT-IR spectra to further verify participation of NH$_4^+$ in the SCR reaction. As seen in Figure 7, vibrational bands for NH$_4^+$, P-O, and Mo-O-Mo in the Keggin structure are observed at 1402, 1062, and 906 cm^{-1}, respectively. After adsorption of NO, the peak at 1402 cm^{-1} disappeared. NH$_3$ was added after cutting off NO + O$_2$, and the peak at 1402 cm^{-1} reappeared. Moreover, the peaks at 1062 and 906 cm^{-1} remained stable during the whole reaction process, and this suggests that HPAs did not change during the reaction. Thus, it was determined that NH$_4^+$ of the Keggin structure participates in the SCR reaction and was regenerated from ammonia gas. Meanwhile, the Keggin structure (of the HPAs) was unchanged in the reaction. In situ Raman spectra of Cat-A were recorded (Figure S4) to further study the stability of the Keggin structure in the SCR reaction. The results indicate that the peaks corresponding to the Keggin structure did not shift, and this suggests that there was no structural perturbation during the reaction.

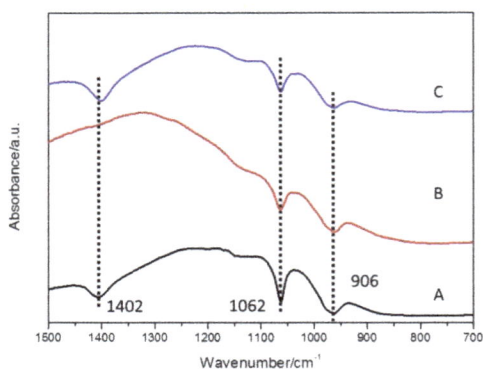

Figure 7. FT-IR spectra of Cat-A with: (A) no adsorption; (B) adsorption of NO + O$_2$; and (C) after cutting off NO + O$_2$, adsorption of NH$_3$.

2.4. Reaction Mechanism

From DRIFT spectra for NH$_3$ adsorption on the two samples, it is observed that NH$_3$ was adsorbed on Brönsted and Lewis acid sites in the forms of NH$_4^+$ and NH$_3$, respectively. After NO + O$_2$ was passed over the catalyst surface that was pretreated with NH$_3$, the adsorbed NH$_3$ species gradually vanished in 10 min, and this indicates that both Brönsted and Lewis acid sites were active centers. As shown in DRIFT spectra of NO + O$_2$ adsorption, NO was adsorbed and oxidized to NO$_2$ and other nitrate species on the two samples. There were more kinds of NO_x species adsorbed on Cat-A than on Cat-B. Different results were observed when NH_3 was passed over the surface of each of the two samples that were pretreated with $NO + O_2$ at 250 °C. Compared with Cat-B, not all of the adsorbed NO_x species participated in the reaction over Cat-A, and most of the adsorbed NO_x species quickly vanished in 1 min, which indicates that both the E-R and L-H mechanisms occurred [35–38]. Noticeably, the peaks of NOH^+ disappeared after 5 min, and this suggests that NO in the secondary structure of the Keggin anions (NOH^+) also participated in reaction. The effect of NH_4^+ in the Keggin structure for the SCR reaction was investigated, and the results reveal that NH_4^+ of the Keggin structure participates in the SCR reaction as an NH_3 pool and can be regenerated from ammonia gas.

Based on the above discussion, a mechanism for improving the low-temperature activity of Cat-A was proposed; the proposed mechanism is shown in Figure 8 and is described as follows.

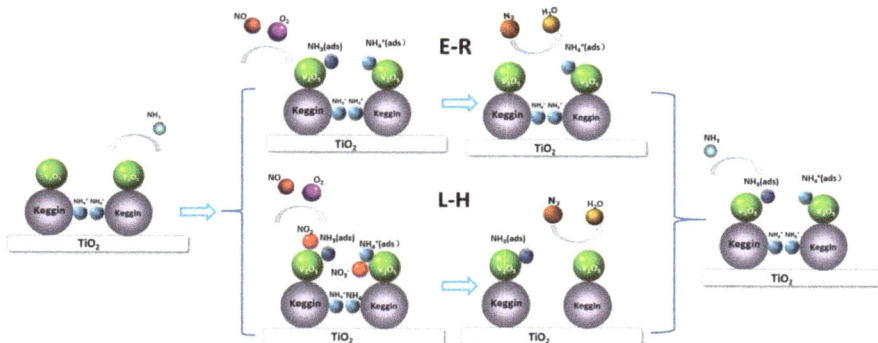

Figure 8. Schematic of the proposed NH_3-SCR reaction mechanism over Cat-A.

Both the E-R and L-H mechanisms occurred in the NH_3-SCR reaction. For the E-R mechanism, NH_3 was adsorbed on Lewis acid sites and reacted directly with gaseous NO and NO_2. This was followed by decomposition into N_2 and H_2O. For the L-H mechanism, NO was adsorbed, and then most of it was oxidized to nitrate species (NO_3^- and NO_2-adsorbed), which reacted with NH_3-adsorbed species. It is noted that the catalyst with Keggin structure had more Brönsted acid Lewis acid sites than the catalyst with oxide phase. In addition, NH_4^+ of the Keggin structure reacted by the L-H mechanism and was regenerated during the reaction; thus, low-temperature SCR activity was promoted via the L-H mechanism.

For Cat-B, the mechanisms were similar to Cat-A. Both the E-R and L-H mechanisms also occurred in the reaction. For the E-R mechanism, gaseous NO and NO_2 could react with NH_3 on Lewis acid sites. For the L-H mechanism, NO could be adsorbed and then oxidized to nitrate species. Most of the nitrate species were monodentate nitrates and NO_2-adsorbed not free NO_3^-, and the nitrate species could react with NH_3-adsorbed species. The results were in agreement with many SCR catalysts with oxide phase [25,26,39].

3. Experimental

3.1. Catalyst Preparation

Two kinds of catalysts were prepared via impregnation, and the main chemical compositions were the same. Specifically, $(NH_4)_3PMo_{12}O_{40}$ loading was 20 wt % and V_2O_5 loading was 1 wt %.

The catalyst with the Keggin structure was prepared as follows. $NH_4H_2PO_4$ (0.6 g, Fuchen, Tianjin, China, 99%) and $(NH_4)_6Mo_7O_{24}$ (11.5 g, Fuchen, Tianjin, China, 99%) were dissolved in distilled water (100 mL). Solution pH was adjusted to a value of about 1. TiO_2 anatase powder (40.0 g, Xinhua,

Chongqing, China, 99%) was added to the precursor solution and solid to liquid ratio (g/mL) was 2:5. The mixture was stirred at 80 °C for 5 h and then dried at 120 °C for 3 h. The solid was ground into a powder and calcined at 400 °C for 5 h. NH_4VO_3 (0.6 g, Fuchen, Tianjin, China, 99%) was then added to a solution of oxalic acid, and the weight of $H_2C_2O_4 \cdot 2H_2O$ (Fuchen, Tianjin, China, 99%) was double that of NH_4VO_3. Desiccation and calcination conditions were the same as above, and the obtained catalyst was denoted as Cat-A.

Catalysts with an oxide phase were prepared via multiple impregnation methods to avoid generating the Keggin structure. $(NH_4)_6Mo_7O_{24}$ (11.5 g) was dissolved in 100 mL of distilled water. TiO_2 anatase powder was impregnated with the precursor solution and stirred at 80 °C for 5 h. The mixture was then dried at 120 °C and was calcined at 400 °C for 5 h. $NH_4H_2PO_4$ (0.6 g) was then dissolved in 100 mL of distilled water. V_2O_5 was loaded using the above conditions, and this sample was denoted as Cat-B.

3.2. Catalytic Activity Test

Catalytic activities of the samples were measured in a fixed bed quartz reactor (9 mm i.d.) with 0.4 mL of catalysts (40–60 mesh). Typical reactant gas was a mixture of 1000 ppm NO, 1000 ppm NH_3, 5% O_2, 5% water vapor, and 350 ppm SO_2 with a balance of N_2 under a flow of 500 mL/min. The space velocity was 40,000 h^{-1}. Concentrations of NO and NH_3 were continuously detected using a Thermo Scientific 17i NO_x chemiluminescence analyzer. N_2O was monitored using a Bruker Tensor 27 FTIR spectrometer. The reaction system was maintained at each reaction temperature for 30 min before analysis. The equations used to calculate NO_x conversion and N_2 selectivity are as follows:

$$\text{NO conversion(\%)} = \frac{[NO]_{in} - ([NO]_{out} + [NO_2]_{out})}{[NO]_{in}} \times 100\% \qquad (1)$$

$$\text{N}_2 \text{ selectivity(\%)} = \left(1 - \frac{[NO_2]_{out} + 2[N_2O]_{out}}{[NH_3]_{in} + [NO]_{in} - [NH_3]_{out} - [NO]_{out}}\right) \times 100\% \qquad (2)$$

4. Conclusions

Two kinds of catalysts (one with and one without the Keggin structure) are synthesized to investigate the effect of the Keggin structure on the NH_3-SCR reaction. Cat-A, which had the Keggin structure, exhibits better low-temperature SCR performance. NH_3 is adsorbed on both Brönsted and Lewis acid sites over the two catalysts but the catalyst with Keggin structure has more Brönsted and Lewis acid sites. Moreover, there are more kinds of adsorbed NO_x species on Cat-A than on Cat-B, and the adsorbed NO_x species are mainly free nitrate ions. The two catalysts follow both E-R and L-H mechanisms. Compared with Cat-B, most of the adsorbed NO_x species for Cat-A react quickly with gaseous NH_3. In addition, NH_4^+ of the Keggin structure participates in the reaction via the L-H mechanism and is recovered by ammonia gas in the flow. Thereby, more acid sites are provided, adsorbed NO_x species activity is improved, and more NH_4^+ ions participate in the L-H mechanism over Cat-A, which had the Keggin structure, thus promoting SCR activity.

Supplementary Materials: The following are available online at http://www.mdpi.com/2073-4344/8/4/143/s1, Characterization methods: XRD and Raman results (Figures S1 and S2); compositions and BET surface areas of Cat-A and Cat-B (Table S1); NH_3-TPD results (Figure S3); in situ sequential Raman spectra results (Figure S4); the time-on-stream stability of the catalysts (Figure S5).

Acknowledgments: The work was supported by the National Key R&D Program of China (2017YFC0210303-02) and the National Natural Science Foundation of China (201577005). The authors thank Ran Liu from Shiyanjia lab for assistance of BET and NH3-TPD analysis (www.Shiyanjia.com).

Author Contributions: Rui Wu and Hong He conceived and designed the experiments; Rui Wu experiments; Rui Wu, Ningqiang Zhang and Wenge Qu analyzed the data; Liyun Song and Lingcong Li contributed analysis tools; Rui Wu wrote the paper.

Conflicts of Interest: The authors declare no conflict of interest.

References

1. Amiridis, M.D.; Duevel, R.V.; Wachs, I.E. The effect of metal oxide additives on the activity of V_2O_5/TiO_2 catalysts for the selective catalytic reduction of nitric oxide by ammonia. *Appl. Catal. B Environ.* **1999**, *20*, 111–122. [CrossRef]

2. Busca, G.; Lietti, L.; Ramis, G.; Berti, F. Chemical and mechanistic aspects of the selective catalytic reduction of NO_x by ammonia over oxide catalysts: A review. *Appl. Catal. B Environ.* **1998**, *18*, 1–36. [CrossRef]

3. Li, P.; Liu, Q.; Liu, Z. Behaviors of NH_4HSO_4 in SCR of NO by NH_3 over different cokes. *Chem. Eng. J.* **2012**, *181–182*, 169–173. [CrossRef]

4. Kamata, H.; Takahashi, K.; Odenbrand, C.I. Surface acid property and its relation to SCR activity of phosphorus added to commercial V_2O_5 $(WO_3)/TiO_2$ catalyst. *Catal. Lett.* **1998**, *53*, 65–71. [CrossRef]

5. Putluru, S.S.R.; Schill, L.; Godiksen, A.; Poreddy, R.; Mossin, S.; Jensen, A.D.; Fehrmann, R. Promoted V_2O_5/TiO_2 catalysts for selective catalytic reduction of NO with NH_3 at low temperatures. *Appl. Catal. B Environ.* **2016**, *183*, 282–290. [CrossRef]

6. Hill, C.L. Foreword: Polyoxometalates in catalysis. *J. Mol. Catal. A Chem.* **1996**, *114*, 1. [CrossRef]

7. Chen, Y.; Wang, Y.; Lu, J.; Wu, C. The viscosity reduction of nano-keggin-$K_3PMo_{12}O_{40}$ in catalytic aquathermolysis of heavy oil. *Fuel* **2009**, *88*, 1426–1434. [CrossRef]

8. Mizuno, N.; Misono, M. Heterogeneous catalysis. *Chem. Rev.* **1998**, *98*, 199–218. [CrossRef] [PubMed]

9. Bélanger, R.; Moffat, J.B. The sorption and reduction of nitrogen oxides by 12-tungstophosphoric acid and its ammonium salt. *Catal. Today* **1998**, *40*, 297–306. [CrossRef]

10. Yang, R.T.; Chen, N. A new approach to decomposition of nitric oxide using sorbent/catalyst without reducing gas: Use of heteropoly compounds. *Ind. Eng. Chem. Res.* **1994**, *33*, 825–831. [CrossRef]

11. Vaezzadeh, K.; Petit, C.; Pitchon, V. The removal of NO_x from a lean exhaust gas using storage and reduction on $H_3PW_{12}O_{40} \cdot 6H_2O$. *Catal. Today* **2002**, *73*, 297–305. [CrossRef]

12. Yoshimoto, R.; Ninomiya, T.; Okumura, K.; Niwa, M. Cooperative effect induced by the mixing of Na-ZSM-5 and $Pd/H_3PW_{12}O_{40}/SiO_2$ in the selective catalytic reduction of NO with aromatic hydrocarbons. *Appl. Catal. B Environ.* **2007**, *75*, 175–181. [CrossRef]

13. Putluru, S.S.R.; Jensen, A.D.; Riisager, A.; Fehrmann, R. Heteropoly acid promoted V_2O_5/TiO_2 catalysts for NO abatement with ammonia in alkali containing flue gases. *Catal. Sci. Technol.* **2011**, *1*, 631–637. [CrossRef]

14. Weng, X.; Dai, X.; Zeng, Q.; Liu, Y.; Wu, Z. DRIFT studies on promotion mechanism of $H_3PW_{12}O_{40}$ in selective catalytic reduction of NO with NH_3. *J. Colloid Interface Sci.* **2016**, *461*, 9–14. [CrossRef] [PubMed]

15. Lietti, L.; Nova, I.; Ramis, G.; Dall'Acqua, L.; Busca, G.; Giamello, E.; Forzatti, P.; Bregani, F. Characterization and Reactivity of V_2O_5–MoO_3/TiO_2 De-NO_x SCR Catalysts. *J. Catal.* **1999**, *187*, 419–435. [CrossRef]

16. Ettireddy, P.R.; Ettireddy, N.; Boningari, T.; Pardemann, R.; Smirniotis, P.G. Investigation of the selective catalytic reduction of nitric oxide with ammonia over Mn/TiO_2 catalysts through transient isotopic labeling and in situ FT-IR studies. *J. Catal.* **2012**, *292*, 53–63. [CrossRef]

17. Song, Z.; Zhang, Q.; Ma, Y.; Liu, Q.; Ning, P.; Liu, X.; Wang, J.; Zhao, B.; Huang, J.; Huang, Z. Mechanism-dependent on the different CeO_2 supports of phosphotungstic acid modification CeO_2 catalysts for the selective catalytic reduction of NO with NH3. *J. Taiwan Inst. Chem. E* **2017**, *71*, 277–284. [CrossRef]

18. Yao, X.; Zhang, L.; Li, L.; Liu, L.; Cao, Y.; Dong, X.; Gao, F.; Deng, Y.; Tang, C.; Chen, Z.; et al. Investigation of the structure, acidity, and catalytic performance of $CuO/Ti_{0.95}Ce_{0.05}O_2$ catalyst for the selective catalytic reduction of NO by NH_3 at low temperature. *Appl. Catal. B Environ.* **2014**, *150–151*, 315–329. [CrossRef]

19. Thirupathi, B.; Smirniotis, P.G. Co-doping a metal (Cr, Fe, Co, Ni, Cu, Zn, Ce, and Zr) on Mn/TiO_2 catalyst and its effect on the selective reduction of NO with NH_3 at low-temperatures. *Appl. Catal. B Environ.* **2011**, *110*, 195–206. [CrossRef]

20. You, Y.; Chang, H.; Zhu, T.; Zhang, T.; Li, X.; Li, J. The poisoning effects of phosphorus on CeO_2-MoO_3/TiO_2 DeNOx catalysts: NH_3-SCR activity and the formation of N_2O. *Mol. Catal.* **2017**, *439*, 15–24. [CrossRef]

21. Zeng, Y.; Zhang, S.; Wang, Y.; Zhong, Q. CeO_2 supported on reduced TiO_2 for selective catalytic reduction of NO by NH_3. *J. Colloid Interface Sci.* **2017**, *496*, 487–495. [CrossRef] [PubMed]

22. Li, W.; Oshihara, K.; Ueda, W. Catalytic performance for propane selective oxidation and surface properties of 12-molybdophosphoric acid treated with pyridine. *Appl. Catal. A Gen.* **1999**, *182*, 357–363. [CrossRef]

23. Boningari, T.; Ettireddy, P.R.; Somogyvari, A.; Liu, Y.; Vorontsov, A.; McDonald, C.A.; Smirniotis, P.G. Influence of elevated surface texture hydrated titania on Ce-doped Mn/TiO$_2$ catalysts for the low-temperature SCR of NOx under oxygen-rich conditions. *J. Catal.* **2015**, *325*, 145–155. [CrossRef]

24. Pappas, D.K.; Boningari, T.; Boolchand, P.; Smirniotis, P.G. Novel manganese oxide confined interweaved titania nanotubes for the low-temperature Selective Catalytic Reduction (SCR) of NO$_x$ by NH$_3$. *J. Catal.* **2016**, *334*, 1–13. [CrossRef]

25. Liu, Z.; Zhang, S.; Li, J.; Ma, L. Promoting effect of MoO$_3$ on the NO$_x$ reduction by NH$_3$ over CeO$_2$/TiO$_2$ catalyst studied with in situ DRIFTS. *Appl. Catal. B Environ.* **2014**, *144*, 90–95. [CrossRef]

26. Liu, J.; Li, X.; Zhao, Q.; Ke, J.; Xiao, H.; Lv, X.; Liu, S.; Tadé, M.; Wang, S. Mechanistic investigation of the enhanced NH$_3$-SCR on cobalt-decorated Ce-Ti mixed oxide: In situ FTIR analysis for structure-activity correlation. *Appl. Catal. B Environ.* **2017**, *200*, 297–308. [CrossRef]

27. Valyon, J.; Hall, W. Surface Species Formed from NO on Copper Zeolites. *ChemInform* **1993**, *24*. [CrossRef]

28. Herring, A.M.; McCormick, R.L. In situ infrared study of the absorption of nitric oxide by 12-tungstophosphoric acid. *J. Phys. Chem. B* **1998**, *102*, 3175–3184. [CrossRef]

29. Chen, L.; Li, J.; Ge, M. DRIFT Study on Cerium−Tungsten/Titania Catalyst for Selective Catalytic Reduction of NO$_x$ with NH$_3$. *Environ. Sci. Technol.* **2010**, *44*, 9590–9596. [CrossRef] [PubMed]

30. Li, Z.; Li, J.; Liu, S.; Ren, X.; Ma, J.; Su, W.; Peng, Y. Ultra hydrothermal stability of CeO$_2$-WO$_3$/TiO$_2$ for NH$_3$-SCR of NO compared to traditional V$_2$O$_5$-WO$_3$/TiO$_2$ catalyst. *Catal. Today* **2015**, *258*, 11–16. [CrossRef]

31. Toops, T.J.; Smith, D.B.; Epling, W.S.; Parks, J.E.; Partridge, W.P. Quantified NO$_x$ adsorption on Pt/K/gamma-Al$_2$O$_3$ and the effects of CO$_2$ and H$_2$O. *Appl. Catal. B Environ.* **2005**, *58*, 255–264. [CrossRef]

32. Büchel, R.; Strobel, R.; Baiker, A.; Pratsinis, S.E. Flame-Made Pt/K/Al$_2$O$_3$ for NO$_x$ Storage–Reduction (NSR) Catalysts. *Top. Catal.* **2009**, *52*, 1799–1802. [CrossRef]

33. Kamasamudram, K.; Currier, N.W.; Chen, X.; Yezerets, A. Overview of the practically important behaviors of zeolite-based urea-SCR catalysts, using compact experimental protocol. *Catal. Today* **2010**, *151*, 212–222. [CrossRef]

34. Schmieg, S.J.; Oh, S.H.; Kim, C.H.; Brown, D.B.; Lee, J.H.; Peden, C.H.F.; Kim, D.H. Thermal durability of Cu-CHA NH$_3$-SCR catalysts for diesel NO$_x$ reduction. *Catal. Today* **2012**, *184*, 252–261. [CrossRef]

35. Ramis, G.; Bregani, F.; Forzatti, P. Fourier transform-infrared study of the adsorption and coadsorption of nitric oxide, nitrogen dioxide and ammonia on vanadia-titania and mechanism of selective catalytic reduction. *Appl. Catal.* **1990**, *64*, 259–278. [CrossRef]

36. Topsøe, N.-Y. Characterization of the nature of surface sites on vanadia-titania catalysts by FTIR. *J. Catal.* **1991**, *128*, 499–511. [CrossRef]

37. Lietti, L.; Svachula, J.; Forzatti, P.; Ramis, G.; Bregani, P. Surface and catalytic properties of Vanadia-Titania and Tungsta-Titania systems in the Selective Catalytic Reduction of nitrogen oxides. *Catal. Today* **1993**, *17*, 131–139. [CrossRef]

38. Topsøe, N.-Y. Mechanism of the selective catalytic reduction of nitric oxide by ammonia elucidated by in situ on-line Fourier transform infrared spectroscopy. *Science* **1994**, *265*, 1217–1219. [CrossRef] [PubMed]

39. Peng, Y.; Li, K.; Li, J. Identification of the active sites on CeO$_2$–WO$_3$ catalysts for SCR of NO$_x$ with NH$_3$: An in situ IR and Raman spectroscopy study. *Appl. Catal. B Environ.* **2013**, *140*, 483–492. [CrossRef]

catalysts

MDPI

Article

The Role of Fe$_2$O$_3$ Species in Depressing the Formation of N$_2$O in the Selective Reduction of NO by NH$_3$ over V$_2$O$_5$/TiO$_2$-Based Catalysts

Moon Hyeon Kim * and Ki Hyuck Yang

Department of Environmental Engineering, Daegu University 201 Daegudae-ro, Jillyang, Gyeongsan 38453, Korea; wiws3@naver.com
* Correspondence: moonkim@daegu.ac.kr; Tel.: +82-53-850-6693; Fax: +82-53-850-6699

Received: 28 February 2018; Accepted: 27 March 2018; Published: 30 March 2018

Abstract: Promotion of 2.73% Fe$_2$O$_3$ in an in-house-made V$_2$O$_5$-WO$_3$/TiO$_2$ (VWT) and a commercial V$_2$O$_5$-WO$_3$/TiO$_2$ (c-VWT) has been investigated as a cost effective approach to the suppression of N$_2$O formation in the selective catalytic reduction of NO by NH$_3$ (NH$_3$-SCR). The promoted VWT and c-VWT catalysts all gave a significantly decreased N$_2$O production at temperatures >400 °C compared to the unpromoted samples. However, such a promotion led to the loss in high temperature NO conversion, mainly due to the oxidation of NH$_3$ to N-containing gases, particularly NO. Characterization of the unpromoted and promoted catalysts using X-ray diffraction (XRD), NH$_3$ adsorption-desorption, and Raman spectroscopy techniques could explain the reason why the promotion showed much lower N$_2$O formation levels at high temperatures. The addition of Fe$_2$O$_3$ to c-VWT resulted in redispersion of the V$_2$O$_5$ species, although this was not visible for 2.73% Fe$_2$O$_3$/VWT. The iron oxides exist as a highly-dispersed noncrystalline α-Fe$_2$O$_3$ in the promoted catalysts. These Raman spectra had a new Raman signal that could be tentatively assigned to Fe$_2$O$_3$-induced tetrahedrally coordinated polymeric vanadates and/or surface V-O-Fe species with significant electronic interactions between the both metal oxides. Calculations of the monolayer coverage of each metal oxide and the surface total coverage are reasonably consistent with Raman measurements. The proposed vanadia-based surface polymeric entities may play a key role for the substantial reduction of N$_2$O formed at high temperatures by NH$_3$ species adsorbed strongly on the promoted catalysts. This reaction is a main pathway to greatly suppress the extent of N$_2$O formation in NH$_3$-SCR reaction over the promoted catalysts.

Keywords: NH$_3$-SCR reaction; V$_2$O$_5$-WO$_3$/TiO$_2$ catalyst; N$_2$O formation; Fe$_2$O$_3$ promotion; NH$_3$ oxidation; Raman spectra

---

1. Introduction

Commercially-available, anatase-type TiO$_2$-supported V$_2$O$_5$ catalysts with either WO$_3$ or MoO$_3$ as a structure stabilizer of the support and a surface acidity enhancer are typical for selective catalytic reduction (SCR) of NO$_x$ from relatively large scale stationary and mobile sources in the presence of N-containing reductants, such as gaseous NH$_3$ and aqueous urea [1,2]. They are usually formulated to 0.1–3% V$_2$O$_5$ and 7–10% WO$_3$ or 6–10% MoO$_3$ [3–8], depending on industrial application target, and V$_2$O$_5$-WO$_3$/TiO$_2$ systems are prevailed for such deNO$_x$ processes. The overall NH$_3$-SCR reaction with V$_2$O$_5$/TiO$_2$-based catalysts could be adequately described by [5–7]:

$$4NO + 4NH_3 + O_2 \rightarrow 4N_2 + 6H_2O, \tag{1}$$

$$NO + NO_2 + 2NH_3 \rightarrow 2N_2 + 3H_2O. \tag{2}$$

The standard Reaction (1) takes place in the presence of excess oxygen, while Reaction (2) has been known as the fast SCR pathway in an equimolar mixture of NO and NO_2 and is known to be much faster than the Reaction (1) at low temperatures [6,9,10].

In addition to the main SCR reactions over V_2O_5-WO_3 (or MoO_3)/TiO_2 catalysts, many side reactions can occur, and among them, one is the production of N_2O that is a greenhouse gas with a global warming potential of 310 at a 100-year time horizon, and the extent of its emissions can greatly depend on the loading and crystallinity of V_2O_5, the secondary component, reaction temperature, concentrations of H_2O and O_2, and so forth [3,6,11–15]. Such a formation of N_2O in the NH_3-SCR reaction is proposed by the following major routes [6,12–15]:

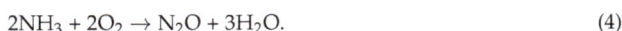

$$4NO + 4NH_3 + 3O_2 \rightarrow 4N_2O + 6H_2O \tag{3}$$

$$2NH_3 + 2O_2 \rightarrow N_2O + 3H_2O. \tag{4}$$

These reactions predominantly occur at high temperatures when V_2O_5-WO_3/TiO_2 catalysts are employed for deNO$_x$ SCR reaction and are associated with a decrease in high-temperature NO conversion and N_2 selectivity [12–15].

An attempt to minimize the formation of N_2O in NH_3-SCR processes has been reported: a direct coating of Fe-ZSM-5 onto a commercial V_2O_5-WO_3/TiO_2 catalyst [16], a sequential configuration of V_2O_5-WO_3/TiO_2 and Fe-ZSM-5 [17], and a modification of TiO_2 using Fe_2O_3 before V_2O_5 and WO_3 loadings [18,19]. These studies were started from the fact that Fe-exchanged zeolites, representatively Fe-ZSM-5, are a commercially proven catalyst not only for the NH_3-SCR reaction but also for direct N_2O decomposition; thereby, the decomposition of N_2O produced in the SCR reaction into N_2 on Fe ions and/or its reduction by residual NH_3 species [20–23] is expected, and from that a nanosized iron oxide, γ-Fe_2O_3 (maghemite), is active for the NH_3-SCR reaction, although N_2O production levels were not addressed for this system [24]. The Fe-ZSM-5-coated V_2O_5-WO_3/TiO_2 catalysts could greatly depress the formation of N_2O in NH_3-SCR reaction, depending on the coating content [16], unlike the series configuration systems [17]. Samples of coprecipitated, Fe_2O_3-TiO_2-supported 1% V_2O_5, and 1% V_2O_5-10% WO_3/TiO_2, had similar NO conversions at 150–400 °C but at higher temperatures, lower ones, depending on the temperature, were indicated for the former sample [18]. This showed a better N_2 selectivity at high temperatures $\geq$400 °C [18]; however, this observation may be apparent because of the difference in NO conversion at the temperature region between the catalysts. All 1% V_2O_5-10% WO_3 catalysts dispersed on 1–5% Fe_2O_3/TiO_2 gave an improvement to N_2 selectivity at temperatures >450 °C, compared to the bare TiO_2-based catalyst; in addition to that, they showed higher NO conversion below 350 °C but comparable NO conversion at higher temperatures [19].

Unsupported and supported Fe_2O_3 are still suspicious of the tolerance to SO_2 existing in flue gases, depending strongly on their preparation techniques [19,24–26], but this point can be avoided for no sulfur applications, such as natural gas-fired plants. The earlier approaches to the reduction in N_2O emissions from NH_3-SCR reaction could advise us of an efficient way of promoting V_2O_5-WO_3/TiO_2 catalysts using Fe_2O_3. Such a utilization of Fe_2O_3 is cost effective, compared to Fe-zeolites whose preparation requires more complicated, expensive processes. Therefore, we have studied a promotional effect of Fe_2O_3 in samples of laboratory-made V_2O_5-WO_3/TiO_2 and commercial V_2O_5-WO_3/TiO_2 on the suppression of N_2O production in NH_3-SCR reaction. The Fe_2O_3 as a promoter was added to the V_2O_5/TiO_2-based catalysts using the well-known impregnation technique, and this is much simpler than the utilization of a coprecipitated Fe_2O_3-TiO_2 and a Fe_2O_3-coated TiO_2 to support V_2O_5 and WO_3 reported in the literature.

2. Results and Discussion

2.1. Physicochemical Properties of Fe₂O₃-Promoted V₂O₅-WO₃/TiO₂ Catalysts

2.1. Physicochemical Properties of Fe_2O_3-Promoted V_2O_5-WO_3/TiO_2 Catalysts

X-ray diffraction (XRD) patterns for VWT, 2.73% Fe_2O_3/VWT, c-VWT, and 2.73% Fe_2O_3/c-VWT at a 2θ value of 10–80° are shown in Figure 1. All the catalysts gave a predominant peak at 2θ = 25.31° with much smaller diffractions at higher 2θ values, as displayed in Figure 1a–d, and all these peaks were the same as those existing in the pattern for anatase TiO_2 shown in Figure 1e. No diffraction due to the crystalline WO_3 was shown for all the samples, indicating that WO_3 existing in them is amorphous WO_x species [27]. There was also no presence of crystalline V_2O_5 phases, except for c-VWT, in which a weak reflection near 2θ = 21.64° appeared when the diffraction was magnified, as indicated in Figure 1c, which is assigned to the crystallographic (101) plane of a polycrystalline α-V_2O_5 (JCPDS card # 41-1426) by comparing it with the pure reference V_2O_5 with an orthorhombic structure as provided in Figure 1f [27,28]. However, the XRD peak disappeared in 2.73% Fe_2O_3/c-VWT, as shown in the magnified pattern in Figure 1d. It should be mentioned that all of the other catalysts gave no peak at the 2θ value, even in similar magnified spectra (not shown here).

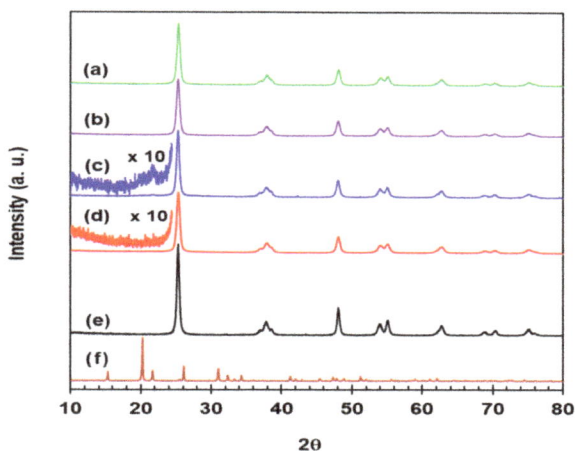

Figure 1. XRD patterns for: (**a**) VWT; (**b**) 2.73% Fe_2O_3/VWT; (**c**) c-VWT; (**d**) 2.73% Fe_2O_3/c-VWT; (**e**) pure anatase TiO_2; and (**f**) α-phase orthorhombic V_2O_5.

Comparing hematite (α-Fe_2O_3), maghemite (γ-Fe_2O_3), magnetite (Fe_3O_4), goethite (α-FeOOH), akaganeite (β-FeOOH), lepidocrocite (γ-FeOOH), and feroxyhyte (δ-FeOOH) in the corresponding JCPDS card #s 33-0664, 39-1346, 02-1035, 29-0713, 34-1266, 08-98, and 29-712, neither of them were indicated in XRD patterns for 2.73% Fe_2O_3-promoted catalysts, as disclosed in Figure 1b,d. Even a sample of 8% Fe_2O_3/VWT after calcination at 500 °C gave no crystalline peaks due to the iron compounds (not shown here). Magnetite nanoparticles are easily transformed to γ-Fe_2O_3 and α-Fe_2O_3 when, respectively, calcined at 300 and 500 °C in air [29], and all the FeOOH phases can be altered to α-Fe_2O_3 even at relatively lower thermal energy [30]. Our XRD measurements, and the thermal stability of the iron oxides and oxyhydroxides, reasonably indicate that FeO_x species existing in the promoted catalysts are in the form of α-Fe_2O_3 as a highly-dispersed noncrystalline particle, and that the Fe_2O_3 could significantly interact with V_2O_5 in c-VWT, thereby resulting in redispersion of the V_2O_5.

2.2. Textural Features of Fe₂O₃-Promoted V₂O₅-WO₃/TiO₂ Catalysts

2.2. Textural Features of Fe_2O_3-Promoted V_2O_5-WO_3/TiO_2 Catalysts

N_2 sorption isotherm of 2.73% Fe_2O_3/VWT was similar to that measured for VWT and WT, as provided in Figure 2a–c. This case was the same for c-VWT and 2.73% Fe_2O_3/c-VWT (Figure 2d,e).

All the isotherms showed a typical character of mesoporous materials because of the desorption hysteresis at P/P_o = 0.53–0.67 [31], depending on the sample.

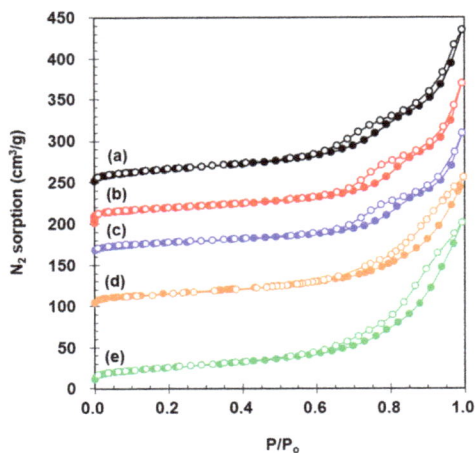

Figure 2. N_2 sorption isotherms on: (**a**) WT; (**b**) VWT; (**c**) 2.73% Fe_2O_3/VWT; (**d**) c-VWT; and (**e**) 2.73% Fe_2O_3/c-VWT.

Values for the specific BET surface area (S_{BET}), the mesopore size (d_m), and the total pore volume (V_t) are listed in Table 1. The S_{BET} value of WT decreased upon 1.6% V_2O_5 loading, which is due to some blockage of relatively small pores by the vanadia, but a further decrease after 2.73% Fe_2O_3 addition to the VWT was insignificant. These are consistent with changes in values for d_m and V_t. On the other hand, the promotion of c-VWT using 2.73% Fe_2O_3 led to an increased S_{BET} compared to the bare sample, and yielded somewhat smaller d_m and larger V_t values (Table 1). This might be caused by rearrangement of mechanical additives existing in c-VWT, such as glass fibers, during the sample preparation using an aqueous solution of the iron precursor [6,32,33].

Table 1. Chemical compositions and textural properties of V_2O_5/TiO_2-based catalysts.

Catalyst	Amount (%)		S_{BET} (m²/g)	d_m (Å) [a]	V_t (cm³/g) [b]
	V_2O_5	WO_3			
WT	-	10	91	127	0.30
VWT	1.6	10	67	142	0.26
2.73% Fe_2O_3/VWT	1.6	10	60	139	0.23
c-VWT	1.44	9.42	70	144	0.27
2.73% Fe_2O_3/c-VWT	1.44	9.42	89	126	0.31

Note. "-": not applicable or measured; S_{BET}: specific BET surface area; d_m: mesopore size; V_t: total pore volume. [a] Using the Barrett-Joyner-Halenda (BJH) mesopore model. [b] Calculated using N_2 sorption amounts at $P/P_o \approx 0.994$.

2.3. Effect of Fe_2O_3 Species on NH_3-SCR Reaction and N_2O Formation

Conversions of NO and NH_3 in deNO$_x$ reaction with VWT, 2.73% Fe_2O_3/VWT, c-VWT, and 2.73% Fe_2O_3/c-VWT and N_2O production are shown in Figures 3 and 4, respectively. The VWT-only exhibited 100% NO conversion at 300–400 °C, while at higher temperatures, it decreased depending on the temperature, as provided in Figure 3a. This shape of activity loss is a common feature of NH_3-SCR reaction over V_2O_5/TiO_2-based catalysts due to some side reactions giving N_2O, N_2, and NO [1,13,34]. The commercial unpromoted catalyst, i.e., c-VWT, basically showed a similar temperature vs. deNO$_x$ activity. These behaviors are consistent with those reported for 1.7–3.5% V$_2$O$_5$ dispersed on a commercial 10% WO$_3$/TiO$_2$ [1], 1.5% V$_2$O$_5$-10% WO$_3$/TiO$_2$ [35], and 1–5% V$_2$O$_5$ on four different commercial WO$_3$/TiO$_2$ supports with a WO$_3$ content of 4.7–6.8% [36]. After addition of 2.73% Fe$_2$O$_3$ to VWT and c-VWT, these all gave a significant decrease in NO conversion above 400 °C. As an example, the 2.73% Fe$_2$O$_3$/VWT had a NO conversion of 60% at 480 °C, which is lower, by 20%, than that indicated over the unpromoted catalyst. In case the 2.73% Fe$_2$O$_3$/c-VWT was used, the extent of such a decrease was much smaller.

Figure 3b shows NH$_3$ conversions over the unpromoted and promoted catalysts. Each value for NH$_3$ conversion below 400 °C could be comparable to that indicated for NO (Figure 3a), which is consistent with the overall reaction described by Equations (1) and (2). However, at higher temperatures the value was 100% regardless of the catalyst employed, and there is a difference in conversion between NH$_3$ and NO. This discrepancy depended not only on the catalyst employed but also on the presence of the iron oxide. The extent of the difference was smaller over the laboratory-made VWT than over the c-VWT, but this trend was reversed when 2.73% Fe$_2$O$_3$ was promoted to VWT and c-VWT (Figure 3a,b). Consequently, it is definitely represented that besides the general SCR mechanism, NH$_3$ would be consumed via undesired pathways at high temperatures >400 °C.

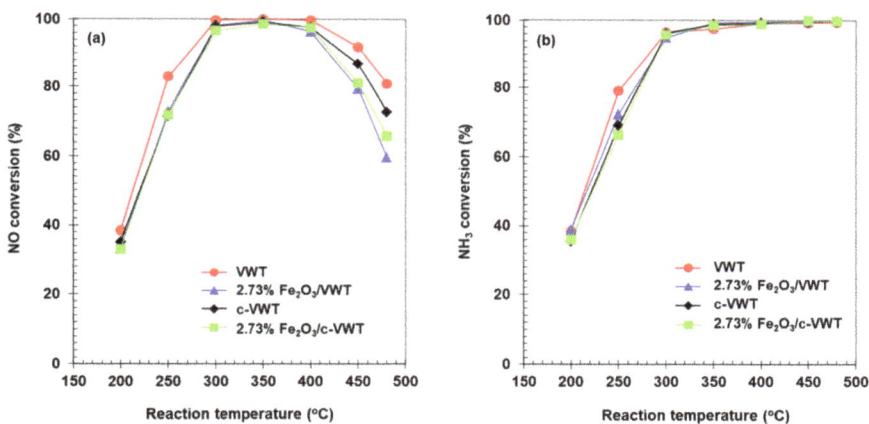

Figure 3. Conversions of (a) NO and (b) NH$_3$ in the reduction of NO by NH$_3$ over unpromoted and Fe$_2$O$_3$-promoted V$_2$O$_5$/TiO$_2$-based catalysts.

N$_2$O could be produced over V$_2$O$_5$-WO$_3$ (or MoO$_3$)/TiO$_2$ catalysts widely used for stationary and mobile applications and it can approach about 750 ppm depending strongly on the catalyst formulation and reaction conditions [8,13,37]. Whether or not the formation of N$_2$O in NH$_3$-SCR reaction with Fe$_2$O$_3$-promoted V$_2$O$_5$-WO$_3$/TiO$_2$ catalysts can be significantly depressed is of particular interest. Results are provided in Figure 4. A comparison between VWT and 2.73% Fe$_2$O$_3$/VWT indicated that the Fe$_2$O$_3$ promotion can greatly suppress N$_2$O production. That is, the VWT catalyst had, respectively, ca. 60 and 110 ppm N$_2$O at 450 and 480 °C, but the respective values decreased to about 25 and 45 ppm over the promoted VWT sample, corresponding to a reduction by 60% regardless of the temperature. This catalyst lost a NO conversion by 12–20% at 450–480 °C, compared to that observed for the VWT-only (Figure 3a). Although the indicated difference in NO conversion between c-VWT and 2.73% Fe$_2$O$_3$/c-VWT at temperatures >400 °C was less than 10% (Figure 3a), a depression effect on N$_2$O formation was similar to that of the promoted VWT catalyst (Figure 4). It represents that the Fe$_2$O$_3$ species in the VWT and c-VWT can be responsible for significant reduction in N$_2$O emissions from NH$_3$-SCR reaction at high temperatures.

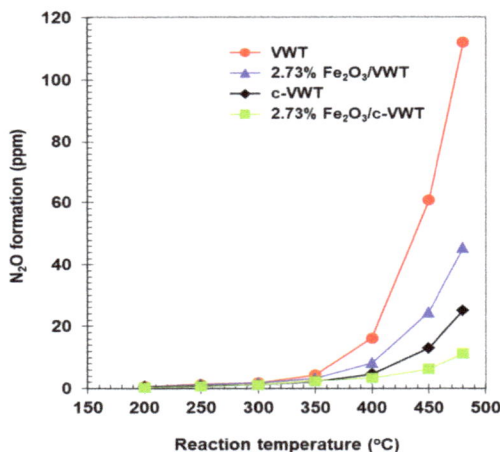

Figure 4. Formation of N_2O in the reduction of NO by NH_3 over unpromoted and Fe_2O_3-promoted V_2O_5/TiO_2-based catalysts.

The small changes in the textural properties of the Fe_2O_3-promoted catalysts (Table 1) might not contribute to the observed decrease in both NO removal and N_2O formation above 400 °C, because these reactions are predominantly determined by chemical compositions rather than textural features [8,38]. A commercial V_2O_5-WO_3/TiO_2 catalyst after prolonged usage at industrial deNO$_x$ plants could yield a large amount of N_2O via Equation (3) at high temperatures [6]. However, low N_2O production over the Fe_2O_3-promoted catalysts may not be due to significant inactivation of Equation (3), since in this circumstance, all NO and NH_3 conversions shall decrease by Equation (1). If the observed decrease in the NO conversion were because of Equation (4), the extent of the N_2O formed should increase. Based on the discussion, and the measured data for NO and NH_3 conversions, it is proposed that the oxidation of NH_3 into NO over the promoted catalysts at high temperatures,

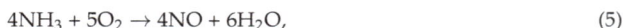

$$4NH_3 + 5O_2 \rightarrow 4NO + 6H_2O, \tag{5}$$

could occur. Thus, the NO gives rise to the decrease in deNO$_x$, activity while NH_3 conversion is still high. Such a NO generation even at 300 °C can take place over Fe_2O_3-containing mixed metal oxides, such as Fe_2O_3-TiO_2, which has been also highly active for NH_3-SCR reaction [39]. A 3% V_2O_5-9% WO_3/TiO_2 catalyst shows the formation of NO in NH_3 oxidation at temperatures >425 °C, depending on concentrations of O_2 in a feed gas stream [14]. Consequently, the addition of Fe_2O_3 to V_2O_5/TiO_2-based catalysts depresses the formation of N_2 in NH_3-SCR reaction at high temperatures, but unfortunately this approach can accompany the undesired pathway.

Another interest to us is whether or not different N_2O production levels of unpromoted and Fe_2O_3-promoted catalysts could result in a difference in absolute NO conversion between them. VWT had a NO conversion near 80% at 480 °C, and a similar value could be exhibited over 2.73% Fe_2O_3/VWT at 450 °C (Figure 3a). Therefore, both samples should give 50 ppm N_2O because of 100% conversion for NH_3 at each temperature (Figure 3b) and the fact that all the NH_3 was consumed according to Equation (4). However, the latter catalyst generated ca. 25 ppm N_2O at 450 °C (Figure 4), which is only a half of the concentration of N_2O expected using Equation (4), while the unpromoted one did ca. 110 ppm N_2O at 480 °C, which is over twice as high as the expected N_2O (Figure 4). These points represent that the VWT-only could produce N_2O via Equation (3) and that Fe_2O_3 species added to VWT can depress N_2O emissions in NH_3-SCR reaction. Such a role is probable, with an assumption not only that the promoted VWT could have more abundant NH_3 species strongly adsorbed on the surface but also that the NH_3 could readily reduce N_2O produced [24]. This surface reaction can be successfully described as the overall stoichiometry [16,20,21],

$$4N_2O + 4NH_3 + O_2 \rightarrow 6N_2 + 6H_2O, \tag{6}$$

and it is consistent with earlier studies that N_2O has been easily reduced by NH_3 over Fe-zeolites [20,21,40]. Furthermore, this pathway to decrease N_2O production was very similar to that reported in our previous work for a commercial V_2O_5-WO_3/TiO_2 catalyst coated by Fe-ZSM-5 [16], indicating that Fe_2O_3 is promised as a much cheaper, simpler promoter for significant reduction in N_2O emissions in NH_3-SCR reaction at high temperatures. On the other hand, 2.73% Fe_2O_3/c-VWT and c-VWT both showed much lower N_2O concentrations than maximum values that can be reached by only a single pathway of Equation (4) at similar a NO conversion to each other (Figures 3a and 4). Besides this, all the catalysts above 350 °C gave lesser N_2O production than those over the unpromoted and promoted VWT (Figure 4). It suggests that the c-VWT-based catalysts may possess much greater ability to catalyze Equation (6). Surface chemistry regarding this reaction occurring on the Fe_2O_3-promoted VWT and c-VWT catalysts will be evident in NH_3 TPD (temperature-programmed desorption) measurements below.

2.4. Role of Fe_2O_3 Species for the Suppression of N_2O Formation

In the TPD of NH_3 adsorbed on VWT, 2.73% Fe_2O_3/VWT, c-VWT, and 2.73% Fe_2O_3/c-VWT, mass spectra for the releasing NH_3 and gaseous products such as H_2O, N_2O, NO, and N_2 are provided in Figures 5 and 6. Predominant desorption of the adsorbed NH_3 occurred at temperatures ranging from 225 to 275 °C with bumps at 370–425 °C and high-temperature peaks at 650–690 °C, depending on the catalyst, as shown in spectra S1 to S4 in Figure 5a. Both a downward-shift of a maximum desorption temperature by 20–35 °C and an appreciable increase of the 650–690 °C NH_3 occurred on the Fe_2O_3-promoted samples (Figure 5a(S2,S4)). This implies that although the Fe_2O_3 species weakened the acid strength of sites for NH_3 desorbing at relatively low temperature region, they could significantly increase very stable NH_3 species. This agrees well with the discussion above. Figure 5b shown for H_2O generation is indicative of the oxidation of the adsorbed NH_3 to the N-containing products. The H_2O gave peaks with maxima around 250–260, 450–470, and 645–660 °C, depending on the sample (Figure 5b(S1–S4)). A relatively much higher H_2O production at the 645–660 °C region was detected for c-VWT and 2.73% Fe_2O_3/c-VWT (Figure 5b(S3,S4)), proposing that these catalysts can, to a much greater extent, oxidize the adsorbed NH_3 species into the N-containing gases, as shown below.

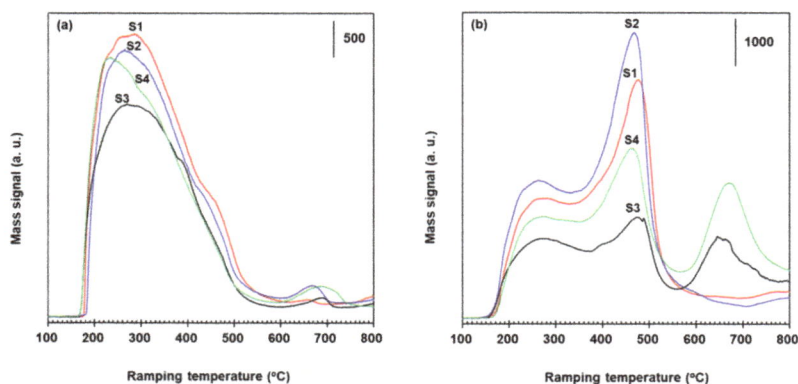

Figure 5. Mass spectra for NH_3 desorbed and H_2O produced during NH_3 TPD runs with unpromoted and Fe_2O_3-promoted catalysts: (**a**) NH_3 and (**b**) H_2O. In (**a**) and (**b**), S1: VWT; S2: 2.73% Fe_2O_3/VWT; S3: c-VWT; S4: 2.73% Fe_2O_3/c-VWT.

Mass spectra for N_2O, NO, and N_2 in the NH_3 TPD runs with the bare and Fe_2O_3-promoted catalysts are provided in Figure 6. All these catalysts gave major NO peaks having a maximum at 240–280 °C with shoulders at 375–415 and 480–495 °C, and relatively small peaks at temperatures >570 °C (Figure 6a). The last peaks are indirect evidence of NH_3-related species strongly adsorbed on the catalyst surface. It is clear that NH_3 adsorbed on the catalysts reacts with labile surface oxygen atoms existing in VO_x, FeO_x, and WO_x to create gaseous NO [41,42]. The most appropriate overall stoichiometry of this surface reaction can be described using Equation (5). Of course, the extent of the NO formation is small when the indicated size of the vertical solid bar in Figures 5 and 6 was compared each other.

Figure 6b shows N_2O production whose peaks at 270–285, 445–490, and 650–680 °C appear. Their intensity greatly depended on the catalyst, as seen in spectra S1–S4. c-VWT had an additional peak at 780 °C and showed a N_2O production profile similar to that reported for another commercial 1.68% V_2O_5-7.6% WO_3/TiO_2 catalyst [16]. The 780 °C N_2O peak disappeared after promotion with 2.73% Fe_2O_3 (Figure 6b(S4)). All these N_2O peaks might come from NH_4^+-like species on Bronsted acid sites such as W^{5+}-OH and WO_3-induced V^{4+}-OH, and from NH_x moieties (x = 1–3) adsorbed on Lewis acid sites, VO_x, WO_x, and FeO_x [43–45]. Both α-Fe_2O_3 and γ-Fe_2O_3 have no Bronsted acidity [46,47]. An intensity of the 650–680 °C N_2O peak in VWT and 2.73% Fe_2O_3/VWT (S1 and S2) was very weak compared to that in the other samples (S3 and S4), which is in excellent agreement with the H_2O production levels (Figure 5b). All the unpromoted catalysts (spectra S1 and S3 in Figure 6b) gave N_2O peaks at 490 °C. This temperature was shifted to 445 °C over Fe_2O_3-promoted samples (S2 and S4), implying that the Fe_2O_3 allows a decrease in the activation energy for the oxidation of adsorbed NH_3 to N_2O. The promoted catalysts showed lesser N_2O emissions in NH_3-SCR reaction (Figure 4). It is well-known that direct oxidation of NH_3 to N_2 and N_2O, not via adsorbed NH_x species (x = 1 or 2), is improbable and NH species on V^{5+}=O sites reacts with gas-phase NO to form N_2O and V^{4+}-OH [11,39,48]. This represents that gaseous NO is essential for such N_2O production, suggesting an involvement of NO molecules shown in Figure 6a because of no feed of NO upon NH_3 TPD experiments.

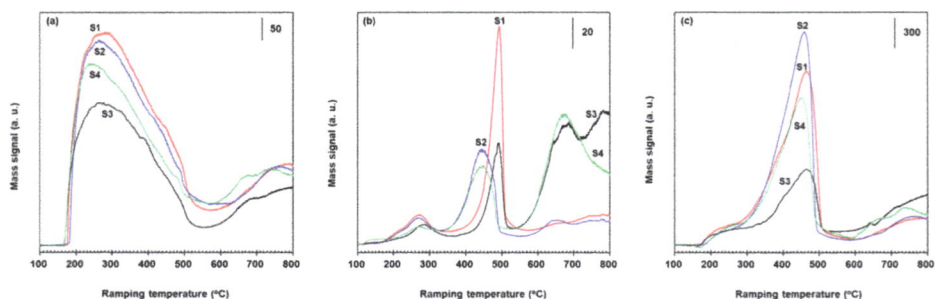

Figure 6. Mass spectra for the production of (**a**) NO, (**b**) N_2O, and (**c**) N_2 in NH_3 TPD runs with unpromoted and promoted VWT-based catalysts. In (**a–c**), S1: VWT; S2: 2.73% Fe_2O_3/VWT; S3: c-VWT; S4: 2.73% Fe_2O_3/c-VWT.

Mass spectra for N_2 evolved in NH_3 TPD runs with the unpromoted and promoted catalysts are shown in Figure 6c. Each catalyst had a predominant peak at 450–465 °C with a shoulder at 280–350 °C and a broad peak at temperatures >600 °C, depending on the sample, as seen in profiles S1–S4. As for NO and N_2O in Figure 6a,b, all those N_2 peaks would be associated with ammonium ions and NH_x on Bronsted and Lewis acid centers [43–45]. The NH_x species are oxidized to N_2 via NO-assistant route, as discussed [11,39,48]. The presence of Fe_2O_3 in VWT and c-VWT caused not only a shift of the major N_2 peak to lower temperatures, by 5–15 °C, but also a significant increase in their intensity, and such changes were greater for the commercial-based catalysts (Figure 6c(S3,S4)). These are in good accordance with the observed trend in the water production of H_2O as an indicator of N-associated side reactions (Figure 5b). It is thought that the promotion by amorphous iron oxide particles can decrease the activation energy for the oxidation of the adsorbed NH_3 species into N_2 and significantly enhance this reaction.

Based on the results in NH_3 TPD experiments with the unpromoted and 2.73% Fe_2O_3-promoted catalysts, and the previous discussion, the iron oxide promotion could significantly increase surface NH_3 species, producing the 650–690 °C peak (Figure 5a), and this ammonia species can be oxidized to NO, N_2O, and N_2 at high temperatures >550 °C (Figure 6). The presence of the Fe_2O_3 is, to some extent, responsible for the loss in $deNO_x$ activity at temperatures >400 °C due to the NO formation (Figure 3a). The NO formed via Equation (5) and/or NO in the feed stream may act as an intermediate for the N_2O and N_2 production [11,39,48]. Thus, Equation (3) rather than Equation (4) may be a main route for the formation of N_2O in NH_3-SCR reaction at temperatures >350 °C, and according to Equation (6), the N_2O can be easily reduced by the strongly-adsorbed NH_3 species on Fe_2O_3 and/or its related surface complex interacting with surface V_2O_5 species, thereby producing less N_2O in NH_3-SCR reaction with the promoted catalysts (Figure 4). This proposal is similar to that reported for great reduction in N_2O emissions over Fe-zeolites alone and Fe-ZSM-5-coated V_2O_5-WO_3/TiO_2 catalysts [16,20,21,23,40]. Consequently, Fe_2O_3 species existing in V_2O_5/TiO_2-based catalysts play a significant role for the suppression of N_2O formation in NH_3-SCR reaction at high temperature side, and this approach is a much simpler and more cost-effective compared to Fe-zeolite promotion techniques.

2.5. Surface Structure of Fe_2O_3, V_2O_5, and WO_3 Species

Figure 7 shows Raman spectra for reference materials and WT whose original signals have been reduced by 1/10–1/1000 for an easier comparison, but all bands have been kept unchanged even after such a data processing. α-phase V_2O_5 polycrystallites yielded a sharp peak at 990 cm^{-1} with subsequent signals near 400, 470, 515, and 700 cm^{-1}, as seen in Figure 7a. Crystalline WO_3 had characteristic sharp bands around 800 and 715 cm^{-1} (Figure 7b), and no Raman signals at frequencies >800 cm^{-1} even in a magnified spectrum existed as indicated by the dashed line. The measured spectra for the V_2O_5 and monoclinic γ-WO_3 ($P2_1$/n) are in agreement with earlier studies [49,50]. Bands due to highly-dispersed amorphous WO_x in a sample of WT that has been employed for preparing VWT and 2.73% Fe_2O_3/VWT catalysts can be differentiated by comparing with those existing in a bare anatase TiO_2 (Figure 7c,d). This support had Raman signals at 397, 515, and 638 cm^{-1} (Figure 7c) with a predominant band at 145 cm^{-1} and a weak one at 197 cm^{-1} (not shown here). The 397, 515 and 145, and 197 and 638 cm^{-1} bands correspond to the respective B_{1g}, A_{1g} + B_{1g} and E_g phonon vibrations of anatase-type TiO_2 [49,51–53]. A weak signal at 793 cm^{-1} (see the magnified spectrum in Figure 7c) is due to the first overtone of the 395 cm^{-1} [49,52,53]. Besides these peaks, bands at 882 and 980 cm^{-1} existed in the WT sample, as provided in the magnified spectrum in Figure 7d. The 980 cm^{-1} peak is assigned to two-dimensional polytungstate species with distorted octahedrally-coordinated environments that have been reported for calcined WO_3/TiO_2 samples not exceeding a monolayer coverage [53,54]. The 882 cm^{-1} is associated with asymmetric W-O-W vibrations in polymeric surface WO_x species [53]. It is represented that the WT sample has two different surface WO_x species. The absence of the 800 and 715 cm^{-1} bands reveals no crystalline WO_3 particles, indicating a very good dispersion of the tungsten oxide on the titania.

Raman spectra for VWT, 2.73% Fe_2O_3/VWT, c-VWT, and 2.73% Fe_2O_3/c-VWT are provided in Figure 8. All the catalysts had bands near 394, 515, 636, and 790 cm^{-1}, that are due to the TiO_2 support (Figure 7c), and they all also produced peaks at 880 and 981 cm^{-1} (Figure 7a–d). Any of the catalysts indicated no band near 1020–1030 cm^{-1} regarding isolated surface VO_x species [35,55]. The 981 and 880 cm^{-1} peaks had appeared even in the WT-only sample (Figure 7d). Regardless, these two bands may not be originated from the W=O and W-O-W structures in the surface tungsten oxide species. Symmetric V=O and W=O stretching vibrations occur at 950–990 cm^{-1}, making it difficult to clearly distinguish between both species, but the V=O species have an interaction cross section that is almost four times greater than that of the W=O groups [53,55–57]; thus, a Raman signal by the V=O structures in surface vanadium oxide species would give much stronger intensity at the frequency region. Furthermore, it has been reported not only for no appreciable mutual effect on the surface structures of VO_x and WO_x species in V_2O_5-WO_5/TiO_2 systems with low total VO_x and WO_x coverages but also for Raman bands at 940–987 cm^{-1}, due to two-dimensional VO_x species in 1–4.5% V_2O_5/TiO_2 below a monolayer coverage, whose positions are usually higher, by 10–30 cm^{-1}, than that of surface WO_x species in 1–7% WO_3/TiO_2 [55,58]. Consequently, the 880 and 981 cm^{-1} bands correspond to the respective bridging V-O-V and terminal V=O structures in distorted octahedrally coordinated VO_6 species, even in the TiO_2-supported binary and ternary component systems employed here, and indicate the presence of polymerized mono-oxo vanadate species [49,53,55,56,58].

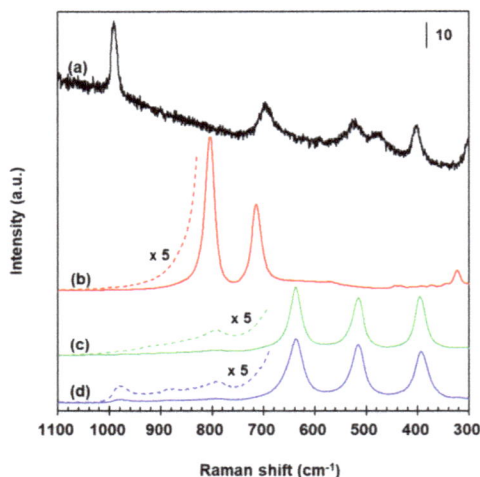

Figure 7. Raman spectra for reference chemicals and a supported WO_3-only sample: (**a**) V_2O_5; (**b**) WO_3; (**c**) TiO_2; (**d**) 10% WO_3/TiO_2 (WT).

All the iron oxides discussed in the XRD measurements, except for magnetite, exhibit Raman bands at 1000–1600 cm^{-1} in which no Raman signals by TiO_2 appear [30,49,51,59]. However, all the 2.73% Fe_2O_3-promoted catalysts gave no bands in the frequency region as well as at 800–1100 cm^{-1} (Figure 8b,d). The latter is because of the absence of Fe=O bonds in the iron oxides. Of course, bands below 800 cm^{-1} by Fe-O in surface iron oxide species could not be visible due to very intense signals of the titania support itself. All the promoted samples also gave no peaks near 1320, and 1375 and 1580 cm^{-1} for crystalline α- and γ-Fe_2O_3, respectively [30]. The intensity of all the characteristic vibration modes of the support was maintained unchanged even after the addition of 2.73% Fe_2O_3 to VWT and c-VWT, proposing that no great interaction of the iron oxide species with the titania surface occurred. There is no indication of the formation of V_2O_5, WO_3, and Fe_2O_3 crystals even in the titania-supported ternary systems, consistent with the previous discussion in the XRD measurements (Figure 1). This indicates that they all consist of submonolayer coverage of the metal oxides species. Table 2 lists the surface density of each metal oxide existing in the unpromoted and promoted catalysts that was estimated using its amount and their S_{BET} value given in Table 1. Calculations of the surface coverage based on monolayer coverage of each metal oxide reported in the literature [60–66] are also provided in Table 2. Both VO_x and FeO_x were below monolayer coverage irrespective of the catalyst, but WO_x in VWT, 2.73% Fe_2O_3/VWT and c-VWT could exceed the monolayer coverage, which is mainly due to a rather wide range of experimentally-determined WO_x monolayer values, such as

4.6–7.0 µmol WO_x/m^2 [60,61,63,65,66]. The upper limit can give us a value near monolayer coverage. The calculated coverages are in reasonable agreement with the Raman measurements. The total surface coverage expressed as MeO_x in Table 2 can be similarly explained, even though it was somewhat high for 2.73% Fe_2O_3/VWT. Finally, c-VWT after Fe_2O_3 promotion indicated a decrease in the surface coverages of each component, as well as of MeO_x, consistent with the redispersion of V_2O_5 in the promoted catalyst (Figure 1d).

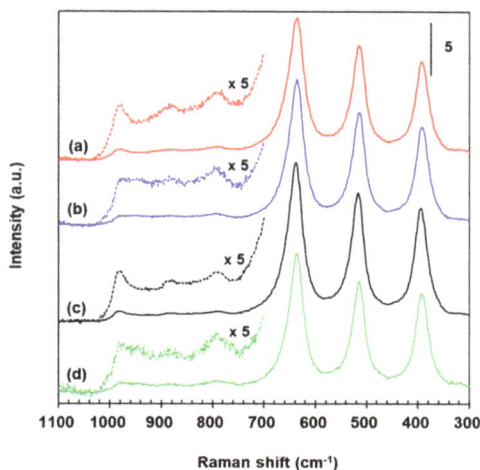

Figure 8. Raman spectra for unpromoted and promoted V_2O_5/TiO_2-based catalysts: (**a**) VWT; (**b**) 2.73% Fe_2O_3/VWT; (**c**) c-VWT; (**d**) 2.73% Fe_2O_3/c-VWT.

The noticeable difference between the unpromoted samples and the promoted ones is a new shoulder around 945 cm^{-1}, broadening of the 981 cm^{-1} band, and appreciable weakening of the phonon intensity of the 880 cm^{-1} signal associated with polymeric surface VO_x species (see the magnified spectra in Figure 8b,d). These strongly suggest an alteration of the molecular structure of surface vanadia and/or tungsta species when 2.73% Fe_2O_3 was introduced to VWT and c-VWT. However, it does not seem that there was a great structural interaction between the surface VO_x species and the FeO_x species, since the position of the 880 and 981 cm^{-1} bands is the same even after the promotion [61]. The terminal V=O band broadening is because of small interactions between the surface metal oxides, perhaps VO_x and FeO_x [55,60,63]. This was indicated by the disappearance of the XRD peak due to V_2O_5 in 2.73% Fe_2O_3/c-VWT (see Figure 1c,d), although such a change was not visible for VWT and 2.73% Fe_2O_3/VWT. The interaction can increase the extent of the disorder of the catalyst surface, and this becomes more energetically heterogeneous, suggesting that an enhancement to the main NH_3-SCR routes and/or the side reactions over the promoted catalysts is probable. The new signal at 945 cm^{-1} may be associated with rearrangement of the polymeric surface VO_x species, because it appeared with the weakening of the 880 cm^{-1} band. All samples of V_2O_5/TiO_2 and V_2O_5-WO_3/TiO_2 with low total surface coverage exhibited signals near 945 cm^{-1} in their Raman spectra collected at ambient conditions, and these were assigned to surface metavanadate species [55]. On the other hand, not only a significant electronic interaction between Fe^{3+} species and V^{5+} ones in titania-supported iron vanadates that are highly active for NH_3-SCR reaction could be evident from the Fe and V K-edge X-ray absorption, but a reduced oxidation capability of the vanadium oxide in the V-O-Fe species was alsoproposed to give a better N_2 selectivity at high temperatures >300 °C [67]. Combining our Raman results with the earlier reports suggests that the tetrahedrally coordinated polymeric vanadates adjacent to Fe_2O_3 species play a role in the reduction of N_2O formed in NH_3-SCR reaction at high temperatures by strongly-adsorbed NH_3 residues.

Table 2. Surface density and surface coverages of the active ingredients in V_2O_5/TiO_2-based catalysts.

Catalyst	Surface Density (μmol/m^2)				Surface Coverage			
	VO_x	WO_x	FeO_x	MeO_x [a]	VO_x [b]	WO_x [c]	FeO_x [d]	MeO_x [e]
WT	-	4.74	-	4.74	-	0.86 ± 0.17	-	0.86 ± 0.17
VWT	2.63	6.44	-	9.07	0.22 ± 0.02	1.16 ± 0.24	-	1.38 ± 0.26
2.73% Fe_2O_3/VWT	2.93	7.19	5.69	15.81	0.24 ± 0.03	1.30 ± 0.26	0.87	1.54 ± 0.29
c-VWT	2.26	5.80	-	8.06	0.19 ± 0.02	1.05 ± 0.21	-	1.24 ± 0.23
2.73% Fe_2O_3/c-VWT	1.78	4.57	3.84	10.19	0.15 ± 0.01	0.82 ± 0.17	0.59	0.97 ± 0.18

Note. "-": not applicable or measured. [a] Sum of the surface density values for VO_x, WO_x, and FeO_x. [b] Based on monolayer coverage of 10.9–13.1 μmol VO_x/m^2 [60–64]. [c] Based on monolayer coverage of 4.6–7.0 μmol WO_x/m^2 [60,61,63,65,66]. [d] Based on monolayer coverage of ca. 6.5 μmol FeO_x/m^2 [61,63]. [e] Sum of the surface coverages of VO_x, WO_x, and FeO_x.

3. Experimental

3.1. Preparation of Catalyst Samples

A commercial powder-type 10% WO_3/TiO_2 (Tronox Ltd., formerly Kerr-McGee Corp., Stamford, CT, USA), hereafter designated to "WT", was employed to prepare a V_2O_5-WO_3/TiO_2 catalyst. An appropriate amount of the WT (ca. 10 g) was calcined at 500 °C for 4 h in flowing 21% O_2/79% N_2 (Praxair, 99.999%, Changwon, Korea) at a total flow rate of 300 cm^3/min. A 1.6% V_2O_5/WT catalyst was prepared by impregnating the calcined WT with an aqueous NH_4VO_3 (Aldrich, 99.99%, Saint Louis, MO, USA) solution with the corresponding vanadium content, which had been obtained by dissolving it in an aqueous solution of oxalic acid (Aldrich, ≥99%) with a pH near 2.5 dissolved in distilled, deionized water (DDI), referred to as "VWT". A part of this sample was used for the preparation of approximately 5 g 2.73% Fe_2O_3/VWT that was made by impregnating the VWT with a aqueous solution of $Fe(NO_3)_3 \cdot 9H_2O$ (Aldrich, ≥99.95%) dissolved in DDI water. A commercial extruded V_2O_5-WO_3/TiO_2 honeycomb was supplied from a domestic coal-fired power plant and crushed, finely ground, and calcined as for the WT, denoted to "c-VWT" to differentiate it from the VWT. A 2.73% Fe_2O_3/c-VWT catalyst was prepared using the calcined c-VWT in a fashion similar to that described for the VWT-supported iron oxide. All the catalysts used were dried at 110 °C overnight in an oven and then calcined as for the WT prior to using them for NH_3-SCR reaction and characterization. The N_2 and O_2 used were further purified by passing them through moisture trap and Oxytraps (Alltech Assoc., Deerfield, IL, USA). The amounts of V_2O_5 and WO_3 existing in the c-VWT were determined using ICP (inductively-coupled plasma) measurements.

3.2. NH_3-SCR deNO$_x$ Reaction and Determination of N_2O Formation

Details of the modified Model MARS 0.75 L/8.0 V White gas cell (Zemini Scientific Instr., Buena Park, CA, USA) combined with a Thermo Electron Nicolet 7600 FT-IR spectrophotometer (Thermo Fisher Scientific, Waltham, MA, USA) used in the present study to measure conversions of NO and NH_3 and the extent of N_2O production in NH_3-SCR reaction have been described earlier [6,68]. A continuous purge of this system was allowed by flowing a compressed air at a rate of 15 L/min that had passed through a train of large volume silica traps to remove H_2O in the air. The NH_3-SCR reaction with catalyst samples was conducted in a continuous flow type I-shaped 3/8″ OD quartz reactor placed in a three-independent temperature adjustable electric furnace (Lindberg/Blue M Model HTF55347C, Thermo Electron Corp., Asheville, NC, USA) coupled with a Lindberg/Blue M Model CC584343PC PID controller [6,16,27,68]. All gas feed lines were maintained at a temperature near 180 °C to prevent the homogeneous reaction between NO and O_2, and the condensation of H_2O produced in the reaction.

Catalysts 2018, 8, 134

A flowing mixture of 21% O_2/79% N_2 at a total rate of 1000 cm^3/min was flowed through the reactor with typically 0.5 g catalyst sample to calcine it at 500 °C for 1 h, and then the temperature and oxygen concentration were changed to 200 °C and 5%, respectively. After this, the downstream was switched to the upstream, at which point the gas cell was fully purged prior to recording a background interferogram that was used for Fourier-transforming sample interferograms. Then NO and NH_3 were added to the gas flow so as to be 500 ppm, respectively. An interferogram before reaction was collected at their steady-state concentration. Following this, the gas mixture was flowed over the catalyst bed at chosen temperatures, corresponding to a gas hourly space velocity of 76,200 h^{-1}, and a sample interferogram was obtained after ca. 30 min. All interferograms were collected with a resolution of 0.5 cm^{-1} and a scan number of 100. The NO (Omega grade, 99.99%, Scott Specialty Gases, South Plainfield, NJ, USA) and NH_3 (Scott Specialty Gases, Electronic grade, 99.999%) were used without any purification, while the N_2 was flowed through an Alltech moisture trap. All gas flows were accurately controlled by using a Model 5850E mass flow controller (Brooks Instr., Hatfield, PA, USA) and a Model F-200CV one (Bronkhorst High-Tech, Ruurlo, The Netherland). Details of the standard procedures to collect the gas-phase spectra have been provided elsewhere [6,68].

3.3. Characterization of Catalyst Samples

A Model D/MAX2500 PC diffractometer (Rigaku, Tokyo, Japan) with a Cu Kα (λ = 1.5405 Å) radiation source was employed for XRD measurements in which an X-ray tube voltage and current were 40 kV and 20 mA, respectively. Each calcined sample charged in a thin quartz holder was scanned from a 2θ value of 10 to 80° at a scanning rate of 0.1°/min to allow an accurate resolution.

A Model 3 Flex Version 3.01 system (Micromeritics Instr., Norcross, GA, USA) was used to determine textural properties of the catalysts, such as S_{BET}, d_m, and V_t. A cell containing ca. 60 mg of each sample was directly connected to the system, evacuated at 300 °C overnight (under a high dynamic vacuum below 10^{-7} Torr (1 Torr = 133.3 Pa)) and allowed for a further evacuation at room temperature. Following this, N_2 was introduced into the sample cell at a liquid nitrogen temperature (−196 °C).

Temperature-programmed desorption (TPD) studies were conducted using a Model HPR-20 QIC quadrupole mass (Hiden Analytical, Warrington, UK) spectrometer system described in detail elsewhere [16,69,70]. For this, an adsorption cell with 160 mg of each sample was coupled with a gas handling system to calcine at 500 °C for 1 h in a flowing mixture of 21% O_2/79% He (Praxair, 99.9999%) at a total flow rate of 100 cm^3/min. Then, 2% NH_3 in flowing He at the same flow rate was admitted to the cell at 100 °C for 1 h prior to fully purging it using the pure He flow. After the NH_3 approached a background level, it was heated to 800 °C at 10 °C/min in a He flow at a total rate of 20 cm^3/min. During this process, N-containing products, i.e., N_2O, NO, and N_2, were monitored at each corresponding m/z, but NH_3 desorbed was monitored at m/z = 16, because H_2O formed upon the surface reaction, producing the nitrogenous products that could be fragmentized to OH [16,69]. All the gases were controlled using a Model 5850E mass flow controller (Brooks Instr., Hatfield, PA, USA). The He used has been further purified in a similar fashion as described above.

Ex situ Raman spectra for fresh catalysts calcined in a fashion similar to that described for the activity measurements were recorded in the range of 50–3400 cm^{-1} with a 2 cm^{-1} spectral resolution. V_2O_5 (Aldrich, 99.99%), WO_3 (99.995%), TiO_2 (DT51D, Millennium Inorganic Chemicals, Henderson, Australia), and the WT were used as a reference material. The measurements were conducted using a Thermo Scientific DXR 2xi Raman spectrometer (Thermo Fisher Scientific, Waltham, MA, USA) equipped with a liquid N_2-cooled EM-CCD detector around −120 °C. A 532-nm diode laser was employed to excite the samples. A power of the laser at a surface of each sample was applied to be ca. 2 mW in order to minimize laser heating effects. Exposure of each sample to the laser beam was approximately 0.01–0.03 s with 500 to 800 averaged signal accumulations, depending on the sample. All spectra were collected with a powder under ambient conditions.

71

4. Conclusions

Fe_2O_3-promoted V_2O_5/TiO_2-based catalysts show no great changes in the textural features, and no XRD peaks due to crystalline Fe_2O_3 phases are indicated. Significant depression of N_2O emissions in NH_3-SCR reaction over V_2O_5-WO_3/TiO_2 catalysts at high temperatures could be successfully established by their promotion using Fe_2O_3 particles. This approach results in a decrease in NO conversion at high temperatures, which is mainly because of the oxidation of NH_3 to NO. NH_3 TPD measurements suggest that the Fe_2O_3 existing in VWT and c-VWT can significantly increase strongly-adsorbed NH_3 and NH_x moieties, and these species participate in the reduction of N_2O formed at high temperatures. Raman spectra for the promoted catalysts propose the presence of Fe_2O_3-induced, tetrahedrally coordinated polyvanadates and/or surface V-O-Fe species that are probably responsible for the N_2O reduction. The reaction between NO and NH_3 may predominantly take place to form N_2O that can be readily reduced by the strongly adsorbed NH_3 over the Fe_2O_3-promoted catalysts.

Acknowledgments: A partial grant-in-aid for this study was provided by Basic Science Research Program through the National Research Foundation of Korea (NRF) via Grant # 2017080772.

Author Contributions: The key approach to this study was designed by Moon Hyeon Kim who also determined a significance of all data and wrote this manuscript. Ki Hyuck Yang performed the experiments regarding activity measurements and instrumental characterization. He also prepared a draft version of all Figures included here. The co-authors have made an approval of the final version of this manuscript.

Conflicts of Interest: The authors declare no conflict of interest.

References

1. Marberger, A.; Elsener, M.; Ferri, D.; Krocher, O. VO_x surface coverage optimization of V_2O_5/WO_3-TiO_2 SCR catalysts by variation of the V loading and by aging. *Catalysts* **2015**, *5*, 1704–1720. [CrossRef]
2. He, Y.; Ford, M.E.; Zhu, M.; Liu, Q.; Tumuluri, U.; Wu, Z.; Wachs, I.E. Influence of catalyst synthesis method on selective catalytic reduction (SCR) of NO by NH_3 with V_2O_5-WO_3/TiO_2 catalysts. *Appl. Catal. B* **2016**, *193*, 141–150. [CrossRef]
3. Madia, G.; Elsener, M.; Koebel, M.; Raimondi, F.; Wokaun, A. Thermal stability of vanadia-tungsta-titania catalysts in the SCR process. *Appl. Catal. B* **2002**, *39*, 181–190. [CrossRef]
4. Forzatti, P. Present status and perspectives in de-NO_x SCR catalysis. *Appl. Catal. A* **2001**, *222*, 221–236. [CrossRef]
5. Nova, I.; dall'Acqua, L.; Lietti, L.; Giamello, E.; Forzatti, P. Study of thermal deactivation of a de-NO_x commercial catalyst. *Appl. Catal. B* **2001**, *35*, 31–42. [CrossRef]
6. Kim, M.H.; Ham, S.W. Determination of N_2O emissions levels in the selective reduction of NO_x by NH_3 over an on-site-used commercial V_2O_5-WO_3/TiO_2 catalyst using a modified gas cell. *Top. Catal.* **2010**, *53*, 597–607. [CrossRef]
7. Kompio, P.G.W.A.; Bruckner, A.; Hipler, F.; Auer, G.; Loffler, E.; Grunert, W. A new view on the relations between tungsten and vanadium in V_2O_5-WO_3/TiO_2 catalysts for the selective reduction of NO with NH_3. *J. Catal.* **2012**, *286*, 237–247. [CrossRef]
8. Lietti, L.; Nova, I.; Ramis, G.; Dall'Acqua, L.; Busca, G.; Giamello, E.; Forzatti, P.; Bregani, F. Characterization and reactivity of V_2O_5–MoO_3/TiO_2 de-NO_x SCR catalysts. *J. Catal.* **1999**, *187*, 419–435. [CrossRef]
9. Nova, I.; Ciardelli, C.; Tronconi, E.; Chatterjee, D.; Weibel, M. Unifying redox kinetics for standard and fast NH_3-SCR over a V_2O_5-WO_3/TiO_2 catalyst. *AIChE J.* **2009**, *55*, 1514–1529. [CrossRef]
10. Nova, I.; Ciardelli, C.; Tronconi, E.; Chatterjee, D.; Bandl-Konrad, B. NH_3-NO/NO_2 chemistry over V-based catalysts and its role in the mechanism of the fast SCR reaction. *Catal. Today* **2006**, *114*, 3–12. [CrossRef]
11. Xiong, S.; Xiao, X.; Liao, Y.; Dang, H.; Shan, W.; Yang, S. Global kinetic study of NO reduction by NH_3 over V_2O_5–WO_3/TiO_2: Relationship between the SCR performance and the key factors. *Ind. Eng. Chem. Res.* **2015**, *54*, 11011–11023. [CrossRef]
12. Koebel, M.; Madia, G.; Elsener, M. Selective catalytic reduction of NO and NO_2 at low temperatures. *Catal. Today* **2002**, *73*, 239–247. [CrossRef]

13. Madia, G.; Koebel, M.; Elsener, M.; Wokaun, A. Side reactions in the selective catalytic reduction of NO_x with various NO_2 fractions. *Ind. Eng. Chem. Res.* **2002**, *41*, 4008–4015. [CrossRef]

14. Djerad, S.; Crocoll, M.; Kureti, S.; Tifouti, L.; Weisweiler, W. Effect of oxygen concentration on the NO_x reduction with ammonia over V_2O_5–WO_3/TiO_2 catalyst. *Catal. Today* **2006**, *113*, 208–214. [CrossRef]

15. Kim, M.H.; Lee, H.S. Effect of Fe-zeolite on formation of N_2O in selective reduction of NO by NH_3 over V_2O_5-WO_3/TiO_2 catalyst. *Res. Chem. Intermed.* **2016**, *42*, 171–184. [CrossRef]

16. Kim, M.H.; Park, S.W. Selective reduction of NO by NH_3 over Fe-zeolite-promoted V_2O_5-WO_3/TiO_2-based catalysts: Great suppression of N_2O formation and origin of NO removal activity loss. *Catal. Commun.* **2016**, *86*, 82–85. [CrossRef]

17. Krocher, O.; Elsener, M. Combination of V_2O_5/WO_3-TiO_2, Fe-ZSM5, and Cu-ZSM5 catalysts for the selective catalytic reduction of nitric oxide with ammonia. *Ind. Eng. Chem. Res.* **2008**, *47*, 8588–8593. [CrossRef]

18. Yang, S.; Wang, C.; Ma, L.; Peng, Y.; Qu, Z.; Yan, N.; Chen, J.; Chang, H.; Li, J. Substitution of WO_3 in V_2O_5/WO_3-TiO_2 by Fe_2O_3 for selective catalytic reduction of NO with NH_3. *Catal. Sci. Technol.* **2013**, *3*, 161–168. [CrossRef]

19. Gao, R.; Zhang, D.; Liu, X.; Shi, L.; Maitarad, P.; Li, H.; Zhang, J.; Cao, W. Enhanced catalytic performance of V_2O_5-WO_3/Fe_2O_3/TiO_2 microspheres for selective catalytic reduction of NO by NH_3. *Catal. Sci. Technol.* **2013**, *3*, 191–199. [CrossRef]

20. Devadas, M.; Krocher, O.; Elsener, M.; Wokaun, A.; Mitrikas, G.; Soger, N.; Pfeifer, M.; Demel, Y.; Mussmann, L. Characterization and catalytic investigation of Fe-ZSM5 for urea-SCR. *Catal. Today* **2007**, *119*, 137–144. [CrossRef]

21. Qi, G.; Yang, R.T. Ultra-active Fe/ZSM-5 catalyst for selective catalytic reduction of nitric oxide with ammonia. *Appl. Catal. B* **2005**, *60*, 13–22. [CrossRef]

22. Rivallan, M.; Ricchiardi, G.; Bordiga, S.; Zecchina, A. Adsorption and reactivity of nitrogen oxides (NO_2, NO, N_2O) on Fe-zeolites. *J. Catal.* **2009**, *264*, 104–116. [CrossRef]

23. Coq, B.; Mauvezin, M.; Delahay, G.; Butet, J.B.; Kieger, S. The simultaneous catalytic reduction of NO and N_2O by NH_3 using an Fe-zeolite-beta catalyst. *Appl. Catal. B* **2000**, *27*, 193–198. [CrossRef]

24. Mou, X.; Zhang, B.; Li, Y.; Yao, L.; Wei, X.; Su, D.S.; Shen, W. Rod-shaped Fe_2O_3 as an efficient catalyst for the selective reduction of nitrogen oxide by ammonia. *Angew. Chem. Int. Ed.* **2012**, *51*, 2989–2993. [CrossRef] [PubMed]

25. Qu, Z.; Miao, L.; Wang, H.; Fu, Q. Highly dispersed Fe_2O_3 on carbon nanotubes for low-temperature selective catalytic reduction of NO with NH_3. *Chem. Commun.* **2015**, *51*, 956–958. [CrossRef] [PubMed]

26. Liu, F.; Asakura, K.; He, H.; Shan, W.; Shi, X.; Zhang, C. Influence of sulfation on iron titanate catalyst for the selective catalytic reduction of NO_x with NH_3. *Appl. Catal. B* **2011**, *103*, 369–377. [CrossRef]

27. Kim, M.H.; An, T.H. A commercial V_2O_5-WO_3/TiO_2 catalyst used at an NH_3-SCR deNO$_x$ process in an oil-fired power plant: Cause of an increase in deNO$_x$ing and NH_3 oxidation performances at low temperatures. *Res. Chem. Intermed.* **2011**, *37*, 1333–1344. [CrossRef]

28. Chen, Y.; Yang, G.; Zhang, Z.; Yang, X.; Hou, W.; Zhu, J.J. Polyaniline-intercalated layered vanadium oxide nanocomposites—One-pot hydrothermal synthesis and application in lithium battery. *Nanoscale* **2010**, *2*, 2131–2138. [CrossRef] [PubMed]

29. Xu, J.; Yang, H.; Fu, W.; Du, K.; Sui, Y.; Chen, J.; Zeng, Y.; Li, M.; Zou, G. Preparation and magnetic properties of magnetite nanoparticles by sol–gel method. *J. Magn. Magn. Mater.* **2007**, *309*, 307–311. [CrossRef]

30. De Faria, D.L.A.; Silva, S.V.; de Oliveira, M.T. Raman microspectroscopy of some iron oxides and oxyhydroxides. *J. Raman Spectrosc.* **1997**, *28*, 873–878. [CrossRef]

31. Thommes, M.; Smarsly, B.; Groenewolt, M.; Ravikovitch, P.I.; Neimark, A.V. Adsorption hysteresis of nitrogen and argon in pore networks and characterization of novel micro- and mesoporous silicas. *Langmuir* **2006**, *22*, 756–764. [CrossRef] [PubMed]

32. Xia, B.; Li, W.; Zhang, B.; Xie, Y. Low temperature vapor-phase preparation of TiO_2 nanopowder. *J. Mater. Sci.* **1999**, *34*, 3505–3511. [CrossRef]

33. Alemany, L.J.; Berti, F.; Busca, G.; Ramis, G.; Robba, D.; Toledo, G.P.; Trombetta, M. Characterization and composition of commercial V_2O_5-WO_5-TiO_2 SCR catalysts. *Appl. Catal. B* **1996**, *10*, 299–311. [CrossRef]

34. Gutierrez, M.J.F.; Baxter, D.; Hunter, C.; Svoboda, K. Nitrous Oxide (N_2O) emissions from waste and biomass to energy plants. *Waste Manag. Res.* **2005**, *23*, 133–147. [CrossRef] [PubMed]

35. Kompio, P.G.W.A.; Bruckner, A.; Hipler, F.; Manoylova, O.; Auer, G.; Mestl, G.; Grunert, W. V_2O_5-WO_3/TiO_2 catalysts under thermal stress: Responses of structure and catalytic behavior in the selective catalytic reduction of NO by NH_3. *Appl. Catal. B* **2017**, *217*, 365–377. [CrossRef]

36. Wang, C.; Yang, S.; Chang, H.; Peng, Y.; Li, J. Dispersion of tungsten oxide on SCR performance of V_2O_5-WO_3/TiO_2: Acidity, surface species and catalytic activity. *Chem. Eng. J.* **2013**, *225*, 520–527. [CrossRef]

37. Lietti, L.; Nova, I.; Forzatti, P. Selective catalytic reduction (SCR) of NO by NH_3 over TiO_2-supported V_2O_5-WO_3 and V_2O_5-MoO_3 catalysts. *Top. Catal.* **2000**, *11*, 111–122. [CrossRef]

38. Busca, G.; Lietti, L.; Ramis, G.; Berti, F. Chemical and mechanistic aspects of the selective catalytic reduction of NO_x by ammonia over oxide catalysts: A review. *Appl. Catal. B* **1998**, *18*, 1–36. [CrossRef]

39. Long, R.Q.; Yang, R.T. Selective catalytic oxidation of ammonia to nitrogen over Fe_2O_3-TiO_2 prepared with a sol–gel method. *J. Catal.* **2002**, *207*, 158–165. [CrossRef]

40. Zhang, X.; Shen, Q.; He, C.; Ma, C.; Cheng, J.; Hao, Z. N_2O catalytic reduction by NH_3 over Fe-zeolites: Effective removal and active site. *Catal. Commun.* **2012**, *18*, 151–155. [CrossRef]

41. Went, G.T.; Leu, L.J.; Rosin, R.R.; Bell, A.T. The effects of structure on the catalytic activity and selectivity of V_2O_5/TiO_2 for the reduction of NO by NH_3. *J. Catal.* **1992**, *134*, 492–505. [CrossRef]

42. Usberti, N.; Jablonska, M.; Blasi, M.D.; Forzatti, P.; Lietti, L.; Beretta, A. Design of a "high-efficiency" NH_3-SCR reactor for stationary applications. A kinetic study of NH_3 oxidation and NH_3-SCR over V-based catalysts. *Appl. Catal. B* **2015**, *179*, 185–195. [CrossRef]

43. Giraud, F.; Geantet, C.; Guilhaume, N.; Loridant, S.; Gros, S.; Porcheron, L.; Kanniche, M.; Bianchi, D. Experimental microkinetic approach of de-NO_x by NH_3 on V_2O_5/WO_3/TiO_2 catalysts. 3. Impact of superficial WO_z and V_xO_y/WO_z groups on the heats of adsorption of adsorbed NH_3 species. *J. Phys. Chem. C* **2015**, *119*, 15401–15413. [CrossRef]

44. Kantcheva, M.M.; Hadjiivanov, K.I.; Klissurski, D.G. An IR spectroscopy study of the state and localization of vanadium-oxo species adsorbed on TiO_2 (anatase). *J. Catal.* **1992**, *134*, 299–310. [CrossRef]

45. Topsoe, N.Y. Characterization of the nature of surface sites on vanadia-titania catalysts by FTIR. *J. Catal.* **1991**, *128*, 499–511. [CrossRef]

46. Lorenzelli, V.; Busca, G. Infrared studies of the surface of α-Fe_2O_3. *Mater. Chem. Phys.* **1985**, *13*, 261–281. [CrossRef]

47. Ramis, G.; Yi, L.; Busca, G.; Turco, M.; Kotur, E.; Willey, R.J. Adsorption, activation, and oxidation of ammonia over SCR catalysts. *J. Catal.* **1995**, *157*, 523–535. [CrossRef]

48. Jung, S.M.; Grange, P. DRIFTS investigation of V=O behavior and its relations with the reactivity of ammonia oxidation and selective catalytic reduction of NO over V_2O_5 catalyst. *Appl. Catal. B* **2002**, *36*, 325–332. [CrossRef]

49. Went, G.T.; Oyama, S.T.; Bell, A.T. Laser Raman spectroscopy of supported vanadium oxide catalysts. *J. Phys. Chem.* **1990**, *94*, 4240–4246. [CrossRef]

50. Boulova, M.; Lucazeau, G. Crystallite nanosize effect on the structural transitions of WO_3 studied by Raman spectroscopy. *J. Solid State Chem.* **2002**, *167*, 425–434. [CrossRef]

51. Ohsaka, T.; Izumi, F.; Fujiki, Y. Raman spectrum of anatase, TiO_2. *J. Raman Spectrosc.* **1978**, *7*, 321–324. [CrossRef]

52. Frank, O.; Zukalova, M.; Laskova, B.; Kurti, J.; Koltai, J.; Kavan, L. Raman spectra of titanium dioxide (anatase, rutile) with identified oxygen isotopes (16, 17, 18). *Phys. Chem. Chem. Phys.* **2012**, *14*, 14567–14572. [CrossRef] [PubMed]

53. Kim, D.S.; Ostromecki, M.; Wachs, I.E. Surface structures of supported tungsten oxide catalysts under dehydrated conditions. *J. Mol. Catal. A* **1996**, *106*, 93–102. [CrossRef]

54. Engweiler, J.; Harf, J.; Baiker, A. WO_x/TiO_2 catalysts prepared by grafting of tungsten alkoxides: Morphological properties and catalytic behavior in the selective reduction of NO by NH_3. *J. Catal.* **1996**, *159*, 259–269. [CrossRef]

55. Vuurman, M.A.; Wachs, I.E.; Hirt, A.M. Structural determination of supported V_2O_5-WO_3/TiO_2 catalysts by in situ Raman spectroscopy and X-ray photoelectron spectroscopy. *J. Phys. Chem.* **1991**, *95*, 9928–9937. [CrossRef]

56. Reiche, M.A.; Burgi, T.; Baiker, A.; Scholz, A.; Schnyder, B.; Wokaun, A. Vanadia and tungsta grafted on TiO_2: Influence of the grafting sequence on structural and chemical properties. *Appl. Catal. A* **2000**, *198*, 155–169. [CrossRef]

57. Wachs, I.E.; Roberts, C.A. Monitoring surface metal oxide catalytic active sites with Raman spectroscopy. *Chem. Soc. Rev.* **2010**, *39*, 5002–5017. [CrossRef] [PubMed]

58. Amiridis, M.D.; Duevel, R.V.; Wachs, I.E. The effect of metal oxide additives on the activity of V_2O_5/TiO_2 catalysts for the selective catalytic reduction of nitric oxide by ammonia. *Appl. Catal. B* **1999**, *20*, 111–122. [CrossRef]

59. Colomban, P.; Cherifi, S.; Despert, G. Raman identification of corrosion products on automotive galvanized steel sheets. *J. Raman Spectrosc.* **2008**, *39*, 881–886. [CrossRef]

60. Bourikas, K.; Fountzoula, C.; Kordulis, C. Monolayer transition metal supported on titania catalysts for the selective catalytic reduction of NO by NH_3. *App. Catal. B* **2004**, *52*, 145–153. [CrossRef]

61. Dunn, J.P.; Stenger, H.G., Jr.; Wachs, I.E. Oxidation of SO_2 over supported metal oxide catalysts. *J. Catal.* **1999**, *181*, 233–243. [CrossRef]

62. Amiridis, M.; Wachs, I.E.; Deo, G.; Jehng, J.M.; Kim, D.S. Reactivity of V_2O_5 catalysts for the selective catalytic reduction of NO by NH_3: Influence of vanadia loading, H_2O, and SO_2. *J. Catal.* **1996**, *161*, 247–253. [CrossRef]

63. Wachs, I.E. Raman and IR studies of surface metal oxide species on oxide supports: Supported metal oxide catalysts. *Catal. Today* **1996**, *27*, 437–455. [CrossRef]

64. Bond, G.C.; Bruckman, K. Selective oxidation of *o*-xylene by monolayer V_2O_5-TiO_2 catalysts. *Faraday Discuss. Chem. Soc.* **1981**, *72*, 235–246. [CrossRef]

65. Vermaire, D.C.; van Berge, P.C. The preparation of WO_3/TiO_2 and WO_3/Al_2O_3 and characterization by temperature-programmed reduction. *J. Catal.* **1989**, *116*, 309–317. [CrossRef]

66. Yu, X.F.; Wu, N.Z.; Huang, H.Z.; Xie, Y.C.; Tang, Y.Q. A study on the monolayer dispersion of tungsten oxide on anatase. *J. Mater. Chem.* **2001**, *11*, 3337–3342. [CrossRef]

67. Liu, F.; He, H.; Lian, Z.; Shan, W.; Xie, L.; Asakura, K.; Yang, W.; Deng, H. Highly dispersed iron vanadate catalyst supported on TiO_2 for the selective catalytic reduction of NO_x with NH_3. *J. Catal.* **2013**, *307*, 340–351. [CrossRef]

68. Kim, D.W.; Kim, M.H.; Ham, S.W. An on-line infrared spectroscopic system with a modified multipath White cell for direct measurements of N_2O from NH_3-SCR reaction. *Korean J. Chem. Eng.* **2010**, *27*, 1730–1737. [CrossRef]

69. Kim, M.H.; Cho, I.H.; Park, J.H.; Choi, S.O.; Lee, I.S. Adsorption of CO_2 and CO on H-zeolites with different framework topologies and chemical compositions and a correlation to probing protonic sites using NH_3 adsorption. *J. Porous Mater.* **2016**, *23*, 291–299. [CrossRef]

70. Kim, M.H.; Cho, I.H.; Choi, S.O.; Lee, I.S. Surface energetic heterogeneity of nanoporous solids for CO_2 and CO adsorption: The key to an adsorption capacity and selectivity at low pressures. *J. Nanosci. Nanotechnol.* **2016**, *16*, 4474–4479. [CrossRef] [PubMed]

catalysts

MDPI

Article

Experimental Research of an Active Solution for Modeling In Situ Activating Selective Catalytic Reduction Catalyst

Tuo Ye [1], Donglin Chen [1,*], Yanshan Yin [1], Jing Liu [2] and Xi Zeng [1]

[1] School of Energy and Power Engineering, Changsha University of Science and Technology, Changsha 410114, China; Ye_Tuo@csust.edu.cn (T.Y.); yanshan.yin@csust.edu.cn (Y.Y.); zx199077@163.com (X.Z.)
[2] School of Electric Power, South China University of Technology, Guangzhou 510640, China; msliujing@mail.scut.edu.cn
*   Correspondence: Chendl_02@sina.com; Tel.: +86-139-7483-7965

Received: 21 July 2017; Accepted: 29 August 2017; Published: 31 August 2017

Abstract: The effect of active solutions suitable for the in situ activation of selective catalytic reduction (SCR) catalysts was experimentally investigated using a designed in situ activation modeling device. To gain further insight, scanning electron microscopy (SEM), specific surface area analysis (BET), Fourier transform infrared spectroscopy (FT-IR), X-ray diffraction (XRD), and energy dispersive spectroscopy (EDS) analyses were used to investigate the effects of different reaction conditions on the characteristics of the deactivated catalysts. The activation effect of loading V_2O_5, WO_3 and MoO_3 on the surface of the deactivated catalysts was analyzed and the correlation to the denitrification activity was determined. The results demonstrate that the prepared activating solution of 1 wt % vanadium (V), 9 wt % tungsten (W), and 6 wt % molybdenum (Mo) has a beneficial effect on the deactivation of the catalyst. The activated catalyst resulted in a higher NO removal rate when compared to the deactivated catalyst. Furthermore, the NO removal rate of the activated catalyst reached a maximum of 32%. The activity of the SCR catalyst is closely linked to the concentration of the active ingredients. When added in optimum amounts, the active ingredients helped to restore the catalytic activity. In particular, the addition of active ingredients, the availability of labile surface oxygen, and the presence of small pores improved the denitrification efficiency. Based on these results, active solutions can effectively solve the problem of denitrification catalyst deactivation. These findings are a reference for the in-situ activation of the selective catalytic reduction of nitrogen oxides (SCR-DeNOx) catalyst.

Keywords: in situ; activating solution; reactivation; denitrification; catalyst

1. Introduction

Selective catalytic reduction of nitrogen oxides (SCR-DeNO$_x$) has been widely used for the treatment of flue gas streams in coal-fired power plants in China, owing to its superior denitrification efficiency and lower ammonia emissions [1]. However, when the flue gas contacts the catalyst, the adsorption and desorption of the reactants are affected, the specific surface area of the catalyst is reduced, and thus the number of active catalyst sites derease, leading to the deactivation of the catalyst. SO_2, SO_3, fly ash, and alkali metals contained in the flue gas may decrease the denitrification capacity of the catalyst. SO_3 reacts with NH_3 to form $(NH_4)_2SO_4$ [2], NH_4HSO_4 [3], and $CaSO_4$ [4]. These small particles, with a particle size <10 μm [5], with sticky ammonium sulfate particles, especially ammonium bisulfate, may clog the micropores of the catalyst surface [6,7] and stain and corrode the downstream devices of the SCR, such as the air preheater [8,9]. The combined presence of metal sites and SO_x may also result in irreversible loss of active sites via metal sulfates formation [10–12].

Fly ash particles hitting the catalyst surface can degrade the catalyst. The particles can also block the microporous channels on the catalyst surface [13], and the deposition of fly ash on the surface of the catalyst may contaminate and shield the catalyst [14], which may prevent the NO_x, NH_3, and O_2 in the flue gas from reaching the active site of the catalyst [5]. The alkali metal oxide decreases the number of active sites on the catalyst and poisons the catalyst by combining with the active acid sites of V_2O_5, which also reduces the ammonia adsorption on the catalyst surface, decreasing the denitrification activity of the catalyst [15–17].

Two main technical methods are generally accepted for addressing the problem of denitrification capacity decrease of the SCR-DeNO$_x$ catalyst caused by the flue gas. The first involves the use of fresh catalyst in the overall replacement of deactivated catalyst when the denitrification catalyst activity decreases to a certain level. The cost of this method is prohibitive for the boiler users. The second method is returning the deactivated catalyst back to the plant and then reinstalling it in the denitrification reactor. However, the deactivated catalyst must be removed from the boiler's denitrification reactor. This requires long-distance transport and secondary installation, which not only requires time and effort, but also causes secondary mechanical damage to the catalyst.

With the aim of solving these problems by increasing the activity of the deactivated catalysts, a method of in situ activation of the denitrified catalyst based on the activating solution is proposed in this study. The activating solution was used to activate the deactivated catalyst in the SCR denitrification reactor to effectively load the active ingredients. It in the in-situ activation process, it is important that the activating solution be efficient, cost effective, and does not cause secondary pollution. The acidity of an in situ activating solution cannot be too strong as it may damage the denitrification reactor. The solution should not only result in the recovery of the denitrification catalyst activity, but should be able to be recycled. Several studies were carried out that resulted in important achievements. For example, Li et al. [18] studied the effect of 1-hydroxy ethylidene-1 and 1-diphosphonic acid solution on the removal of $CaWO_4$, caused by the high content of CaO, from the catalyst surface to activate the denitrified catalyst. However, this activation method was not able to recover the strong acid sites and supplement the active sites. In addition, the effect of flue gas was not considered, and a higher vanadium content must be maintained to ensure the denitrification efficiency. Dong et al. [19] investigated the effect of the acidity of the precursor solution on the activity of the denitrified catalyst. The results showed that with increasing acidity of the precursor solution, more vanadium species and active sites would form on the surface of the catalyst. However, the optimum acidity of the activating solution and the vanadium content in the catalyst activity recovery have not been reported. Although these investigations provided valuable information about activating the SCR denitrification catalyst, only a few studies have reported in situ activation of the catalyst, and the activating solution suitable for in situ activation has not been investigated. Not only is there no activating solution suitable for in situ activation, but no research on the effect of the flue gas in in situ activation process exists.

This study aimed to find an efficient and inexpensive activation solution suitable for in situ activation. An in-situ activation modeling device was designed to activate the deactivated catalyst. The denitrification activity of the catalyst was tested using an activity measuring device. First, the effects of the active ingredients, including vanadium, tungsten, molybdenum, and oxalic acid, and the flue gas conditions, including space velocity, oxygen concentration, ammonia to nitrogen ratio, and the initial concentration of NO, on the in-situ activation of the catalyst were studied. In addition, the effects of these different reaction conditions on the deactivated catalysts were analyzed by scanning electron microscopy (SEM), specific surface area analysis (BET), Fourier transform infrared spectroscopy (FT-IR), X-ray diffraction (XRD), and energy dispersive spectroscopy (EDS). Finally, an efficient activating solution for in situ activation was obtained. The in situ activating modeling experiment provides a powerful platform for activating the deactivated catalysts and is broadly applicable to a variety of systems and experimental conditions. We expect that these results can pave the way for the application of the above catalysts to in situ denitrification situations.

77

2. Experimental Materials and Methods

2.1. Equipment

The experimental system includes cleaning system, activation system, drying and roasting system. As shown in Figure 1.

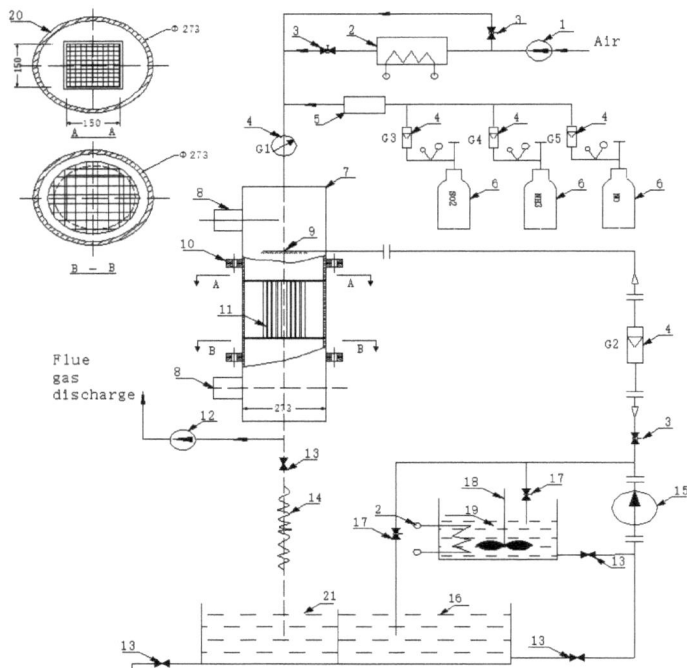

Figure 1. Schematic diagram of the experimental device used for modeling in-situ activating catalysts, where 1 is the induced draft fan, 2 is the air heater, 3 is the regulating valve, 4 is the flow meter, 5 is the mixer, 6 is the gas cylinder, 7 is the DeNO$_x$ reaction tower, 8 is the thief hole, 9 is the nozzle, 10 is the flange, 11 is the catalyst, 12 is the induced draft fan, 13 is the regulating valve, 14 is a hose, 15 is a pump, 16 is the clean water tank, 17 is the valve for circulating, 18 is the stirrer, 19 is the activating solution tank, 20 is insulation, and 21 is the sewage tank.

2.1.1. Cleaning

The in-situ denitrification modeling system for the activation process included a clean water tank, sewage tank, valves, flowmeters, and other devices used in the cleaning process of deactivated catalysts. The valve on the clean water tank was adjusted and the flowmeter was used to adjust the rinsing water to appropriate flow volume. The outlet valve of the clean water tank was opened and the pump was started to ensure the water was completely sprayed on the catalyst.

2.1.2. Activation

The in-situ denitrification modeling system for the activation process included activation pools, pumps, valves, and other devices used in the activation process of deactivated catalysts. The inlet valve of the sewage pool and the outlet valve of the clean-water reservoir were closed, while the inlet and outlet valves of the activation pool were opened. Subsequently, the pump was started and the flowmeter was used to adjust the regulating valve of the activation pool until the activate fluid reached a certain flow volume and rate and completely covered the surface of the catalyst.

2.1.3. Drying and Roasting

The in-situ denitrification modeling system for the activation process included air heaters, induced draft fans, regulating valves, and other devices used in the drying and roasting process of catalysts. The air volume and air temperature were adjusted by controlling the heaters, induced draft fans, and regulating valves, to achieve the drying and roasting of the catalysts.

2.2. Materials

2.2.1. Catalyst Preparation

The fresh and deactivated commercial honeycomb catalysts were obtained from a coal-fired power plant. The type of the prepared catalyst samples was ZERONOX1831K, and the catalyst parameters are listed in Table 1. The catalyst adopts TiO_2 as the carrier, and the main active ingredients are V_2O_5, WO_3, and MoO_3. The fresh and deactivated catalysts were labeled as C and C0, respectively.

Table 1. Catalyst parameters.

Parameter	Value
Channel number	18×18
Square width	150 mm
Channel width	7.09 mm
Specific geometry area	$408 \ m^2/m^3$
Hardened front length	25 mm

2.2.2. Activating Solution Preparation

Commercial vanadium catalysts usually contain ~1 wt % vanadium and ~9 wt % tungsten-molybdenum on a titanium (anatase) substrate. The activating solution composition was calculated according to the mass percentages of active precursors, and were weighed using an electronic balance (JA3003N, accuracy of ±1 mg). In this experiment, a single factor method was used to prepare the activating solution. The main ingredients of the prepared activating solution are listed in Table 2.

Table 2. Mass percentage of activating ingredients.

Number	Ammonium Vanadate (wt %)		Ammonium Molybdate (wt %)		Ammonium Tungstate (wt %)		Oxalic Acid (wt %)		Notes
L0–L4	L0	0.8	9		6		7.6		
	L1	0.9							
	L2	1.0							
	L3	1.1							
	L4	1.2							
L5–L8, Li_F	Content of vanadate as in solution Li_F		L5	7	6		7.6		Li_F is the number of the optimal solution in L0–L4
			L6	8					
			L7	10					
			L8	11					
			Li_F	9					
L9–L12, Li_W	Content of vanadate and molybdate as in solution Li_W				L9	7	7.6		Li_W is the number of the optimal solution in L5–L8, Li_F
					L10	8			
					L11	4			
					L12	5			
					Li_W	6			
L13–L16, Li_M	Content of vanadate, molybdate and tungstate as in solution Li_M						L13	6	Li_M is the number of the optimal solution in L9–L12, Li_W
							L14	7	
							L15	8	
							L16	9	
							Li_M	7.6	

2.2.3. Catalyst Activation

To improve the denitrification efficiency of the deactivated catalyst, in situ activation of the deactivated catalyst was performed using an in-situ activation modeling device. First, a water flow of 2500 L/h was used to clean the deactivated catalyst for 1.0 min; the removal of the toxic ingredients with the water improved the micromorphology of the deactivated catalyst. After the cleaning was completed, a 2500 L/h flow of the activation solution was used to spray the deactivation catalyst for 6 min, completely covering the surface of the catalyst and penetrating into it. Then, regulating the frequency of a draft fan and blower, the cold air was blown over the catalyst for 10 min to take away the dust and sulfuric acid, adhering to the surface of the catalyst. Then, an electric heater was turned on to heat the cold air to $100 \pm 10\,^{\circ}C$ to dry the catalyst for 1 h, reducing the water content and the competitive adsorption of H_2O and NH_3 and increasing the denitrification efficiency of the catalyst. Finally, the frequency of the draft fan and the blower was adjusted to 45 Hz, the temperature of the electric heater was adjusted to $380 \pm 10\,^{\circ}C$, and the catalyst was reacted in the hot air for 1 h, forming V_2O_5, WO_3, MoO_3, and other active ingredients.

During the calcination of the catalyst, the chemical reactions in Equations (1) to (3) occur between the active ingredients:

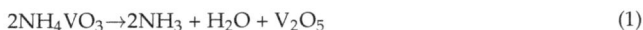

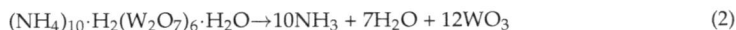

$$2NH_4VO_3 \rightarrow 2NH_3 + H_2O + V_2O_5 \tag{1}$$

$$(NH_4)_{10} \cdot H_2(W_2O_7)_6 \cdot H_2O \rightarrow 10NH_3 + 7H_2O + 12WO_3 \tag{2}$$

$$(NH_4)_6 \cdot (Mo_7O_{24})_6 \cdot 4H_2O \rightarrow 6NH_3 + 7H_2O + 7MoO_3 \tag{3}$$

The main active ingredient (V_2O_5) would evenly precipitate and attach to the surface of the catalyst to supplement the active ingredient lost by the deactivated catalyst, which is either lost, poisoned, or abraded at high temperature. At the same time, a chemical reaction also occurs on the surface of the catalyst to produce WO_3 and MoO_3, which are evenly distributed on the surface of the catalyst.

2.3. Methodology

2.3.1. Catalytic Performance Test

Figure 2 shows the schematic diagram of the experimental device, used to test the denitrification efficiency of catalysts. Industrial gas flow was simulated by flowing a test gas with the following composition: NH_3 (1000 ppm), NO (1000 ppm), SO_2 (1200 ppm), CO_2 (13%), O_2 (5%), and H_2O (5%). N_2 was used as the equilibrium gas. The feed gases were mixed to preheat in a chamber before entering the reactor. The total gas flow rate was 600 mL/min at a space velocity of $6000\ h^{-1}$. The gas flow was controlled using a mass flow meter, and the NO_x concentration before and after the reaction was measured using an in-situ flue gas analyzer (T-350, Testo Company, Selangor, Malaysia). In addition, the temperature range of the flue gases was varied from 300 to 400 $^{\circ}C$. The NO removal rate was calculated according to Equation (4):

$$\eta_{NOx} = \frac{C_{NOxin} - C_{NOxout}}{C_{NOxin}} \times 100\% \tag{4}$$

where η_{NOx} is the nitrogen oxide removal rate, C_{NOxin} is the nitrogen oxide inlet concentration, and C_{NOxout} is the nitrogen oxide outlet concentration.

The gas flow at the outlet was controlled by adjusting the valve, and the concentrations of NO and NH_3 were the same. The change in NO would change the concentration of NH_3. The concentration of NH_3 in the flue gas should be <5% to prevent explosion due to NH_3 combustion. The inlet and outlet temperatures, and the concentration of NO in the denitrification reactor, were measured using a flue gas analyzer. The denitrification rate of the experimental flue gas, under different activating solution formulations and different flue gas conditions, was calculated, and the catalyst activity (K) was also calculated.

2.3.2. Activity Evaluations

In order to evaluate the catalytic activity, kinetic parameters were calculated by using Equation (5):

$$\kappa = -\frac{V_{fg}}{V_{cat}} Ln(1 - \frac{\eta_{NOx}}{M}) = -S_V Ln(1 - \frac{\eta_{NOx}}{M}) \qquad (5)$$

where κ is the catalyst activity coefficient, V_{fg} is the flue gas flow (m^3/h), V_{cat} is the catalyst volume (m^3), η_{NOx} is the nitrogen oxide removal rate, M is the NH$_3$/NO molar ratio, and S_V is the space velocity. Based on the experimental conditions, M is always 1.0, and S_V is always 6000 h^{-1}. The denitrification efficiency and activity of the activated catalyst was calculated by using Equations (4) and (5), respectively.

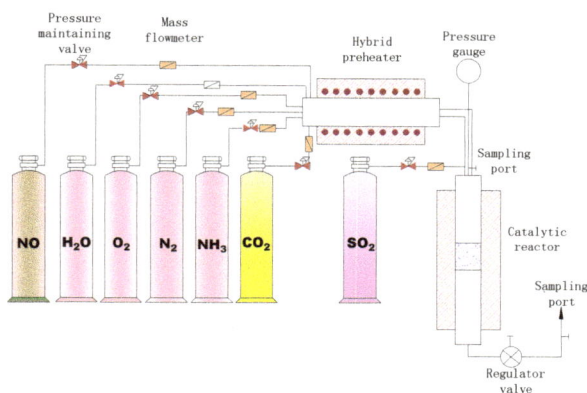

Figure 2. Schematic diagram of the experimental device used for denitrified catalyst activity evaluation.

2.3.3. Catalyst Characteristics

Pore Structure and Morphology

The morphology of the samples was analyzed by the SEM (Sirion200, FEI, Eindhoven, The Netherlands). The magnification was 20,000 times when the photos were analyzed. The specific surface area, pore volume, and pore size distribution of the samples were determined by the BET (V-Sorb2800, Gold APP Instruments, Beijing, China), which yielded important information about the structural features. Crystal structures of the samples were established by using an X-ray diffractometer (XRD, D/MAX-2200, Rigaku, Tokyo, Japan). The XRD patterns of all samples were recorded on a Phillips X-pert (50 KV, 40 mA, 10° min^{-1} from 10° to 80°). An Energy Dispersive Spectrometer (EDS) was used to identify the elemental mapping of the samples.

Carbon Surface Chemistry

The main functional groups of the active substances' surfaces were determined by FT-IR (Nicolet iS10, Thermo Fisher, Waltham, MA, USA); the spectral range was 400 to 4000 cm^{-1}, and the resolution was 4.0 cm^{-1}.

3. Results and Discussion

3.1. Effect of Active Ingredients on NO Removal Rate

3.1.1. Effect of Vanadium

The effect of the activating solution with vanadium on the NO removal rate is displayed in Figure 3a. The NO removal rate of the activated catalyst increased significantly with increasing vanadium concentration in the activating solution. From Figure 3a, Catalyst L2 had the greatest impact on NO removal rate. The NO removal rate of the active catalyst increased by 32% compared to the deactivated catalyst. The addition of vanadium increases the concentration of Brønsted acid sites and ammonia adsorption, and the Brønsted acid site concentration is directly correlated to the rate of NO removal. NH_3 adsorbs on V-OH, the Brønsted acid site, and acts in conjunction with an oxidation state shift of the V=O site (+4 to +5) to reduce the gas phase NO [20]. Nevertheless, the NO removal rate showed a downward trend when the vanadium concentration exceeded 1%. The reason for this may be that with increasing V_2O_5 content, the number of vanadium active sites increases, and the surface adsorption of SO_2 increases. This increasing ammonium sulfate formation after the adsorption of NH_3 affects the NO removal rate. Vanadium will also accelerate the formation of anatase TiO_2 crystals, decreasing the anatase surface area, forming low-activity vanadium species and serious pore blockages, reducing the performance of the catalyst. At the same time, the presence of vanadium affects the sintering of TiO_2, and the transformation from anatase to rutile is faster with high vanadium content [21]. The experimental consequences of different vanadium loadings on the catalyst are consistent with the experimental results of Giakoumelou et al. [22].

3.1.2. Effect of Tungsten

The activating effect of tungsten on the catalyst is shown in Figure 3b. The presence of WO_3 effectively improved the NO removal rate, reaching the maximum removal rate at a WO_3 concentration of 9 wt % (Figure 3b). Tungsten contributes significantly to the number of Brønsted acid sites and hence the activity of vanadia-based catalysts [21]. Tungsten also plays an important role in maintaining the structure and morphology properties of the samples and can stabilize the anatase and disperse the vanadium oxide on the surface [21]. In addition, WO_3 (9–10 wt %) has been applied to improve the activity of surface acidity, as well as significantly to mitigate the effects of alkali and alkaline soil poisoning [23]. Tungsten oxide delays the decrease in the BET surface area [24] and hinders the conversion of monomer vanadium to crystalline V_2O_5 [25]. However, with increasing the loading amounts of tungsten oxides, the NO removal rate initially increases then subsequently decreases, indicating that increasing the WO_3 loading will not always promote the catalytic activity and a maximum is reached at an appropriate tungsten loading [26]. The activation effect of the activator gradually decreased with increasing WO_3 content because the denitrification efficiency of WO_3 is less than that of V_2O_5 and MoO_3, and the excess WO_3 loading will occupy the active acid sites of the original V_2O_5 and MoO_3. The activity increase of the WO_3 load is less than the activity loss value, thus decreasing the catalyst denitrification efficiency with excess WO_3 loading. WO_3 is generally believed to affect the stalline V_2O_5 growth. Detailed experimental studies of tungsten supported on titania have been reported [27–29].

3.1.3. Effect of Molybdenum Composition

Figure 3c shows the activation effect of molybdenum composition on the NO removal rate. The presence of MoO_3 effectively improved the NO removal rate, reaching a maximum removal at a MoO_3 concentration of 6 wt % (Figure 3c), because MoO_3 promotes V_2O_5 and WO_3 activity in the catalyst. The Lewis acid site formed by Mo^{6+} can enhance the acidity of the catalyst surface and strengthen the adsorption process of NH_3 on the catalyst surface. Previous studies [30,31] have shown that Mo could promote the dispersion of VO_x species on the surface of TiO_2. Importantly, V and W

Catalysts 2017, 7, 258

have a synergistic effect [32]. The NO removal rate increases slowly when the MoO$_3$ concentration is <6 wt %, which may be because the activating solution cannot supplement the loss of molybdenum. In addition, excessive molybdenum does not contribute to the recovery of the catalyst activity when the molybdenum concentration in the activating solution is >6 wt %. In addition, some researchers proposed that the structural and morphological characteristics of MoO$_3$ and WO$_3$ on TiO$_2$ are similar [33]. In addition, TiO$_2$-loaded MoO$_3$ and WO$_3$ catalysts have the same metal oxide loading, but exhibit different catalytic behavior [34]. Moreover, MoO$_3$-containing catalysts produce more N$_2$O than WO$_3$-containing catalysts [35]. Previous studies demonstrated that the catalysts' overall reduction ability of the dramatically improved with the addition of MoO$_3$ [36].

3.1.4. Effect of Oxalic Acid

The effect of oxalic acid on the NO removal rate of catalysts is shown in Figure 3d. The experimental results demonstrated that using the appropriate amount of oxalic acid is advantageous for NO removal (Figure 3d). The addition of oxalic acid in the activating solution creates an acidic environment, which promotes the dissolution of ammonia, while maintaining a definite reduction property on the catalyst surfaces. The results indicate that the highest activity was observed at 7.6 wt % oxalic acid, and the NO removal rate decreased with increasing oxalic acid. Based on these results, we concluded that the excessive oxalic acid concentration increases the acidity of the activating solution, which increases the solubility of V$_2$O$_5$, thus decreasing the NO removal rate [37]. Oxalic acid is mainly used as an auxiliary solvent in the activating solution, and is critical for successfully dissolving the main active ingredients.

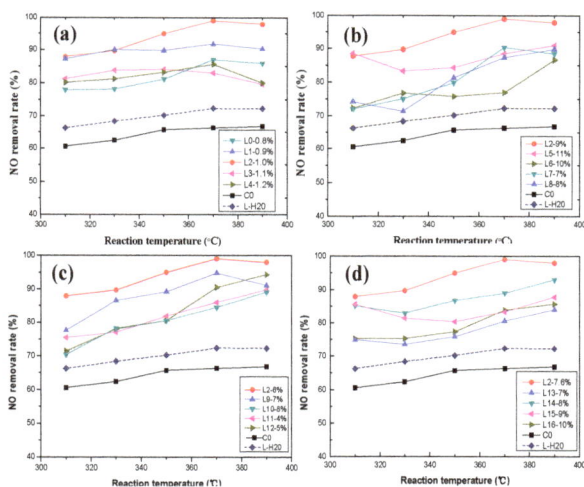

Figure 3. Effect of active ingredients on NO removal rate: (**a**) vanadium; (**b**) tungsten; (**c**) molybdenum; and (**d**) oxalic acid.

3.2. Activity Analyses of the Activated Catalyst

The activity of the catalyst can be calculated according to Equation (2), where κ_0 is the catalyst activity coefficient of the fresh catalyst. The κ/κ_0 of the deactivated catalyst C0 is 0.332. The κ/κ_0 values of the activated catalysts, demonstrating the activation properties of the activated catalyst, are presented in Figure 4. The results show that the activity of the catalysts, activated by different formulations of the activating solutions, increases at 370 °C. This result demonstrates that the activity of the catalyst is significantly restored by activation treatment and was an improvement of up to 83.4%

83

over the fresh catalyst. From Figure 4, we concluded that the activation of the catalyst was affected by the ingredients in the activating solution.

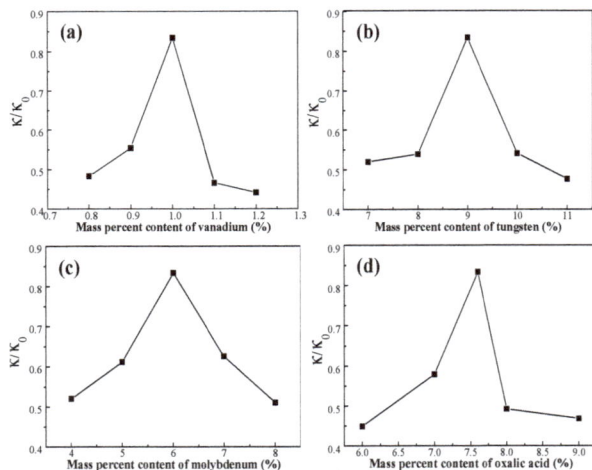

Figure 4. Effect of active ingredients on activity of the catalyst: (**a**) W = 9 wt %, Mo = 6 wt %, oxalic acid = 7.6 wt %; (**b**) V = 1 wt %, Mo = 6 wt %, oxalic acid = 7.6 wt %; (**c**) V = 1 wt %, W = 9 wt %, oxalic acid = 7.6 wt %; and (**d**) V = 1 wt %, W = 9 wt %, Mo = 6 wt %.

3.3. Effects of Activated Catalysts on the NO Removal Rate under Flue Gas Conditions

3.3.1. Space Velocity

Figure 5a shows the effect of space velocity on the activation of the NO removal rate of catalyst L2. As shown in Figure 5a, the catalyst has a high NO removal rate when the space velocity is low, but the trend gradually slows, reaching its maximum when the reaction temperature is 370 °C. However, the NO removal rate can reach >90%, when the velocity is <9000 h^{-1}. Moreover, the removal rate reached 99% when the space velocity was 3000 and 6000 h^{-1}. The NO removal rate also decreases with increasing space velocity, especially at a space velocity of 12,000 h^{-1}. At a high space velocity, the decrease in the NO removal rate may be due to the following the contact time between the reactants and the catalyst may be insufficient or insufficient numbers of active sites exist to accommodate the high flux in reactants. The effect of space velocity on the NO removal rate in the high temperature zone (>370 °C) is not as obvious as that in the low temperature zone (<370 °C). Figure 5a shows the NO removal rate of Catalyst L2 varies slightly when the space velocity is <9000 h^{-1}.

3.3.2. Oxygen Concentration

The oxygen in flue gas has an important effect on the catalysts' NO removal rate. As shown in Figure 5b, the catalysts' NO removal rate gradually increased as the oxygen increased from 2% to 4%. The NO removal rate of the catalyst increased gradually with increasing oxygen content. Oxygen can regenerate the active site [20] and increase the reaction rate of SCR; it is first transformed into oxygen free radicals during the transformation of SO$_2$ [38]. Ammonia adsorbs on V=O sites by interacting with V instead of O in the V=O bond. Gas-phase NO reacts with adsorbed ammonia on the V=O sites and produces N$_2$ and H$_2$O. Gas-phase oxygen then oxidizes the V=O site, preparing the site for another cycle of SCR reactions [39]. The increase in the oxygen content plays an important role in the formation of N$_2$O by ammonia oxidation [40]. The NO removal rate tends to be stable at an oxygen content of >4%. This shows that the effect of oxygen on the NO removal rate can be eliminated at an oxygen content of >4%.

3.3.3. Ratio of Ammonia to Nitrogen

Figure 5c shows the effect of the ammonia to nitrogen ratio on the NO removal rate. The NO removal rate of the catalyst increases sharply with increasing ammonia concentration when the ratio of ammonia to nitrogen is <1, which may be explained by the fact insufficient ammonia exists to encounter the acidic sites on the catalyst surface at this time. Higher ammonia to nitrogen ratios increase the diffusion pressure of ammonia in the inner and outer channels, thus accelerating the diffusion speed. Therefore, increasing the NH_3 volume fraction in the reaction gas can increase the amount of NH_3 adsorbed on the catalyst surface. NH_3 adsorption on the Brønsted acid sites of the catalyst surface is the first step in a SCR reaction, so NH_3 adsorption would increase the SCR reaction rate [41]. This is consistent with the results of Lei et al. [42], however, the NO removal rate tends to be constant when the ratio of ammonia to nitrogen is >1.0. This effect may be because the number of acid sites attached to NH_3 on the catalyst surface is finite, and the catalyst surface acidity is limited at this time. Moreover, the acidity of the catalyst and ammonia essentially reached dynamic equilibrium. The maximum denitrification capacity of catalysts was theoretically attained.

3.3.4. Initial Concentration of NO

Figure 5d shows the effect of the initial NO concentration on the NO removal rate. The NO concentration in the flue gas inlet was in the range 600–1400 ppm, indicating that the change in the initial NO concentration slightly affects the NO removal rate. Moreover, this reoxidation of the vanadium phase was highly affected by the presence of the second phase of WO_x and MoO_x. Figure 5d shows that the NO removal rate of the catalyst increased with increasing NO content. The NO removal rate of Catalyst L2 remained >90%. This result illustrates that Catalyst L2 is adaptable and is not sensitive to the NO concentration on its entrance into the flue gas.

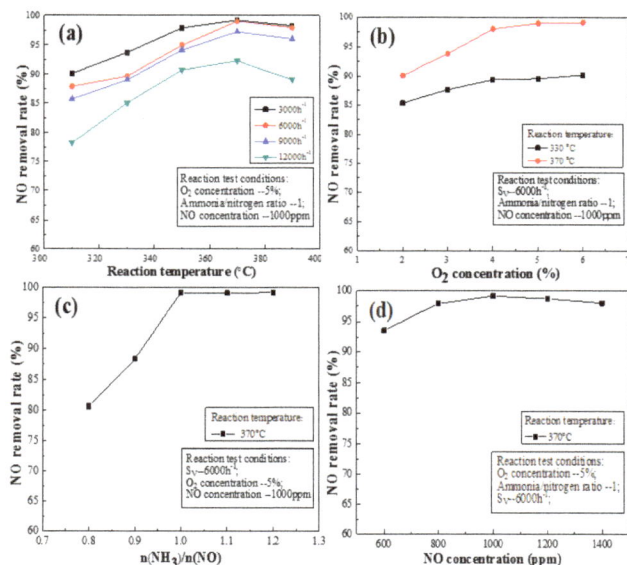

Figure 5. Effect of flue gas conditions on the NO removal rate of activated catalyst: (**a**) space velocity; (**b**) O_2 concentration; (**c**) ammonia/nitrogen ratio; and (**d**) NO concentration.

3.4. Characterization of the Catalyst

3.4.1. SEM

Figure 6 shows the representative SEM images of the deactivated, activated, and fresh samples. As shown in Figure 6a, the surface of the deactivated catalyst C0 is covered with fine particles and the spaces between the particles are clogged by fine particles. Fouling and masking may prevent the reactant NO_x and ammonia from reaching the active catalyst sites. Figure 6a,c show that the pores disappeared in the deactivated catalyst sample, suggesting that the sediments and sintering blocked and eventually removed the catalyst pores, reducing the pore aperture and volume [43]. The gap between the particles on the catalyst surface decreases, so it is difficult for the flue gas and the active ingredient to come into contact, reducing the catalyst denitrification efficiency (Figure 2). Thus, the denitrification activity of catalyst C0 is much lower than that of the fresh catalyst. After activation (Figure 6b), the gaps between the catalyst particles increased owing to the washing and activation by the activating solution. The toxic substance was washed away [44] and the internal structure became more evident, improving the micromorphology of the catalyst surface. Simultaneously, the activating solution ingredients may partially assemble on the surface of the catalyst, forming a novel porous surface. The clear particle distribution increases the porosity, while high porosity increases the number of active sites on the catalyst, which in turn increases the denitrification efficiency of SCR catalyst, as reported by Reddy et al. [44]. Therefore, the activation process is beneficial for catalyst activation.

(a)   (b)   (c)

Figure 6. SEM images of (**a**) C0 catalyst; (**b**) activated catalyst; and (**c**) fresh catalyst.

3.4.2. BET

The BET surface area, pore volume, and pore size of the various catalysts are shown in Table 3. The results indicate that the specific surface area of the deactivated catalysts is relatively small ($41.85 \ m^2/g$), whereas the specific surface area and pore size of the activated catalyst increased significantly. This shows that the activation process increases the specific surface area and total pore volume of the catalyst. Comparing the structural characteristics results of catalyst C and catalyst C0 indicates that the specific surface area of catalyst C0 decreases 20% more than that of catalyst C, which may be because of the catalyst being sintered, abraded, blocked, contaminated, or obscured [43]. Comparing the structural characteristics results of L8, L2 and L5, the surface area decreased with increasing ammonium molybdate. We attributed the excess Mo loading to the decrease in surface area [36]. Comparing the structural characteristics results of L2 and L11, the surface area increased with increasing ammonium tungstate. Tungsten oxide was reported to delay the loss of BET surface area [24]. The effective contact area between the catalyst and the flue gases increased, and the NO removal rate increased, coinciding with the result obtained by the SEM studies (Figure 6). The pore size distribution and the catalytic activity show that the small pore active sites may be the important factor for improving the denitrification activity (Figure 6), which can be ascribed to the longer effective contact time [42].

Table 3. Structural characteristics of catalysts before and after activation.

Catalyst Sample	Specific Surface Area (m^2/g)	Total Pore Volume (cc/g)	Aperture (nm)
C	52.31	0.8022	18.6
C0	41.85	0.3434	20.0
L2	47.55	0.6884	17.0
L5	50.61	0.7502	18.1
L8	46.80	0.6710	16.7
L11	47.03	0.6725	16.8

3.4.3. FT-IR

Figure 7 shows the FT-IR spectra of the deactivated and activated catalysts. For the activated catalysts, the peaks at 2332.31 and 2360.49 cm^{-1} correspond to CO_2, the absorption peak at 1400.18 cm^{-1} corresponds to the deformation vibration of the N–H bond from NH^{4+} adsorbed on the Brönsted acid sites [45]. The bands in the range of 1080.30–1180.95 cm^{-1} are close to the vibration frequency of V=O, which is associated with V_2O_5. The number and strength of V=O bonds are one of the most important factors adsorbing NH_3 onto the V=O sites, as NH^{4+} reacts with NO_x [20]. The infrared absorption peak of TiO_2 appeared at ~519.86 and 671.07 cm^{-1} [46]. However, in the FT-IR spectrum of the deactivated catalyst, these active material absorption peaks almost vanished, indicating the decrease in active material content. Thus, the active ingredient on the surface of the activated catalysts increased significantly. This shows that the active ingredient can be effectively loaded, and the activation process is effective.

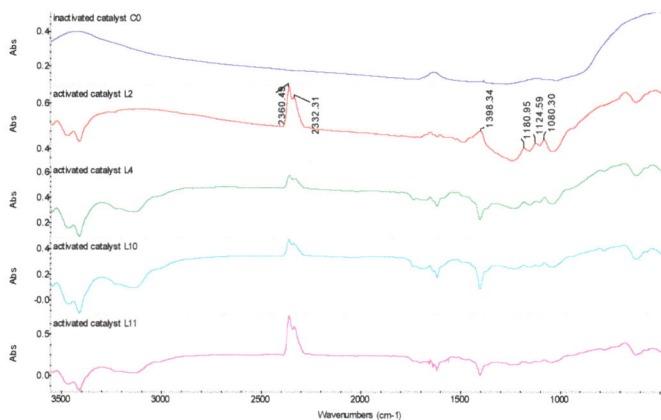

Figure 7. Infrared spectra of catalysts before and after activation.

3.4.4. XRD

The crystal structure of the catalyst determines the catalytic performance to some extent; therefore, the deactivated and activated catalysts were characterized by XRD. Figure 8 shows that the deactivated sample exhibits the diffraction peaks of anatase TiO_2. The crystal type of the TiO_2 carrier does not change. The diffraction peaks of V_2O_5 were not detected in any of the samples. There are numerous possible reasons for this observation. Firstly, the depletion of the active ingredient (V_2O_5) could be the cause, the vanadium oxide could have been disseminated on the catalyst in an amorphous or highly dispersed form (Figure 8), or the molybdenum improved the dispersion of monolayer VO_x

species on TiO_2 and the tungsten on the catalyst surface suppressed the progressive transformation of monomeric vanadyl species into crystalline V_2O_5 [31,47]. At the same time, the diffraction peak of WO_3 appeared in the sample, but the diffraction peaks of WO_3 did not appear, indicating that the dispersion of WO_3 was slightly less, and WO_3 did not sufficiently contact TiO_2. The aggregation of WO_3 occurs, to a certain extent, due to the inter molecular forces, resulting in the formation of a small amount of crystalline phase [27]. MoO_3 had good dispersion on the support. The diffraction peaks of $BaSO_4$ and $CaWO_4$ were detected, corresponding to the byproduct of the reaction with fly ash particles and a catalyst in the process of denitrification [18,48]. The $BaSO_4$ and $CaWO_4$ diffraction peaks of the activated catalyst were obviously weakened, which may be the effect of the active ingredient's adhesion and the washing away or covering of the $BaSO_4$ and $CaWO_4$.

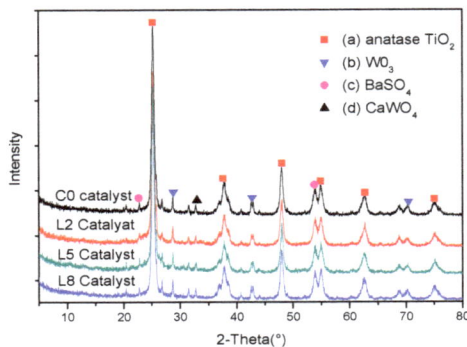

Figure 8. X-ray diffraction (XRD) spectra of the catalysts: (**a**) anatase TiO_2; (**b**) WO_3; (**c**) $BaSO_4$; and (**d**) $CaWO_4$.

3.4.5. EDS

The weight percentage of each element in the fresh, deactivated, and activated catalysts are listed in Table 4, indicating that the relative mass fractions of Ti, W, Mo, and V of the deactivated catalyst decreased significantly, and those of Al, Si, S, and Ca increased, reaching 1.66 wt %, 6.74 wt %, 6.37 wt %, and 4.93 wt %, respectively. The active ingredient of the catalyst reduced during the operation of the SCR system because of various physical and chemical effects including poisoning, abrasion, sintering, blockage, etc. The results showed that the dirt on the deactivated catalyst surface was mainly due to Al, Si, S, and Ca. The S, which might react with Al_2O_3 to form $Al_2(SO_4)_3$, led to the distinct decrease in the specific surface area and total pore volume (Figure 6, Table 1) and this result is consistent with the study by Lisi et al. [49]. In addition, Al and Ca are derived from fly ash particles in the flue gas, in accordance with the study of Zheng et al. [50]. The active ingredient of the activated catalyst changed significantly, and W, Mo, and V content increased by 7.78 wt %, 8.11 wt %, and 0.55 wt %, respectively. Moreover, Al, Si, S, and Ca decreased significantly. The results showed that active solution L2 could effectively increase the active ingredients of deactivated catalyst.

Table 4. Statistical table of the weight percentage of catalyst element.

Element	C		C0		L2	
	wt %	at %	wt %	at %	wt %	at %
O	4.3	13.38	9.25	22.94	6.87	21.33
Al	0	0	1.66	2.44	0	0
Si	0.7	1.25	6.74	9.53	2.5	4.43
W	13.72	3.72	8.96	1.93	18.64	5.04
Mo	5.52	2.87	2.48	1.03	10.59	5.49
S	0	0	6.37	7.89	0.10	0.15
Ca	0	0	4.93	4.88	0	0
Ti	74.08	77.84	58.97	48.86	60.10	62.38
V	0.96	0.94	0.65	0.5	1.2	1.17

4. Conclusions

In this study, the performance of several activating solutions, during the in situ activation of SCR-DeNO$_x$ catalysts, was investigated. Remarkable changes were observed in the deactivated catalysts after the activation process. The texture, morphology, and surface chemistry of the activated catalysts were investigated. We drew the following conclusions from the results of this study:

(1) The deactivated catalysts were effectively activated by the in-situ activation method, and the active material was found to be highly loaded. The active ingredients, V_2O_5, WO_3, and MoO_3, of the activating solution effectively increased the NO removal rate of the activated catalysts; the activation effect of the L2 solution, which contained 1.0% V, 9% W, and 6% Mo, was best, increasing the NO removal rate by 32%.

(2) The result of the flue gas conditions of the NO removal rate of the activated catalyst exhibited that the effect of space velocity on the NO removal rate in the high temperature zone (>370 °C) was not as obvious as that in the low temperature zone (<370 °C). The NO removal rate effect at a low space velocity was stronger than that at a high space velocity. When the oxygen concentration was <4%, the NO removal rate of the catalyst was effectively promoted. The NO removal rate of the catalyst increased sharply with increasing ammonia concentration at the ammonia to nitrogen ratio of <1. The initial concentration of NO had a slight effect on the NO removal rate.

(3) The fresh, deactivated, and activated catalysts were characterized by SEM, BET, FTIR, XRD, and EDS analyses. The results showed that the gaps between the activated catalyst particles increased owing to the washing and activation by the activating solution, which removed the toxic substances, resulting in a more evident internal structure. The activation process is beneficial for increasing the specific surface area and total pore volume of the catalyst. The active ingredient can be effectively loaded, and the vanadium oxide was disseminated on the catalyst in an amorphous or highly dispersed form. Activating solutions could effectively increase the active ingredients of the deactivated catalyst and enhance the NO removal rate as a result.

Our findings suggest that the in-situ activation method of denitrified catalysts, based on an activation solution, has the advantages of having a rapid activation period and prevents the disassembly and removal of the catalyst, which can solve the deactivation problem of denitrified catalysts as well. We expect that the in-situ activation and activation device have potential applications to SCR-DeNO$_x$. In fact, it should be noted that very little optimization work has been performed on these devices. Further mechanistic studies and development of durability, activating process, and corrosion behavior are ongoing in our laboratory.

Acknowledgments: This work was financially supported by the National Natural Science Foundation of China (Grant No. 51206012), the Scientific Research Foundation of the Hunan Education Department (10A004), the Hunan Province 2011 Collaborative Innovation Center of Clean Energy and Smart Grid, the Southern Power Grid Science and Technology Project (K-GD2012-395).

Author Contributions: Tuo Ye, Donglin Chen and Xi Zeng conceived of and designed the experiments. Tuo Ye, Yanshan Yin, Xi Zeng and Jing Liu performed the experiments. Tuo Ye, Xi Zeng and Donglin Chen analyzed the data. Tuo Ye and Donglin Chen wrote the paper.

Conflicts of Interest: The authors declare no conflict of interest.

References

1. Qi, C.; Bao, W.; Wang, L.; Li, H. Study of the V_2O_5-WO_3/TiO_2 Catalyst Synthesized from Waste Catalyst on Selective Catalytic Reduction of NO_x by NH_3. *Catalysts* **2017**, *7*, 110. [CrossRef]
2. Liang, Z.; Ma, X.; Lin, H.; Tang, Y. The energy consumption and environmental impacts of SCR technology in China. *Appl. Energy* **2011**, *88*, 1120–1129. [CrossRef]
3. Bao, J.; Mao, L.; Zhang, Y.; Fang, H.; Shi, Y.; Yang, L.; Yang, H. Effect of Selective Catalytic Reduction System on Fine Particle Emission Characteristics. *Energy Fuels* **2016**, *30*, 1325–1334. [CrossRef]
4. Schwaemmle, T.; Heidel, B.; Brechtel, K.; Scheffknecht, G. Study of the effect of newly developed mercury oxidation catalysts on the $DeNO_x$-activity and SO_2-SO_3-conversion. *Fuel* **2012**, *101*, 179–186. [CrossRef]
5. Guo, X. Poisoning and Sulfation on Vanadia SCR Catalyst. Ph.D. Thesis, Brigham Young University, Provo, UT, USA, 2006.
6. Peng, Y.; Li, J.; Si, W.; Luo, J.; Wang, Y.; Fu, J.; Li, X.; Crittenden, J.; Hao, J. Deactivation and regeneration of a commercial SCR catalyst: Comparison with alkali metals and arsenic. *Appl. Catal. B Environ.* **2015**, *168–169*, 195–202. [CrossRef]
7. Zheng, Y.J.; Jensen, A.D.; Johnsson, J.E.; Thøgersen, J.R. Deactivation of V_2O_5-WO_3-TiO_2, SCR catalyst at biomass fired power plants: Elucidation of mechanisms by lab- and pilot-scale experiments. *Appl. Catal. B Environ.* **2008**, *83*, 186–194. [CrossRef]
8. Benson, S.A.; Laumb, J.D.; Crocker, C.R.; Pavlish, J.H. SCR catalyst performance in flue gases derived from subbituminous and lignite coals. *Fuel Process. Technol.* **2005**, *86*, 577–613. [CrossRef]
9. Xu, W.; He, H.; Yu, Y. Deactivation of a Ce/TiO_2 Catalyst by SO_2 in the Selective Catalytic Reduction of NO by NH_3. *J. Phys. Chem. C* **2009**, *113*, 4426–4432. [CrossRef]
10. Yu, J.; Guo, F.; Wang, Y.; Xu, G. Sulfur poisoning resistant mesoporous Mn-based catalyst for low-temperature SCR of NO with NH_3. *Appl. Catal. B Environ.* **2010**, *95*, 160–168. [CrossRef]
11. Shen, B.X. Deactivation of MnO-CeO/ACF Catalysts for Low-Temperature NH-SCR in the Presence of SO. *Acta Phys. Chim. Sin.* **2010**, *26*, 3009–3016.
12. Aguilar-Romero, M.; Camposeco, R.; Castillo, S.; Marín, J.; Glez, V.R.; García-Serrano, L.A.; Mejía-Centeno, I. Acidity, surface species, and catalytic activity study on V_2O_5-WO_3/TiO_2 nanotube catalysts for selective NO reduction by NH_3. *Fuel* **2017**, *198*, 123–133. [CrossRef]
13. Wilburn, R.T.; Wright, T.L. SCR Ammonia Slip Distribution in Coal Plant Effluents and Dependence upon SO. *Powerpl. Chem.* **2004**, *6*, 295–314.
14. Tang, F.; Xu, B.; Shi, H.; Qiu, J.; Fan, Y. The poisoning effect of Na^+ and Ca^{2+} ions doped on the V_2O_5/TiO_2 catalysts for selective catalytic reduction of NO by NH_3. *Appl. Catal. B Environ.* **2010**, *94*, 71–76. [CrossRef]
15. Chen, L.; Li, J.; Ge, M. The poisoning effect of alkali metals doping over nano V_2O_5-WO_3/TiO_2 catalysts on selective catalytic reduction of NO_x by NH_3. *Chem. Eng. J.* **2011**, *170*, 531–537. [CrossRef]
16. Nicosia, D.; Czekaj, I.; Kröcher, O. Chemical deactivation of V_2O_5/WO_3-TiO_2 SCR catalysts by additives and impurities from fuels, lubrication oils and urea solution: Part II. Characterization study of the effect of alkali and alkaline earth metals. *Appl. Catal. B Environ.* **2008**, *77*, 228–236. [CrossRef]
17. Xiong, Z.B.; Hu, Q.; Liu, D.Y.; Wu, C.; Zhou, F.; Wang, Y.Z.; Jin, J.; Lu, C.M. Influence of partial substitution of iron oxide by titanium oxide on the structure and activity of iron–cerium mixed oxide catalyst for selective catalytic reduction of NO_x, with NH_3. *Fuel* **2016**, *165*, 432–439. [CrossRef]
18. Li, X.; Li, X.; Chen, J.; Hao, J. An efficient novel regeneration method for Ca-poisoning V_2O_5-WO_3/TiO_2 catalyst. *Catal. Commun.* **2016**, *87*, 45–48. [CrossRef]
19. Dong, G.; Zhang, Y.; Zhao, Y. Effect of the pH value of precursor solution on the catalytic performance of VO-WO/TiO in the low temperature NH-SCR of NO. *J. Fuel Chem. Technol.* **2014**, *42*, 1455–1463. [CrossRef]
20. Topsøe, N.Y.; Topsøe, H.; Dumesic, J.A. Vanadia-Titania Catalysts for Selective Catalytic Reduction (SCR) of Nitric-Oxide by Ammonia 1. Combined Temperature-Programmed In-Situ FTIR and Online Mass-Spectroscopy Studies. *J. Catal.* **1995**, *151*, 226–240. [CrossRef]
21. Djerad, S.; Tifouti, L.; Crocoll, M.; Weisweiler, W. Effect of vanadia and tungsten loadings on the physical and chemical characteristics of V_2O_5-WO_3/TiO_2 catalysts. *J. Mol. Catal. A Chem.* **2004**, *208*, 257–265. [CrossRef]

22. Giakoumelou, I.; Fountzoula, C.; Kordulis, C.; Boghosian, S. Molecular structure and catalytic activity of V_2O_5/TiO_2, catalysts for the SCR of NO by NH_3: In situ Raman spectra in the presence of O_2, NH_3, NO, H_2, H_2O and SO_2. *J. Catal.* **2006**, *239*, 1–12. [CrossRef]

23. Kobayashi, M.; Hagi, M. V_2O_5-WO_3/TiO_2-SiO_2-SO_4^{2-} catalysts: Influence of active components and supports on activities in the selective catalytic reduction of NO by NH_3 and in the oxidation of SO_2. *Appl. Catal. B Environ.* **2006**, *63*, 607–612. [CrossRef]

24. Oliveri, G.; Ramis, G.; Busca, G.; Escribano, V.S. Thermal stability of vanadia–titania catalysts. *J. Mater. Chem.* **1993**, *3*, 1239–1249. [CrossRef]

25. Alemany, L.J.; Lietti, L.; Ferlazzo, N.; Forzatti, P.; Busca, G.; Giamello, E.; Bregani, F. Reactivity and Physicochemical Characterization of V_2O_5-WO_3/TiO_2 De-NO_x Catalysts. *J. Catal.* **1995**, *155*, 117–130. [CrossRef]

26. Sun, C.; Dong, L.; Yu, W.; Liu, L.; Li, H.; Gao, F.; Dong, L.; Chen, L. Promotion effect of tungsten oxide on SCR of NO with NH_3, for the V_2O_5-$WO_3/Ti_{0.5}$ $Sn_{0.5}$ O_2, catalyst: Experiments combined with DFT calculations. *J. Mol. Catal. A Chem.* **2011**, *346*, 29–38. [CrossRef]

27. Wachs, I.E.; Kim, T.; Ross, E.I. Catalysis science of the solid acidity of model supported tungsten oxide catalysts. *Catal. Today* **2006**, *116*, 162–168. [CrossRef]

28. Fierro, J.L.G. (Ed.) *Metal Oxides: Chemistry and Applications*; CRC Press: Boca Raton, FL, USA, 2006; pp. 1–30.

29. Kobayashi, M.; Miyoshi, K. WO_3-TiO_2 monolithic catalysts for high temperature SCR of NO by NH_3: Influence of preparation method on structural and physico-chemical properties activity and durability. *Appl. Catal. B Environ.* **2007**, *72*, 253–261. [CrossRef]

30. Peng, Y.; Si, W.; Li, X.; Luo, J.; Li, J.; Crittenden, J.; Hao, J. Comparison of MoO_3 and WO_3 on arsenic poisoning V_2O_5/TiO_2 catalyst: DRIFTS and DFT study. *Appl. Catal. B Environ.* **2016**, *181*, 692–698. [CrossRef]

31. Yang, S.; Xiong, S.; Liao, Y.; Xiao, X.; Qi, F.; Peng, Y.; Fu, Y.; Shan, W.; Li, J. Mechanism of N_2O formation during the low-temperature selective catalytic reduction of NO with NH_3 over Mn-Fe Spinel. *Environ. Sci. Technol.* **2014**, *48*, 10354–10362. [CrossRef] [PubMed]

32. Shan, W.; Liu, F.; He, H.; Shi, X.; Zhang, C. ChemInform Abstract: Novel Cerium-Tungsten Mixed Oxide Catalyst for the Selective Catalytic Reduction of NO_x with NH_3. *Chem. Commun.* **2011**, *47*, 8046–8048. [CrossRef] [PubMed]

33. Liu, Z.; Zhang, S.; Li, J.; Ma, L. Promoting effect of MoO_3 on the NO_x reduction by NH_3 over CeO_2/TiO_2 catalyst studied with in situ DRIFTS. *Appl. Catal. B Environ.* **2014**, *144*, 90–95. [CrossRef]

34. Nova, I.; Lietti, L.; Casagrande, L.; Dall'Acquab, L.; Giamellob, E.; Forzattia, P. Characterization and reactivity of TiO_2-supported MoO_3 De-Nox SCR catalysts. *Appl. Catal. B Environ.* **1998**, *17*, 245–258. [CrossRef]

35. Zhao, B.; Liu, X.; Zhou, Z.; Shao, H.; Wang, W.; Si, J.; Xu, M. Effect of molybdenum on mercury oxidized by V_2O_5-MoO_3/TiO_2 catalysts. *Chem. Eng. J.* **2014**, *253*, 508–517. [CrossRef]

36. Huang, X.; Peng, Y.; Liu, X.; Li, K.; Deng, Y.; Li, J. The promotional effect of MoO_3 doped V_2O_5/TiO_2 for chlorobenzene oxidation. *Catal. Commun.* **2015**, *69*, 161–164. [CrossRef]

37. Khodayari, R.; Odenbrand, C.U.I. Regenaration of commercial TiO_2-V_2O_5-WO_3 SCR catalysts used in biofuel plants. *Appl. Catal. B Environ.* **2001**, *30*, 87–99. [CrossRef]

38. Dunn, J.P.; Koppula, P.R.; Stenger, H.G.; Wachs, I.E. Oxidation of sulfur dioxide to sulfur trioxide over supported vanadia catalysts. *Appl. Catal. B Environ.* **1998**, *19*, 103–117. [CrossRef]

39. Busca, G.; Lietti, L.; Ramis, G.; Berti, F. Chemical and mechanistic aspects of the selelctive catalytic reduction of NO_x by ammonia over oxide catalysts: A Review. *Appl. Catal. B Environ.* **1998**, *18*, 1–36. [CrossRef]

40. Djerad, S.; Crocoll, M.; Kureti, S.; Weisweiler, W. Effect of oxygen concentration on the NO_x reduction with ammonia over V_2O_5-WO_3/TiO_2 catalyst. *Catal. Today* **2006**, *113*, 208–214. [CrossRef]

41. Lee, C. Modeling urea-selective catalyst reduction with vanadium catalyst based on NH_3 temperature programming desorption experiment. *Fuel* **2016**, *173*, 155–163. [CrossRef]

42. Lei, Z.; Liu, X.; Jia, M. Modeling of Selective Catalytic Reduction (SCR) for NO Removal Using Monolithic Honeycomb Catalyst. *Energy Fuels* **2009**, *23*, 6146–6151. [CrossRef]

43. Kling, Å.; Andersson, C.; Myringer, Å.; Eskilsson, D.; Järås, S.G. Alkali deactivation of high-dust SCR catalysts used for NO_x, reduction exposed to flue gas from 100 MW-scale biofuel and peat fired boilers: Influence of flue gas composition. *Appl. Catal. B Environ.* **2007**, *69*, 240–251. [CrossRef]

44. Reddy, B.M.; Ganesh, I.; Chowdhury, B. Design of stable and reactive vanadium oxide catalyst supported on binary oxides. *Catal. Today* **1999**, *49*, 115–121. [CrossRef]

45. Cai, Y.P. Investigation of the Reaction Network and Catalytic Sites in Selective Catalytic Reduction Ognitric Oxide with Ammonia over Vanadia Catalysts. Ph.D. Dissertation, The Ohio State University, Columbus, OH, USA, 1993.
46. Liu, F.D.; He, H.; Zhang, C.B.; Shan, W.; Shi, X. Selective catalytic reduction of NO with NH_3 over iron titanate catalyst: Catalytic performance and characterization. *Appl. Catal. B Environ.* **2010**, *96*, 408–420. [CrossRef]
47. Peng, Y.; Li, J.; Si, W.; Luo, J.; Dai, Q.; Luo, X.; Liu, X.; Hao, J. Insight into deactivation of commercial SCR catalyst by arsenic: An experiment and DFT study. *Environ. Sci. Technol.* **2014**, *48*, 13895. [CrossRef] [PubMed]
48. Choo, S.T.; Yim, S.D.; Nam, I.S.; Ham, S.W.; Lee, J.B. Effect of promoters including WO and BaO on the activity and durability of VO/sulfated TiO catalyst for NO reduction by NH. *Appl. Catal. B Environ.* **2003**, *44*, 237–252. [CrossRef]
49. Lisi, L.; Lasorella, G.; Malloggi, S.; Russo, G. Single and combined deactivating effect of alkali metals and HCl on commercial SCR catalysts. *Appl. Catal. B Environ.* **2004**, *50*, 251–258. [CrossRef]
50. Zheng, Y.; Jensen, A.D.; Johnsson, J.E. Deactivation of V_2O_5-WO_3-TiO_2 SCR catalyst at a biomass-fired combined heat and power plant. *Appl. Catal. B Environ.* **2005**, *60*, 253–264. [CrossRef]

catalysts

MDPI

Article

Investigation of Iron Vanadates for Simultaneous Carbon Soot Abatement and NH$_3$-SCR

Marzia Casanova *, Sara Colussi and Alessandro Trovarelli

Dipartimento Politecnico di Ingegneria e Architettura, Università degli Studi di Udine, via Cotonificio 108, 33100 Udine, Italy; sara.colussi@uniud.it (S.C.); trovarelli@uniud.it (A.T.)
* Correspondence: marzia.casanova@uniud.it; Tel.: +39-0432-558-755

Received: 7 March 2018; Accepted: 22 March 2018; Published: 26 March 2018

Abstract: FeVO$_4$ and Fe$_{0.5}$Er$_{0.5}$VO$_4$ were prepared and loaded over standard Selective Catalytic Reduction (SCR) supports based on TiO$_2$-WO$_3$-SiO$_2$ (TWS) and redox active supports like CeO$_2$ and CeZrO$_2$ with the aim of finding a suitable formulation for simultaneous soot abatement and NH$_3$-SCR and to understand the level of interaction between the two reactions. A suitable bi-functional material was identified in the composition FeVO$_4$/CeZrO$_2$ where an SCR active component is added over a redox active support, to increase carbon oxidation properties. The influence of the presence of ammonia in soot oxidation and the effect of the presence of soot on SCR reaction have been addressed. It is found that the addition of NO and NO/NH$_3$ mixtures decreases at different levels the oxidation temperature of carbon soot, while the presence of carbon adversely affects the NH$_3$-SCR reaction by increasing the oxidation of NH$_3$ to NO, thus lowering the NO removal efficiency.

Keywords: NH$_3$-SCR; soot oxidation; de-soot/deNO$_x$; FeVO$_4$; Titania; Ceria

1. Introduction

The study of Diesel exhaust after-treatment technologies is focusing on one side in finding novel formulations for active soot oxidation [1] and de-NO$_x$ catalysts [2–4] and on the other side in looking for suitable strategies to limit the number of different process units and combine more functionalities in one single step [5]. Nowadays, the mobile exhaust gas after-treatment is comprised of a series of stand-alone devices, starting from an oxidation catalyst (DOC), a diesel particulate filter (DPF) and a de-NO$_x$ reactor which includes a NO$_x$ storage/reduction system or, alternatively, a NH$_3$-SCR (Selective Catalytic Reduction) unit. Among the NO$_x$ removal technologies, Selective Catalytic Reduction with NH$_3$ (NH$_3$-SCR) has been successfully employed both in stationary and heavy duty mobile applications [6].

Increasingly restrictive limitations concerning emissions of nitrogen oxides and soot particulates from vehicle engines will demand the simultaneous application of a particulate filter and a SCR de-NO$_x$ system. The motivation for the integration of the two reactions is to obtain a reduction of cost and volume of the engine after-treatment system via inclusion of SCR and DPF functionalities into a single entity. Different bi-functional configurations are possible with DPF and SCR catalyst placed in the same housing [7], or with a DPF filter coated with an SCR catalyst formulation [8–10].

The aim of this work is to study the behaviour of a suitable catalyst with acceptable performances both in soot oxidation and in the NH$_3$-SCR reaction conducted simultaneously. For this purpose, an active SCR catalyst formulation has been loaded over supports showing different soot oxidation functionalities. SCR catalyst materials were selected among metal/rare earth vanadates that have been shown to act as efficient catalyst precursors for NH$_3$-SCR reaction [11–16], while among the different supports, standard TiO$_2$-WO$_3$-SiO$_2$ (TWS) was used and compared with more efficient carbon oxidation carriers like CeO$_2$ and Ce$_{0.75}$Zr$_{0.25}$O$_2$ [17]. In particular, this work investigates FeVO$_4$ and

$Fe_{0.5}Er_{0.5}VO_4$ supported on TWS, CeO_2, $Ce_{0.75}Zr_{0.25}O_2$ (CeZrO$_2$), and Al_2O_3 in the simultaneous soot oxidation and NH_3-SCR reactions. Specific attention is given to the influence of all the components of the complex atmosphere under which these catalysts operate and to the study of the interactions between the two reactions. A promising catalytic system is proposed as a suitable candidate for both reactions.

2. Results and Discussion

2.1. Catalysts Characterization

All the samples were characterized by B.E.T. (Brunauer-Emmett-Teller) surface area measurements (reported in Table 1 and Table S1) and powder XRD analysis. As it can be observed, the specific surface area of bare supports is negatively affected by the loading of iron vanadate (B.E.T. = 4 m^2/g) and iron erbium vanadate (13 m^2/g). XRD patterns of samples made with $FeVO_4$ are reported in Figure 1A, whereas XRD diffraction patterns of supported $Fe_{0.5}Er_{0.5}VO_4$ are reported in Figure 1B. In the diffraction profile of 8.4 wt % $FeVO_4$/TWS only diffraction lines due to anatase-TiO_2 and Fe_2O_3 were observed. This is due to partial decomposition of $FeVO_4$ on TiO_2-based supports following thermal treatments [14,18]. This decomposition generates Fe_2O_3 plus VO_x species which at low coverage are not detected by XRD, while the remaining $FeVO_4$ could not be clearly identified because of superposition of its main diffraction line to the anatase main diffraction peak. In the XRD pattern of 8.4 wt % $FeVO_4$/CeO_2 instead, diffraction lines due to CeO_2, Fe_2O_3, and $CeVO_4$ are identified. In this case, the precursor $FeVO_4$ partially decomposes releasing Fe_2O_3 and VO_x species, which then react with CeO_2 already at 650 °C, resulting in the formation of $CeVO_4$ [19]. A very similar situation was detected for the 8.4 wt % $FeVO_4$/$Ce_{0.75}Zr_{0.25}O_2$ catalyst. The formation of surface VO_x species was confirmed by Raman spectroscopy on a representative 8.4 wt % $FeVO_4$/TWS sample calcined at 650 °C (Figure 2). We observed Raman shifts at ca. 1000 cm^{-1} that have been attributed to monomeric or dimeric species formed upon decomposition of $FeVO_4$ [20–22]. By carefully looking at different spots of the sample, also on the remaining $FeVO_4$, that were not observed by XRD, was easily detected by Raman as highlighted in Figure 2. $FeVO_4$ decomposition takes place also on alumina as revealed by Fe_2O_3 peaks in 8.4 wt % $FeVO_4$/Al_2O_3 diffraction pattern. Also in this case this is accompanied by the release of non-detectable VO_x species. Residual $FeVO_4$ species, not detectable by XRD, could be identified by Raman spectroscopy also on 8.4 wt % $FeVO_4$/Al_2O_3, as shown in Figure S1.

XRD spectra of supported $Fe_{0.5}Er_{0.5}VO_4$ (Figure 1B) are characterized by the presence of $ErVO_4$ and Fe_2O_3. It was previously highlighted that $Fe_{0.5}Er_{0.5}VO_4$ does not form a solid solution and the two vanadates ($FeVO_4$ and $ErVO_4$) are present as an intimate physical mixture [13]. The $FeVO_4$ fraction of the mixed vanadate partially decomposes releasing Fe_2O_3 and VO_x, while $ErVO_4$, which is more thermally stable than $FeVO_4$ [23], can still be detected on the various supports.

For a better understanding of the effect of the vanadate-support interaction on catalytic activity, some additional investigations were conducted also on a family of catalysts made by loading 50 wt % of vanadate on the support. Their XRD diffraction profiles are reported in Figure S2 of the supplementary material; as already noticed for 8.4 wt % $FeVO_4$ loading, the presence of Fe_2O_3 in the XRD patterns and of VO_x in Raman spectrum (Figure S3) indicates that part of $FeVO_4$ is decomposed.

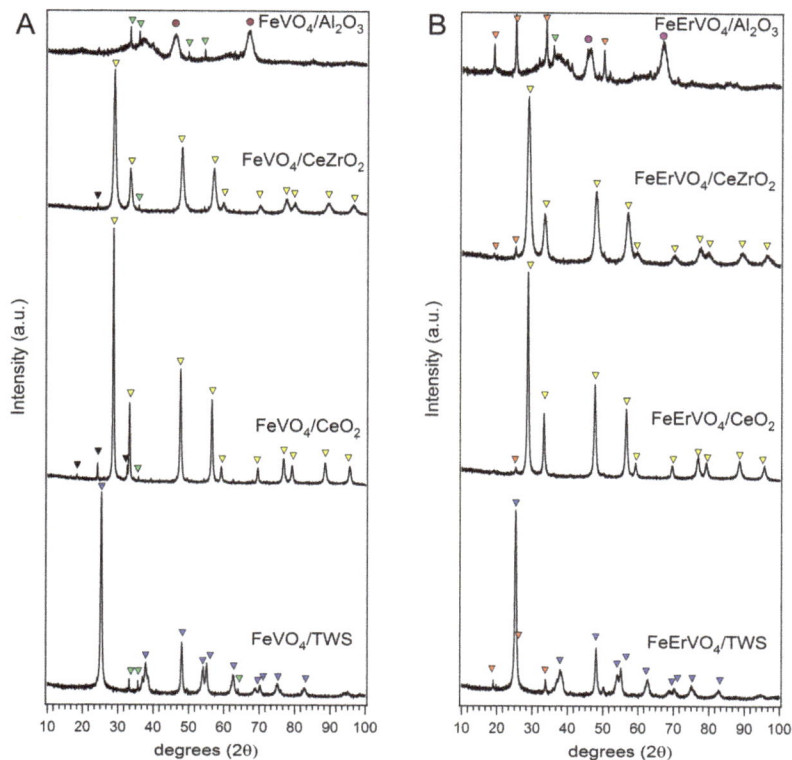

Figure 1. XRD patterns of (**A**) 8.4 wt % FeVO$_4$ and (**B**) 8.4 wt % Fe$_{0.5}$Er$_{0.5}$VO$_4$ on different supports. ▼ TiO$_2$-anatase, ▼ Fe$_2$O$_3$, ▼ CeO$_2$ or CeZrO$_2$, ▼ CeVO$_4$, ▼ ErVO$_4$, ● Al$_2$O$_3$.

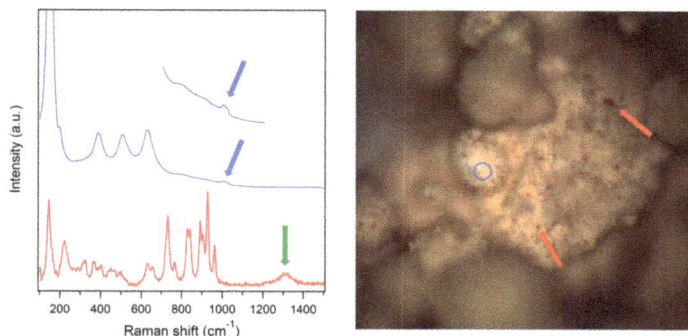

Figure 2. Raman spectra of 8.4 wt % FeVO$_4$/TWS (TiO$_2$-WO$_3$-SiO$_2$). The blue line represents Raman spectrum of a light spot on the sample (blue circle in the optical image) where anatase and surface VO$_x$ (blue arrows) were detected; the red line is the spectrum of a red spot (red arrows in the optical image) showing mainly FeVO$_4$ and a band belonging to Fe$_2$O$_3$ (green arrow in the Raman spectrum).

Catalyst reducibility was measured by temperature-programmed reduction under H_2 atmosphere (TPR). Since many reducible species are involved, complex reduction profiles were obtained, as shown in Figure 3. From XRD results it is clear that $FeVO_4$ undergoes decomposition over all supports, with the formation of surface Fe_2O_3, VO_x and, on ceria-containing supports, $CeVO_4$. On Al_2O_3, which is a non-reducible support, all contributions to reduction profile are related to Fe_2O_3, VO_x, and $FeVO_4$ species. The attribution of the single TPR peaks is not straightforward, but similarly to what reported for $FeVO_4$ supported on TWS [24], the low temperature feature can be related mostly to the reduction of Fe_2O_3 and VO_x species, whereas at increasing temperature H_2 consumption due to $FeVO_4$ reduction is detected. At higher loading, an increase in hydrogen consumption is observed, which has been tentatively attributed to the residual part of $FeVO_4$, based on quantitative analysis obtained also from XRD spectra. The onset reduction temperature is independent from $FeVO_4$ initial loading (Table 1). TWS reduction profile (Figure 3A) is characterized by three reduction steps of WO_3 with the main hydrogen consumption taking place in the 800–950 °C range [25]. This feature is maintained also in the supported catalyst. Hydrogen consumption at lower temperature for 8.4 wt % $FeVO_4$/TWS (400–700 °C) is due to reduction of the supported species VO_x, Fe_2O_3, and $FeVO_4$. As for Al_2O_3 samples, it is noteworthy that the onset temperature of reduction is not related to the vanadate loading (Table 1).

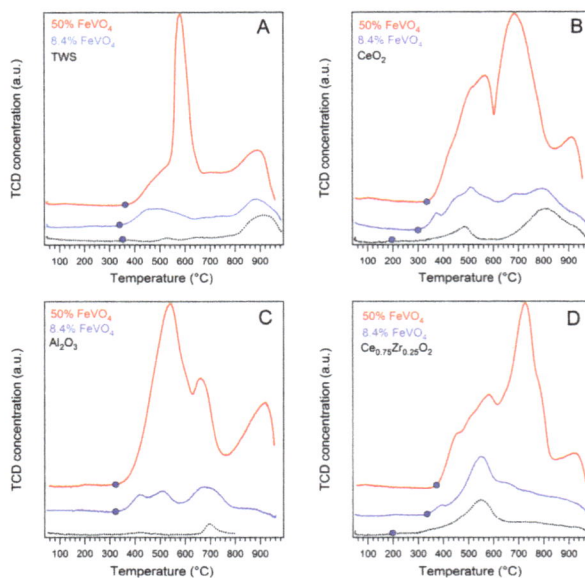

Figure 3. H_2-TPR temperature-programmed reduction under H_2 atmosphere profiles of (**A**) $FeVO_4$/TWS; (**B**) $FeVO_4$/CeO_2; (**C**) $FeVO_4$/Al_2O_3; (**D**) $FeVO_4$/$Ce_{0.75}Zr_{0.25}O_2$.

TPR profile of CeO_2 shows a low-T reduction feature due to surface species and a higher-T peak due to bulk reduction, as expected for pure ceria [26]. For $Ce_{0.75}Zr_{0.25}O_2$ one single hydrogen consumption peak is detected in the medium temperature range, in agreement with reported TPR profiles of ceria-zirconia solid solutions [27]. In the case of CeO_2-containing supports, a difference in the onset of the reduction can be detected at increasing $FeVO_4$ loading, in contrast to what observed for TWS and Al_2O_3-supported vanadates, indicating a strong influence of vanadate loading on sample reducibility. The temperatures of the onset of reduction are reported in Table 1 together with the overall hydrogen consumption and the H/V atomic ratio. As it can be observed from Table 1 and Figure 3, when increasing the vanadate loading up to 50 wt % the reduction peaks become more intense, the overall H_2 consumption increases but with a lower H/V ratio, indicating a better reducibility of the low-loaded samples.

Table 1. B.E.T. surface areas, hydrogen consumption in TPR, H/V atomic ratio and onset temperature of reduction for each sample.

Catalyst	B.E.T. (m^2/g)	H_2 Consumption (mmol/g_{cat})	H/V Atomic Ratio [1]	Onset of Reduction (°C)
TWS	89	0.953	/	350
8.4% FeVO$_4$/TWS	68	2.184	8.88	344
50% FeVO$_4$/TWS	21	5.786	3.95	361
CeO$_2$	26	0.997	/	190
8.4% FeVO$_4$/CeO$_2$	12	1.926	7.83	285
50% FeVO$_4$/CeO$_2$	6	6.379	4.36	338
Ce$_{0.5}$Zr$_{0.5}$O$_2$ (CZ)	52	1.404	/	186
8.4% FeVO$_4$/CZ	24	2.448	9.95	320
50% FeVO$_4$/CZ	8	6.694	4.57	348
Al$_2$O$_3$	168	0.080	/	/
8.4% FeVO$_4$/Al$_2$O$_3$	166	1.022	4.15	320
50% FeVO$_4$/Al$_2$O$_3$	89	5.240	3.58	320

[1] Calculated on the basis of H_2 consumption from TPR profiles and total vanadium content of the catalyst.

2.2. Soot Oxidation Activity

Table 2 shows soot oxidation efficiency of the samples investigated under air atmosphere. For 8.4 wt % loaded catalysts activity under $NO/N_2/O_2$ and $NO/N_2/O_2/NH_3$ mixtures is also reported.

Soot oxidation values (T50) of the vanadates supported on TWS indicate the low efficiency of TiO_2-based materials in this reaction, as reported in the literature [17]. Compared to the oxidation potential of pure vanadates (FeVO$_4$ and Fe$_{0.5}$Er$_{0.5}$VO$_4$) supported catalysts show an increase of T50. These formulations were compared with those obtained by loading vanadates on supports that are known to be more active for soot oxidation like CeO$_2$ and Ce$_{0.75}$Zr$_{0.25}$O$_2$ [28].

As expected, CeO$_2$-based materials show lower values of T50. This can be explained through the so called "active oxygen mechanism" where, thanks to the redox ability of ceria, oxygen can be activated over a vacancy and spill over the carbon soot for oxidation [17,29,30]. Conversely, the use of redox inactive supports like Al$_2$O$_3$, results in catalyst with the highest T50. A comparison of the TPR profiles reported in Figure 3 with the corresponding onset temperatures of reduction (Table 1) evidences the correlation between sample reducibility and soot oxidation behavior. Bare CeO$_2$ or Ce$_{0.75}$Zr$_{0.25}$O$_2$ are the most active materials thanks to their low-temperature surface reducibility. Even with 8.4 wt % FeVO$_4$ a change in the onset temperature of reducibility of ceria is observed and, consequently, soot oxidation activity is lowered. For 50 wt % FeVO$_4$/CeO$_2$, the reduction is further shifted towards higher temperatures resulting in a higher soot oxidation temperature (Table 2). The opposite is obtained for TWS materials, for which an increase in FeVO$_4$ loading favours activity as no detrimental effect is displayed on reducibility in the temperature range of soot oxidation.

Catalysts **2018**, _8_, 130

Table 2. Soot oxidation characteristic temperatures of investigated catalysts.

Catalyst	T50 (°C) [1] air	Tp (°C) [2] NO/O$_2$/N$_2$	Tp (°C) [3] NO/O$_2$/N$_2$/NH$_3$
8.4% FeVO$_4$/TWS	520	490	480
8.4% Fe$_{0.5}$Er$_{0.5}$VO$_4$/TWS	567	548	541
8.4% FeVO$_4$/CeO$_2$	473	459	462
8.4% Fe$_{0.5}$Er$_{0.5}$VO$_4$/CeO$_2$	460	448	440
8.4% FeVO$_4$/CZ	459	459	447
8.4% Fe$_{0.5}$Er$_{0.5}$VO$_4$/CZ	423	419	411
8.4% FeVO$_4$/Al$_2$O$_3$	602	-	-
8.4% Fe$_{0.5}$Er$_{0.5}$VO$_4$/Al$_2$O$_3$	603	-	-
FeVO$_4$	450	-	-
Fe$_{0.5}$Er$_{0.5}$VO$_4$	450	-	-
TWS	580	-	-
50% FeVO$_4$/TWS	474	-	-
50% Fe$_{0.5}$Er$_{0.5}$VO$_4$/TWS	449	-	-
CeO$_2$	390	-	-
50% FeVO$_4$/CeO$_2$	493	-	-
50% Fe$_{0.5}$Er$_{0.5}$VO$_4$/CeO$_2$	493	-	-
CeVO$_4$	495	-	-
Ce$_{0.75}$Zr$_{0.25}$O$_2$ (CZ)	385	-	-
50% FeVO$_4$/CZ	486	-	-
50% Fe$_{0.5}$Er$_{0.5}$VO$_4$/CZ	488	-	-
Al$_2$O$_3$	612	-	-
50% FeVO$_4$/Al$_2$O$_3$	546	-	-
50% Fe$_{0.5}$Er$_{0.5}$VO$_4$/Al$_2$O$_3$	505	-	-

[1] Temperature of 50% weight loss measured in thermogravimetric temperature programmed ramps from RT up to 800 °C on catalyst samples mixed in tight contact with soot. [2] Peak temperature calculated from CO$_2$ profiles during TPO (Temperature Programmed Oxidation) runs in a flow-trough microreactor under 500 ppm NO, 10% O$_2$, (balance N$_2$) atmosphere. [3] Peak temperature calculated from CO$_2$ profiles during TPO runs in a flow-trough microreactor under 500 ppm NO, 500 ppm NH$_3$, 10% O$_2$, 5% H$_2$O, (balance N$_2$) atmosphere.

These results clearly show that the mechanisms involved in soot oxidation on ceria and ceria-zirconia are different than those observed over TWS and alumina. This is better understood by looking at Figure 4 which shows the correlation between soot oxidation activity of pure and supported vanadates against loading. Similar results were obtained with FeErVO$_4$ (Supplementary material Figure S4).

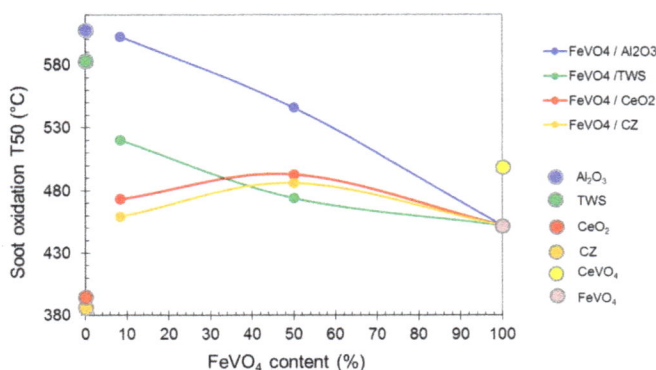

Figure 4. Effect of vanadate loading on soot oxidation activity.

For FeVO$_4$ supported on either Al$_2$O$_3$ or TWS the addition of FeVO$_4$ progressively increases activity. The effect is much more pronounced on TWS, where a strong influence is already observed at low loading. On Al$_2$O$_3$, a simple dilution effect seems to prevail, with a linear decrease in T50 going towards pure FeVO$_4$. From XRD analyses of these compounds (Figure 1 and Figure S2) it can be observed that both in Al_2O_3 and TWS samples the decomposition of $FeVO_4$ is confirmed by the presence of Fe_2O_3 diffraction lines (corresponding VO_x were detected by Raman spectroscopy on TWS support—Figure 2 and Figure S3 in supplementary material). If on TWS the role of VO_x formed is essential for soot oxidation enhancement, $FeVO_4$ decomposition on Al_2O_3 does not have the same beneficial effect and activity is governed simply by the dilution of the catalysts components. An opposite trend is observed with CeO_2 and $Ce_{0.75}Zr_{0.25}O_2$-based materials, where the highest activity is obtained with the two bare supports; the activity then decreases reaching a minimum for the 50 wt % mixtures to increase again in correspondence of the pure vanadates. This catalytic behaviour is due to formation of $CeVO_4$ which is less active than either CeO_2 or $FeVO_4$ (Table 2). For the catalyst with a higher vanadate loading, the higher amount of $CeVO_4$ formed (Figure S2A in supplementary material), contributes to a further decrease in activity.

The effect of addition of NO and NO/NH_3 to the gas feed stream for soot oxidation was also evaluated (Table 2). In comparison to O_2/N_2 atmosphere, the presence of NO enhances soot oxidation performances of the catalysts by lowering the peak temperature due to the formation of NO_2 with its better oxidation properties compared to O_2 [31]. In a subsequent step, 500 ppm of ammonia were added to the feed gas mixture with the overall effect of slightly lowering Tp temperature for almost all catalysts (Table 2). The positive effect of the presence of NH_3 on soot combustion efficiency using NO_2 was observed by Tronconi et al. for Cu-Zeolite catalysts [32].

2.3. NH_3-SCR Activity

NO conversion (defined as $\{[NO_{in}] - [NO_{out}]/[NO_{in}]\}$) in the NH_3-SCR reaction was investigated for the vanadates on the different supports. Figure 5 compares their activity in the range of temperature 200–500 °C. The NO reduction profile shows a typical volcano-type shape with NO conversion increasing with temperature up to a maximum value which is reached at ca. 300–450 °C depending on the catalyst composition. At this point, the activity starts to decrease due to the ammonia oxidation reaction which is prevailing over NO reduction to N_2. The activity of supported $FeVO_4$ is higher than that of $Fe_{0.5}Er_{0.5}VO_4$ irrespective of the support used, in agreement with our earlier findings [13]. Among the different carriers, the highest activity is found with TWS, for which maximum conversions in the range 80–100% are reported. The best performing de-soot catalysts show lower activity in the standard NH_3-SCR reaction with NO conversion values in the range 45–60% at medium SCR temperatures. Al_2O_3-based catalysts are almost inactive up to 250 °C. Overall, the NO_x conversion values coupled with the better efficiency in carbon soot oxidation at lower temperature makes vanadates supported on CeO_2 and $Ce_{0.75}Zr_{0.25}O_2$ as promising catalysts to investigate the interaction between the two reactions occurring simultaneously.

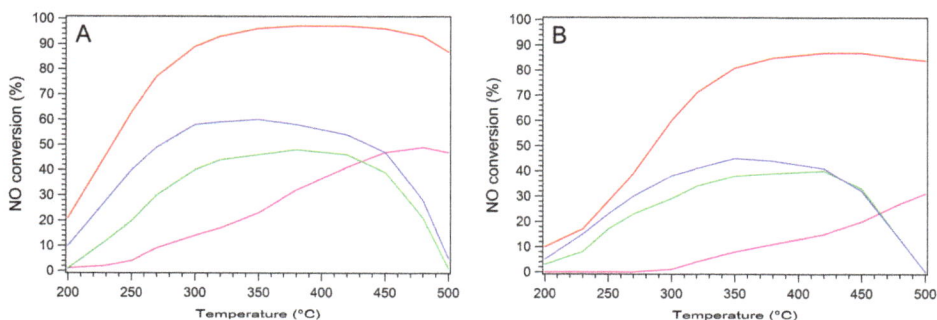

Figure 5. NO conversion in NH_3-SCR over (**A**) $FeVO_4$ (8.4 wt %) and (**B**) $Fe_{0.5}Er_{0.5}VO_4$ (8.4 wt %) loaded on different supports. Red lines: TWS, blue lines: $Ce_{0.75}Zr_{0.25}O_2$, green lines: CeO_2, magenta lines: Al_2O_3.

2.4. Simultaneous NH₃-SCR and Soot Oxidation Activity

To study the influence of soot oxidation on SCR reaction, a different set up configuration was used employing 20 mg of powder catalyst mixed in tight contact with soot. Under these conditions, the SCR reaction that is carried out at a GHSV of ca. 900,000 h^{-1} results in a lower NO$_x$ conversion than that observed previously. This downscale of the catalytic bed was necessary to reduce pressure drop issues originating by mixing a higher amount of catalyst with soot in powder form. Figure 6 shows that for 8.4 wt % catalysts, the comparison of NO conversion obtained under the conditions described above in SCR and SCR/de-soot reaction. In addition, the CO$_2$ concentration profile due to soot combustion is also reported. In agreement with the results obtained with lower space velocity (Figure 5), SCR only reaction (blue curve in Figure 6), shows a typical volcano shape with a maximum conversion in the range 350–400 °C. Iron vanadates over TWS support are the best performing catalytic materials with a maximum NO conversion in the range 40–60% while a drop to lower conversion values is observed using CeO$_2$ and CeZrO$_2$ as supports (16–18%). In correspondence of CO$_2$ evolution peak, a decrease of NO conversion is observed, which is much more evident for CZ based supports. Similar results are obtained with CeO$_2$ (supplementary material Figure S5). NH$_3$ evolution profiles (not shown) reveal a decrease in NH$_3$ concentration in correspondence to soot oxidation that indicates a large consumption of ammonia to form NO.

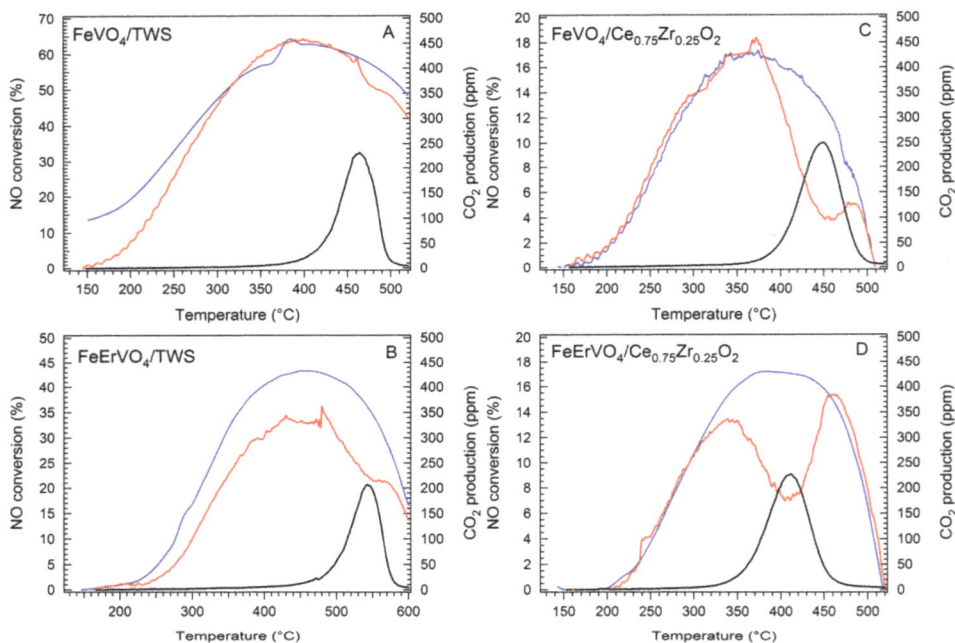

Figure 6. Comparison of 8.4 wt % loaded catalysts under the following conditions: catalyst 20 mg, NO 500 ppm, NH$_3$ 500 ppm, O$_2$ 10%, H$_2$O 5%, N$_2$ balance, and GHSV 900,000 h^{-1}. NO conversion in SCR (blue line), NO conversion in SCR combined with soot oxidation (red line) CO$_2$ evolution (black lines) (**A**) FeVO$_4$/TWS, (**B**) FeErVO$_4$/TWS, (**C**) FeVO$_4$/Ce$_{0.75}$Zr$_{0.25}$O$_2$, (**D**) FeErVO$_4$/Ce$_{0.75}$Zr$_{0.25}$O$_2$.

On pure FeVO$_4$ and Ce$_{0.75}$Zr$_{0.25}$O$_2$, negative NO conversion values are obtained (Figure 7). This indicates that NO can be generated through ammonia oxidation, and this occurs in correspondence to carbon soot oxidation. Monitoring catalyst temperature during heating ramp allowed to exclude that a local increase of temperature during carbon oxidation might affect reaction mechanism (Figure S6,

Supplementary). A similar result, that better highlights the decrease of NO conversion in SCR, is found when running the fast-SCR with NO_2 present in the feed gas on bare $Ce_{0.75}Zr_{0.25}O_2$ mixed with soot. In this case, the high NO conversion observed already at 150 °C, drops in correspondence of soot oxidation and CO_2 release (Figure 7D).

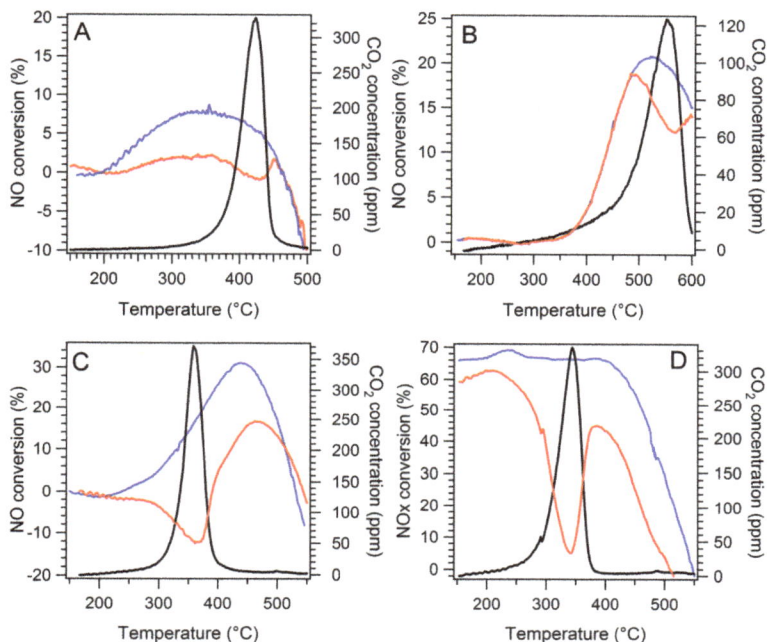

Figure 7. SCR runs on single components of the catalysts. Blue lines represent SCR of the bare samples, red lines represent SCR activity of the samples in tight contact with soot, black lines represent CO_2 production during soot combustion. (**A**) $FeVO_4$; (**B**) TWS; (**C**) $Ce_{0.75}Zr_{0.25}O_2$ standard SCR; (**D**) $Ce_{0.75}Zr_{0.25}O_2$ fast SCR.

Interestingly, a small drop in NO conversion in correspondence to soot oxidation is observed also on pure $FeVO_4$ and TWS (Figure 7A,B) which indicates that soot oxidation slightly affects NH_3-SCR also on these compounds. For this reason, ammonia oxidation reaction was studied for the single catalyst components. In Figure 8, NH_3 oxidation results over $FeVO_4$, TWS, CeO_2, $Ce_{0.75}Zr_{0.25}O_2$ are shown. The four different profiles compare reaction carried out with empty reactor, and with reactor filled with soot, catalyst and soot/catalyst mixtures. Compared to the empty reactor, soot alone has a minimal effect towards ammonia oxidation, as already observed by Mehring et al. [33]. $FeVO_4$ is the most active in the oxidation of NH_3, followed by ceria and ceria-zirconia. For $FeVO_4$, there is no large difference in NH_3 oxidation activity with or without the presence of soot. For CeO_2 and $Ce_{0.75}Zr_{0.25}O_2$ instead, the tight contact with soot (Figure 8C,D—red lines) leads to an additional NH_3 consumption in correspondence of CO_2 release due to soot combustion. This consumption is higher than the sum of contributions of catalyst and soot alone, showing a synergistic effect of the two components in NH_3 oxidation reaction and indicating that the presence of carbon soot on CeO_2 and $CeZrO_2$ produces a synergic effect increasing NH_3 oxidation to NO. No significant effect is displayed in NH_3 oxidation reaction by $CeVO_4$ which is formed on these catalysts when $FeVO_4$ is supported on CeO_2 as highlighted in Figure S7 in supplementary material. This effect is the responsible of the large decrease in NO conversion observed over ceria-containing materials in NH_3-SCR reaction.

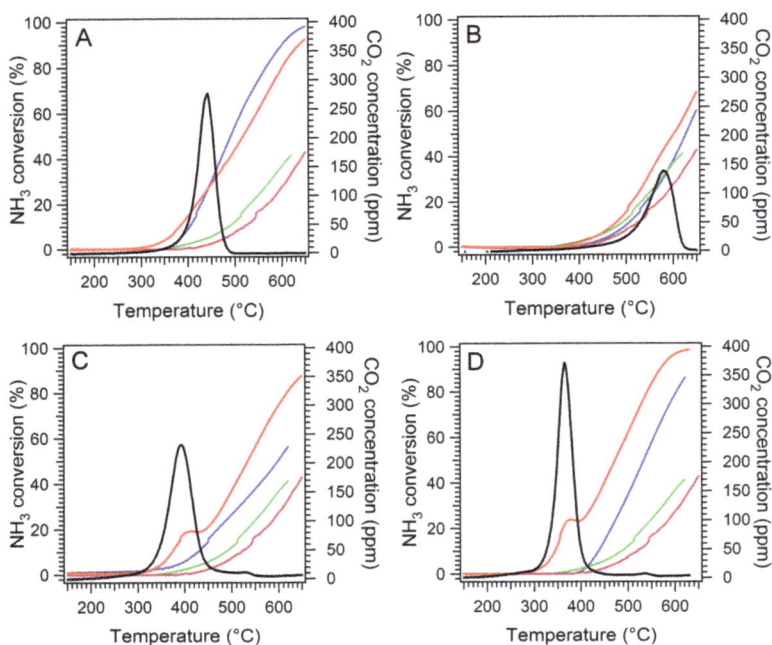

Figure 8. NH_3 oxidation carried out on single catalyst components and/or soot under the following conditions: catalyst 20 mg, NH_3 500 ppm, O_2 10%, N_2 balance, GHSV 900,000 h^{-1}. Magenta lines: NH_3 conversion curve over empty reactor; green lines: NH_3 conversion curve of bare soot; blue lines: NH_3 conversion curve over catalyst (**A**) $FeVO_4$—(**B**) TWS—(**C**) CeO_2—(**D**) $Ce_{0.75}Zr_{0.25}O_2$, red lines: NH_3 conversion over catalyst in tight contact with soot, black lines: CO_2 released during soot combustion with ongoing NH_3 oxidation reaction.

3. Experimental Section

3.1. Catalyst Preparation

$FeVO_4$ and $Fe_{0.5}Er_{0.5}VO_4$ were prepared by co-precipitation of metal nitrates ($Fe(NO_3)_3 \cdot 9H_2O$ and $Er(NO_3)_3 \cdot 6H_2O$, Treibacher Industrie AG, Althofen, Austria)) and ammonium metavanadate (99+%, Sigma-Aldrich, St. Louis, MO, USA). Aqueous solutions of the precursors were separately prepared at 80 °C. After unification, the pH was adjusted by means of ammonia (28% aqueous solution, Sigma-Aldrich) to obtain the respective precipitate which was filtered, washed and dried at 120 °C overnight. These vanadates were subsequently used to make the supported catalysts. All support used in this study were commercial materials (CeO_2 from Treibacher Industrie AG, $CeZrO_2$ from W.R. Grace (Columbia, MD, USA), TWS from Cristal (Hong Kong, China), Al_2O_3 from Sasol Germany GmbH (Hamburg, Germany)). Supported catalysts were prepared by mixing two separate aqueous slurries containing, respectively, the vanadate and the support. Vanadate/support weight ratio was 8.4/91.6 or 50/50. Combined slurries were evaporated to dryness. The resulting solids were dried at 120 °C overnight followed by calcination at 650 °C/2 h in a muffle furnace to obtain the final catalyst.

3.2. Catalytic Tests

Soot combustion efficiency under air was measured by means of thermogravimetry using soot (Printex U-Degussa)/catalyst mixtures under tight contact conditions obtained by mixing the two components in an agate mortar for 10 min. The soot/catalyst weight ratio was 1:20. The samples were placed on Pt crucibles licked by air flow (60 mL/min) and heated at a constant rate of 10 °C/min from RT up to 800 °C. T50 values, corresponding to the temperature at which 50 wt % of carbon is lost, was used as a measure of soot oxidation activity. Soot oxidation under $NO/O_2/N_2$ and $NO/O_2/N_2/NH_3$ was carried out in a laboratory set-up equipped with a quartz microreactor for powder testing. In these experiments 20 mg of sample in powder form (soot/catalyst ratio of 20:1, mixed in tight contact) were exposed to a 300 mL/min feed stream of composition 500 ppm NO, 500 ppm NH_3 (when present), 10% O_2, 5% H_2O, balance N_2 with a GHSV of 900,000 h^{-1}.

NH_3-SCR activity was measured in the same laboratory set-up described above on 100 mg of catalyst pressed and sieved (350 < Ø < 425 µm) under a 300 mL/min feed stream of composition 500 ppm NO, 500 ppm NH_3, 10% O_2, 5% H_2O, balance N_2 with a GHSV of 180,000 h^{-1}.

NH_3 oxidation experiments were conducted on 20 mg catalytic beds under a reactive atmosphere of composition 500 ppm NH_3, 10% O_2, balance N_2 with a GHSV of 900,000 h^{-1}. Ammonia concentration was monitored during a temperature ramp from 150 °C up to 650 °C. When soot was present, also CO_2 release due to carbon oxidation was monitored.

For all experiment conducted in the gas bench, the composition of the feed stream (reagents and products) was analyzed with an on line MKS 2030 MultiGas Analyzer FT-IR instrument.

3.3. Characterization of the Catalysts

Specific surface areas of catalysts were measured by nitrogen adsorption at -196 °C according to the B.E.T. method on a Tristar 3000 gas analyzer (Micromeritics, Norcross, GA, USA). Prior to the measurement all samples were outgassed at 150 °C for 1.5 h. Powder X-ray diffraction patterns were collected on a Philips X'Pert diffractometer (PANalytical B.V., Almelo, The Netherlands) operated at 40 kV and 40 mA using Ni filtered Cu Kα radiation. Data were recorded from 10 to 100° 2θ using a step size of 0.02° and a counting time of 40 s per angular abscissa. Phase identification was conducted with the Philips X'Pert Highscore software (PANalytical B.V., Almelo, The Netherlands, 2002). Reducibility of the samples was measured by temperature programmed reduction (TPR) runs; approximately 50 mg of samples were heated at 10 °C/min in a U-shaped quartz reactor from room temperature up to 960 °C under a 4.5 vol % H_2/N_2 flow (35 mL/min). The samples were pretreated at 500 °C for 1 h under air flow. Hydrogen consumption was monitored with a thermal conductivity detector (TCD) in an Autochem II 2920 instrument (Autochem II 2920, Micromeritics, Norcross, GA, USA). Raman spectra were collected with a Xplora Plus Micro-Raman system (Horiba, Kyoto, Japan) equipped with a cooled CCD detector (−60 °C) and Edge filter. The samples were excited with the 512 nm radiation in a Linkam CR1000 in situ cell at 200 °C under flowing air. The spectral resolution was 1 cm^{-1} and the spectra acquisition was of 5 accumulations of 10 or 20 s with a 50× LWD objective. The optical images were collected with an integrated microscope Olympus BX43 (Olympus, Tokyo, Japan) with a 10× objective.

4. Conclusions

A systematic screening to evaluate the interaction between soot oxidation and NH_3-SCR has been carried out investigating the behavior of $FeVO_4$ and $Fe_{0.5}Er_{0.5}VO_4$ loaded on standard SCR supports like TiO_2-WO_3-SiO_2, redox active carriers like CeO_2 or $Ce_{0.75}Zr_{0.25}O_2$ and Al_2O_3.

It is shown by combined XRD, TPR, and Raman spectroscopy that vanadates partially decompose over TiO_2-based materials and Al_2O_3 with formation of Fe_2O_3 and surface VO_x species. In the presence of $CeO_2/CeZrO_2$, formation of crystalline $CeVO_4$ along with Fe_2O_3 is reported.

Al$_2$O$_3$-based materials are almost inactive either in SCR and soot oxidation while, as expected, TiO$_2$-based materials are highly active in NH$_3$-SCR and CeO$_2$/CeZrO$_2$ are active in carbon soot oxidation. The addition of ammonia is found to slightly influence soot oxidation by reducing its combustion temperature by a few degrees while the presence of soot is found to affect SCR reaction, especially in the presence of CeO$_2$ and CeZrO$_2$. This is due to the combined effect of the catalyst and carbon soot that results in an enhancement of NH$_3$ oxidation reaction.

Supplementary Materials: The following are available online at http://www.mdpi.com/2073-4344/8/4/130/s, Figure S1: Raman spectra of 8.4 wt % FeVO$_4$/Al$_2$O$_3$. In the blue line (grating 1800 L/mm with 5 s per accumulation) only Fe$_2$O$_3$ and Al$_2$O$_3$ Raman shifts are visible. Surface VO$_x$ could not be detected. In the red line Raman shifts mainly attributable to FeVO$_4$ and Fe$_2$O$_3$ are shown, Figure S2: XRD patterns of (A) 50 wt % FeVO$_4$ and (B) 50 wt % FeErVO$_4$ on different supports. Figure S3: Raman spectra of 50 wt % FeVO$_4$/TWS. The optical image of the sample shows that the dark spots of FeVO$_4$ and Fe$_2$O$_3$ are uniformly distributed and all Raman spectra show the Raman shifts of FeVO$_4$. Figure S4: Effect of FeErVO$_4$ loading on soot oxidation activity, Figure S5: Comparison of SCR catalytic activity of 20 mg catalytic beds of bare catalysts and with soot oxidation conducted simultaneously, Figure S6: Comparison between theoretical heating ramp and temperature measured in the catalyst bed during soot oxidation reaction on CeO$_2$ sample (tight contact, soot/catalyst weight ratio of 1:20), Figure S7: NH$_3$ oxidation carried out on single catalyst components under the following conditions: catalyst 20 mg, NH$_3$ 500 ppm, O$_2$ 10%, N$_2$ balance, GHSV 900,000 h^{-1}, Table S1: B.E.T. surface areas of Fe$_{0.5}$Er$_{0.5}$VO$_4$ catalysts.

Acknowledgments: Authors would like to thank Interreg V Italy-Austria project Coat4Cata (project n. ITAT 1019) for funding, for research support and for covering the costs to publish in open access.

Author Contributions: M.C. and A.T. conceived and designed the experiments; M.C. performed the experiments; M.C., S.C., and A.T. analyzed the data; A.T. contributed reagents/materials/analysis tools; M.C. prepared tables and figures and wrote the first draft of the manuscript and S.C. and A.T. contributed to subsequent revisions. All authors agreed to the final version of the paper.

Conflicts of Interest: The authors declare no conflict of interest.

References

1. Bassou, B.; Guilhaume, N.; Iojoiu, E.E.; Farrusseng, D.; Lombaert, K.; Bianchi, D.; Mirodatos, C. High-throughput approach to the catalytic combustion of diesel soot II: Screening of oxide-based catalysts. *Catal. Today* **2011**, *159*, 138–143. [CrossRef]
2. Granger, P.; Parvulescu, V.I. Catalytic NO$_x$ abatement systems for mobile sources: From three-way to lean burn after-treatment technologies. *Chem. Rev.* **2011**, *111*, 3155–3207. [CrossRef] [PubMed]
3. Liu, Z.M.; Woo, S.I. Recent advances in catalytic deNO$_{(x)}$ science and technology. *Catal. Rev.* **2006**, *48*, 43–89. [CrossRef]
4. Brandenberger, S.; Krocher, O.; Tissler, A.; Althoff, R. The state of the art in selective catalytic reduction of NO$_x$ by ammonia using metal-exchanged zeolite catalysts. *Catal. Rev.* **2008**, *50*, 492–531. [CrossRef]
5. Boger, T. Integration of SCR functionality into diesel particulate filters. In *Urea-Scr Technology for Denox after Treatment of Diesel Exhausts*; Nova, I., Tronconi, E., Eds.; Springer: New York, NY, USA, 2014; pp. 623–655.
6. Johnson, T.V. Review of selective catalytic reduction (SCR) and related technologies for mobile applications. In *Urea-SCR Technology for deNO$_x$ after Treatment of Diesel Exhausts*; Nova, I., Tronconi, E., Eds.; Springer: New York, NY, USA, 2014; pp. 3–31.
7. Weibel, M.; Waldbusser, N.; Wunsch, R.; Chatterjee, D.; Bandl-Konrad, B.; Krutzsch, B. A novel approach to catalysis for NO$_x$ reduction in diesel exhaust gas. *Top. Catal.* **2009**, *52*, 1702–1708. [CrossRef]
8. Watling, T.C.; Ravenscroft, M.R.; Avery, G. Development, validation and application of a model for an SCR catalyst coated diesel particulate filter. *Catal. Today* **2012**, *188*, 32–41. [CrossRef]
9. Song, X.; Johnson, J.H.; Naber, J.D. A review of the literature of selective catalytic reduction catalysts integrated into diesel particulate filters. *Int. J. Engine Res.* **2015**, *16*, 738–749. [CrossRef]
10. Cavataio, G.; Warner, J.R.; Girard, J.W.; Ura, J.; Dobson, D.; Lambert, C.K. Laboratory study of soot, propylene and diesel fuel impact on zeolite-based SCR filter catalysts. *SAE Int. J. Fuels Lubr.* **2009**, *2*, 342–368. [CrossRef]
11. Casanova, M.; Schermanz, K.; Llorca, J.; Trovarelli, A. Improved high temperature stability of NH$_3$-SCR catalysts based on rare earth vanadates supported on TiO$_2$-WO$_3$-SiO$_2$. *Catal. Today* **2012**, *184*, 227–236. [CrossRef]

12. Vargas, M.A.L.; Casanova, M.; Trovarelli, A.; Busca, G. An IR study of thermally stable V_2O_5-WO_3-TiO_2 SCR catalysts modified with silica and rare-earths (Ce, Tb, Er). *Appl. Catal. B Environ.* **2007**, *75*, 303–311. [CrossRef]

13. Casanova, M.; Llorca, J.; Sagar, A.; Schermanz, K.; Trovarelli, A. Mixed iron-erbium vanadate NH_3-SCR catalysts. *Catal. Today* **2015**, *241*, 159–168. [CrossRef]

14. Marberger, A.; Elsener, M.; Ferri, D.; Sagar, A.; Schermanz, K.; Krocher, O. Generation of NH_3 selective catalytic reduction active catalysts from decomposition of supported $FeVO_4$. *ACS Catal.* **2015**, *5*, 4180–4188. [CrossRef]

15. Liu, F.; He, H.; Lian, Z.; Shan, W.; Xie, L.; Asakura, K.; Yang, W.; Deng, H. Highly dispersed iron vanadate catalyst supported on TiO_2 for the selective catalytic reduction of NO_x with NH_3. *J. Catal.* **2013**, *307*, 340–351. [CrossRef]

16. Gillot, S.; Tricot, G.; Vezin, H.; Dacquin, J.-P.; Dujardin, C.; Granger, P. Development of stable and efficient $CeVO_4$ systems for the selective reduction of NO_x by ammonia: Structure-activity relationship. *Appl. Catal. B Environ.* **2017**, *218*, 338–348. [CrossRef]

17. Bueno-Lopez, A. Diesel soot combustion ceria catalysts. *Appl. Catal. B Environ.* **2014**, *146*, 1–11. [CrossRef]

18. Casanova, M.; Nodari, L.; Sagar, A.; Schermanz, K.; Trovarelli, A. Preparation, characterization and NH_3-SCR activity of $FeVO_4$ supported on TiO_2-WO_3-SiO_2. *Appl. Catal. B Environ.* **2015**, *176*, 699–708. [CrossRef]

19. Wu, Z.L.; Rondinone, A.J.; Ivanov, I.N.; Overbury, S.H. Structure of vanadium oxide supported on ceria by multiwavelength raman spectroscopy. *J. Phys. Chem. C* **2011**, *115*, 25368–25378. [CrossRef]

20. Went, G.T.; Leu, L.J.; Bell, A.T. Quantitative structural-analysis of dispersed vanadia species in TiO_2(anatase)-supported V_2O_5. *J. Catal.* **1992**, *134*, 479–491. [CrossRef]

21. Wachs, I.E. Raman and IR studies of surface metal oxide species on oxide supports: Supported metal oxide catalysts. *Catal. Today* **1996**, *27*, 437–455. [CrossRef]

22. Haber, J.; Machej, T.; Serwicka, E.M.; Wachs, I.E. Mechanism of surface spreading in vanadia-titania system. *Catal. Lett.* **1995**, *32*, 101–114. [CrossRef]

23. Gaur, K.; Lal, H.B. Electrical transport in heavy rare-earth vanadates. *J. Mater. Sci.* **1986**, *21*, 2289–2296. [CrossRef]

24. Casanova, M.; Sagar, A.; Schermanz, K.; Trovarelli, A. Enhanced stability of Fe_2O_3-doped $FeVO_4$/TiO_2-WO_3-SiO_2 SCR catalysts. *Top. Catal.* **2016**, *59*, 996–1001. [CrossRef]

25. Vermaire, D.C.; Vanberge, P.C. The preparation of WO_3/TiO_2 and WO_3/Al_2O_3 and characterization by temperature-programmed reduction. *J. Catal.* **1989**, *116*, 309–317. [CrossRef]

26. Giordano, F.; Trovarelli, A.; de Leitenburg, C.; Giona, M. A model for the temperature-programmed reduction of low and high surface area ceria. *J. Catal.* **2000**, *193*, 273–282. [CrossRef]

27. Daturi, M.; Finocchio, E.; Binet, C.; Lavalley, J.-C.; Fally, F.; Perrichon, V.; Vidal, H.; Hickey, N.; Kašpar, J. Reduction of high surface area CeO_2-ZrO_2 mixed oxides. *J. Phys. Chem. B* **2000**, *104*, 9186–9194. [CrossRef]

28. Gallert, T.; Casanova, M.; Puzzo, F.; Strazzolini, P.; Trovarelli, A. SO_2 resistant soot oxidation catalysts based on orthovanadates. *Catal. Commun.* **2017**, *97*, 120–124. [CrossRef]

29. Machida, M.; Murata, Y.; Kishikawa, K.; Zhang, D.J.; Ikeue, K. On the reasons for high activity of CeO_2 catalyst for soot oxidation. *Chem. Mater.* **2008**, *20*, 4489–4494. [CrossRef]

30. Soler, L.; Casanovas, A.; Escudero, C.; Pérez-Dieste, V.; Aneggi, E.; Trovarelli, A.; Llorca, J. Ambient pressure photoemission spectroscopy reveals the mechanism of carbon soot oxidation in ceria-based catalysts. *ChemCatChem* **2016**, *8*, 2748–2751. [CrossRef]

31. Zouaoui, N.; Issa, M.; Kehrli, D.; Jeguirim, M. CeO_2 catalytic activity for soot oxidation under NO/O_2 in loose and tight contact. *Catal. Today* **2012**, *189*, 65–69. [CrossRef]

32. Tronconi, E.; Nova, I.; Marchitti, F.; Koltsakis, G.; Karamitros, D.; Maletic, B.; Markert, N.; Chatterjee, D.; Hehle, M. Interaction of NO_x reduction and soot oxidation in a DPF with Cu-Zeolite SCR coating. *Emiss. Control Sci. Technol.* **2015**, *1*, 134–151. [CrossRef]

33. Mehring, M.; Elsener, M.; Kröcher, O. Selective catalytic reduction of NO_x with ammonia over soot. *ACS Catal.* **2012**, *2*, 1507–1518. [CrossRef]

catalysts

MDPI

Article

Nb-Modified Ce/Ti Oxide Catalyst for the Selective Catalytic Reduction of NO with NH$_3$ at Low Temperature

Jawaher Mosrati [1], Hanan Atia [2], Reinhard Eckelt [2], Henrik Lund [2], Giovanni Agostini [2], Ursula Bentrup [2], Nils Rockstroh [2], Sonja Keller [2], Udo Armbruster [2,*] and Mourad Mhamdi [1,3]

[1] Laboratoire de Chimie des Matériaux et Catalyse, Departement de Chimie, Faculté des Sciences de Tunis, Campus Universitaire El Manar, 2092 El Manar Tunis, Tunisia; Jawahermosrati98@gmail.com (J.M.); mourad.mhamdi@gmail.com (M.M.)

[2] Leibniz-Institut für Katalyse e.V., Albert-Einstein-Strasse 29a, 18059 Rostock, Germany; hanan.atia@catalysis.de (H.A.); reinhard.eckelt@catalysis.de (R.E.); henrik.lund@catalysis.de (H.L.); giovanni.agostini@catalysis.de (G.A.); ursula.bentrup@catalysis.de (U.B.); nils.rockstroh@catalysis.de (N.R.); sonja.keller@catalysis.de (S.K.)

[3] Institut Supérieur des Technologies Médicales de Tunis, Université de Tunis El Manar, 9 Rue Zouhaier Essafi 1006 Tunis, Tunisia

* Correspondence: udo.armbruster@catalysis.de; Tel.: +49-381-1281-257

Received: 9 March 2018; Accepted: 23 April 2018; Published: 26 April 2018

Abstract: Recently, great attention has been paid to Ceria-based materials for selective catalytic reduction (SCR) with NH$_3$ owing to their unique redox, oxygen storage, and acid-base properties. Two series of bimetallic catalysts issued from Titania modified by Ce and Nb were prepared by the one-step sol-gel method (SG) and by the sol-gel route followed by impregnation (WI). The resulting core-shell and bulk catalysts were tested in NH$_3$-SCR of NO$_x$. The impregnated Nb5/Ce40/Ti100 (WI) catalyst displayed 95% NO$_x$ conversion at 200 °C (GHSV = 60,000 mL·g^{-1}·h^{-1}, 1000 ppm NO$_x$, 1000 ppm NH$_3$, 5% O$_2$/He) without forming N$_2$O. The catalysts were characterized by various methods including ICP-OES, N$_2$-physisorption, XRD, Raman, NH$_3$-TPD, DRIFTS, XPS, and H$_2$-TPR. The results showed that the introduction of Nb decreases the surface area and strengthens the surface acidity. This behavior can be explained by the strong interaction between Ceria and Titania which generates Ce-O-Ti units, as well as a high concentration of amorphous or highly dispersed Niobia. This should be the reason for the excellent performance of the catalyst prepared by the sol-gel method followed by impregnation. Furthermore, Nb5/Ce40/Ti100 (WI) has the largest NH$_3$ adsorption capacity, which is helpful to promote the NH$_3$-SCR reaction. The long-term stability and the effect of H$_2$O on the catalysts were also evaluated.

Keywords: low-temperature SCR; Ceria-based catalysts; Niobia; sol-gel method

1. Introduction

Nitrogen oxides (NO$_x$), emitted from the industrial combustion of fossil fuels and automobile exhaust gas, are major air pollutants leading to various environmental problems: acid rain, photochemical smog, ozone depletion, and the greenhouse effect [1–3]. The steady increase of NO$_x$ emissions necessitates the improvement of abatement technologies. The selective catalytic reduction (SCR) is the most promising way to remove NO$_x$ from stationary sources and vehicles exhaust gas [4–6]. A reducing agent is needed to convert NO$_x$ into inert components like water and nitrogen. For this purpose, mostly urea is used for fixed stations (nitric acid plants); however, it is also used in catalytic reactors fitted to diesel engines and some vessels because of the high NO$_x$ conversions that can be achieved at a high space velocity [4]. Nowadays, TiO$_2$ (anatase) supported V$_2$O$_5$-WO$_3$ or V$_2$O$_5$-MoO$_3$

oxides have been commercialized for stationary applications [2,7]. V_2O_5 is the main redox active site, whereas WO_3 (or MoO_3) promotes the surface acidity [2,8]. However, some drawbacks of vanadium-based catalysts were observed during industrial application, such as the biological toxicity and volatility of V_2O_5, the narrow operating temperature window (300–400 °C), and high activity for the oxidation of SO_2 to SO_3 [9–11]. As an alternative, low-temperature catalysts—if available—could be placed downstream of electrostatic precipitator and desulfurization units. Therefore, it is crucial to develop novel, environment-friendly, and non-vanadium-based catalysts with high NO_x conversion, high N_2 selectivity, nontoxic species, and high resistance to H_2O for NH_3-SCR.

Ceria-based materials are of particular interest for NH_3-SCR owing to their redox cycle between Ce^{3+} and Ce^{4+}, ability for oxygen storage, and acid-base properties [12,13]. Ceria is not very toxic and comparatively cheap. Therefore, Ceria has already been widely used as an indispensable component in automotive three-way catalysts. Recently, many researchers studied the effect of introducing solid acid components on NH_3-SCR catalysts such as CeO_2-TiO_2 [14,15], CeO_2-WO_3 [16–22], CeO_2-$SO_4{}^{2-}$ [23–25], CeO_2-Nb_2O_5 [26–30], CeO_2-$PO_4{}^{3-}$ [31,32], and CeO_2-MoO_3 [33]. Additionally, CeO_2 was reported to have the ability to resist SO_2 and alkali poisoning [34].

Among them, CeO_2-Nb_2O_5-based catalysts offer high potential for diesel exhaust gas treatment due to their multi-functionality, such as the hydrolysis of urea to NH_3, the SCR of NO_x with NH_3, and the oxidation of soot [28–30]. Binary oxides of Cerium and Niobium were reported to improve the acidity of catalysts, which was suggested to be an important feature in the NH_3-SCR reaction [29]. Niobium modification in general appears to be an effective strategy in selective catalytic reduction of NO_x with ammonia or hydrocarbons [29,30]. The Nb–OH bond is the source for Brønsted acid sites and the Nb=O bond for Lewis acid sites, both of which are essential in the NH_3-SCR reaction [29]. The positive effect of modifying the acidic properties with NbO_x and the generation of structural defects in Ce-Nb (oxygen vacancies) increase the electron conductivity [29,30,35,36]. So, the combination of the Nb_2O_5 acid properties and the redox behavior of CeO_2 may lead to a good SCR performance. Despite the availability of some data from these studies on SCR performance in different reaction conditions, an understanding of the structure-activity relationship of CeO_2-Nb_2O_5-TiO_2 catalysts is still missing, e.g., the correlation of De-NO_x activity to surface acidity and redox ability.

Regarding the successful application of TiO_2 as a support in V_2O_5-WO_3/TiO_2 and V_2O_5-MoO_3/TiO_2 catalysts for NH_3-SCR and the potential of the CeO_2-Nb_2O_5 catalyst for the SCR of NO_x, it is of great interest to study a combination of them. The NH_3-SCR performance of NO_x on this type of supported metal oxide has not yet been studied.

In the present study, we synthesized Ce40/Ti100 (numbers indicate elemental ratios) and Nb5/Ce40/Ti100 catalysts (Nb given in weight percent) by the one-step sol-gel method or by the impregnation of support Ce40/Ti100 with Niobia and their efficiency in the low-temperature NH_3-SCR was elucidated. The beneficial effect of Nb and Ce on NO conversion and N_2 selectivity will be demonstrated and explained.

2. Results

2.1. NH_3-SCR Performance

The results from NH_3-SCR runs with different catalysts are given in Figure 1a,b, including some blank tests. The reference materials TiO_2 (support) and Nb5/Ti100 (Ceria-free catalyst, 5 wt % Nb) were inactive below 300 °C; first upon raising the temperature to 400 °C, the conversion increased to 15 and 30%, respectively. On the other hand, the NO_x conversion with the Nb-free Ce40/Ti100 (SG) catalyst showed a remarkable increase with temperature, reaching 78% at 200 °C and 99% at 300 °C. Presence of Ceria is essential for high SCR activity.

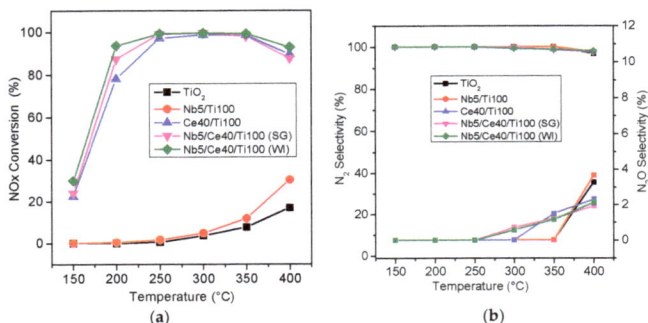

Figure 1. (**a**) NO_x conversion and (**b**) N_2-N_2O selectivity for different catalysts (1000 ppm NO_x, 1000 ppm NH_3, 5% O_2/He, GHSV = 60,000 mL·g^{-1}·h^{-1}).

Both catalysts Nb5/Ce40/Ti100 (SG) and Nb5/Ce40/Ti100 (WI) outnumbered all these reference materials and showed almost complete conversion of NO_x at 250 °C. At 200 °C, the Nb5/Ce40/Ti100 (WI) catalyst was more active and exhibited 95% NO_x conversion, while Nb5/Ce40/Ti100 (SG) showed 87%. Even at 150 °C, the Nb5/Ce40/Ti100 (WI) showed higher conversion (30%) than Nb5/Ce40/Ti100 (SG) (24%) and Ce40/Ti100 (22%). A similar system with 20% CeO_2/TiO_2 prepared by an impregnation method was reported elsewhere, which affords 95% conversion in the range of 250–375 °C at a GHSV of 25,000 h^{-1} [15]. However, on setting the GHSV to 50,000 h^{-1}, still being lower than the value applied in the present paper (60,000 mL·g^{-1}·h^{-1}), a much higher temperature of 350 °C was needed to reach 97% conversion. A comparison of three preparation methods for CeO_2/TiO_2 catalysts revealed that only the catalyst prepared by a single step sol-gel method showed complete conversion at 225 °C and GHSV = 50,000 h^{-1} [34]. In addition, the Ce-Nb system with a ratio of 1:1 prepared with the co-precipitation method presented 80% NO conversion in a temperature range of 200–450 °C with a GHSV = 120,000 mL·g^{-1}·h^{-1} [29]. A ternary oxide system Mn/Ce/Nb/Ti exhibited complete NO_x conversion at 150 °C at a high GHSV of 180,000 h^{-1} [37]. Mn-Ce oxide catalysts were found to allow complete conversions at 100–150 °C at a space velocity of 42,000 h^{-1} but showed considerable deactivation in the presence of 19% H_2O [38]. The present catalyst does not reach such high values at these temperatures. However, Mn-based catalysts, due to their oxidative power, promoted the formation of undesired products N_2O and NO_2 [37]. It can be first pointed out that the addition of Nb to Ce-TiO_2 improves the NO_x conversion particularly at temperatures below 250 °C and in the absence of H_2O, and a clear synergistic effect exists between Nb, Ce, and Ti species. Regarding these data, the presented catalyst Nb5/Ce40/Ti100 (WI) appears to be an attractive intermediate temperature catalyst as it avoids the over-oxidation as known from Mn catalysts. Furthermore, it shows higher activity than conventional V_2O_5/TiO_2. In conclusion, Nb5/Ce40/Ti100 (WI) showed highly competitive results comparable to the state-of-art. Second, the applied preparation methods seem to have a significant effect on the catalyst efficiencies.

The N_2 selectivity with Nb5/Ce40/Ti100 (WI) reaches 100% below 250 °C and exceeds 99% within the whole temperature range, as shown in Figure 1b; this indicates that the interaction of Nb species with Ce40/Ti100 (SG) suppresses the unselective catalytic oxidation of NH_3 by O_2 towards N_2O and NO_2 at high temperature, as it was observed elsewhere [29,37]. The N_2O concentration recorded with these catalysts is especially low (<2.5%) at 300–400 °C.

Water vapor is an inevitable compound in automotive exhaust gases and always significantly influences the catalytic activity. To investigate the long-term stability and the effect of water (8 vol %) on the performance of the catalysts, runs were made at 200 °C to apply technically relevant conditions for low-temperature SCR.

When adding 8 vol % of water to the gas flow during catalyst testing, the conversion drops by 22% for Nb5/Ce40/Ti100 (WI) and by 27% for both Ce40/Ti100 and Nb5/Ce40/Ti100 (SG). Within the first 4 h on stream Nb5/Ce40/Ti100 (WI) showed a slight decrease in conversion from 65 to 62% (Figure 2). Upon switching off the water dosage, the NO_x conversion rapidly recovered to 85%, indicating that the inhibition effect of water is largely reversible. However, the initial conversion of 95% (Figure 1a) was not recovered probably due to some minor structural changes which could occur in the presence of water, causing a deactivation of the catalyst. Nevertheless, on switching on H_2O again, the conversion recovered, reached its initial value of 62%, and then decreased slowly to 60% over 12 h. Surprisingly, Ce40/Ti100 displayed during the first 4 h on stream comparable activity as Nb5/Ce40/Ti100 (SG) and even appeared more stable over 12 h. In general, the catalysts deactivate in the presence of H_2O, but Nb5Ce40Ti100 (WI) is less pronounced deactivated compared to the others. This can be explained by the competitive adsorption between NH_3/NO_x and H_2O on the active sites, leading to an inhibiting effect of H_2O on the low-temperature SCR reaction [39,40].

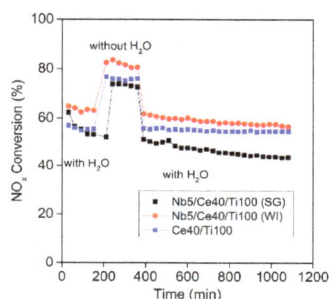

Figure 2. Long-term activity and effect of water on the SCR activity at 200 °C (1000 ppm NO_x, 1000 ppm NH_3, 5 vol % O_2, 8 vol % H_2O, balance He, GHSV = 60,000 mL·g^{-1}·h^{-1}).

2.2. Physicochemical Characterization

2.2.1. Structural and Textural Properties (ICP-OES, BET, XRD, and Raman)

The chemical compositions of the catalysts were analyzed by the ICP-OES technique and the weight percentage of each element is listed in Table 1. The values for Nb are lower than expected for all the catalysts, while for Ce, the content is slightly higher than expected. This can be explained by the method of drying the solvent under supercritical conditions, e.g., by partial leaching of Nb species by this solvent.

The textural properties of the samples were investigated by N_2 adsorption-desorption at 77 K (Table 1). Figure 3a–e show the corresponding N_2 adsorption-desorption isotherms and BJH pore size distributions for the catalysts, respectively. All the catalysts exhibited typical type IV isotherms with an obvious type H3 hysteresis loop as classified by IUPAC, which is always assigned to capillary condensation in mesopores ascribed to the interstices between the particles [41]. The specific BET surface area of the chosen TiO_2 support was 141 m^2·g^{-1} and a loss of 15% was observed upon impregnating with $NbCl_5$ (WI). After sol-gel preparation with CeO_2, the TiO_2 supported sample lost 36% of its surface. Addition of Nb during the sol-gel (SG) technique did not cause any decrease in surface area, while by impregnation, the surface area decreased to 56 m^2·g^{-1} as expected due to the blocking of pores. While the addition of Ce to Ti in the sol-gel formation of the support leads to a loss in BET surface area, the modification of the preparation method by adding Nb as a third component seems to enlarge the BET surface area, as well as the pore volumes and diameters. This might be explained by the incorporation of Nb in the framework formed by Ce and Ti species. In contrast, after impregnation of the sol-gel-based Ce40/Ti100 with Nb, parts of the pores are blocked. The average pore diameters of all samples are between 5 and 10 nm (Figure 3b), which confirms that mesopores account for the dominant role in the catalysts (Table 1).

Table 1. Catalyst composition, specific BET surface area, and pore volume (SG = sol-gel route; WI = wet impregnation).

Catalyst	Element Composition (wt %)			S_{BET} $(m^2 \cdot g^{-1})$	Pore Volume $(cm^3 \cdot g^{-1})$	Average Pore Diameter (nm)
	Nb	Ce	Ti			
TiO$_2$	-	-	-	141	0.26	5.4
Nb5/Ti100 (WI)	4.3	-	49.1	121	0.26	7.3
Ce40/Ti100 (SG)	-	35.7	28.2	90	0.22	6.9
Nb5/Ce40/Ti100 (SG)	4.1	35.0	26.7	94	0.32	9.8
Nb5/Ce40/Ti100 (WI)	4.6	34.4	26.8	56	0.17	7.1

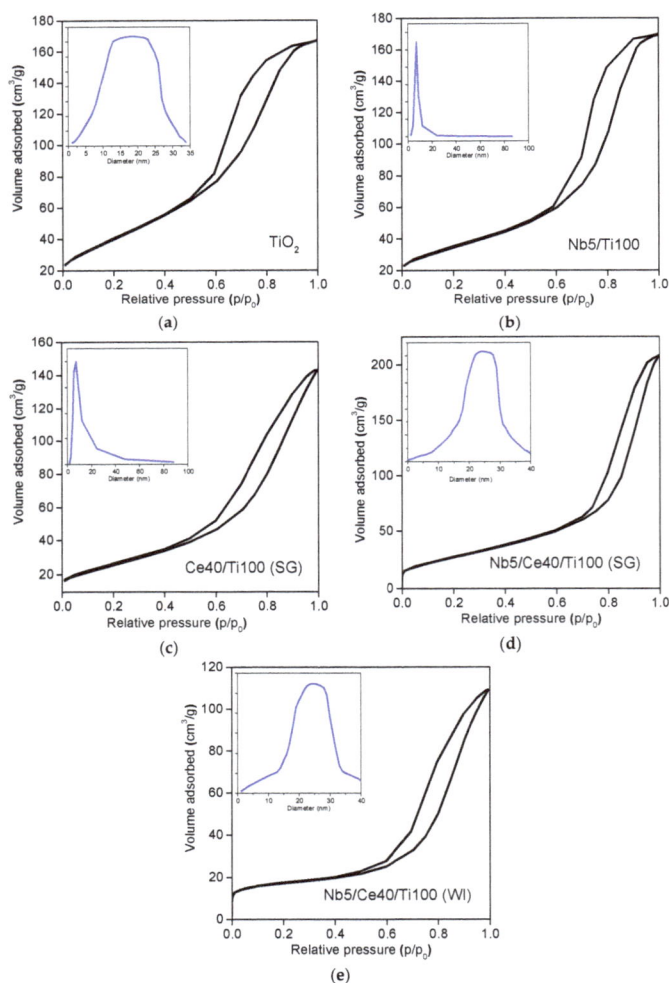

Figure 3. N$_2$ adsorption-desorption isotherms and BJH pore distributions for all fresh catalysts: (**a**) TiO$_2$, (**b**) Nb5/Ti100, (**c**) Ce40/Ti100 (SG), (**d**) Nb5/Ce40/Ti100 (SG), (**e**) Nb5/Ce40/Ti100 (WI).

The XRD patterns of the catalysts are presented in Figure 4. The powder patterns of the catalysts synthesized from the pure Ti(IV) isopropoxide indicate the presence of anatase modification of Titania (ICDD PDF-2 2017). After the impregnation of TiO_2 with an Nb precursor, no additional reflexes which can be assigned to a Nb-containing phase were observed. Furthermore, no change of the unit cell parameters of anatase was seen. This suggests that the Nb species are highly dispersed on the surface or in an amorphous state. In contrast, the impregnation with Ce precursor leads to the formation of cubic Ceria (Figure 4, triangles) and a non-stoichiometric brannite-like structured Ceria-Titania-oxide ($Ce_{0.97}Ti_2O_{5.95}$, Figure 4, hereinafter called $CeTi_2O_6$). Further treatment after impregnation with Niobium species (Nb5/Ce40/Ti100 (SG), Nb5/Ce40/Ti100 (WI)) leads to the loss of crystallinity, as indicated by the additional peak broadening (cf. calculated crystallite sizes, Table 2). In general, the catalyst prepared by the impregnation method was more crystalline than the catalysts obtained via the sol-gel method. As Nb^{5+} has a smaller ionic radius (0.064 nm) compared to Ce^{4+} (0.087 nm), Nb can diffuse into the lattice of CeO_2 and decrease its crystallinity [42]. This may explain the decrease in Nb5/Ce40/Ti100 (SG) crystallinity compared to the impregnated sample, as due to the homogeneous distribution of the elements it is more likely that Nb interacts with Ce and Ti. This behavior is also proved by the decrease in the crystallite size of CeO_2 and $CeTi_2O_6$ (Table 2). It is important to note that upon impregnating Ce40/Ti100 with Nb, there was a loss in surface area which enhances the agglomeration and causes an increase in the crystallite size of $CeTi_2O_6$. These effects were closely related to the catalyst preparation methods. However, the presence of Titania (anatase) can be discussed in all of the three samples which contain Ce as well.

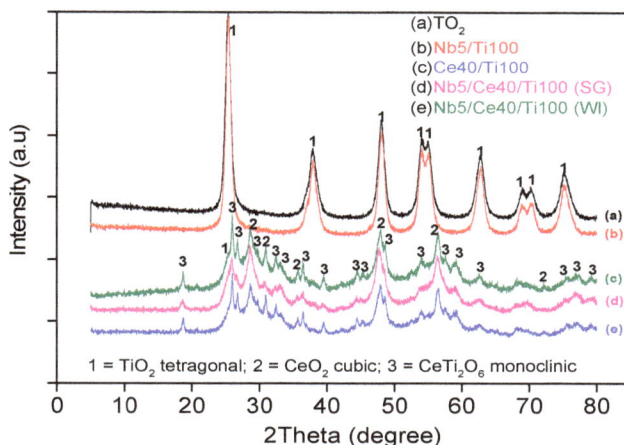

Figure 4. XRD patterns of the fresh catalysts.

Table 2. Crystallite sizes of prepared catalysts calculated with Scherrer's formula [43].

Catalysts	Crystallite Size (nm)		
	$CeTi_2O_6$	TiO_2	CeO_2
Ce40/Ti100	23.5	-	5.6
Nb5/Ce40/Ti100 (SG)	12.6	6.2	4.3
Nb5/Ce40/Ti100 (WI)	18.8	-	7.4

These results are also backed by Raman spectra (Figure 5), wherein all the catalysts show typical bands of TiO_2 (anatase) at 143, 396, 515, and 637 cm^{-1} [44–46]. These results are in good agreement with XRD data and further confirm that the anatase structure of the TiO_2 support is retained. However, preparing the mixture of Ce and TiO$_2$ with the sol-gel technique creates several new bands at 162, 191, 270, 332, 371, 571, and 641 cm^{-1}, which can be assigned to CeTi$_2$O$_6$ [47,48]. The formation of such Ce-Ti phases is probably responsible for the improved catalyst performance. The band at 463 cm^{-1} can be attributed to CeO$_2$ and/or CeTi$_2$O$_6$ phase. No bands for Nb$_2$O$_5$ were observed in all of the catalysts, suggesting the high dispersion of Nb species on the surface of Ce40/Ti100. Additionally, it is noticeable for the SG catalyst that the band intensities at 193, 332, 368, 571, and 640 cm^{-1} are lower compared to the WI sample, which proves that the addition of Niobium decreases crystallinity, thereby confirming the XRD results.

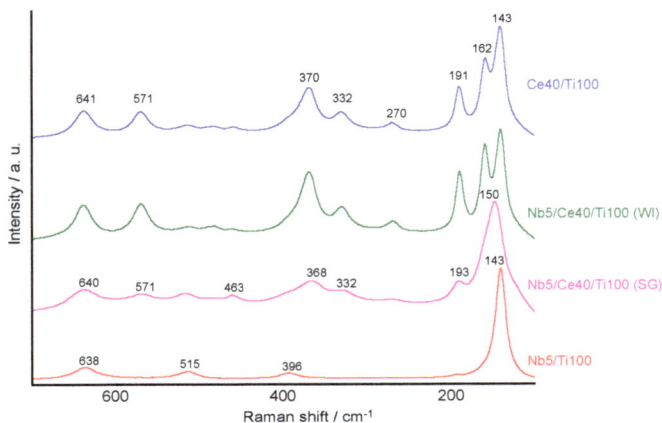

Figure 5. Raman spectra of all fresh samples.

2.2.2. Acid Properties (NH$_3$-TPD and DRIFTS)

NH$_3$-TPD runs were carried out to analyse the amount and strength of the acid sites of the materials. These parameters correlate with the position and area of the desorption peak, respectively [49]. As shown in Figure 6, all of the catalysts possess one broad peak ranging from 150 to 500 °C with a maximum around 300 °C. TiO$_2$ has the highest number of acidic centers (605.7 µmol/g) but they are weak ones (Table 3). Upon the addition of Nb to TiO$_2$ (WI), the amount of acidity decreased to 556.7 µmol/g due to the Nb coverage of the TiO$_2$ surface. However, a shoulder is observed at 410 °C, indicating the generation of stronger acidic sites. Furthermore, a comparison of both WI and SG samples reveals that in the WI sample, the shoulder is shifted to 385 °C and for the SG sample to 320 °C, and the overall acidity is higher (212 µmol/g) than in SG (146.2 µmol/g). In the case of the WI sample, Nb is mostly located on the surface and provides higher and stronger acidic sites than the SG sample, where a larger fraction of the Nb is expected to be located in the bulk. According to Table 3, the total amount of acidity for these catalysts follows the sequence TiO$_2$ > Nb5/Ti100 > Ce40/Ti100 > Nb5/Ce40/Ti100 (WI) > Nb5/Ce40/Ti100 (SG).

In situ DRIFT spectroscopy was applied to the two catalysts prepared with SG and WI methods, as well as with Ce40/Ti100 to study the interaction with the feed components NO/O$_2$ and NH$_3$. The spectra obtained after exposure to NO/O$_2$/He at 200 °C and flushing with He are depicted in Figure 7a. On all samples, a couple of distinct bands were observed and assigned to adsorbed NO$_2$ (1606–1616 cm^{-1}), bidentate nitrate (1570 and 1220 cm^{-1}), and monodentate nitrate (around 1550 and 1370 cm^{-1}) species [50,51]. All samples have in common that the bidentate nitrate species are only stable under NO/O$_2$ flow and vanish after He flushing. The band intensities of the remaining monodentate nitrate species on the SG and WI samples differ from those observed for the Ce40/Ti100 sample. This suggests that a part of the adsorption sites available on the latter are blocked by doping with Nb, which is more pronounced for the WI sample. While under NO/O_2 flow, significantly more nitrate species are formed on SG compared to the WI sample, the amount of remaining adsorbates after He flushing is comparable for both catalysts. A small moiety of NO_2 is more strongly adsorbed on the SG sample and therefore also detectable after He flushing.

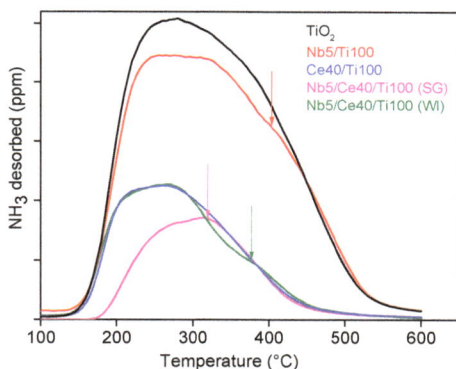

Figure 6. NH_3-TPD plots of supported Ceria-based catalysts and reference materials.

Table 3. Total amount of NH_3 desorbed in from the fresh samples.

Catalysts	Amount of NH_3 Desorbed (μmol/g)
TiO_2	605.7
Nb5/Ti100	556.7
Ce40/Ti100	213.4
Nb5/Ce40/Ti100 (SG)	146.2
Nb5/Ce40/Ti100 (WI)	212.7

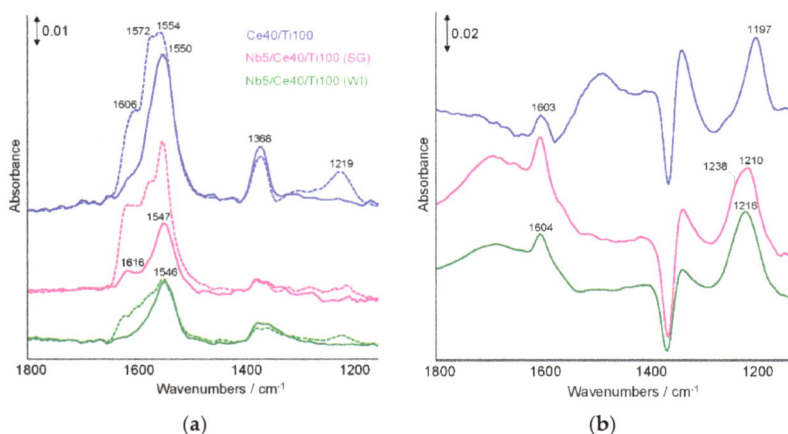

(a)

(b)

Figure 7. In situ DRIFT spectra of Ce40/Ti100, Nb5/Ce40/Ti100 (SG), and Nb5/Ce40/Ti100 (WI) measured at 200 °C after (**a**) NO/O_2/He adsorption (dashed lines) and subsequent flushing with He (solid lines) and (**b**) exposure to NH_3/He and subsequent flushing with He.

In another experiment, the catalysts were exposed to NH_3/He at 200 °C followed by flushing with He (Figure 7b). The bands at 1604 cm^{-1} and around 1200 cm^{-1} were assigned to asymmetric and symmetric bending vibrations of NH_3, respectively, coordinatively bound to Lewis acid sites [52]. Compared to the Ce40/Ti100 sample, the symmetric bending vibration (1197 cm^{-1}) is shifted to higher wavenumbers for the SG and WI samples (around 1216 cm^{-1}), while the band intensities slightly increase. This points to a change of the strength and amount of acidic sites caused by Nb doping. The intensities of the bands around 1216 cm^{-1} are similar for the SG and WI catalysts, indicating a comparable amount of Lewis sites. However, for the SG catalyst, the band positions slightly differ. Here, a shoulder at 1238 cm^{-1} is observed besides the main band at 1209 cm^{-1}, which suggests the appearance of Lewis sites with different strengths. Both experiments show that the introduction of Nb enhances the amount and strength of acid sites and therefore the ability for NH_3 activation, as well as for NO oxidation, which are beneficial for good NH_3-SCR activity in the temperature range below 250 °C.

2.2.3. Redox Properties (XPS and H_2-TPR)

It is well known that the surface composition and oxidation states of catalysts are very important for the NH_3-SCR reaction, as they can remarkably influence the adsorption and activation of reactant molecules. Therefore, the surface properties of these catalysts were investigated by XPS and the deconvoluted spectra are shown in Figure 8 and the atomic ratio percentage of Ce, Nb, and oxygen species are listed in Table 4.

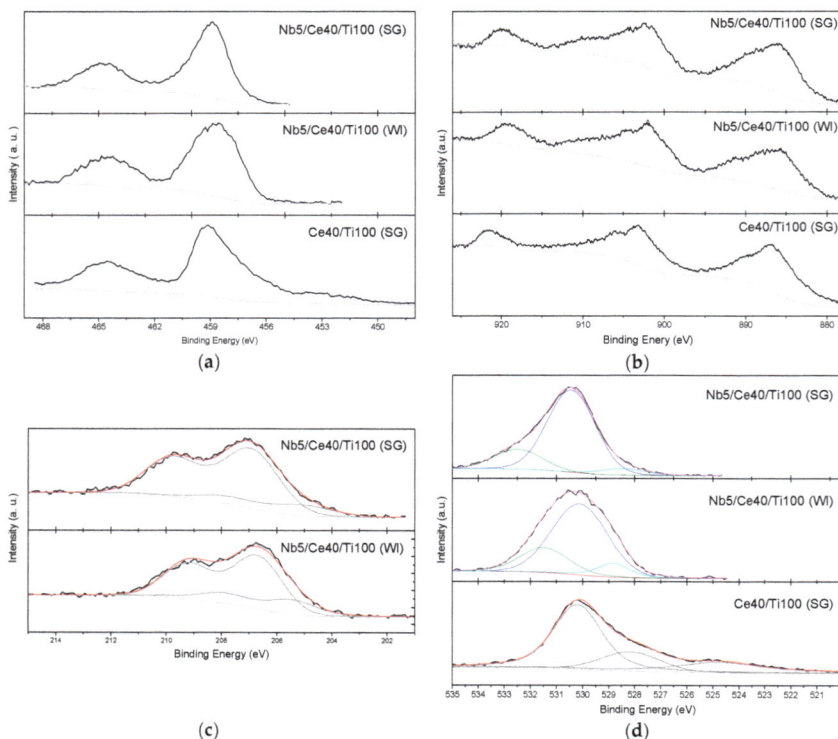

Figure 8. XPS data of the Ceria-containing catalysts: (**a**) Ti 2p, (**b**) Ce 3d, (**c**) Nb 3d, and (**d**) O 1s.

Table 4. Surface composition and element ratios of the supported Ceria-containing catalysts calculated from XPS.

Catalysts	Concentration (atom%)				Atomic Ratio (%)		
	Ce	Nb	Ti	O	Nb/(Ce + Ti)	Ce/(Ce + Ti)	$O_\alpha/(O_\alpha + O_\beta)$
Nb5/Ce40/Ti100 (WI)	5.2	3.5	16.7	74.7	15.9	23.6	24.6
Nb5/Ce40/Ti100 (SG)	7.7	3.2	17.1	72.0	12.7	31.2	19.6
Ce40/Ti100 (SG)	9.8	-	21.1	69.1	-	31.6	-

The Ti 2p region is dominated by two main peaks around 459 and 464 eV due to $2p_{3/2}$ and $2p_{1/2}$ contribution (Figure 8a). The spectra exhibit different symmetries and full widths at half maximum, reflecting the different local environments of Ti atoms according to the preparation method used.

The Ce 3d XPS spectra (Figure 8b) exhibit a very broad signal in the region with binding energies between 880 and 917 eV. The XPS spectra for fully oxidized Ce^{4+} and fully reduced Ce^{3+} reveal six respective four peaks at similar energy values. In the literature, the quantitative estimation of relative amounts of Ce^{4+} and Ce^{3+} is reported by fitting a linear combination of reference materials. But in the present study, the peaks are too broad for reliable deconvolution [53].

Nb 3d XPS spectra are dominated by two contributions around 206.9 and 209.7 eV that can be ascribed to Nb $3d_{5/2}$ and Nb $3d_{3/2}$, respectively (Figure 8c). With respect to the $3d_{5/2}$ peak, a second contribution at a lower binding energy was detected. The energy shift is 1.9 eV for the SG sample and 1.3 eV for the WI sample, testifying to the role of different preparation methods on Nb electronic structure.

The O 1s spectra show a main signal for Ce40/Ti100 in all the samples with a maximum in the range of 530.1–530.4 eV (Figure 8d), which could be caused by lattice oxygen species named as O_β. The other signal at 528.1–528.9 eV was assigned to the oxygen ions in CeO_2 [37,54]. The signals at 531.5 and 532.1 eV for Nb5/Ce40Ti100 (WI) and (SG), respectively, are denoted as O_α, representing chemisorbed oxygen fixed on defective sites such as oxygen vacancies. The relative ratio of $O_\alpha/(O_\alpha + O_\beta)$ is quantified based on the areas of O_α and O_β (Table 4). Accordingly, the amount of chemisorbed oxygen was higher in the WI sample (24.6%) than in the SG sample (19.6%). This could be due to the formation of Ce^{3+} as a result of the Ce-O-Nb unit, which forms charge imbalance and oxygen vacancies. These sites can promote the oxidation of NO to NO_2, resulting in high SCR activity. That is the reason why the WI sample showed more activity compared to the SG sample. It must be noted that this activation effect of Nb is only relevant for the oxygen atoms in Ceria next to Nb [29,37]. This can also explain why upon the addition of Nb via WI preparation, the amount of reducible surface species is increased, as will be discussed later by means of TPR results.

Since the redox properties are crucial for the activity of the NH_3-SCR catalyst, H_2-TPR was carried out to compare the reducibility of the prepared catalysts (Figure 9). The H_2-TPR profiles of all Ce, Ti, and Nb containing catalysts exhibit three reduction peaks between 400 and 900 °C. The low-temperature peak around 500 °C and the high-temperature peak around 800 °C can be attributed to the reduction of surface dispersed CeO_2 and bulk-like CeO_2, respectively [55,56]. Previous research [37,42] indicated that NbO_x reduction partly overlaps with CeO_2 reduction and therefore it is hard to distinguish the reduction peaks of NbO_x at high temperatures. The mid-reduction peak at 584 °C may be assigned to the reduction of Ce-O-Ti species, which form via the interaction of CeO_2 and TiO_2. Comparing Ce40Ti100 (SG) with both Nb containing catalysts suggests that Nb addition increases the reduction temperature. On the other hand, the catalyst Nb/Ce40/Ti100 (WI) shows enhanced reduction of the surface Ce^{4+} species as the related peak intensity in the impregnated sample is highest. This indicates that a synergetic effect between Ce and Nb increases the amount of oxygen vacancies due to the charge imbalance between Nb^{5+} and Ce^{4+} on Nb–O–Ce interfaces, as reported in literature [30,42], generating more active oxygen for the SCR reaction, as has been discussed by means of XPS data above.

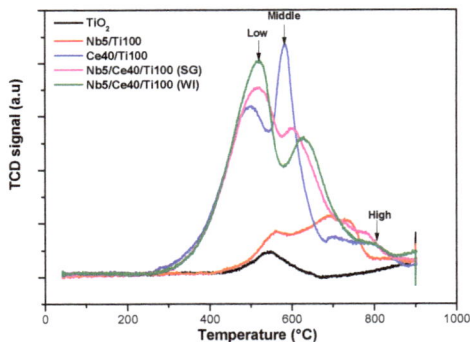

Figure 9. H_2-TPR profiles of the fresh samples.

The H_2 consumption of these supported Ceria-based catalysts is summarized in Table 5. The actual H_2 consumption of the Ce40/Ti100, Nb5/Ce40/Ti100 (SG), and (WI) catalysts is slightly higher than the theoretical H_2 consumption. It is possible that part of TiO_2 is also reduced to form Ti^{3+}. All the above features are considered beneficial for the excellent SCR activity.

Table 5. H_2 consumption calculated from temperature-programmed reduction (TPR).

Catalysts	H_2 Consumption (mmol·g^{-1})	
	Experimental	Theoretical
Ce40/Ti100	1548	1401
Nb5/Ce40/Ti100 (SG)	1509	1343
Nb5/Ce40/Ti100 (WI)	1455	1319

3. Materials and Methods

3.1. Catalyst Preparation

All the chemicals used in this study were supplied by Sigma-Aldrich Chemie GmbH, Munich, Germany with purities of 98–99%. Pure TiO_2 support was synthesized by a sol-gel method using Ti(IV) isopropoxide and anhydrous ethanol as the precursor. Next, ethyl acetoacetate (EAcAc) used as a complexing agent was added to the solution under stirring at room temperature. Finally, a solution of HNO_3 (0.1 M) was added drop-wise and stirred until a yellow gel was formed. The gel was extracted with supercritical ethanol in an autoclave (T = 245 °C, p = 60 bar). The resulting product was calcined under O_2 flow at 500 °C for 3 h (heating rate 3 K/min).

The Ce40/Ti100 sample (molar ratio 0.4:1) was prepared by adding $(NH_4)_2Ce(NO_3)_6$ dissolved in anhydrous ethanol to the above Ti precursor solution at 60 °C until complete dissolution. For preparation of Nb5/Ce40/Ti100 (SG), $NbCl_5$ dissolved in anhydrous ethanol was added. Further processing was conducted as described for TiO_2.

The Nb5/Ti100 (WI) and Nb5/Ce40/Ti100 (WI) catalysts were synthesized by impregnating the as-received pure TiO_2 and Ce40/Ti100 powders, respectively, with $NbCl_5$ and anhydrous ethanol. The nominal loadings of Nb were 5 wt %.

3.2. Activity Measurement

The NH_3-SCR activity measurement was carried out in a fixed-bed quartz reactor (i. d. = 10 mm, length = 150 mm) from 150 to 400 °C at atmospheric pressure with 100 mg of catalyst (sieved in the range 315–710 μm). The feed gas consisted of 1000 ppm of NO, 1000 ppm of NH_3, 5 vol % of O_2

(8 vol % of water if used), and helium as the balance at a 100 mL/min total flow rate. The modified space velocity (GHSV) for activity measurement was kept at 60,000 mL·g^{-1}·h^{-1}. Before each test, the catalyst was pretreated with 5% O_2 in helium at 500 °C for 30 min. The analysis of N_2, N_2O, and O_2 in the product mixture was performed using on-line gas chromatography (HP 6890) employing a molecular sieve 5A column. Simultaneously, NO, NO_2, and NH_3 concentrations in the product stream were continuously monitored by a multigas sensor (Limas 11HW, ABB, Mannheim, Germany). The concentrations of NO, N_2O, and NO_2 were measured in ppm. The continuous reactor was kept for 1 h in steady state at each chosen reaction temperature. The NH_3-SCR activity (NO_x conversion) of catalysts was calculated using Equation (1) and the N_2 selectivity was calculated using Equation (2):

$$NOx_{Conv.}\% = \frac{(NO + NO_2)_{in} - (NO + NO_2)_{out}}{(NO + NO_2)_{in}} \times 100 \tag{1}$$

$$S_{N_2}\% = 1 - \frac{2 * N_2O}{\left(((NO + NO_2)_{in} - (NO + NO_2)_{out}) + \left(NH_{3_{in}} - NH_{3_{out}}\right)\right)} \times 100 \tag{2}$$

As all nitrogen-containing gaseous reactants were quantified, it was possible to calculate the nitrogen balance. Typically, the values ranged from 94 to 100%. To investigate the long-term stability and the effect of water (8 vol %) on the catalyst performance, the temperature was set to 200 °C.

3.3. Catalyst Characterization

The elemental compositions were determined by inductively coupled plasma optical emission spectroscopy (ICP-OES) using a 715-ES ICP emissions spectrometer (Varian, Palo Alto, CA, USA). The samples were digested in a mixture of HF and aqua regia and then treated in a microwave assisted sample preparation apparatus at 200 °C and 60 bar.

BET surface area and pore volume of the prepared catalysts were measured from nitrogen adsorption isotherms measured at −196 °C (Micromeritics ASAP 2010). Before the measurement, each sample was degassed at 200 °C for 4 h. The average pore diameters were calculated from the desorption branch of the isotherm using the BJH method.

XRD powder patterns were recorded on a X'Pert diffractometer (Panalytical, Almelo, The Netherlands) equipped with a Xcelerator detector, used with automatic divergence slits and Cu kα1/α2 radiation (40 kV, 40 mA; λ = 0.015406 nm, 0.0154443 nm). Cu beta-radiation was excluded by using nickel filter foil. The measurements were performed in 0.0167° steps and 25 s of data collecting time per step. The samples were mounted on silicon zero background holders. The collected data were converted from automatic divergence slits to fixed divergence slits (0.25°) before data analysis to obtain the correct intensities. Peak positions and profile were fitted with the Pseudo-Voigt function using the HighScore Plus software package (Panalytical, Almelo, The Netherlands). Phase identification was done by using the PDF-2 database of the International Center of Diffraction Data (ICDD). Crystallite size was calculated by applying the Scherrer equation using the integral breadth under the assumption of spherically shaped crystallites. K was set to 1.0747.

The Raman spectra of the samples were collected on a in Via Raman microscope (Renishaw, Wotton-under-edge, UK) using a 633 nm laser with a laser power between 0.81 and 1.61 mW. The samples were mounted onto object slides and an objective with a magnification of 50 was applied. To prove the homogeneity of the materials, spectra were acquired at different points of the sample.

X-ray photoelectron spectra were performed with a Thermo ESCALAB 220 iXL spectrometer (ThermoFisher, Walham, MA, USA) at room temperature using monochromatic AlKα radiation. Binding energies were corrected to C-C contribution at 284.8 eV in the C1s region. Signal intensities were normalized using the sensitivity factors of Scofield [57] and the transmission function of the spectrometer.

In situ DRIFT spectra were measured on a Nicolet 6700 FTIR spectrometer using a high-temperature reaction cell (Harrick) equipped with a temperature programmer (Eurotherm) and connected to a gas-dosing device with mass flow controllers (Bronkhorst). The catalyst samples were pretreated for 1 h at 300 °C in air, cooled to a reaction temperature of 200 °C in He, and subsequently exposed to a flow (30 mL/min) of 1000 ppm NO, 4% O_2, balanced with He or 1000 ppm NH_3, balanced with He, respectively, for 45 min. Then, the cell was flushed with He for 20 min. Generally, subtracted spectra were evaluated, obtained by subtraction of the spectrum recorded after pretreatment and He flushing at reaction temperature from the respected adsorbate spectra.

TPR measurement was done using a Autochem II 2920 instrument (Micromeritics, Aachen, Germany). A 70 mg sample was loaded into a U-shaped quartz reactor and heated from RT to 400 °C with 10 K/min in 5% O_2/He (50 mL/min) for 30 min at 400 °C to remove any adsorbed species. Then, the system cooled down to RT in a flow of 5% O_2/He. The TPR run was carried out from RT to 900 °C in a 5% H_2/Ar flow (50 mL/min) with a heating rate of 10 K/min. The hydrogen consumption was continuously recorded using a thermal conductivity detector and quantitatively analyzed.

Temperature-programmed desorption of ammonia (TPD-NH_3) was performed by inserting a 230 mg sample into a home-made quartz-tube reactor. The samples were flushed in He (100 mL/min) for 1 h at 100 °C. Then, the sample was exposed to 1000 ppm NH_3/He (100 mL/min) for 1 h at 100 °C. Afterwards, physisorbed ammonia was removed by flushing the sample with He (100 mL/min) for 1 h at 100 °C. The sample was then ramped to 600 °C at a heating rate of 10 K/min in He (100 mL/min). The temperature was held at 600 °C for 30 min. The amount of ammonia desorbed with temperature was continuously monitored by a multigas sensor (Limas 11HW, ABB, Mannheim, Germany). The amount of acidity was calculated based on peak area integration.

4. Conclusions

The combination of three metal oxides (CeO_2, TiO_2, and Nb_2O_5) was chosen as the catalyst system for the SCR reaction and the effects of structure (core-shell vs bulk catalyst) and Nb doping on the catalytic activity were studied. Nb doped Ce40/Ti100 catalysts prepared by two methods WI and SG were used in the NH_3-SCR reaction. Nb5/Ce40/Ti100 (WI) showed the highest SCR activity (95%) at 200 °C in the absence of H_2O at GHSV = 60,000 h^{-1}. To our knowledge, this is among the best results for catalysts based on this ternary system for an intermediate temperature SCR. The stability of the catalysts was tested in the presence of H_2O, and the catalysts showed a loss of 22–27% in conversion. Nb5/Ce40/Ti100 (WI) exhibited higher conversions compared to the SG sample and showed a less pronounced deactivation over 12 h. It is noteworthy that at temperatures below 250 °C, byproduct formation is completely suppressed, which is advantageous compared to Mn-based systems. The latter tend to over-oxidize NO. On the other side, the Nb/Ce/Ti system could be an alternative to the V_2O_5/TiO_2 system, due to higher activity, particularly if it would be able to avoid the oxidation of SO_2 to SO_3 (as known for V_2O_5/TiO_2). The catalyst might be preferably applicable for intermediate temperature SCR. The results of BET, XRD, Raman, XPS, H_2-TPR, in situ DRIFT, and TPD-NH_3 studies clearly evidenced that the addition of Nb via wet impregnation leads to a high Nb concentration on the surface, which has a prominent effect on acidity, as well as the amount of adsorbed oxygen, with the latter resulting in a good redox ability. Second, the addition of Niobium improved the dispersion of Cerium, and third the coexistence of Nb and Ce/Ti species was of essential importance for a high NH_3-SCR performance. These are the key factors for the good performance of the Nb-modified Ce-based catalyst in SCR. This means that both the preparation method (core-shell versus bulk catalyst) and metal addition are essential for generating a good catalyst which is active and selective.

Author Contributions: J.M., H.A., M.M., and U.A. conceived and designed the experiments; J.M. and R.E. performed the catalyst tests and analyzed the data; H.A. performed the TPR and NH_3-TPD analyses and analyzed the data; R.E. performed the BET measurements; H.L. performed the XRD measurements; G.A. performed the XPS measurements; N.R., U.B., and S.K. performed the RAMAN and DRIFTS experiments and analyzed the data; J.M., H.A., and U.A. wrote the paper.

Catalysts **2018**, *8*, 175

Acknowledgments: The authors are grateful to Karin Struve for ICP measurements and acknowledge the staff of LIKAT and LCMC for all the help in discussing the results. J.M. acknowledges the staff of LIKAT and LCMC for all the efforts and help they have given. The publication of this article was funded by the Open Access Fund of the Leibniz Association.

Conflicts of Interest: The authors declare no conflict of interest.

References

1. Skalska, K.; S Miller, J.; Ledakowicz, S. Trends in NO$_x$ abatement: A review. *Sci. Total Environ.* **2010**, *408*, 3976–3989. [CrossRef] [PubMed]
2. Busca, G.; Lietti, L.; Ramis, G.; Berti, F. Chemical and mechanistic aspects of the selective catalytic reduction of NO$_x$ by ammonia over oxide catalysts: A review. *Appl. Catal. B Environ.* **1998**, *18*, 1–36. [CrossRef]
3. Lietti, L.; Nova, I.; Forzatti, P. Selective catalytic reduction (SCR) of NO by NH$_3$ over TiO$_2$-supported V$_2$O$_5$-WO$_3$ and V$_2$O$_5$-MoO$_3$ catalysts. *Top. Catal.* **2000**, *11*, 111–122. [CrossRef]
4. Long, R.Q.; Yang, R.T. Fe-ZSM-5 for Selective Catalytic Reduction of NO with NH$_3$: A Comparative Study of Different Preparation Techniques. *Catal. Lett.* **2001**, *74*, 201–205. [CrossRef]
5. Delahay, G.; Valade, D.; Guzmán-Vargas, A.; Coq, B. Selective catalytic reduction of nitric oxide with ammonia on Fe-ZSM-5 catalysts prepared by different methods. *Appl. Catal. B Environ.* **2005**, *55*, 149–155. [CrossRef]
6. Ayari, F.; Mhamdi, M.; Álvarez-Rodríguez, J.; Ruiz, A.R.G.; Delahay, G.; Ghorbel, A. Selective catalytic reduction of NO with NH$_3$ over Cr-ZSM-5 catalysts: General characterization and catalysts screening. *Appl. Catal. B Environ.* **2013**, *134–135*, 367–380. [CrossRef]
7. Forzatti, P. Present status and perspectives in de-NO$_x$ SCR catalysis. *Appl. Catal. A Gen.* **2001**, *222*, 221–236. [CrossRef]
8. Michalow-Mauke, K.A.; Lu, Y.; Kowalski, K.; Graule, T.; Nachtegaal, M.; Kröcher, O.; Ferri, D. Flame-Made WO$_3$/CeO$_x$-TiO$_2$ Catalysts for Selective Catalytic Reduction of NO$_x$ by NH$_3$. *ACS Catal.* **2015**, *5*, 5657–5672. [CrossRef]
9. Ding, S.; Liu, F.; Shi, X.; Liu, K.; Lian, Z.; Xie, L.; He, H. Significant Promotion Effect of Mo Additive on a Novel Ce-Zr Mixed Oxide Catalyst for the Selective Catalytic Reduction of NO$_x$ with NH$_3$. *ACS Appl. Mater. Interfaces* **2015**, *7*, 9497–9506. [CrossRef] [PubMed]
10. Liu, J.; Li, X.; Zhao, Q.; Ke, J.; Xiao, H.; Lv, X.; Liu, S.; Tadé, M.; Wang, S. Mechanistic investigation of the enhanced NH$_3$-SCR on cobalt-decorated Ce-Ti mixed oxide: In situ FTIR analysis for structure-activity correlation. *Appl. Catal. B Environ.* **2017**, *200*, 297–308. [CrossRef]
11. Liu, C.; Chen, L.; Li, J.; Ma, L.; Arandiyan, H.; Du, Y.; Xu, J.; Hao, J. Enhancement of Activity and Sulfur Resistance of CeO$_2$ Supported on TiO$_2$-SiO$_2$ for the Selective Catalytic Reduction of NO by NH$_3$. *Environ. Sci. Technol.* **2012**, *46*, 6182–6189. [CrossRef] [PubMed]
12. Krishna, K.; B F Seijger, G.; M van den Bleek, C.; Calis, H.P.A. Very active CeO$_2$-zeolite catalysts for NO$_x$ reduction with NH$_3$. *Chem. Commun.* **2002**, *18*, 2030–2031. [CrossRef]
13. Jin, R.; Liu, Y.; Wu, Z.; Wang, H.; Gu, T. Relationship between SO$_2$ poisoning effects and reaction temperature for selective catalytic reduction of NO over Mn–Ce/TiO$_2$ catalyst. *Catal. Today* **2010**, *153*, 84–89. [CrossRef]
14. Xu, W.; He, H.; Yu, Y. Deactivation of a Ce/TiO$_2$ Catalyst by SO$_2$ in the Selective Catalytic Reduction of NO by NH$_3$. *J. Phys. Chem. C* **2009**, *113*, 4426–4432. [CrossRef]
15. Xu, W.; Yu, Y.; Zhang, C.; He, H. Selective catalytic reduction of NO by NH$_3$ over a Ce/TiO$_2$ catalyst. *Catal. Commun.* **2008**, *9*, 1453–1457. [CrossRef]
16. Chen, L.; Li, J.; Ge, M.; Zhu, R. Enhanced activity of tungsten modified CeO$_2$/TiO$_2$ for selective catalytic reduction of NO$_x$ with ammonia. *Catal. Today* **2010**, *153*, 77–83. [CrossRef]
17. Peng, Y.; Li, K.; Li, J. Identification of the active sites on CeO$_2$–WO$_3$ catalysts for SCR of NO$_x$ with NH$_3$: An in situ IR and Raman spectroscopy study. *Appl. Catal. B Environ.* **2013**, *140–141*, 483–492. [CrossRef]
18. Ma, Z.; Weng, D.; Wu, X.; Si, Z. Effects of WO$_x$ modification on the activity, adsorption and redox properties of CeO$_2$ catalyst for NO$_x$ reduction with ammonia. *J. Environ. Sci.* **2012**, *24*, 1305–1316. [CrossRef]
19. Shan, W.; Liu, F.; He, H.; Shi, X.; Zhang, C. A superior Ce-W-Ti mixed oxide catalyst for the selective catalytic reduction of NO$_x$ with NH$_3$. *Appl. Catal. B Environ.* **2012**, *115–116*, 100–106. [CrossRef]
20. Shan, W.; Liu, F.; He, H.; Shi, X.; Zhang, C. Novel Cerium-tungsten mixed oxide catalyst for the selective catalytic reduction of NO$_x$ with NH$_3$. *Chem. Commun.* **2011**, *47*, 8046–8048. [CrossRef] [PubMed]

21. Chen, L.; Li, J.; Ge, M. DRIFT Study on Cerium-Tungsten/Titiania Catalyst for Selective Catalytic Reduction of NO_x with NH_3. *Environ. Sci. Technol.* **2010**, *44*, 9590–9596. [CrossRef] [PubMed]

22. Peng, Y.; Liu, Z.; Niu, X.; Zhou, L.; Fu, C.; Zhang, H.; Li, J.; Han, W. Manganese doped CeO_2-WO_3 catalysts for the selective catalytic reduction of NO_x with NH_3: An experimental and theoretical study. *Catal. Commun.* **2012**, *19*, 127–131. [CrossRef]

23. Gu, T.; Liu, Y.; Weng, X.; Wang, H.; Wu, Z. The enhanced performance of Ceria with surface sulfation for selective catalytic reduction of NO by NH_3. *Catal. Commun.* **2010**, *12*, 310–313. [CrossRef]

24. Si, Z.; Weng, D.; Wu, X.; Yang, J.; Wang, B. Modifications of CeO_2–ZrO_2 solid solutions by nickel and sulfate as catalysts for NO reduction with ammonia in excess O_2. *Catal. Commun.* **2010**, *11*, 1045–1048. [CrossRef]

25. Si, Z.; Weng, D.; Wu, X.; Ma, Z.; Ma, J.; Ran, R. Lattice oxygen mobility and acidity improvements of NiO–CeO_2–ZrO_2 catalyst by sulfation for NO_x reduction by ammonia. *Catal. Today* **2013**, *201*, 122–130. [CrossRef]

26. Casapu, M.; Kröcher, O.; Elsener, M. Screening of doped MnO_x-CeO_2 catalysts for low-temperature NO-SCR. *Appl. Catal. B Environ.* **2009**, *88*, 413–419. [CrossRef]

27. Casapu, M.; Kröcher, O.; Mehring, M.; Nachtegaal, M.; Borca, C.; Harfouche, M.; Grolimund, D. Characterization of Nb-Containing MnO_x-CeO_2 Catalyst for Low-Temperature Selective Catalytic Reduction of NO with NH_3. *J. Phys. Chem. C* **2010**, *114*, 9791–9801. [CrossRef]

28. Casapu, M.; Bernhard, A.; Peitz, D.; Mehring, M.; Elsener, M.; Kröcher, O. A Niobia-Ceria based multi-purpose catalyst for selective catalytic reduction of NO_x, urea hydrolysis and soot oxidation in diesel exhaust. *Appl. Catal. B Environ.* **2011**, *103*, 79–84. [CrossRef]

29. Qu, R.; Gao, X.; Cen, K.; Li, J. Relationship between structure and performance of a novel Cerium-Niobium binary oxide catalyst for selective catalytic reduction of NO with NH_3. *Appl. Catal. B Environ.* **2013**, *142–143*, 290–297. [CrossRef]

30. Ma, Z.; Weng, D.; Wu, X.; Si, Z.; Wang, B. A novel Nb-Ce/WO_x-TiO_2 catalyst with high NH_3-SCR activity and stability. *Catal. Commun.* **2012**, *27*, 97–100. [CrossRef]

31. Si, Z.; Weng, D.; Wu, X.; Ran, R.; Ma, Z. NH3-SCR activity, hydrothermal stability, sulfur resistance and regeneration of $Ce_{0.75}Zr_{0.25}O_2$-PO43-catalyst. *Catal. Commun.* **2012**, *17*, 146–149. [CrossRef]

32. Yu, J.; Si, Z.; Chen, L.; Wu, X.; Weng, D. Selective catalytic reduction of NO_x by ammonia over phosphate-containing $Ce_{0.75}Zr_{0.25}O_2$ solids. *Appl. Catal. B Environ.* **2015**, *163*, 223–232. [CrossRef]

33. Liu, Z.; Zhang, S.; Li, J.; Ma, L. Promoting effect of MoO_3 on the NO_x reduction by NH_3 over CeO_2/TiO_2 catalyst studied with in situ DRIFTS. *Appl. Catal. B Environ.* **2014**, *144*, 90–95. [CrossRef]

34. Gao, X.; Jiang, Y.; Fu, Y.; Zhong, Y.; Luo, Z.; Cen, K. Preparation and characterization of CeO_2/TiO_2 catalysts for selective catalytic reduction of NO with NH_3. *Catal. Commun.* **2010**, *11*, 465–469. [CrossRef]

35. Brayner, R.; Ciuparu, D.; da Cruz, G.M.; Fiévet-Vincent, F.; Bozon-Verduraz, F. Preparation and characterization of high surface area Niobia, Ceria–Niobia and Ceria–Zirconia. *Catal. Today* **2000**, *57*, 261–266. [CrossRef]

36. Yashiro, K.; Suzuki, T.; Kaimai, A.; Matsumoto, H.; Nigara, Y.; Kawada, T.; Mizusaki, J.; Sfeir, J.; Van herle, J. Electrical properties and defect structure of Niobia-doped Ceria. *Solid State Ion.* **2004**, *175*, 341–344. [CrossRef]

37. Zhao, B.; Ran, R.; Guo, X.; Cao, L.; Xu, T.; Chen, Z.; Wu, X.; Si, Z.; Weng, D. Nb-modified Mn/Ce/Ti catalyst for the selective catalytic reduction of NO with NH_3 at low temperature. *Appl. Catal. A Gen.* **2017**, *545*, 64–71. [CrossRef]

38. Qi, G.; Yang, R.T. A superior catalyst for low-temperature NO reduction with NH_3. *Chem. Commun.* **2003**, *7*, 848–849. [CrossRef]

39. Shen, Q.; Zhang, L.; Sun, N.; Wang, H.; Zhong, L.; He, C.; Wei, W.; Sun, Y. Hollow MnO_x-CeO_2 mixed oxides as highly efficient catalysts in NO oxidation. *Chem. Eng. J.* **2017**, *322*, 46–55. [CrossRef]

40. Lian, Z.; Liu, F.; He, H.; Liu, K. Nb-doped VO_x/CeO_2 catalyst for NH_3-SCR of NO_x at low temperatures. *RSC Adv.* **2015**, *5*, 37675–37681. [CrossRef]

41. Li, Y.; Li, Y.; Wan, Y.; Zhan, S.; Guan, Q.; Tian, Y. Structure-performance relationships of MnO_2 nanocatalyst for the low-temperature SCR removal of NO_x under ammonia. *RSC Adv.* **2016**, *6*, 54926–54937. [CrossRef]

42. Ma, Z.; Wu, X.; Si, Z.; Weng, D.; Ma, J.; Xu, T. Impacts of Niobia loading on active sites and surface acidity in NbO_x/CeO_2–ZrO_2 NH_3–SCR catalysts. *Appl. Catal. B Environ.* **2015**, *179*, 380–394. [CrossRef]

43. Langford, J.I.; Wilson, A.J.C. Scherrer after sixty years: A survey and some new results in the determination of crystallite size. *J. Appl. Crystallogr.* **1978**, *11*, 102–113. [CrossRef]

44. Yao, X.; Zhang, L.; Li, L.; Liu, L.; Cao, Y.; Dong, X.; Gao, F.; Deng, Y.; Tang, C.; Chen, Z.; et al. Investigation of the structure, acidity, and catalytic performance of $CuO/Ti_{0.95}Ce_{0.05}O_2$ catalyst for the selective catalytic reduction of NO by NH_3 at low temperature. *Appl. Catal. B Environ.* **2014**, *150–151*, 315–329. [CrossRef]

45. Liu, L.; Zhao, H.; Andino, J.M.; Li, Y. Photocatalytic CO_2 Reduction with H_2O on TiO_2 Nanocrystals: Comparison of Anatase, Rutile, and Brookite Polymorphs and Exploration of Surface Chemistry. *ACS Catal.* **2012**, *2*, 1817–1828. [CrossRef]

46. Balachandran, U.; Eror, N.G. Raman spectra of Titanium dioxide. *J. Solid State Chem.* **1982**, *42*, 276–282. [CrossRef]

47. Kong, L.; Gregg, D.J.; Karatchevtseva, I.; Zhang, Z.; Blackford, M.G.; Middleburgh, S.C.; Lumpkin, G.R.; Triani, G. Novel Chemical Synthesis and Characterization of $CeTi_2O_6$ Brannerite. *Inorg. Chem.* **2014**, *53*, 6761–6768. [CrossRef] [PubMed]

48. Fu, M.; Wei, L.; Li, Y.; Zhou, X.; Hao, S.; Li, Y. Surface charge tuning of Ceria particles by Titanium doping: Towards significantly improved polishing performance. *Solid State Sci.* **2009**, *11*, 2133–2137. [CrossRef]

49. Fang, C.; Zhang, D.; Cai, S.; Zhang, L.; Huang, L.; Li, H.; Maitarad, P.; Shi, L.; Gao, R.; Zhang, J. Low-temperature selective catalytic reduction of NO with NH_3 over nanoflaky MnO_x on carbon nanotubes in situ prepared via a chemical bath deposition route. *Nanoscale* **2013**, *5*, 9199–9207. [CrossRef] [PubMed]

50. Dines, T.J.; Rochester, C.H.; Ward, A.M. Infrared and Raman study of the adsorption of NH_3, pyridine, NO and NO_2 on anatase. *J. Chem. Soci. Faraday Trans.* **1991**, *87*, 643–651. [CrossRef]

51. Hadjiivanov, K.I. Identification of Neutral and Charged N_xO_y Surface Species by IR Spectroscopy. *Catal. Rev.* **2000**, *42*, 71–144. [CrossRef]

52. Davydov, A. The Nature of Oxide Surface Centers. In *Molecular Spectroscopy of Oxide Catalyst Surfaces*; John Wiley & Sons: Hoboken, NJ, USA, 2003.

53. Mullins, D.R. The surface chemistry of Cerium oxide. *Surf. Sci. Rep.* **2015**, *70*, 42–85. [CrossRef]

54. Reddy, B.M.; Khan, A. Nanosized CeO_2-SiO_2, CeO_2-TiO_2, and CeO_2-ZrO_2 Mixed Oxides: Influence of Supporting Oxide on Thermal Stability and Oxygen Storage Properties of Ceria. *Catal. Surv. Asia* **2005**, *9*, 155–171. [CrossRef]

55. Li, P.; Xin, Y.; Li, Q.; Wang, Z.; Zhang, Z.; Zheng, L. Ce-Ti Amorphous Oxides for Selective Catalytic Reduction of NO with NH3: Confirmation of Ce-O-Ti Active Sites. *Environ. Sci. Technol.* **2012**, *46*, 9600–9605. [CrossRef] [PubMed]

56. Neri, G.; Pistone, A.; Milone, C.; Galvagno, S. Wet air oxidation of p-coumaric acid over promoted Ceria catalysts. *Appl. Catal. B Environ.* **2002**, *38*, 321–329. [CrossRef]

57. Scofield, J.H. Hartree-Slater subshell photoionization cross-sections at 1254 and 1487 eV. *J. Electron Spectrosc. Relat. Phenom.* **1976**, *8*, 129–137. [CrossRef]

catalysts

MDPI

Review

Sulfur and Water Resistance of Mn-Based Catalysts for Low-Temperature Selective Catalytic Reduction of NO$_x$: A Review

Chen Gao, Jian-Wen Shi *, Zhaoyang Fan, Ge Gao and Chunming Niu *

Center of Nanomaterials for Renewable Energy, State Key Laboratory of Electrical Insulation and Power Equipment, School of Electrical Engineering, Xi'an Jiaotong University, Xi'an 710049, China; rick.gcgc@stu.xjtu.edu.cn (C.G.); z.y.fan@stu.xjtu.edu.cn (Z.F.); gaoge5022@stu.xjtu.edu.cn (G.G.)
* Correspondence: jianwen.shi@mail.xjtu.edu.cn (J.-W.S.); cniu@mail.xjtu.edu.cn (C.N.)

Received: 6 December 2017; Accepted: 3 January 2018; Published: 7 January 2018

Abstract: Selective catalytic reduction (SCR) with NH$_3$ is the most efficient and economic flue gas denitrification technology developed to date. Due to its high low-temperature catalytic activity, Mn-based catalysts present a great prospect for application in SCR de-NO$_x$ at low temperatures. However, overcoming the poor resistance of Mn-based catalysts to H$_2$O and SO$_2$ poison is still a challenge. This paper reviews the recent progress on the H$_2$O and SO$_2$ resistance of Mn-based catalysts for the low-temperature SCR of NO$_x$. Firstly, the poison mechanisms of H$_2$O and SO$_2$ are introduced in detail, respectively. Secondly, Mn-based catalysts are divided into three categories—single MnO$_x$ catalysts, Mn-based multi-metal oxide catalysts, and Mn-based supported catalysts—to review the research progress of Mn-based catalysts for H$_2$O and SO$_2$ resistance. Thirdly, several strategies to reduce the poisonous effects of H$_2$O and SO$_2$, such as metal modification, proper support, the combination of metal modification and support, the rational design of structure and morphology, are summarized. Finally, perspectives and future directions of Mn-based catalysts for the low-temperature SCR of NO$_x$ are proposed.

Keywords: selective catalytic reduction; Mn-based catalysts; H$_2$O and SO$_2$ resistance; low-temperature; de-NO$_x$

1. Introduction

Nitrogen oxides (NO$_x$, x = 1,2) emitted from power plants and diesel engines are major air pollutants that can cause acid rain, photochemical smog, ozone depletion, and other severe environmental problems [1–5]. Selective catalytic reduction (SCR) with NH$_3$ is the most efficient and economic method for post-NO$_x$ abatement, and V$_2$O$_5$–WO$_3$(MoO$_3$)/TiO$_2$ has been the most popular commercial SCR catalyst since the 1970s [6,7]. However, V$_2$O$_5$-based catalysts have drawbacks, such as the toxicity of vanadium, SO$_2$ oxidation to SO$_3$, over-oxidation of NH$_3$ to N$_2$O, and a high working temperature [8]. Because of the high working temperature window (300–400 °C), V$_2$O$_5$-based catalysts have to be placed upstream of the dust removal system and desulfurization units to avoid costly heating of the flue gas, where the catalysts are susceptible to deactivation by dust accumulation and SO$_2$ poison. Therefore, SCR catalysts that are environmentally friendly and can work at low temperatures (around 250 °C or even lower) urgently need to be developed [9–11].

Due to its high low-temperature catalytic activity, manganese oxide (MnO_x) has been intensively studied in recent decades [4,12–14]. Recently, our research group has also made a series of progress in the low-temperature SCR of NO with NH_3 over Mn-based catalysts [15–21]. However, several problems, including thermal instability, narrow operation window, and poor resistance to H_2O and SO_2 poison, remain. Among these drawbacks, the poor tolerance to H_2O and SO_2 is one of the most significant disadvantages, which limits the practical application of Mn-based catalysts [22]. Researchers have done numerous studies to develop Mn-based catalysts with good tolerance to water and sulfur and to uncover the deactivation mechanism of Mn-based catalysts in the presence of water and sulfur. In this review, we focused on the recent progress on the water and sulfur resistance of Mn-based catalysts. To make the organization clear, catalysts were introduced by the following three categories, single MnO_x catalysts, Mn-based multi-metal oxide catalysts and Mn-based supported catalysts. Table 1 summarizes Mn-based catalysts reported in the literature that have exhibited good performance in the presence of water and sulfur.

Table 1. Summary of the current status of H_2O and SO_2 tolerance study on Mn-based catalysts in the literature.

Catalyst	Reaction Conditions	T/°C	X_{NO}	X_{NO}-U	X_{NO}-A	References
MnOx	500 ppm NH₃, 500 ppm NO, 3% O₂, 10% H₂O, 100 ppm SO₂ GHSV at 47,000 h⁻¹	80	98%	70%	90%	[13]
MnOx	500 ppm NH₃, 500 ppm NO, 5% O₂, 11% H₂O, 100 ppm SO₂ GHSV at 50,000 h⁻¹	120	100%	94%	100%	[23]
Mn-Ce	1000 ppm NH₃, 1000 ppm NO, 2% O₂, 2.5% H₂O, 100 ppm SO₂ GHSV at 42,000 h⁻¹	120	100%	95%	100%	[24]
Mn-Ce	500 ppm NH₃, 500 ppm NO, 5% O₂, 5% H₂O, 50 ppm SO₂ GHSV at 64,000 h⁻¹	150	~98%	~95%	/	[25]
Mn-Ce	500 ppm NH₃, 500 ppm NO, 5% O₂, 5% H₂O, 100 ppm SO₂ 60,000 mL g⁻¹ h⁻¹	200	~97%	~70%	~85%	[26]
Mn-Fe	1000 ppm NH₃, 1000 ppm NO, 2% O₂, 2.5% H₂O, 37.5 ppm SO₂ GHSV at 15,000 h⁻¹	160	100%	~98%	/	[27]
Mn-Fe	1000 ppm NH₃, 1000 ppm NO, 3% O₂, 5% H₂O, 100 ppm SO₂ GHSV at 30,000 h⁻¹	120	100%	87%	93%	[28]
Mn-Co	500 ppm NH₃, 500 ppm NO, 3% O₂, 8% H₂O, 200 ppm SO₂ GHSV at 38,000 h⁻¹	175	100%	90%	100%	[29]
Mn-Co	500 ppm NH₃, 500 ppm NO, 5% O₂, 5% H₂O, 100 ppm SO₂ GHSV at 50,000 h⁻¹	200	100%	80%	~90%	[30,31]
Mn-Cu	500 ppm NH₃, 500 ppm NO, 5% O₂, 11% H₂O, 100 ppm SO₂ GHSV at 50,000 h⁻¹	125	95%	64%	~90%	[32]
Mn-Sm	500 ppm NH₃, 500 ppm NO, 5% O₂, 2% H₂O, 100 ppm SO₂ GHSV at 49,000 h⁻¹	100	100%	91%	97%	[33]
Mn-Eu	600 ppm NH₃, 600 ppm NO, 5% O₂, 5% H₂O, 100 ppm SO₂ GHSV at 108,000 h⁻¹	350	100%	90%	95%	[34]
Mn-Fe-Ce	1000 ppm NH₃, 1000 ppm NO, 2% O₂, 2.5% H₂O, 100 ppm SO₂ GHSV at 42,000 h⁻¹	150	98%	95%	98%	[35]
Mn-Ce-Fe	1000 ppm NH₃, 1000 ppm NO, 3% O₂, 10% H₂O, 100 ppm SO₂ GHSV at 30,000 h⁻¹	120	100%	75%	95%	[36]
Mn-Sn-Ce	1000 ppm NH₃, 1000 ppm NO, 2% O₂, 12% H₂O, 100 ppm SO₂ GHSV at 35,000 h⁻¹	110	100%	70%	90%	[37,38]
Mn-Ce-Ni	500 ppm NH₃, 500 ppm NO, 5% O₂, 10% H₂O, 150 ppm SO₂ GHSV at 48,000 h⁻¹	175	~90%	~78%	~90%	[39]
Mn-Ce-Co	500 ppm NH₃, 500 ppm NO, 5% O₂, 10% H₂O, 150 ppm SO₂ GHSV at 48,000 h⁻¹	175	~90%	~72%	~90%	[39]
Mn-W-Zr	500 ppm NH₃, 500 ppm NO 5% O₂, 5% H₂O, 50 ppm SO₂ GHSV at 128,000 h⁻¹	300	100%	~90%	100%	[40]
Mn-Fe/TiO₂	1000 ppm NH₃, 1000 ppm NO, 2% O₂, 2.5% H₂O, 100 ppm SO₂ GHSV at 15,000 h⁻¹	150	100%	90%	100%	[41]
Mn/Fe-TiO₂	500 ppm NH₃, 500 ppm NO, 2% O₂, 8% H₂O, 60 ppm SO₂ GHSV at 12,000 h⁻¹	200	100%	83%	100%	[42]
Mn-Ce/TiO₂	1000 ppm NH₃, 1000 ppm NO, 3% O₂, 3% H₂O, 100 ppm SO₂ GHSV at 30,000 h⁻¹	150	100%	84%	/	[9]
Mn-Fe-Ce/TiO₂	600 ppm NH₃, 600 ppm NO, 3% O₂, 3% H₂O, 100 ppm SO₂ GHSV at 50,000 h⁻¹	180	100%	84%	90%	[43]
Mn-Ce/Ti-PILC	600 ppm NH₃, 600 ppm NO, 3% O₂, 3% H₂O, 100 ppm SO₂ GHSV at 50,000 h⁻¹	200	~95%	~90%	~90%	[44]
Mn-Ce/TiO₂	220 ppm NH₃, 200 ppm NO, 8% O₂, 6% H₂O, 100 ppm SO₂ GHSV at 60,000 h⁻¹	180	100%	62%	70%	[45]
Mn-Ce/TiO₂	500 ppm NH₃, 500 ppm NO, 5% O₂, 5% H₂O, 50 ppm SO₂ GHSV at 64,000 h⁻¹	200	~95%	~90%	~93%	[46]
Ni–Mn/TiO₂	1000 ppm NH₃, 1000 ppm NO, 3% O₂, 15% H₂O, 100 ppm SO₂ GHSV at 40,000 h⁻¹	240	100%	~95%	100%	[47]
MnCe@CNTs	500 ppm NH₃, 500 ppm NO, 3% O₂, 4% H₂O, 100 ppm SO₂ GHSV at 10,000 h⁻¹	300	100%	87%	90%	[48]
Fe₂O₃@MnOx@CNTs	550 ppm NH₃, 550 ppm NO, 5% O₂, 10% H₂O, 100 ppm SO₂ GHSV at 20,000 h⁻¹	240	97%	91%	95%	[49]
Mn-Ce/TiO₂-graphene	500 ppm NH₃, 500 ppm NO, 7% O₂, 10% H₂O, 200 ppm SO₂ GHSV at 67,000 h⁻¹	180	95%	95%	100%	[50,51]
MnOx(0.6)/Ce₀.₅Zr₀.₅O	600 ppm NH₃, 600 ppm NO, 3% O₂, 3% H₂O, 200 ppm SO₂ GHSV at 30,000 h⁻¹	180	100%	~92%	~98%	[52]
WySnMnCeOx	500 ppm NH₃, 500 ppm NO, 5% O₂, 5% H₂O, 100 ppm SO₂ 60,000 mL g⁻¹ h⁻¹	200	~97%	~90%	~95%	[53]
MnO₂/3DOMC	1000 ppm NH₃, 1000 ppm NO, 5% O₂, 5% H₂O, 200 ppm SO₂ GHSV at 36,000 h⁻¹	190	100%	~87%	~95%	[54]
W₀.₂₅-Mn₀.₂₅-Ti₀.₅	1000 ppm NH₃, 1000 ppm NO, 5% O₂, 10% H₂O, 100 ppm SO₂ GHSV at 25,000 h⁻¹	/	~100%	~100%	/	[55]

X_{NO}, X_{NO}-U, and X_{NO}-A represent NOx conversion of regular SCR reaction, NOx conversion under tolerance test and after tolerance test, respectively.

2. The Poisoning Mechanism of Mn-Based Catalysts

2.1. The Poisoning Mechanism of H_2O

Water vapor has a negative effect on SCR reaction mainly because of the loss of available active sites on the surface of catalysts [43,56–58]. Even under dry conditions, the catalysts can be affected by the water vapor produced in the SCR reaction [56,59]. It is believed that the poisonous effects of H_2O can be generally divided into two aspects: reversible and irreversible deactivation. As shown in Figure 1, the competitive adsorption between H_2O and NH_3 (or NO) is generally considered as the cause of reversible deactivation. Less adsorption of reacting agents on the surface leads to a decrease in NO_x conversion. Fortunately, this effect generally disappears if H_2O vapors are removed [60]. The formation of additional surface hydroxyls (–OH) caused by dissociative adsorption and decomposition of H_2O on the catalyst surface is likely to be the reason of irreversible deactivation, and this effect can occur at a relatively low temperature (below 200 °C) [61]. Because of the good thermal stability of hydroxyls (in the 250–500 °C range), the NO_x conversion cannot be recovered even shutting the H_2O stream down at such a low temperature, thus resulting in an irreversible deactivation [62].

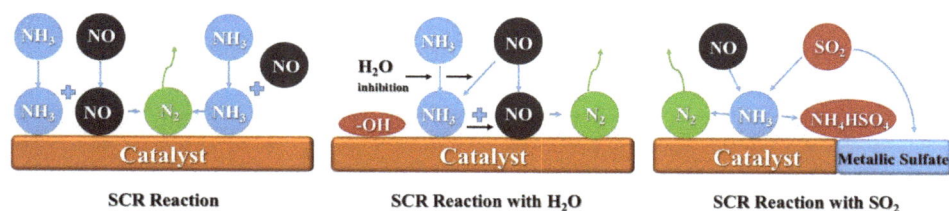

Figure 1. Scheme of the regular selective catalytic reduction (SCR) reaction, the H_2O poisoning effect, and the SO_2 poisoning effect.

2.2. The Poisoning Mechanism of SO_2

The presence of a significant amount of SO_2 in flue gas has a critical influence on the catalyst for SCR reaction at low temperatures. The poisonous effects of SO_2 can be generally classified in two categories: reversible and irreversible deactivation. For reversible deactivation, as displayed in Figure 1, SO_2 is easily oxidized to SO_3, which will easily react with NH_3 to generate ammonia sulfate. The ammonia sulfates (NH_4HSO_4 and $(NH_4)_2SO_4$) could cover on the active sites of catalysts and lead to a decrease in NO_x conversion [63–65]. In addition, the competitive adsorption between SO_2 and NO on the active sites of the catalysts also contributes to the poisoning effect of SO_2 on the SCR reaction [66]. However, the reversible effect can be eliminated by washing with water or acid solution, or high temperature treatment of catalyst. For the irreversible case, as illustrated in Figure 1, SO_2 (or SO_3) can directly react with active components and form metallic sulfate, which leads to surface active site loss. Hence, the conversion decreases. Due to the high thermal stability of metallic sulfate, washing with water or high temperature treatment cannot bring much recovery of NO_x conversion. When H_2O and SO_2 are introduced simultaneously, water will make the poisoning effect of SO_2 severer, leading to a great decrease in the NO_x conversion. Mn-based catalysts can work at low temperatures, which means that the SCR unit can be installed downstream of the dust removal system and desulfurization units. According to the Chinese Standard (GB13233-2011) and the EU Standard (BREF), residual SO_2 in flue gas after desulfurization (35–150 ppm depending on different fuels and desulfurization methods) is allowed. The remaining SO_2 and H_2O in flue gas still have inevitable effects [59,67]. Thus, developing Mn-based catalysts with good tolerance to water and sulfur is crucial for commercial applications.

2.3. The Effect on N_2 Selectivity

The N_2 selectivity is another indicator for the evaluation of SCR catalysts, which is closely related to the yield of N_2O. During NH_3–SCR reaction, N_2O can be produced together with N_2, especially at high temperatures. However, for Mn-based catalysts, some N_2O can also be formed even at low temperatures due to side reactions resulting from the oxidative properties of manganese oxides, whether the NH_3–SCR reaction follows the Eley–Rideal (E–R) mechanism or the Langmuir–Hinshelwood (L–H) mechanism. It has been reported that H_2O presents a positive effect on N_2 selectivity. Xiong et al. found that the formation of N_2O over the Mn–Fe spinel and MnO_x–CeO_2 catalysts following the E-R mechanism was notably restrained by H_2O due to the decrease in the oxidation ability of MnO_x, the suppression of NH_3 adsorption and the inhibition of the interface reaction. Furthermore, the generation of N_2O through the L–H mechanism was completely suppressed by H_2O due to the fact that the formation of NH_4NO_3 was inhibited or the decomposition of generated NH_4NO_3 was promoted [68,69]. As regards the effect of SO_2 on N_2 selectivity, there is a lack of research in this area at the moment.

3. Research Progress of Mn-Based Catalysts for Water and Sulfur Resistance

3.1. Single MnO_x Catalysts

It has been proven that pure MnO_x has terrific catalytic activity for the SCR of NO_x with NH_3 but poor resistance to the poison of water and sulfur at low temperatures [4,70–72]. It has been reported that several factors, such as the preparation method and the specific surface area, have a great influence on the tolerance of MnO_x. Tang et al. prepared a series of amorphous MnO_x catalysts using three methods, the solid phase reaction method (SP), the co-precipitation method (CP), and the rheological phase reaction method (RP) [13], and they found that the MnO_x (CP) exhibited the best sulfur and water resistance, but the MnO_x (SP) presented a larger surface area (150 m^2g^{-1} for MnO_x (SP) and 96 m^2g^{-1} for MnO_x (CP)). As shown in Figure 2, the NO_x conversion at 80 °C over MnO_x (CP) decreased from 98 to 73% in 3 h. After turning off SO_2 and H_2O, the NO_x conversion was quickly restored to 90%. After the MnO_x (CP) was heated for 1–2 h in N_2 at 280 °C, its activity was restored to the initial level. Kang et al. prepared two MnO_x catalysts using sodium carbonate (SC) and ammonia (AH) as precipitants [23]. They found that an MnO_x-SC catalyst showed better SCR activity and great sulfur and water tolerance, and they ascribed this to the larger surface area (173.3 m^2/g for MnO_x-SC and 18.7 m^2/g for MnO_x-AH). As displayed in Figure 3, the NO_x conversion over the MnO_x-SC catalyst was decreased from 100 to 94% after both SO_2 (100 ppm) and H_2O (11 vol %) were fed into the reaction system with aspace velocity of 50,000 h^{-1}, which is still very high de-NO_x activity at 120 °C. Moreover, its activity was rapidly recovered to 100% after the supply of SO_2 and H_2O was cut off.

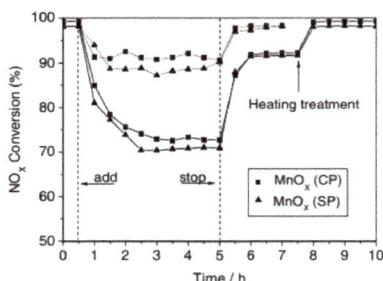

Figure 2. The effect of SO_2 and H_2O on NO_x conversion over MnO_x (CP) and MnO_x (SP) (dotted line: only added 10% H_2O; solid line: added 10% H_2O + 100 ppm SO_2). (Reproduced with permission from Reference [13], Copyright 2007, Elsevier).

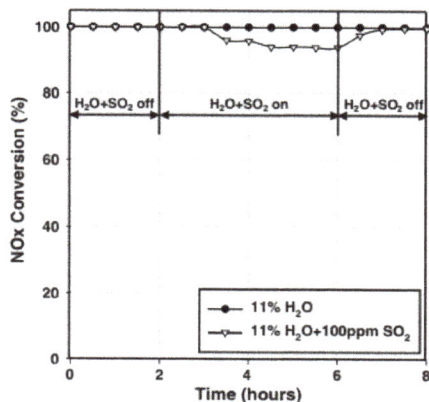

Figure 3. The effects of H_2O and SO_2 on NO_x conversions over MnO_x-SC catalyst at 120 °C. Reactants: 500 ppm NO, 500 ppm NH_3, 5 vol % O_2 in N_2. The gas hourly space velocity (GHSV) was 50,000 h^{-1}. (Reproduced with permission from Reference [23], Copyright 2006, Springer).

3.2. Mn-Based Multi-Metal Oxide Catalysts

It has been widely demonstrated that mixing or doping MnO_x with other metal oxides can greatly improve the water and sulfur resistance of single MnO_x catalysts because of the synergistic effect between them [73]. For Mn-based binary metal oxide catalysts, it has been reported that different dopants have different effects on the improvement of the tolerance to water and sulfur [74,75]. For Mn-based ternary metal oxide catalysts, the modification of a small amount of a third element can enhance the synergistic effect resulted from the changes in both electronic and structural properties.

3.2.1. Mn-Based Binary Metal Oxide Catalysts

Among the metal elements, cerium [24,25], chromium [10], iron [27,28,76], cobalt [29,30], copper [52], nickel [77], and several other elements have drawn the most attention as the mixture or dopant to construct binary metal oxide catalysts with MnO_x. CeO_2 has been studied extensively due to its good characteristics, such as increasing surface acidity after SO_2 poisoning [78,79], high surface area [80,81], good dispersion of MnO_x on the surface [45], and the redox shift between Ce^{4+} and Ce^{3+}. It should be noted that the shift between Ce^{4+} and Ce^{3+} can result in the formation of oxygen vacancies and anincrease in the chemisorbed oxygen on the surface of Mn–Ce binary metal oxide catalysts, which are helpful for the enhancement of water and sulfur resistance [25,82]. Qi and Yang [24] reported that the Mn–Ce catalyst with a proper mole ratio (Mn/(Mn+Ce) = 0.3) showed great tolerance to water and sulfur. As illustrated in Figure 4, the NO conversion over an Mn–Ce catalyst gradually decreased from 100 to 95% within 4 h after 100 ppm SO_2 and 2.5% H_2O were added to the reaction gas at 120 °C. Moreover, the NO conversion was restored after SO_2 + H_2O was stopped. Liu et al. prepared an Mn–Ce catalyst by the surfactant-template method using hexadecyltrimethyl ammonium bromide (CTAB) as the template. The obtained Mn_5Ce_5(ST) catalyst presented a noticeable decrease in the catalytic activity for the NO_x conversion at 100 °C in the presence of H_2O and SO_2 (Figure 5), a slight inhibiting effect was observed from 150 to 200 °C, and the promoting effect was exhibited above 200 °C [25]. Yao et al. successfully prepared a series of Mn/CeO_2 catalysts via impregnation using deionized water, anhydrous ethanol, acetic acid, and oxalic acid as a solvent and found that Mn/Ce–OA (oxalic acid) exhibited the best water and sulfur tolerance among all catalysts (Figure 6) [26]. Chen et al. [10] found that the SO_2 tolerance of MnO_x was dramatically enhanced by the introduction of Cr due to the formation of $CrMn_{1.5}O_4$.

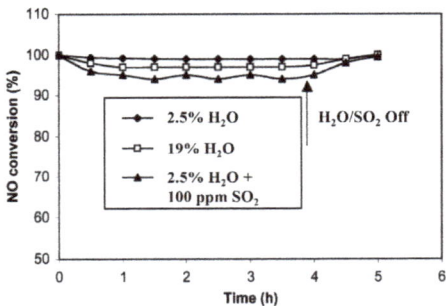

Figure 4. The effect of on-stream time on SCR activity with $H_2O + SO_2$ and without $H_2O + SO_2$ (Reaction conditions: 120 °C, $[NH_3] = [NO] = 1000$ ppm, $[O_2] = 2\%$, GHSV = 42,000 h^{-1}. Catalyst: $MnO_x(0.3)$–CeO_2). (Reproduced with permission from Reference [24], Copyright 2003, The Royal Society of Chemistry).

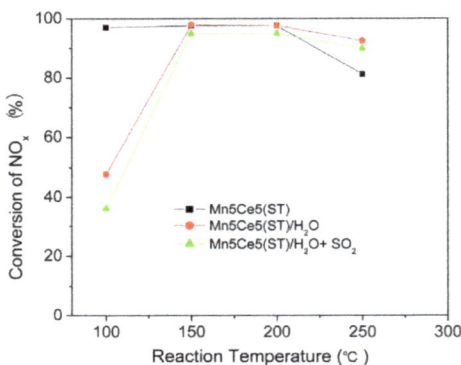

Figure 5. Catalytic performance of $Mn_5Ce_5(ST)$ catalysts in the presence of H_2O and SO_2 (Reaction conditions: NO = 500 ppm, NH_3 = 500 ppm, O_2 = 5%, H_2O = 5%, SO_2 = 50 ppm, GHSV = 64,000 h^{-1}). (Reproduced with permission from Reference [25], Copyright 2013, Elsevier).

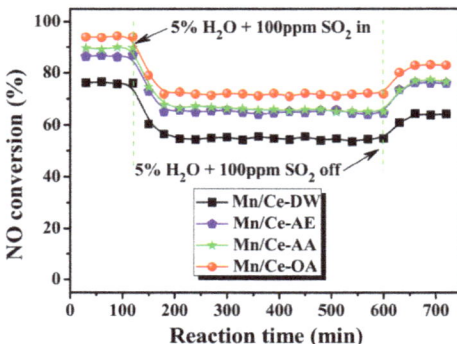

Figure 6. The effect of H_2O and SO_2 on NO conversion for Mn/CeO_2 catalysts (Reaction conditions: $[NO] = [NH_3] = 500$ ppm, $[O_2] = 5$ vol %, $[SO_2] = 100$ ppm, $[H_2O] = 5$ vol %, N_2 balance, T = 200 °C, 60,000 mL g^{-1} h^{-1}). (Reproduced with permission from Reference [26], Copyright 2017, Elsevier).

Iron is another potential element that has been demonstrated to play a positive role in the sulfur and water tolerance of Mn-based catalysts. Long et al. [27] observed that Fe–Mn-based transition metal oxides were resistant to H_2O and SO_2 at 140–180 °C (Figure 7). Chen et al. [28] found that the NO conversion over an Fe–Mn mixed oxide catalyst decreased slightly from 100 to 87% in 4 h at 120 °C in the presence of 5% H_2O and 100 ppm SO_2, which could be restored to 93% after the stopping of both SO_2 and H_2O (Figure 8). They attributed the enhanced resistance to the formed $Fe_3Mn_3O_8$ phase in Fe–Mn mixed oxides. Yang et al. [76] prepared an Mn–Fe spinel catalyst and found that the NO conversion over Mn–Fe spinel catalyst decreased from 100 to about 60% after the addition of H_2O and SO_2 for 100 min, and the NO conversion could be recovered to the original level after washing catalyst with water.

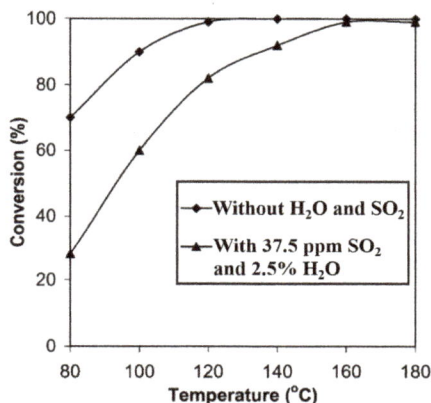

Figure 7. SCR activities on the Fe–Mn-based transition metal oxides in the presence of SO_2 + H_2O (Reaction conditions: 0.5 g catalyst, [NO] = [NH_3] = 1000 ppm, [O_2] = 2%, [SO_2] = 37.5 ppm and [H_2O] = 2.5% (when used), He = balance, total flow rate = 100 mL/min, and GHSV = 15,000 h^{-1}). (Reproduced with permission from Reference [27], Copyright 2002, The Royal Society of Chemistry.).

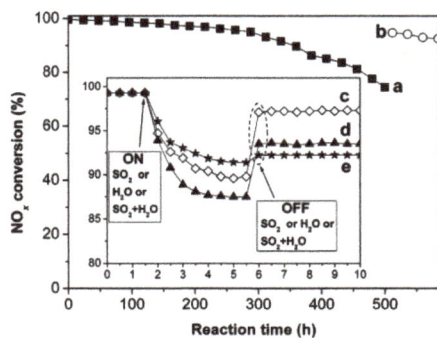

Figure 8. Lifetime, SO_2 tolerance, and water resistance of Fe-(0.4)MnO_x(CA-500) catalyst (Reaction conditions: 120 °C, [NO] = [NH_3] = 1000 ppm, [O_2] = 3%, [SO_2] = 100 ppm, [H_2O] = 5%, N_2 as balance, and GHSV = 30,000 h^{-1}; Plasma treatment conditions: 10 MHz, 25 °C, pure oxygen with 50 mL/min under 2.4 s of residence time, and duration of 6 h): (**a**) Lifetime testing; (**b**) regeneration property; (**c**) SO_2 tolerance; (**d**) water resistance; (**e**) the combined effect of SO_2 and H_2O. (Reproduced with permission from Reference [28], Copyright 2012, American Chemical Society.).

129

It has been reported that cobalt also presents a positive role on the tolerance of Mn-based catalysts to sulfur and water. Zhang et al. [29] found that the $Mn_xCo_{3-x}O_4$ nanocage catalyst exhibited decent SO_2 tolerance due to its hierarchically porous structure, abundant active sites, and strong interaction between Mn and Co oxides (Figure 9). Qiu et al. prepared a mesoporous 3D-$MnCo_2O_4$ catalyst, which exhibited great SCR activity and good tolerance to sulfur and water [30,31]. As illustrated in Figure 10, the NO conversion over $MnCo_2O_4$ was maintained at 86% in the presence of 5 vol % H_2O and 100 ppm SO_2. Futhermore, the NO conversion could be recovered to 93% after the supply of H_2O and SO_2 was cut off.

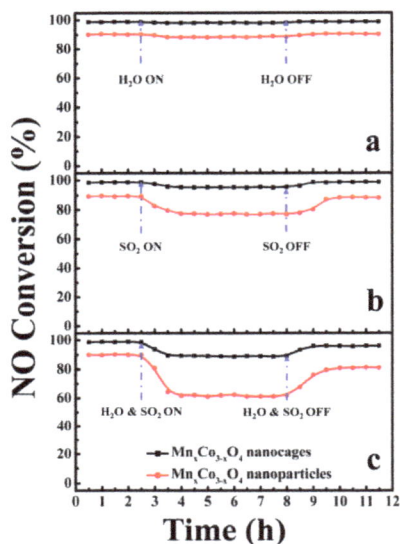

Figure 9. (a) H_2O resistance, (b) SO_2 tolerance, and (c) H_2O and SO_2 synergetic effect study of the catalysts at 175 °C (Reaction conditions: [NO] = [NH_3] = 500 ppm, [O_2] = 3 vol %, [H_2O] = 8 vol % (when used), [SO_2] = 200 ppm (when used), N_2 balance, and GHSV = 38,000 h^{-1}). (Reproduced with permission from Reference [29], Copyright 2014, American Chemical Society.).

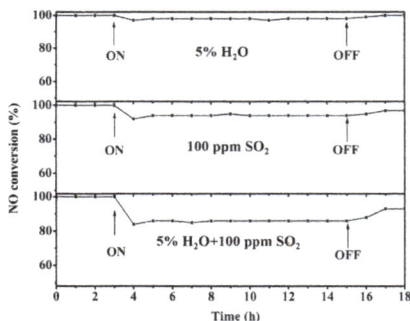

Figure 10. H_2O resistance, SO_2 tolerance, and H_2O and SO_2 synergetic effect study of the $MnCo_2O_4$ catalyst at 200 °C (Reaction conditions: [NO] = [NH_3] = 500 ppm, [O_2] = [H_2O] = 5 vol %, [SO_2] = 100 ppm, N_2 balance, and GHSV = 50,000 h^{-1}). (Reproduced with permission from Reference [30], Copyright 2015, Elsevier).

Copper also presents a positive effect on the tolerance of MnO_x-based catalysts to water and sulfur. Kang et al. reported a Cu–Mn mixed oxide catalyst, which exhibited good tolerance to water and sulfur [32]. When 100 ppm SO_2 and 11 vol % H_2O were added to the reaction gas, the NO_x conversion over Cu–Mn oxides decreased from 95 to 64% at 125 °C after 4 h, and the NO_x conversion was gradually recovered after stopping the supply of SO_2 and H_2O.

Recently, several rare earth elements have been demonstrated to promote the enhanced tolerance to sulfur and water. For instance, Meng et al. developed a Sm-modified MnO_x catalyst [33], and found that a proper Sm modification (Sm/Mn = 1:10) enhanced the sulfur and water tolerance of MnO_x. As shown in Figure 11, the NO_x conversion over the Sm–Mn-0.1 catalyst could be maintained at about 91% at 100 °C when 2% H_2O and 100 ppm SO_2 were added into the feed gas, and the NO_x conversion was recovered to 97% after both H_2O and SO_2 were removed from the feed gas. Sun et al. prepared a Eu-modified MnO_x catalyst [34]. They tested the sulfur and water resistance of this catalyst at a higher temperature (350 °C), instead of a low temperature, such as 100 °C. The deactivation effect of SO_2 and H_2O on $MnEuO_x$-0.1 was weak, and the NO_x conversion over $MnEuO_x$-0.1 kept over 90% in the presence of 100 ppm SO_2 and 5% H_2O. Furthermore, the NO_x conversion nearly recovered its original level after the supply of SO_2 and H_2O was cut off (Figure 12).

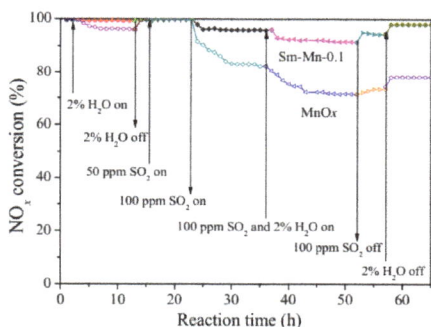

Figure 11. The effect of H_2O and SO_2 on the catalytic activities of the MnO_x and Sm–Mn-0.1 catalysts for the SCR reaction at 100 °C (Reaction conditions: 0.3 g catalyst, 500 ppm NO, 500 ppm NH_3, 5% O_2, Ar to balance, GHSV = 49,000 h^{-1}). (Reproduced with permission from Reference [33], Copyright 2016, American Chemical Society).

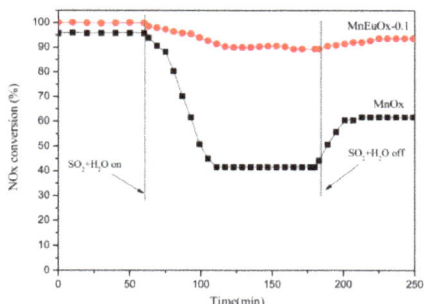

Figure 12. Effect of SO_2 and H_2O on the SCR activities of MnO_x and $MnEuO_x$-0.1 catalysts (Reaction conditions: [NO] = [NH_3] = 600 ppm, [SO_2] = 100 ppm, [H_2O] = [O_2] = 5%, balance Ar, GHSV = 108,000 h^{-1}, reaction temperature = 350 °C). (Reproduced with permission from Reference [34], Copyright 2017, Elsevier).

3.2.2. Mn-Based Ternary Metal Oxide Catalysts

It has been reported that the introduction of a small amount of a third element can enhance the tolerance of Mn-based binary metal oxide catalysts to H_2O and SO_2. Qi et al. successfully prepared Mn–Fe–Ce mixed oxides that performed well under 100 ppm SO_2 and 2.5% H_2O condition (Figure 13) [35]. The NO conversion over Mn–Fe–Ce decreased from 98 to 95% in 3 h in the presence of SO_2 and H_2O and then restored quickly to its original level after the supply of SO_2 and H_2O was cut off. France et al. developed a CeFeMnO$_x$ catalyst that exhibited excellent sulfur and water resistance at a low temperature [36]. As presented in Figure 14, the NO conversion over this catalyst only decreased from 100 to 75% when water and sulfur were introduced, and then recovered to 95% after the supply of H_2O and SO_2 was cut off. Chang et al. found that Sn doping could enhance the sulfur resistance of Mn–Ce catalysts because SO_2 was easier to react with Ce on the surface instead of forming ammonia sulfate; meanwhile, more surface acid sites were introduced due to Sn doping [37,38]. As shown in Figure 15, the NO conversion was kept at around 70% in the presence of SO_2 and H_2O and recovered to almost the original level within less than 3 h after SO_2 and H_2O were removed. Gao et al. [39] found that the SCR pathways over MnO$_x$–CeO$_2$ catalyst are based on the adsorption, activation, and reaction of monodentate nitrite species and coordinated NH_3 species, and these species are significantly inhibited by SO_2 through competitive adsorption. In contrast, over Co- and Ni-doped MnO$_x$–CeO$_2$ catalysts, the primary NO$_x$ adsorbed species are in the form of bidentate nitrate without the influence by SO_2. The NO conversion over Co- and Ni-doped MnO$_x$–CeO$_2$ catalysts decreased 20% after 150 ppm SO_2 and 10% H_2O were introduced, and recovered after the supply of SO_2 and H_2O was cut off (Figure 16). Liu et al. successfully prepared WO_3 promoted Mn–Zr mixed oxide catalyst [40]. As shown in Figure 17, the NO$_x$ conversion over MnWZr catalyst was maintained above 90% in the presence of 50 ppm SO_2 and 5% H_2O, and the conversion quickly recovered after the supply of SO_2 and H_2O was cut off.

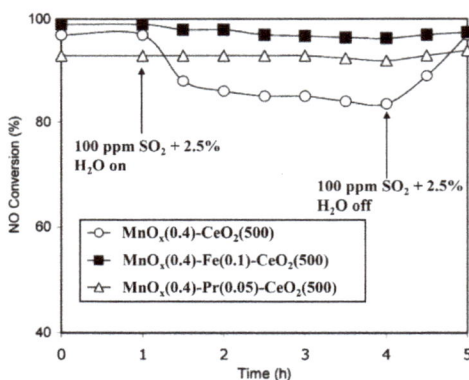

Figure 13. The effects of SO_2 and H_2O on the SCR activities of MnO$_x$(0.4)–CeO$_2$(500)-based catalysts (Reaction conditions: 0.2 g catalyst, [NH$_3$] = [NO] = 1000 ppm, [O$_2$] = 2%, He = balance and total flow rate = 100 mL/min, T = 150 °C). (Reproduced with permission from Reference [35], Copyright 2004, Elsevier).

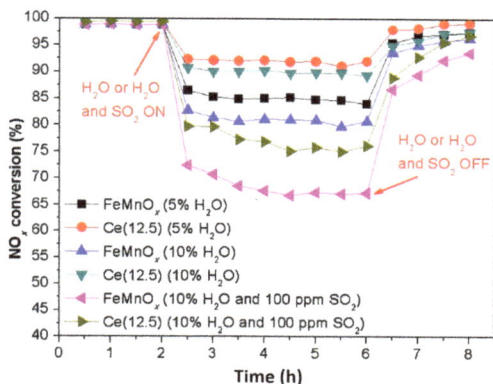

Figure 14. Influence of H_2O and combined H_2O and SO_2 on NO_x conversion of FeMnO$_x$ and Ce(12.5) (Reaction conditions: [NO] = [NH$_3$] = 0.1%, [H$_2$O] = 5% or 10%, [SO$_2$] = 100 ppm, [O$_2$] = 3%, N$_2$ balance, GHSV = 30,000 h^{-1}; reaction temperature = 120 °C). (Reproduced with permission from Reference [36], Copyright 2017, Elsevier).

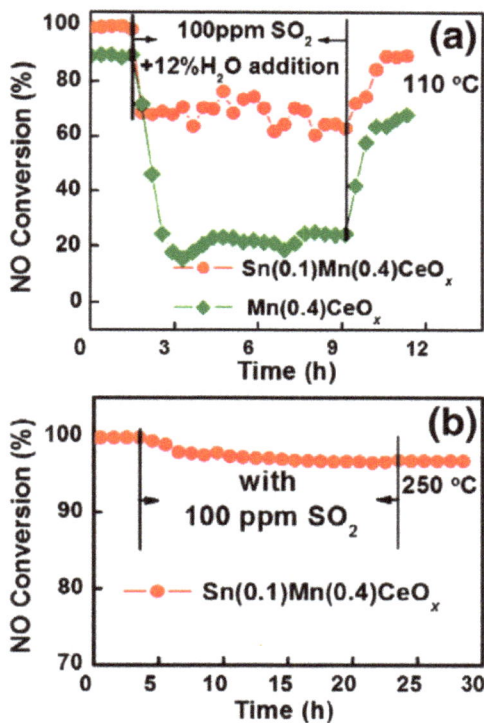

Figure 15. (**a**) NO conversion in the presence of 100 ppm SO$_2$ and 12% H$_2$O at 110 °C; (**b**) NO conversion in the presence of 100 ppm SO$_2$ at 250 °C (Reaction conditions: 0.2 g samples, 1000 ppm NO, 1000 ppm NH$_3$, 2% O$_2$, N$_2$ balance, GHSV = 35,000 h^{-1}). (Reproduced with permission from Reference [37], Copyright 2012, Elsevier).

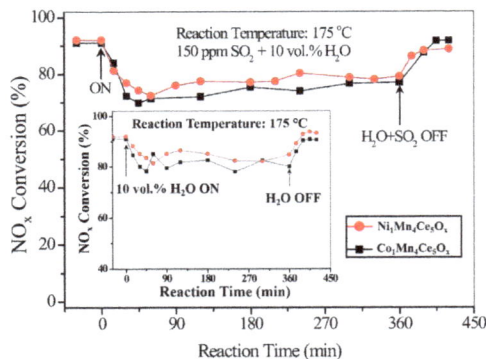

Figure 16. The resistance to H_2O and/or SO_2 over $Co_1Mn_4Ce_5O_x$ and $Ni_1Mn_4Ce_5O_x$ (Reaction conditions: $[NO] = [NH_3] = 500$ ppm, $[O_2] = 5$ vol %, $[SO_2] = 150$ ppm, $[H_2O] = 10$ vol %, N_2 to balance, total flow rate = 200 mL/min, GHSV = 48,000 h^{-1}). (Reproduced with permission from Reference [39], Copyright 2017, Elsevier).

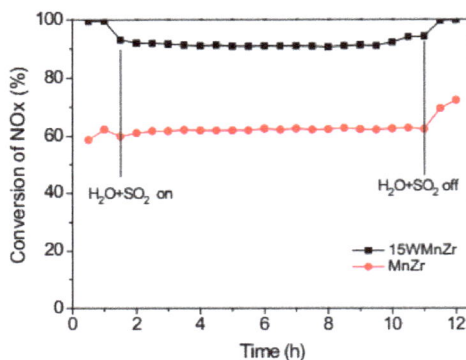

Figure 17. Response of NO_x conversion over MnZr and 15WMnZr catalysts at 300 °C to intermittent feed of H_2O and SO_2 (Reaction conditions: $[NO] = [NH_3] = 500$ ppm, $[O_2] = 5\%$, $[H_2O] = 5\%$, $[SO_2] = 50$ ppm, GHSV = 128,000 h^{-1}). (Reproduced with permission from Reference [40], Copyright 2016, Elsevier).

3.3. Supported Mn-Based Catalysts

Supports play an important role in NH_3–SCR reaction. Proper supports not only can provide a huge surface to disperse the active components and prevent the formation of large crystalline particles but can also affect the sulfur and water tolerance. To date, various materials, such as TiO_2, carbon materials, and Al_2O_3, have been explored as supports to load Mn-based catalysts.

3.3.1. TiO$_2$ Supported Mn-Based Catalysts

TiO_2 is known to be more resistant to sulfur poisoning because of the stability of sulfates on the TiO_2 surface is weaker than that on other oxides [59], which made TiO_2 an ideal support for the loading of Mn-based catalysts.

Qi and Yang [41] prepared a series of MnTi and FeMnTi catalysts. As shown in Figure 18, the NO conversion over Fe–Mn/TiO_2 was decreased from 100 to 90% within 5 h at 150 °C after 100 ppm SO_2 and 2.5% H_2O were added. After the supply of SO_2 and H_2O was cut off, the NO conversion recovered to 100% again quickly. Yang et al. investigated the sulfur and water tolerance of Fe–Ti spinel supported MnO_x catalyst [42]. As shown in Figure 19, the NO_x conversion at 200 °C gradually decreased from 100 to 83% and then kept unchanged after 8% of H_2O and 60 ppm of SO_2 were introduced. After the supply H_2O and SO_2 was shut off, the NO_x conversion rapidly recovered to 100%. Wu et al. found that the sulfur resistance of Mn/TiO_2 can be greatly improved by Ce addition [9]. As displayed in Figure 20, SO_2 presented an obvious poisonous effect on SCR activity of Mn/TiO_2 at low temperatures because the NO conversion over the MnTi catalyst decreased from 93 to 30% in the presence of SO_2 within 6.5 h, while the NO conversion over MnCeTi still maintained at about 84% under the same conditions. As shown in Figure 21, the surface of fresh catalysts was smooth and uniform (Figure 21A,C). After the catalyst was poisoned with 100 ppm SO_2 for 24 h, the significant agglomeration and deposition could be observed from the surface of MnTi–S (Figure 21B), while only a few deposited particles (no agglomeration) appeared on the surface of MnCeTi–S (Figure 21D). Yu et al. [80] developed a mesoporous MnO_2–Fe_2O_3–CeO_2/TiO_2 catalyst. The NO conversion over this catalyst was stable at 80% under astream of SO_2. Shen et al. [43] found that the addition of proper iron enhanced the tolerance of TiO_2-supported Mn–Ce catalyst to water and sulfur. As exhibited in Figure 22, Fe(0.15)–Mn–Ce$/TiO_2$ showed higher resistance under 3 vol % H_2O and 0.01 vol % SO_2 and still provided 83.8% NO conversion over afurther 5 h, an improvement over the Mn–Ce$/TiO_2$ catalyst. Shen et al. found that titanium-pillared clays (Ti–PILCs) presented advantages in sulfur tolerance over traditional TiO_2 supports [44]. It can be seen from Figure 23 that the NO conversion was stable at around 90% without any obvious decrease in the presence of 3 vol % H_2O and 0.01 vol % SO_2, suggesting that Mn–CeO$_x$/Ti–PILC(S) possessed good resistance to H_2O and SO_2. Lee et al. prepared a series of Mn/Ce–TiO_2 catalysts and found that Mn(20)/Ce(4)–TiO_2 showed good H_2O and SO_2 tolerance [45]. As shown in Figure 24, the NO conversion decreased to 60% and it recovered to almost the original level when the SO_2 supply was shut off. Park et al. [83] prepared Mn/Ti catalysts via chemical vapor condensation (CVC) method and claimed that this Mn/Ti not only showed higher activity at low temperature but also exhibited better tolerance to water and sulfur. Only a small NO conversion decrease from 70 to 58% was found under 200 ppm of SO_2 in 250 min at 100 °C. Liu et al. [46] prepared an Mn–Ce–Ti catalyst using the hydrothermal method, and the NO_x conversion over the $Mn_{0.2}Ce_{0.1}Ti_{0.7}O_x$ catalyst under H_2O and SO_2 was further investigated at 200 °C. As shown in Figure 25, the introduction of H_2O and SO_2 induced a slight decrease in NO_x conversion. After H_2O and SO_2 were excluded from the reactant feed, the NO_x conversion completely recovered.

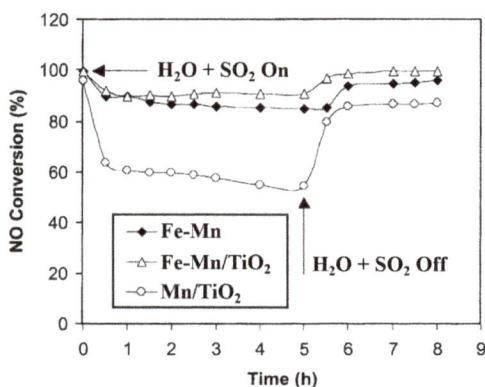

Figure 18. NO conversion on the various Fe–Mn-based catalyst in the presence of SO_2 + H_2O (Reaction conditions: temperature = 150 °C, [NO] = [NH$_3$] = 1000 ppm, [O_2] = 2%, [SO_2] = 100 ppm, [H_2O] = 2.5%, balance He, total flow rate 100 mL/min, catalyst 0.5 g). (Reproduced with permission from Reference [41], Copyright 2003, Elsevier).

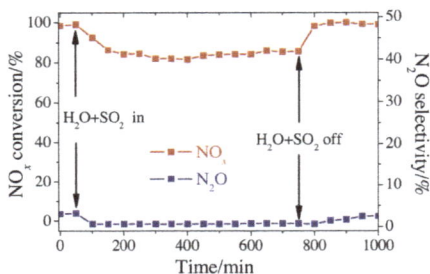

Figure 19. Stability of NO reduction over 10% Mn/Fe–Ti spinel in the presence of H_2O and SO_2 (Reaction conditions: $[NH_3]$ = $[NO]$ = 500 ppm, $[SO_2]$ = 60 ppm, $[H_2O]$ = 8%, catalyst mass = 500 mg, the total flow rate = 100 mL and GHSV = 12000 cm^3 g^{-1} h^{-1}). (Reproduced with permission from Reference [42], Copyright 2016, Elsevier).

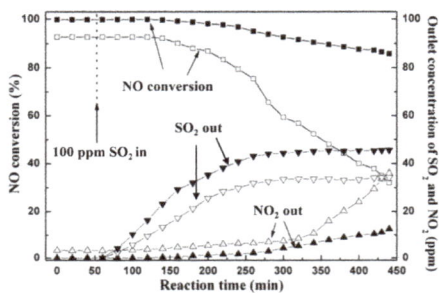

Figure 20. SCR activities of Mn/TiO$_2$ and Ce-doped Mn/TiO$_2$ in the presence of SO_2 (Reaction conditions: $[NH_3]$ = $[NO]$ = 1000 ppm, $[O_2]$ = 3%, $[SO_2]$ = 100 ppm, $[H_2O]$ = 3 vol %, N_2 balance, GHSV = 40,000 h^{-1}, reaction temperature = 150 °C; hollow symbols for MnTi and solid symbols for MnCeTi).

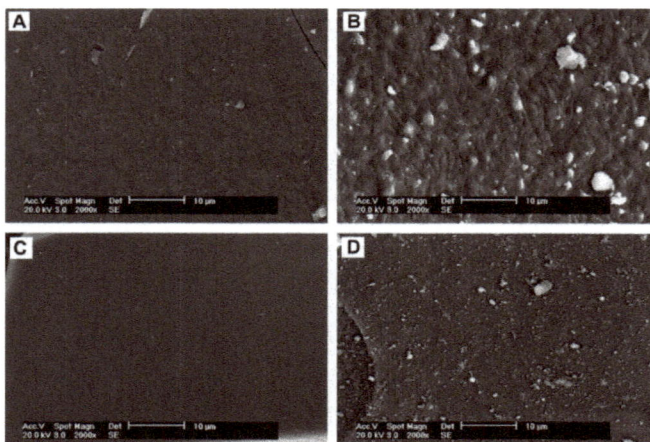

Figure 21. Scanning electron microscope (SEM) micrographs of fresh and SO_2-poisoned catalysts. (**A**) MnTi, (**B**) MnTi-S, (**C**) MnCeTi, and (**D**) MnCeTi-S. (Reproduced with permission from Reference [9], Copyright 2009, Elsevier).

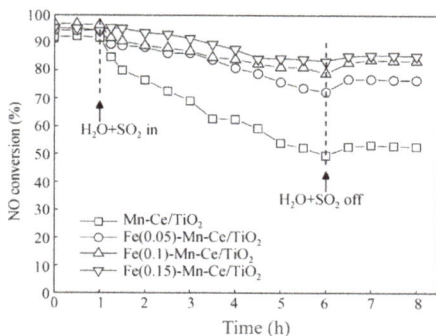

Figure 22. Effect of H_2O and SO_2 on NO conversion over Mn–Ce/TiO_2 and Fe–Mn–Ce/TiO_2 catalysts (Reaction conditions: 0.06 vol % NO, 0.06 vol % NH_3, 3 vol % O_2, 3 vol % H_2O (when used), 0.01 vol % SO_2 (when used), balance N_2, GHSV 50,000 h^{-1}, total flow rate 300 mL/min, tested at 180 °C). (Reproduced with permission from Reference [43], Copyright 2010, Science Direct).

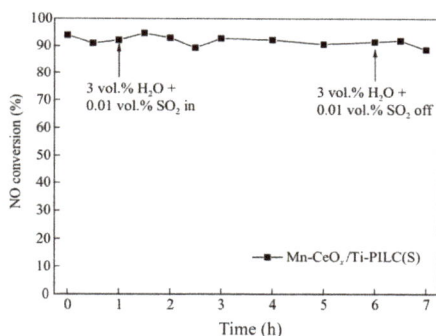

Figure 23. Effect of H_2O and SO_2 on NO conversion over Mn–CeO_x/TiPILC(S) at 200 °C (Reaction conditions: 0.06 vol % NO, 0.06 vol % NH_3, 3 vol % O_2, 3 vol % H_2O (when used), 0.01 vol % SO_2 (when used), balance N_2, GHSV 50,000 h^{-1}, total flow rate 300 mL/min). (Reproduced with permission from Reference [44], Copyright 2012, Science Direct).

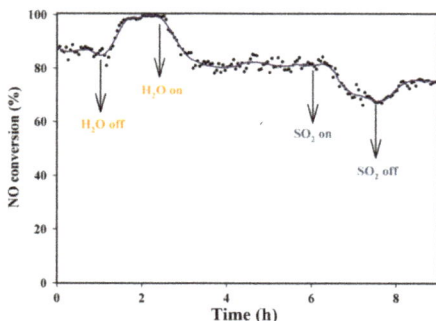

Figure 24. The effects of the H_2O and SO_2 in the Mn/Ce(4)–TiO_2 catalysts with different ratio of Ce/Ti (Reaction condition: 200 ppm NO, 8% O_2, 6% H_2O, 100 ppm SO_2, 0.28 g of sample and 500 cc/min total flow rate, tested at 180 °C). (Reproduced with permission from Reference [45], Copyright 2012, Elsevier).

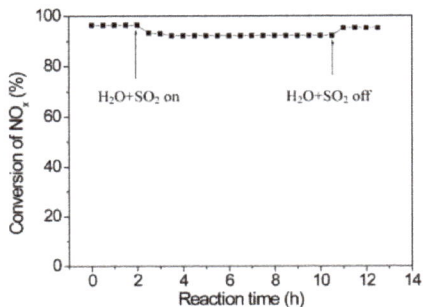

Figure 25. Response of the NO_x conversion over $Mn_{0.2}Ce_{0.1}Ti_{0.7}O_x$ catalyst at 200 °C to the intermittent feed of H_2O and SO_2 (Reaction condition: 500 ppm of NO, 500 ppm of NH_3, 5% O_2, 5% H_2O, 50 ppm of SO_2, balance He, GHSV = 64,000 h^{-1}). (Reproduced with permission from Reference [46], Copyright 2014, American Chemical Society).

3.3.2. Carbon Materials Supported Mn-Based Catalysts

Carbon materials, such as activated carbon (AC), activated carbon fiber (ACF), carbon nanotube (CNT), and graphene (GE), have been widely studied as substrates for supporting low-temperature SCR catalysts due to their high specific surface area, unique pore structure, excellent dispersion of active components, and chemical stability [70,84–86]. Among these carbon materials, CNT and GE have been considered as good supports that can enhance the tolerance of H_2O and SO_2.

Zhang et al. prepared a novel MnCe@CNTs-R catalyst, which exhibited great tolerance to 100 ppm SO_2 and 4 vol % H_2O due to the good dispersion degree of the active components on the surface of CNTs [48]. The coexistence of SO_2 and H_2O induced a 13% NO conversion decrease and the NO conversion was recovered to 90% after the supply of SO_2 and H_2O was cut off. Cai et al. designed a multi-shell Fe_2O_3@MnO_x@CNTs catalyst and found that the Fe_2O_3 shell effectively suppressed the formation of the surface sulfate species, which led to a good tolerance to H_2O and SO_2 (Figure 26) [49]. Lu et al. successfully synthesized a series of TiO_2–graphene-supported Mn and Mn–Ce catalysts with good tolerance to H_2O and SO_2 due to the well dispersed Mn component (Figure 27) [50,51]. Wang et al. investigated the effect of SO_2 on activated carbon honeycomb (ACH)-supported MnO_x and CeO_2–MnO_x catalysts, and the S 2p XPS results are displayed in Figure 28. The peak intensity of Mn/ACH was much higher than that of CeMn/ACH, indicating that Ce doping on ACH had an inhibition of sulfates loading [81].

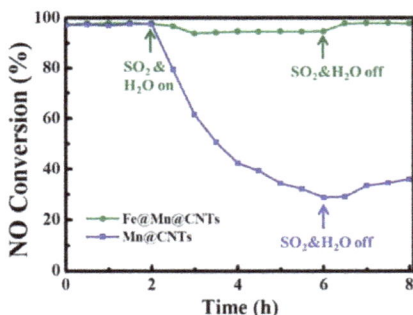

Figure 26. SO_2 + H_2O tolerance test at 240 °C (Reaction conditions: [NH_3] = [NO] = 550 ppm, [O_2] = 5 vol %, [SO_2] = 100 ppm, [H_2O] = 10 vol % (when used), N_2 as balance gas, GHSV = 20,000 h^{-1}). (Reproduced with permission from Reference [49], Copyright 2016, The Royal Society of Chemistry).

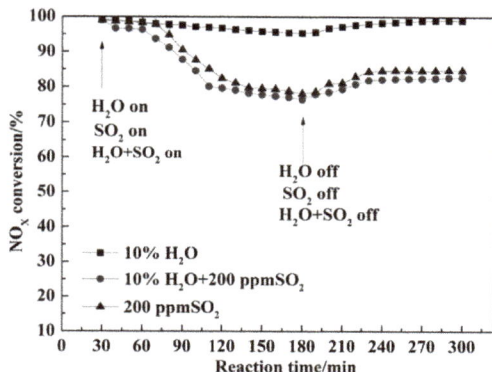

Figure 27. SO$_2$ + H$_2$O tolerance test at 180 °C (Reaction conditions: [NH$_3$] = [NO] = 500 ppm, [O$_2$] = 7 vol %, [SO$_2$] = 200 ppm, [H$_2$O] = 10 vol % (when used), N$_2$ as balance gas, GHSV = 67,000 h^{-1}). (Reproduced with permission from Reference [51], Copyright 2015, Elsevier).

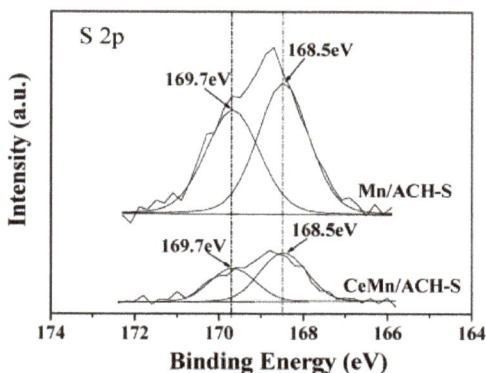

Figure 28. X-ray photoelectron spectroscopy (XPS) spectra of S 2p for poisoned Mn/ACH and CeMn/ACH catalysts. (Reproduced with permission from Reference [81], Copyright 2015, American Chemical Society).

3.3.3. Other Supported Mn-Based Catalysts

Mixed metal oxides and SiO$_2$ have also been studied as substrates for supporting SCR catalysts. Yao et al. prepared MnO$_x$/SiO$_2$, MnO$_x$/Al$_2$O$_3$, MnO$_x$/TiO$_2$, and MnO$_x$/CeO$_2$ catalysts and found that the catalytic activity in the presence of H$_2$O and SO$_2$ was in the order of MnO$_x$/SiO$_2$ < MnO$_x$/TiO$_2$ < MnO$_x$/CeO$_2$ < MnO$_x$/Al$_2$O$_3$ (Figure 29) [87]. Shen et al. also compared the tolerance to H$_2$O and SO$_2$ of MnO$_x$-supported on various substrates including Al$_2$O$_3$, TiO$_2$, CeO$_2$, ZrO$_2$, and Ce$_{0.5}$Zr$_{0.5}$O$_2$. Their results showed that the resistance ability was decreased in the following order: MnO$_x$/Ce$_{0.5}$Zr$_{0.5}$O$_2$ > MnO$_x$/Al$_2$O$_3$ > MnO$_x$/CeO$_2$ > MnO$_x$/TiO$_2$ > MnO$_x$/ZrO$_2$, and they ascribed the excellent toleranceof MnO$_x$/Ce$_{0.5}$Zr$_{0.5}$O$_2$ to the combination of the advantages of the two supports (ZrO$_2$ and CeO$_2$) (Figure 30) [52]. Huang et al. prepared a mesoporous silica-supported Mn–Fe catalyst and found that its SCR activity was suppressed gradually in the presence of SO$_2$ and H$_2$O, and the inhibitory effect was relieved after heating treatment [88].

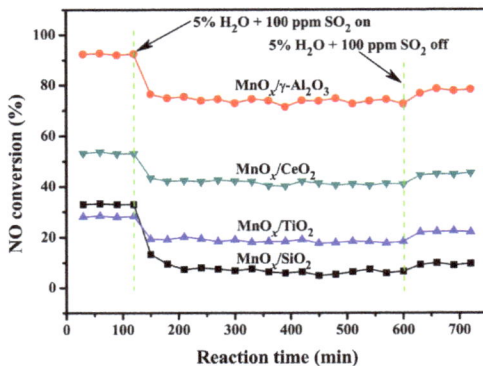

Figure 29. The results of H_2O + SO_2 resistance at 200 °C of these supported Mn-based catalysts with different supports. (Reproduced with permission from Reference [87], Copyright 2017, Elsevier).

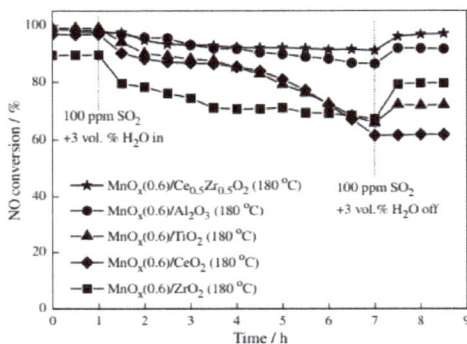

Figure 30. The effect of H_2O and SO_2 on NO conversion for $MnO_x(0.6)/Ce_{0.5}Zr_{0.5}O_2$ (Reaction conditions: [NO] = [NH_3] = 600 ppm, [O_2] = 3 vol %, N_2 balance, T = 180°C, catalyst 0.5 g, GHSV 30,000 h^{-1}). (Reproduced with permission from Reference [52], Copyright 2014, Elsevier).

4. Strategies to Reduce the Poisoning Effect

Although there are many factors that affect the water and sulfur tolerance of Mn-based catalysts, such as the preparation method, the reaction temperature, the gas hourly space velocity (GHSV), and the morphology, structure, and surface area of the catalyst, deactivation can be attributedto three main causes: (1) the competitive adsorption between SO_2 and NO, H_2O, and NH_3 on the active sites, (2) the blocking effect of the NH_4HSO_4 and $(NH_4)_2SO_4$ formed on the surface active sites, and (3) the formation of metallic sulfate, which reduces the active sites on the surface. Hence, suppressing the three negative effects is the key to enhancing resistance against H_2O and SO_2. To date, many strategies have been taken to reduce the poisoning effect on Mn-based catalysts.

4.1. Metal Modification

Metal modification or doping is a common solution to the problem. Most transition metals have been used as dopants to modify Mn-based catalysts for good resistance to SO_2 and H_2O. Cerium has been fully studied, and the mechanism has been uncovered. Cerium reacts more sensitively with SO_2, so the formation of NH_4HSO_4 and $(NH_4)_2SO_4$ is reduced on the surface of Ce-modified Mn-based catalysts [36,79,89]. Furthermore, metallic sulfates formed by cerium and SO_2 are relatively stable and can provide surface acid sites to enhance the adsorption of NH_3 and to inhibit the catalytic oxidization of NH_3 at the same time, thus promoting SCR reactions in the presence of SO_2 and H_2O [68,78,90]. Liu et al. confirmed, using density functional theory, that Ce isable to inhibit the formation of ammonia sulfate on the surface of catalysts, which is believed to be a key factor in improving tolerance [79]. It was also reported that iron is capable of decreasing the formation rate of sulfate species, thus promoting tolerance [41,91]. Furthermore, several reports have shown that the doping of a third metal, such as Sn and W, into Mn-based catalysts can further improve resistance to SO_2 and H_2O [38,92,93]. Zhang et al. found that resistance to SO_2 and H_2O over the W-modified $SnMnCeO_x$ catalysts, in comparison with unmodified $SnMnCeO_x$, was further improved (Figure 31) [53]. They attributed this improvement to the introduction of WO_x species, which prevented the formation of $(NH_4)_2SO_4$ on the catalyst and blocked the interactions between Mn^{n+}, SO_4^{2-}, and gaseous SO_3 [37]. Rare earths have drawn an increasing amount of attention recently, and Sm and Eu doping have been shown to have a positive influence on the tolerance of SO_2 and H_2O [33,34]. However, the mechanism of SO_2 and H_2O resistance still needs to be uncovered.

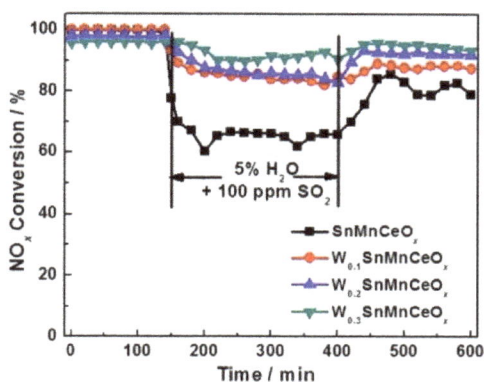

Figure 31. NH_3–SCR activity over $W_ySnMnCeO_x$ catalysts in the presence of SO_2/H_2O at 200 °C (Reaction conditions: [NO] = [NH_3] = 500 ppm, [O_2] = 5%, [SO_2] = 100 ppm (when used), [H_2O] = 5% (when used), N_2 balance, total flow rate 100 mL min^{-1} and GHSV = 60,000 mLg^{-1}h^{-1}). (Reproduced with permission from Reference [53], Copyright 2017, Elsevier).

4.2. Proper Support

It is well believed that loading Mn-based SCR catalysts on a suitable support is an effective measure to enhance the tolerance to SO_2 and H_2O [60,94] because of the high thermal and mechanical stability, large surface area, and highly dispersed active sites. Furthermore, the interaction between support and active components exhibits positive effects on the tolerance to SO_2 and H_2O [95]. Therefore, it is very important for Mn-based catalysts to choose an appropriate support. Among several supports, TiO_2, porous carbon material, and CNTs are considered to be good options. It has been widely reported that TiO_2 can provide a higher specific surface area [58], a higher surface acidity [96], and a good dispersion of active components, all of which effectively enhance SO_2 and H_2O resistance [97]. Gao et al. reported a novel nanocomposite of MnO_x nanoparticles supported on three-dimensionally ordered macroporous carbon (MnO_x/3DOMC). They found that this novel catalyst exhibited good water and sulfur tolerance (Figure 32) [54]. As a special ordered carbon material with unique nanostructure and electronic properties, carbon nanotubes (CNTs) have been reported to be an interesting support for SCR catalysts [98,99]. Zhang et al. proved that active components were well dispersed on the surface of the support such that the blocking effect caused by NH_4HSO_4 and $(NH_4)_2SO_4$ was reduced [48].

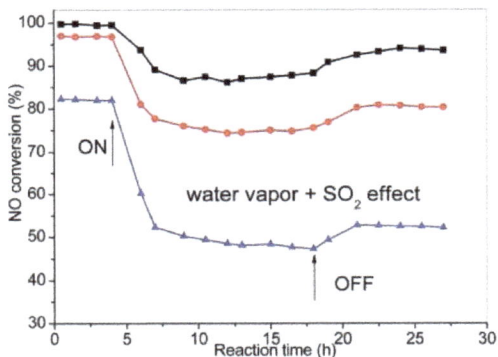

Figure 32. Effect of water vapor and SO_2 on NO conversion over the $MnO_x/3DOMC$ (black), MnO_x/NAC (red), and MnO_x/TiO_2 (blue) catalysts (Reaction conditions: 190 °C, 1000 ppm of NO, 1000 ppm of NH_3, 5% of O_2, 5% of water vapor, and/or 200 ppm of SO_2, He balance). (Reproduced with permission from Reference [54] Copyright 2015, The Royal Society of Chemistry).

4.3. Combination of Metal Modification and Support

The combination of metal modification and support is considered to be a good way of enhancing the water and sulfur tolerance of Mn-based catalysts due to the advantages of both strategies. Compared with non-supported mixed metal oxides, supported catalysts often possess a larger specific surface area and a better dispersion of active components, which facilitates tolerance to water and sulfur. On the other hand, compared with supported single MnO_x catalysts, the synergistic effect introduced by one or more modifiers can reduce the poisonous effect and protect the surface active components. Thus, combining two measures, mixing (or doping) MnO_x with suitable metal oxides and loading active components on a suitable support, is the best way to enhance the tolerance to water and sulfur. Chen et al. prepared an NiMn/Ti catalyst and investigated the effects of H_2O and SO_2 on its SCR performance (Figure 33) [47]. They found that the coexistence of 100 ppm SO_2 and 15 vol % H_2O led to an apparent decrease in NO_x conversion, and the NO_x conversion recovered quickly to 100% after stopping the addition of H_2O. Chen et al. found that W-modified MnO_x/TiO_2 exhibited better tolerance to SO_2 than MnO_x/TiO_2 catalyst due to the fact that W addition inhibited the formation of sulfate species [93]. Wang et al. reported a series of W-modified MnO_x/TiO_2 and found that a W(0.25)–Mn(0.25)–Ti(0.5) catalyst showed the best SCR activity and good tolerance to water and sulfur. As illustrated in Figure 34, the W(0.25)–Mn(0.25)–Ti(0.5) catalyst presented a 100% NO_x conversion from 140 to 260 °C [55]. Our group successfully prepared a europium-modified TiO_2-supported Mn-based catalyst and found that this catalyst showed better tolerance than the Mn–TiO_2 catalyst due to the highly dispersed MnO_x and Eu_2O_3 on the surface of TiO_2 [18]. Zhao et al. synthesized an Nb-modified Mn/Ce/Ti catalyst and tested the water and sulfur tolerance at a high GHSV of 180,000 h^{-1} at 175 °C [100]. They found that the catalyst was deactivated with a decreased NO conversion from 100 to 10% within 4 h, which recovered to almost the original level after regeneration by washing.

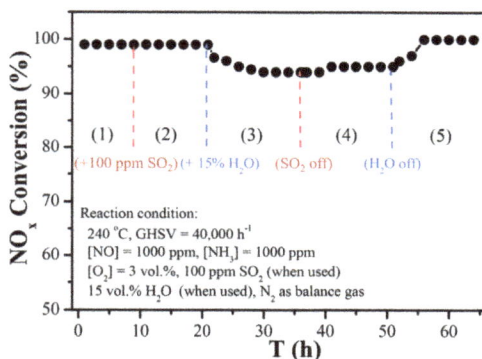

Figure 33. Effects of H_2O and SO_2 on NO_x conversion over the $Ni_{0.4}Mn_{0.6}Ti_{10}$ catalyst. (Reproduced with permission from Reference [47], Copyright 2017,The Royal Society of Chemistry).

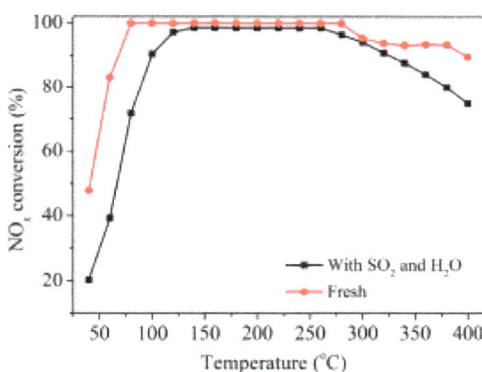

Figure 34. Effects of SO_2 and H_2O on the NO_x conversion of W(0.25)–Mn(0.25)–Ti(0.5) at GHSV of 25,000 h^{-1} (Reaction conditions: [NO] = [NH$_3$] = 1000 ppm, [O$_2$] = 5 vol %, [SO$_2$] = 100 ppm, [H$_2$O] = 10 vol %, in He as balance). (Reproduced with permission from Reference [55], Copyright 2016, Elsevier).

4.4. Rational Design of Structure and Morphology

The rational design of catalyst structure and morphology is another method of reducing the poisoning effect of SO_2 and H_2O. Shen et al. developed a hollow MnO_x–CeO_2 mixed oxide catalyst, which exhibited good SCR performance under water and sulfur poison at a high GHSV of 120,000 h^{-1} due to the hollow structure [101]. Zhang et al. [28] found that an $Mn_xCo_{3-x}O_4$ catalyst with a nanocage structure exhibited much better SO_2 and H_2O tolerance than $Mn_xCo_{3-x}O_4$ without a nanocage structure. Qiu et al. synthesized a mesoporous 3D-$MnCo_2O_4$ catalyst that exhibited great SCR activity and good tolerance to sulfur and water [30,31], and the mesoporous structure enabled a dynamic balance between the formation and decomposition of ammonium sulfate, and thus suppressed the blocking effect during the SCR reaction. Li et al. prepared Mn_2O_3-doped Fe_2O_3 hexagonal microsheet catalyst [102] and found the single H_2O resistance (15%) and the single SO_2 resistance (100 ppm) over this catalyst were good and stable with the NO conversion at around 92% and 85% for 100 h, respectively, because of this special structure.

4.5. Monolithic Catalysts

Preparing monolithic catalysts may be an option to promote the tolerance of Mn-based SCR catalysts to SO_2 and H_2O. As well known, the commercial catalysts (V_2O_5-WO_3(or MoO_3)/TiO_2) used in thermal power plant are in the monolithic form because the honeycomb monoliths are suitable for a high gas flow rate, reduce pressure drop problems, exhibit high tolerance to dust and attrition, and are easy to regenerate [103–106]. Recently, metal foam and wire mesh as novel monolithic support for Mn-based and other vanadium-free SCR catalysts are drawing an increasing amount of attention due to their high porosity, stability, thermal conductivity, and mass transfer ability [107,108]. Xu et al. prepared porous $MnCo_xO_y$ nanocubes on a Ti mesh as a novel monolith de-NO_x catalyst for SCR [109]. They found that this monolithic catalyst exhibited better SCR activity than $MnCo_xO_y$@honeycomb ceramics. Meanwhile, the water resistance test results of this novel monolithic catalyst were promising. Xu et al. successfully synthesized a series of MnO_x–CeO_2/WO_3-ZrO_2 monolithic catalysts that showed good tolerance to water and sulfur [110].

5. Conclusions and Perspectives

Recent progress on the sulfur and water resistance of Mn-based catalysts for the low-temperature selective catalytic reduction of NO_x has been reviewed comprehensively in this work. Although much progress has been made, many questions still need to be answered, and many problems need to be solved:

(1) The exploration of novel Mn-based catalysts with excellent resistance to SO_2 and H_2O is still worthwhile. Resistance to SO_2 and H_2O directly decides whether this catalyst can be commercialized. Up to now, mixing (or doping) MnO_x with suitable metal oxides and loading Mn-based active components on a suitable support are considered an efficient strategy. Discovering new doping elements and novel supports may be promising research directions.

(2) The actual effect of every specific doping element on tolerance promotion needs to be explained. To date, many works have been done to test the tolerance of Mn-based catalysts to H_2O and SO_2. However, the reasons why the tolerance of Mn-based catalysts to H_2O and SO_2 can be enhanced by mixing (or doping) them with other suitable elements need to be further explored in detail.

(3) The role of support ought to be further analyzed. Does support only provide a higher specific surface area and a good dispersion of Mn? Is the support involved in SCR reaction? Such questions need to be answered.

(4) Long-term tolerance tests need to be conducted. Most tests only last for several hours, and it is hard to predict the long-term performance of the catalyst under H_2O and SO_2 poison.

(5) N_2 selectivity is an important indicator for the commercialization of SCR catalysts, which is closely related to the yield of N_2O. However, there is currently a lack of research on the effect of SO_2 on N_2 selectivity over Mn-based catalysts. Therefore, it is necessary to carry out this research in the near future.

(6) Most studies focus on powder catalysts. From a commercial perspective, monolithic catalysts should be given more consideration.

Acknowledgments: This work was sponsored by the National Natural Science Fund Committee-Baosteel Group Corporation Steel Joint Research Fund, China (U1460105), the National Science Foundation of China (51521065), and the Natural Science Foundation of Shaanxi Province, China (2015JM2055).

Conflicts of Interest: There are no conflict of interest.

References

1. Kapteijn, F.; Rodriguez-Mirasol, J.; Moulijn, J.A. Heterogeneous catalytic decomposition of nitrous oxide. *Appl. Catal. B Environ.* **1996**, *9*, 25–64. [CrossRef]
2. Pârvulescu, V.I.; Grange, P.; Delmon, B. Catalytic removal of NO. *Catal. Today* **1998**, *46*, 233–316. [CrossRef]

3. Forzatti, P. Present status and perspectives in de-NO$_x$ SCR catalysis. *Appl. Catal. A Gen.* **2001**, *222*, 221–236. [CrossRef]

4. Kapteijn, F.; Singoredjo, L.; Andreini, A.; Moulijn, J.A. Activity and selectivity of pure manganese oxides in the selective catalytic reduction of nitric oxide with ammonia. *Appl. Catal. B Environ.* **1994**, *3*, 173–189. [CrossRef]

5. Boningari, T.; Smirniotis, P.G. Impact of nitrogen oxides on the environment and human health: Mn-based materials for the NO$_x$ abatement. *Curr. Opin. Chem. Eng.* **2016**, *13*, 133–141. [CrossRef]

6. Busca, G.; Lietti, L.; Ramis, G.; Berti, F. Chemical and mechanistic aspects of the selective catalytic reduction of NO$_x$ by ammonia over oxide catalysts: A review. *Appl. Catal. B Environ.* **1998**, *18*, 1–36. [CrossRef]

7. Boningari, T.; Koirala, R.; Smirniotis, P.G. Low-temperature selective catalytic reduction of NO with NH$_3$ over V/ZrO$_2$ prepared by flame-assisted spray pyrolysis: Structural and catalytic properties. *Appl. Catal. B Environ.* **2012**, *127*, 255–264. [CrossRef]

8. Roy, S.; Hegde, M.S.; Madras, G. Catalysis for NO$_x$ abatement. *Appl. Energy* **2009**, *86*, 2283–2297. [CrossRef]

9. Wu, Z.; Jin, R.; Wang, H.; Liu, Y. Effect of ceria doping on SO$_2$ resistance of Mn/TiO$_2$ for selective catalytic reduction of NO with NH$_3$ at low temperature. *Catal. Commun.* **2009**, *10*, 935–939. [CrossRef]

10. Chen, Z.; Yang, Q.; Li, H.; Li, X.; Wang, L.; Chi Tsang, S. Cr–MnO$_x$ mixed-oxide catalysts for selective catalytic reduction of NO$_x$ with NH$_3$ at low temperature. *J. Catal.* **2010**, *276*, 56–65. [CrossRef]

11. Wu, Z.; Jiang, B.; Liu, Y.; Wang, H.; Jin, R. DRIFT study of Manganese/Titania-based catalysts for low-temperature selective catalytic reduction of NO with NH$_3$. *Environ. Sci. Technol.* **2007**, *41*, 5812–5817. [CrossRef] [PubMed]

12. Kang, M.; Park, E.D.; Kim, J.M.; Yie, J.E. Manganese oxide catalysts for NO$_x$ reduction with NH$_3$ at low temperatures. *Appl. Catal. A Gen.* **2007**, *327*, 261–269. [CrossRef]

13. Tang, X.; Hao, J.; Xu, W.; Li, J. Low temperature selective catalytic reduction of NO$_x$ with NH$_3$ over amorphous MnO$_x$ catalysts prepared by three methods. *Catal. Commun.* **2007**, *8*, 329–334. [CrossRef]

14. Tian, W.; Yang, H.; Fan, X.; Zhang, X. Catalytic reduction of NO$_x$ with NH$_3$ over different-shaped MnO$_2$ at low temperature. *J. Hazard. Mater.* **2011**, *188*, 105–109. [CrossRef] [PubMed]

15. Fan, Z.; Shi, J.W.; Gao, C.; Gao, G.; Wang, B.; Niu, C. Rationally designed porous MnO$_x$–FeO$_x$ nanoneedles for low-temperature selective catalytic reduction of NO$_x$ by NH$_3$. *ACS Appl. Mater. Interfaces* **2017**, *9*, 16117–16127. [CrossRef] [PubMed]

16. Gao, G.; Shi, J.-W.; Fan, Z.; Gao, C.; Niu, C. MnM$_2$O$_4$ microspheres (M = Co, Cu, Ni) for selective catalytic reduction of NO with NH$_3$: Comparative study on catalytic activity and reaction mechanism via in-situ diffuse reflectance infrared Fourier transform spectroscopy. *Chem. Eng. J.* **2017**, *325*, 91–100. [CrossRef]

17. Gao, G.; Shi, J.-W.; Liu, C.; Gao, C.; Fan, Z.; Niu, C. Mn/CeO$_2$ catalysts for SCR of NO$_x$ with NH$_3$: Comparative study on the effect of supports on low-temperature catalytic activity. *Appl. Surf. Sci.* **2017**, *411*, 338–346. [CrossRef]

18. Gao, C.; Shi, J.-W.; Fan, Z.; Yu, Y.; Chen, J.; Li, Z.; Niu, C. Eu–Mn–Ti mixed oxides for the SCR of NO$_x$ with NH$_3$: The effects of Eu-modification on catalytic performance and mechanism. *Fuel Process. Technol.* **2017**, *167*, 322–333. [CrossRef]

19. Xie, C.; Yang, S.; Shi, J.; Li, B.; Gao, C.; Niu, C. MnO$_x$–TiO$_2$ and Sn doped MnO$_x$–TiO$_2$ selective reduction catalysts prepared using MWCNTs as the pore template. *Chem. Eng. J.* **2017**, *327*, 1–8. [CrossRef]

20. Shi, J.-W.; Gao, C.; Liu, C.; Fan, Z.; Gao, G.; Niu, C. Porous MnO$_x$ for low-temperature NH$_3$–SCR of NO$_x$: The intrinsic relationship between surface physicochemical property and catalytic activity. *J. Nanopart. Res.* **2017**, *19*, 194–205. [CrossRef]

21. Liu, C.; Gao, G.; Shi, J.-W.; He, C.; Li, G.; Bai, N.; Niu, C. MnO$_x$–CeO$_2$ shell-in-shell microspheres for NH$_3$–SCR de-NO$_x$ at low temperature. *Catal. Commun.* **2016**, *86*, 36–40. [CrossRef]

22. Loiland, J.A.; Lobo, R.F. Low temperature catalytic NO oxidation over microporous materials. *J. Catal.* **2014**, *311*, 412–423. [CrossRef]

23. Kang, M.; Yeon, T.H.; Park, E.D.; Yie, J.E.; Kim, J.M. Novel MnO$_x$ catalysts for NO reduction at low temperature with ammonia. *Catal. Lett.* **2006**, *106*, 77–80. [CrossRef]

24. Qi, G.; Yang, R.T. A superior catalyst for low-temperature NO reduction with NH$_3$. *Chem. Commun.* **2003**, 848–849. [CrossRef]

25. Liu, Z.; Yi, Y.; Zhang, S.; Zhu, T.; Zhu, J.; Wang, J. Selective catalytic reduction of NO$_x$ with NH$_3$ over Mn–Ce mixed oxide catalyst at low temperatures. *Catal. Today* **2013**, *216*, 76–81. [CrossRef]

26. Yao, X.; Kong, T.; Chen, L.; Ding, S.; Yang, F.; Dong, L. Enhanced low-temperature NH$_3$–SCR performance of MnO$_x$/CeO$_2$ catalysts by optimal solvent effect. *Appl. Surf. Sci.* **2017**, *420*, 407–415. [CrossRef]

27. Long, R.Q.; Yang, R.T.; Chang, R. Low temperature selective catalytic reduction (SCR) of NO with NH$_3$ over Fe–Mn based catalysts. *Chem. Commun.* **2002**, 452–453. [CrossRef]

28. Chen, Z.; Wang, F.; Li, H.; Yang, Q.; Wang, L.; Li, X. Low-temperature selective catalytic reduction of NO$_x$ with NH$_3$ over Fe–Mn mixed-oxide catalysts containing Fe$_3$Mn$_3$O$_8$ Phase. *Ind. Eng. Chem. Res.* **2012**, *51*, 202–212. [CrossRef]

29. Zhang, L.; Shi, L.; Huang, L.; Zhang, J.; Gao, R.; Zhang, D. Rational design of high-performance deNO$_x$ catalysts based on Mn$_x$Co$_{3-x}$O$_4$nanocages derived from metal–organic frameworks. *ACS Catal.* **2014**, *4*, 1753–1763. [CrossRef]

30. Qiu, M.; Zhan, S.; Yu, H.; Zhu, D. Low-temperature selective catalytic reduction of NO with NH$_3$ over ordered mesoporous Mn$_x$Co$_{3-x}$O$_4$ catalyst. *Catal. Commun.* **2015**, *62*, 107–111. [CrossRef]

31. Qiu, M.; Zhan, S.; Yu, H.; Zhu, D.; Wang, S. Facile preparation of ordered mesoporous MnCo$_2$O$_4$ for low-temperature selective catalytic reduction of NO with NH$_3$. *Nanoscale* **2015**, *7*, 2568–2577. [CrossRef] [PubMed]

32. Kang, M.; Park, E.D.; Kim, J.M.; Yie, J.E. Cu–Mn mixed oxides for low temperature NO reduction with NH$_3$. *Catal. Today* **2006**, *111*, 236–241. [CrossRef]

33. Meng, D.; Zhan, W.; Guo, Y.; Guo, Y.; Wang, L.; Lu, G. A highly effective catalyst of Sm–MnO$_x$ for the NH$_3$–SCR of NO$_x$ at low temperature: Promotional role of Sm and its catalytic performance. *ACS Catal.* **2015**, *5*, 5973–5983. [CrossRef]

34. Sun, P.; Guo, R.-T.; Liu, S.-M.; Wang, S.-X.; Pan, W.-G.; Li, M.-Y. The enhanced performance of MnO$_x$ catalyst for NH$_3$–SCR reaction by the modification with Eu. *Appl. Catal. A Gen.* **2017**, *531*, 129–138. [CrossRef]

35. Qi, G.; Yang, R.T.; Chang, R. MnO$_x$–CeO$_2$ mixed oxides prepared by co-precipitation for selective catalytic reduction of NO with NH$_3$ at low temperatures. *Appl. Catal. B Environ.* **2004**, *51*, 93–106. [CrossRef]

36. France, L.J.; Yang, Q.; Li, W.; Chen, Z.; Guang, J.; Guo, D.; Wang, L.; Li, X. Ceria modified FeMnO$_x$—Enhanced performance and sulphur resistance for low-temperature SCR of NO$_x$. *Appl. Catal. B Environ.* **2017**, *206*, 203–215. [CrossRef]

37. Chang, H.; Li, J.; Chen, X.; Ma, L.; Yang, S.; Schwank, J.W.; Hao, J. Effect of Sn on MnO$_x$–CeO$_2$ catalyst for SCR of NO$_x$ by ammonia: Enhancement of activity and remarkable resistance to SO$_2$. *Catal. Commun.* **2012**, *27*, 54–57. [CrossRef]

38. Chang, H.; Chen, X.; Li, J.; Ma, L.; Wang, C.; Liu, C.; Schwank, J.W.; Hao, J. Improvement of activity and SO$_2$ tolerance of Sn-modified MnO$_x$–CeO$_2$ catalysts for NH$_3$–SCR at low temperatures. *Environ. Sci. Technol.* **2013**, *47*, 5294–5301. [CrossRef] [PubMed]

39. Gao, F.; Tang, X.; Yi, H.; Li, J.; Zhao, S.; Wang, J.; Chu, C.; Li, C. Promotional mechanisms of activity and SO$_2$ tolerance of Co- or Ni-doped MnO$_x$–CeO$_2$ catalysts for SCR of NO$_x$ with NH$_3$ at low temperature. *Chem. Eng. J.* **2017**, *317*, 20–31. [CrossRef]

40. Liu, Z.; Liu, Y.; Li, Y.; Su, H.; Ma, L. WO$_3$ promoted Mn–Zr mixed oxide catalyst for the selective catalytic reduction of NO$_x$ with NH$_3$. *Chem. Eng. J.* **2016**, *283*, 1044–1050. [CrossRef]

41. Qi, G.; Yang, R.T. Low-temperature selective catalytic reduction of NO with NH$_3$ over iron and manganese oxides supported on titania. *Appl. Catal. B Environ.* **2003**, *44*, 217–225. [CrossRef]

42. Yang, S.; Qi, F.; Xiong, S.; Dang, H.; Liao, Y.; Wong, P.K.; Li, J. MnO$_x$ supported on Fe–Ti spinel: A novel Mn based low temperature SCR catalyst with a high N$_2$ selectivity. *Appl. Catal. B Environ.* **2016**, *181*, 570–580. [CrossRef]

43. Shen, B.; Liu, T.; Zhao, N.; Yang, X.; Deng, L. Iron-doped Mn–Ce/TiO$_2$ catalyst for low temperature selective catalytic reduction of NO with NH$_3$. *J. Environ. Sci.* **2010**, *22*, 1447–1454. [CrossRef]

44. Shen, B.; Ma, H.; Yao, Y. Mn–CeO$_x$/Ti-PILCs for selective catalytic reduction of NO with NH$_3$ at low temperature. *J. Environ. Sci.* **2012**, *24*, 499–506. [CrossRef]

45. Lee, S.M.; Park, K.H.; Hong, S.C. MnO$_x$/CeO$_2$–TiO$_2$ mixed oxide catalysts for the selective catalytic reduction of NO with NH$_3$ at low temperature. *Chem. Eng. J.* **2012**, *195–196*, 323–331. [CrossRef]

46. Liu, Z.; Zhu, J.; Li, J.; Ma, L.; Woo, S.I. Novel Mn–Ce–Ti mixed-oxide catalyst for the selective catalytic reduction of NO$_x$ with NH$_3$. *ACS Appl. Mater. Interfaces* **2014**, *6*, 14500–14508. [CrossRef] [PubMed]

47. Chen, L.; Li, R.; Li, Z.; Yuan, F.; Niu, X.; Zhu, Y. Effect of Ni doping in $Ni_xMn_{1-x}Ti_{10}$ ($x = 0.1$–0.5) on activity and SO_2 resistance for NH_3–SCR of NO studied with in situ DRIFTS. *Catal. Sci. Technol.* **2017**, *7*, 3243–3257. [CrossRef]
48. Zhang, D.; Zhang, L.; Shi, L.; Fang, C.; Li, H.; Gao, R.; Huang, L.; Zhang, J. In situ supported MnO_x–CeO_x on carbon nanotubes for the low-temperature selective catalytic reduction of NO with NH_3. *Nanoscale* **2013**, *5*, 1127–1136. [CrossRef] [PubMed]
49. Cai, S.; Hu, H.; Li, H.; Shi, L.; Zhang, D. Design of multi-shell Fe_2O_3@MnO_x@CNTs for the selective catalytic reduction of NO with NH_3: Improvement of catalytic activity and SO_2 tolerance. *Nanoscale* **2016**, *8*, 3588–3598. [CrossRef] [PubMed]
50. Lu, X.; Song, C.; Chang, C.-C.; Teng, Y.; Tong, Z.; Tang, X. Manganese oxides supported on TiO_2–graphene nanocomposite catalysts for selective catalytic reduction of NO_x with NH_3 at low temperature. *Ind. Eng. Chem. Res.* **2014**, *53*, 11601–11610. [CrossRef]
51. Lu, X.; Song, C.; Jia, S.; Tong, Z.; Tang, X.; Teng, Y. Low-temperature selective catalytic reduction of NO_x with NH_3 over cerium and manganese oxides supported on TiO_2–graphene. *Chem. Eng. J.* **2015**, *260*, 776–784. [CrossRef]
52. Shen, B.; Wang, Y.; Wang, F.; Liu, T. The effect of Ce–Zr on NH_3–SCR activity over $MnO_x(0.6)/Ce_{0.5}Zr_{0.5}O_2$ at low temperature. *Chem. Eng. J.* **2014**, *236*, 171–180. [CrossRef]
53. Zhang, T.; Qiu, F.; Chang, H.; Peng, Y.; Li, J. Novel W-modified $SnMnCeO_x$ catalyst for the selective catalytic reduction of NO_x with NH_3. *Catal. Commun.* **2017**, *100*, 117–120. [CrossRef]
54. Gao, X.; Li, L.; Song, L.; Lu, T.; Zhao, J.; Liu, Z. Highly dispersed MnO_x nanoparticles supported on three-dimensionally ordered macroporous carbon: A novel nanocomposite for catalytic reduction of NO_x with NH_3 at low temperature. *RSC Adv.* **2015**, *5*, 29577–29588. [CrossRef]
55. Wang, X.; Li, X.; Zhao, Q.; Sun, W.; Tade, M.; Liu, S. Improved activity of W-modified MnO_x–TiO_2 catalysts for the selective catalytic reduction of NO with NH_3. *Chem. Eng. J.* **2016**, *288*, 216–222. [CrossRef]
56. Apostolescu, N.; Geiger, B.; Hizbullah, K.; Jan, M.; Kureti, S.; Reichert, D.; Schott, F.; Weisweiler, W. Selective catalytic reduction of nitrogen oxides by ammonia on iron oxide catalysts. *Appl. Catal. B Environ.* **2006**, *62*, 104–114. [CrossRef]
57. Lee, T.; Bai, H. Low temperature selective catalytic reduction of NO_x with NH_3 over Mn-based catalyst: A review. *AIMS Environ. Sci.* **2016**, *3*, 261–289. [CrossRef]
58. Yao, Y.; Zhang, S.-L.; Zhong, Q.; Liu, X.-X. Low-temperature selective catalytic reduction of NO over manganese supported on TiO_2 nanotubes. *J. Fuel Chem. Technol.* **2011**, *39*, 694–701. [CrossRef]
59. Li, J.; Chang, H.; Ma, L.; Hao, J.; Yang, R.T. Low-temperature selective catalytic reduction of NO_x with NH_3 over metal oxide and zeolite catalysts—A review. *Catal. Today* **2011**, *175*, 147–156. [CrossRef]
60. Liu, F.; He, H.; Zhang, C.; Shan, W.; Shi, X. Mechanism of the selective catalytic reduction of NO_x with NH_3 over environmental-friendly iron titanate catalyst. *Catal. Today* **2011**, *175*, 18–25. [CrossRef]
61. Qu, L.; Li, C.; Zeng, G.; Zhang, M.; Fu, M.; Ma, J.; Zhan, F.; Luo, D. Support modification for improving the performance of MnO_x–CeO_y/γ-Al_2O_3 in selective catalytic reduction of NO by NH_3. *Chem. Eng. J.* **2014**, *242*, 76–85. [CrossRef]
62. Schill, L.; Putluru, S.S.R.; Jensen, A.D.; Fehrmann, R. MnFe/Al_2O_3 catalyst synthesized by deposition precipitation for low-temperature selective catalytic reduction of NO with NH_3. *Catal. Lett.* **2015**, *145*, 1724–1732. [CrossRef]
63. Gao, R.; Zhang, D.; Liu, X.; Shi, L.; Maitarad, P.; Li, H.; Zhang, J.; Cao, W. Enhanced catalytic performance of V_2O_5–$WO_3/Fe_2O_3/TiO_2$ microspheres for selective catalytic reduction of NO by NH_3. *Catal. Sci. Technol.* **2013**, *3*, 191–199. [CrossRef]
64. Zhang, L.; Li, L.; Cao, Y.; Yao, X.; Ge, C.; Gao, F.; Deng, Y.; Tang, C.; Dong, L. Getting insight into the influence of SO_2 on TiO_2/CeO_2 for the selective catalytic reduction of NO by NH_3. *Appl. Catal. B Environ.* **2015**, *165*, 589–598. [CrossRef]
65. Xu, W.; He, H.; Yu, Y. Deactivation of a Ce/TiO_2 catalyst by SO_2 in the selective catalytic reduction of NO by NH_3. *J. Phys. Chem. C* **2009**, *113*, 4426–4432. [CrossRef]
66. Jiang, B.Q.; Wu, Z.B.; Liu, Y.; Lee, S.C.; Ho, W.K. DRIFT study of the SO_2 effect on low-temperature SCR reaction over Fe−Mn/TiO_2. *J. Phys. Chem. C* **2010**, *114*, 4961–4965. [CrossRef]

67. Long, R.Q.; Yang, R.T. Superior Fe-ZSM-5 catalyst for selective catalytic reduction of nitric oxide by ammonia. *J. Am. Chem. Soc.* **1999**, *121*, 5595–5596. [CrossRef]
68. Xiong, S.; Liao, Y.; Xiao, X.; Dang, H.; Yang, S. The mechanism of the effect of H_2O on the low temperature selective catalytic reduction of NO with NH_3 over Mn–Fe spinel. *Catal. Sci. Technol.* **2015**, *5*, 2132–2140. [CrossRef]
69. Xiong, S.; Liao, Y.; Xiao, X.; Dang, H.; Yang, S. Novel Effect of H_2O on the low temperature selective catalytic reduction of NO with NH_3 over MnO_x–CeO_2: Mechanism and kinetic study. *J. Phys. Chem. C* **2015**, *119*, 4180–4187. [CrossRef]
70. Marban, G. Mechanism of low-temperature selective catalytic reduction of NO with NH_3 over carbon-supported Mn_3O_4. Role of surface NH_3 species: SCR mechanism. *J. Catal.* **2004**, *226*, 138–155. [CrossRef]
71. Tang, X.; Li, J.; Sun, L.; Hao, J. Origination of N_2O from NO reduction by NH_3 over β–MnO_2 and α–Mn_2O_3. *Appl. Catal. B Environ.* **2010**, *99*, 156–162. [CrossRef]
72. Wang, C.; Sun, L.; Cao, Q.; Hu, B.; Huang, Z.; Tang, X. Surface structure sensitivity of manganese oxides for low-temperature selective catalytic reduction of NO with NH_3. *Appl. Catal. B Environ.* **2011**, *101*, 598–605. [CrossRef]
73. Liu, C.; Shi, J.-W.; Gao, C.; Niu, C. Manganese oxide-based catalysts for low-temperature selective catalytic reduction of NO_x with NH_3: A review. *Appl. Catal. A Gen.* **2016**, *522*, 54–69. [CrossRef]
74. Chen, L.; Si, Z.; Wu, X.; Weng, D.; Ran, R.; Yu, J. Rare earth containing catalysts for selective catalytic reduction of NO_x with ammonia: A review. *J. Rare Earths* **2014**, *32*, 907–917. [CrossRef]
75. Smirniotis, P.G.; Peña, D.A.; Uphade, B.S. Low-temperature selective catalytic reduction (SCR) of NO with NH_3 by using Mn, Cr, and Cu oxides supported on hombikat TiO_2. *Angew. Chem. Int. Ed.* **2001**, *40*, 2479–2482. [CrossRef]
76. Yang, S.; Wang, C.; Li, J.; Yan, N.; Ma, L.; Chang, H. Low temperature selective catalytic reduction of NO with NH_3 over Mn–Fe spinel: Performance, mechanism and kinetic study. *Appl. Catal. B Environ.* **2011**, *110*, 71–80. [CrossRef]
77. Chen, L.; Niu, X.; Li, Z.; Dong, Y.; Zhang, Z.; Yuan, F.; Zhu, Y. Promoting catalytic performances of Ni–Mn spinel for NH_3–SCR by treatment with SO_2 and H_2O. *Catal. Commun.* **2016**, *85*, 48–51. [CrossRef]
78. Yang, S.; Guo, Y.; Chang, H.; Ma, L.; Peng, Y.; Qu, Z.; Yan, N.; Wang, C.; Li, J. Novel effect of SO_2 on the SCR reaction over CeO_2: Mechanism and significance. *Appl. Catal. B Environ.* **2013**, *136–137*, 19–28. [CrossRef]
79. Liu, Y.; Cen, W.; Wu, Z.; Weng, X.; Wang, H. SO_2 poisoning structures and the effects on pure and Mn doped CeO_2: A first principles investigation. *J. Phys. Chem. C* **2012**, *116*, 22930–22937. [CrossRef]
80. Yu, J.; Guo, F.; Wang, Y.; Zhu, J.; Liu, Y.; Su, F.; Gao, S.; Xu, G. Sulfur poisoning resistant mesoporousMn-based catalyst for low-temperature SCR of NO with NH_3. *Appl. Catal. B Environ.* **2010**, *95*, 160–168. [CrossRef]
81. Wang, Y.; Li, X.; Zhan, L.; Li, C.; Qiao, W.; Ling, L. Effect of SO_2 on activated carbon honeycomb supported CeO_2–MnO_x catalyst for NO removal at low temperature. *Ind. Eng. Chem. Res.* **2015**, *54*, 2274–2278. [CrossRef]
82. Wei, Y.; Sun, Y.; Su, W.; Liu, J. MnO_2 doped CeO_2 with tailored 3-D channels exhibits excellent performance for NH_3–SCR of NO. *RSC Adv.* **2015**, *5*, 26231–26235. [CrossRef]
83. Park, E.; Kim, M.; Jung, H.; Chin, S.; Jurng, J. Effect of sulfur on Mn/Ti catalysts prepared using chemical vapor condensation (CVC) for low-temperature NO reduction. *ACS Catal.* **2013**, *3*, 1518–1525. [CrossRef]
84. Lu, P.; Li, C.; Zeng, G.; He, L.; Peng, D.; Cui, H.; Li, S.; Zhai, Y. Low temperature selective catalytic reduction of NO by activated carbon fiber loading lanthanum oxide and ceria. *Appl. Catal. B Environ.* **2010**, *96*, 157–161. [CrossRef]
85. Sui, M.; Xing, S.; Sheng, L.; Huang, S.; Guo, H. Heterogeneous catalytic ozonation of ciprofloxacin in water with carbon nanotube supported manganese oxides as catalyst. *J. Hazard. Mater.* **2012**, *227–228*, 227–236. [CrossRef] [PubMed]
86. Wang, L.; Huang, B.; Su, Y.; Zhou, G.; Wang, K.; Luo, H.; Ye, D. Manganese oxides supported on multi-walled carbon nanotubes for selective catalytic reduction of NO with NH_3: Catalytic activity and characterization. *Chem. Eng. J.* **2012**, *192*, 232–241. [CrossRef]
87. Yao, X.; Kong, T.; Yu, S.; Li, L.; Yang, F.; Dong, L. Influence of different supports on the physicochemical properties and denitration performance of the supported Mn-based catalysts for NH_3–SCR at low temperature. *Appl. Surf. Sci.* **2017**, *402*, 208–217. [CrossRef]

88. Huang, J.; Tong, Z.; Huang, Y.; Zhang, J. Selective catalytic reduction of NO with NH$_3$ at low temperatures over iron and manganese oxides supported on mesoporous silica. *Appl. Catal. B Environ.* **2008**, *78*, 309–314. [CrossRef]

89. Kwon, D.W.; Nam, K.B.; Hong, S.C. The role of ceria on the activity and SO$_2$ resistance of catalysts for the selective catalytic reduction of NO$_x$ by NH$_3$. *Appl. Catal. B Environ.* **2015**, *166–167*, 37–44. [CrossRef]

90. Xiao, X.; Xiong, S.; Shi, Y.; Shan, W.; Yang, S. Effect of H$_2$O and SO$_2$ on the selective catalytic reduction of NO with NH$_3$ over Ce/TiO$_2$ catalyst: Mechanism and kinetic Study. *J. Phys. Chem. C* **2016**, *120*, 1066–1076. [CrossRef]

91. Tang, X.; Hao, J.; Yi, H.; Li, J. Low-temperature SCR of NO with NH$_3$ over AC/C supported manganese-based monolithic catalysts. *Catal. Today* **2007**, *126*, 406–411. [CrossRef]

92. Kwon, D.W.; Nam, K.B.; Hong, S.C. Influence of tungsten on the activity of a Mn/Ce/W/Ti catalyst for the selective catalytic reduction of NO with NH$_3$ at low temperatures. *Appl. Catal. A Gen.* **2015**, *497*, 160–166. [CrossRef]

93. Chen, Q.-L.; Guo, R.-T.; Wang, Q.-S.; Pan, W.-G.; Wang, W.-H.; Yang, N.-Z.; Lu, C.-Z.; Wang, S.-X. The catalytic performance of Mn/TiWO$_x$ catalyst for selective catalytic reduction of NO$_x$ with NH$_3$. *Fuel* **2016**, *181*, 852–858. [CrossRef]

94. Qi, G.; Yang, R.T. Characterization and FTIR studies of MnO$_x$−CeO$_2$ catalyst for low-temperature selective catalytic reduction of NO with NH$_3$. *J. Phys. Chem. B* **2004**, *108*, 15738–15747. [CrossRef]

95. Guo, F.; Yu, J.; Chu, M.; Xu, G. Interaction between support and V$_2$O$_5$ in the selective catalytic reduction of NO by NH$_3$. *Catal. Sci. Technol.* **2014**, *4*, 2147–2155. [CrossRef]

96. Jin, R.; Liu, Y.; Wu, Z.; Wang, H.; Gu, T. Low-temperature selective catalytic reduction of NO with NH$_3$ over MnCe oxides supported on TiO$_2$ and Al$_2$O$_3$: A comparative study. *Chemosphere* **2010**, *78*, 1160–1166. [CrossRef] [PubMed]

97. Pappas, D.K.; Boningari, T.; Boolchand, P.; Smirniotis, P.G. Novel manganese oxide confined interweaved titania nanotubes for the low-temperature Selective Catalytic Reduction (SCR) of NO$_x$ by NH$_3$. *J. Catal.* **2016**, *334*, 1–13. [CrossRef]

98. Wang, X.; Zheng, Y.; Xu, Z.; Liu, X.; Zhang, Y. Low-temperature selective catalytic reduction of NO over MnO$_x$/CNTs catalysts. *Catal. Commun.* **2014**, *50*, 34–37. [CrossRef]

99. Zhang, Y.; Zheng, Y.; Wang, X.; Lu, X. Preparation of Mn–FeO$_x$/CNTs catalysts by redox co-precipitation and application in low-temperature NO reduction with NH$_3$. *Catal. Commun.* **2015**, *62*, 57–61. [CrossRef]

100. Zhao, B.; Ran, R.; Guo, X.; Cao, L.; Xu, T.; Chen, Z.; Wu, X.; Si, Z.; Weng, D. Nb-modified Mn/Ce/Ti catalyst for the selective catalytic reduction of NO with NH$_3$ at low temperature. *Appl. Catal. A Gen.* **2017**, *545*, 64–71. [CrossRef]

101. Shen, Q.; Zhang, L.; Sun, N.; Wang, H.; Zhong, L.; He, C.; Wei, W.; Sun, Y. Hollow MnO$_x$–CeO$_2$ mixed oxides as highly efficient catalysts in NO oxidation. *Chem. Eng. J.* **2017**, *322*, 46–55. [CrossRef]

102. Li, Y.; Wan, Y.; Li, Y.; Zhan, S.; Guan, Q.; Tian, Y. Low-temperature selective catalytic reduction of NO with NH$_3$ over Mn$_2$O$_3$-doped Fe$_2$O$_3$ hexagonal microsheets. *ACS Appl. Mater. Interfaces* **2016**, *8*, 5224–5233. [CrossRef] [PubMed]

103. Tomašić, V. Application of the monoliths in DeNO$_x$ catalysis. *Catal. Today* **2007**, *119*, 106–113. [CrossRef]

104. Heck, R.M.; Gulati, S.; Farrauto, R.J. The application of monoliths for gas phase catalytic reactions. *Chem. Eng. J.* **2001**, *82*, 149–156. [CrossRef]

105. Shan, W.; Song, H. Catalysts for the selective catalytic reduction of NO$_x$ with NH$_3$ at low temperature. *Catal. Sci. Technol.* **2015**, *5*, 4280–4288. [CrossRef]

106. Ouzzine, M.; Cifredo, G.A.; Gatica, J.M.; Harti, S.; Chafik, T.; Vidal, H. Original carbon-based honeycomb monoliths as support of Cu or Mn catalysts for low-temperature SCR of NO: Effects of preparation variables. *Appl. Catal. A Gen.* **2008**, *342*, 150–158. [CrossRef]

107. Li, H.; Zhang, D.; Maitarad, P.; Shi, L.; Gao, R.; Zhang, J.; Cao, W. In situ synthesis of 3D flower-like NiMnFe mixed oxides as monolith catalysts for selective catalytic reduction of NO with NH$_3$. *Chem. Commun.* **2012**, *48*, 10645–10647. [CrossRef] [PubMed]

108. Shu, Y.; Aikebaier, T.; Quan, X.; Chen, S.; Yu, H. Selective catalytic reaction of NO_x with NH_3 over Ce–Fe/TiO_2-loaded wire-mesh honeycomb: Resistance to SO_2 poisoning. *Appl. Catal. B Environ.* **2014**, *150–151*, 630–635. [CrossRef]
109. Xu, J.; Li, H.; Liu, Y.; Huang, L.; Zhang, J.; Shi, L.; Zhang, D. In situ fabrication of porous $MnCo_xO_y$ nanocubes on Ti mesh as high performance monolith de-NO_x catalysts. *RSC Adv.* **2017**, *7*, 36319–36325. [CrossRef]
110. Xu, H.; Zhang, Q.; Qiu, C.; Lin, T.; Gong, M.; Chen, Y. Tungsten modified MnO_x–CeO_2/ZrO_2 monolith catalysts for selective catalytic reduction of NO_x with ammonia. *Chem. Eng. Sci.* **2012**, *76*, 120–128. [CrossRef]

catalysts

MDPI

Article

Enhanced Low Temperature NO Reduction Performance via MnO$_x$-Fe$_2$O$_3$/Vermiculite Monolithic Honeycomb Catalysts

Ke Zhang [1], Feng Yu [1,*], Mingyuan Zhu [1], Jianming Dan [1,2,3], Xugen Wang [1], Jinli Zhang [1] and Bin Dai [1,*]

[1] Key Laboratory for Green Processing of Chemical Engineering of Xinjiang Bingtuan, School of Chemistry and Chemical Engineering, Shihezi University, Shihezi 832003, China; kezhang1014@163.com (K.Z.); zhuminyuan@shzu.edu.cn (M.Z.); djm_tea@shzu.edu.cn (J.D.); wxgen@shzu.edu.cn (X.W.); zhangjinli@tju.edu.cn (J.Z.)
[2] Key Laboratory of Materials-Oriented Chemical Engineering of Xinjiang Uygur Autonomous Region, Shihezi 832003, China
[3] Engineering Research Center of Materials-Oriented Chemical Engineering of Xinjiang Production and Construction Corps, Shihezi 832003, China
* Correspondence: yufeng05@mail.ipc.ac.cn (F.Y.); db_tea@shzu.edu.cn (B.D.); Tel.: +86-993-205-7272 (F.Y. & B.D.); Fax: +86-993-205-7270 (F.Y. & B.D.)

Received: 31 January 2018; Accepted: 27 February 2018; Published: 28 February 2018

Abstract: Selective catalytic reduction of NO$_x$ by ammonia (NH$_3$-SCR) was the most efficient and economic technology for De-NO$_x$ applications. Therefore, a series of MnO$_x$/vermiculite (VMT) and MnO$_x$-Fe$_2$O$_3$/VMT catalysts were prepared by an impregnation method for the selective catalytic reduction (SCR) of nitrogen oxides (NO$_x$). The MnO$_x$-Fe$_2$O$_3$/VMT catalysts provided an excellent NO conversion of 96.5% at 200 °C with a gas hourly space velocity (GHSV) of 30,000 h^{-1} and an NO concentration of 500 ppm. X-ray photoelectron spectroscopy results indicated that the Mn and Fe oxides of the MnO$_x$-Fe$_2$O$_3$/VMT catalyst were mainly composed of MnO$_2$ and Fe$_2$O$_3$. However, the MnO$_2$ and Fe$_2$O$_3$ components were well dispersed because no discernible MnO$_2$ and Fe$_2$O$_3$ phases were observed in X-ray powder diffraction spectra. Corresponding MnO$_x$-Fe$_2$O$_3$/VMT monolithic honeycomb catalysts (MHCs) were prepared by an extrusion method, and the MHCs achieved excellent SCR activity at low temperature, with an NO conversion greater than 98.6% at 150 °C and a GHSV of 4000 h^{-1}. In particular, the MnO$_x$-Fe$_2$O$_3$/VMT MHCs provided a good SCR activity at room temperature (20 °C), with an NO conversion of 62.2% (GHSV = 1000 h^{-1}). In addition, the NO reduction performance of the MnO$_x$-Fe$_2$O$_3$/VMT MHCs also demonstrated an excellent SO$_2$ resistance.

Keywords: MnO$_x$-Fe$_2$O$_3$/vermiculite; monolithic honeycomb catalyst; room-temperature catalysis; selective catalytic reduction; NO removal efficiency

---

1. Introduction

Nitrogen oxides (NO$_x$) in stationary stack source emissions are strong contributing factors of acid rain, photochemical smog, and ozone depletion and are, therefore, detrimental to the natural environment and human health [1–3]. Therefore, the selective catalytic reduction (SCR) of NO$_x$ by ammonia (NH$_3$-SCR) was developed to reduce the release of NO$_x$ (i.e., De-NO$_x$), and this has thus far been the most economical and effective technology for De-NO$_x$ applications [4]. It is well known that the key component affecting the SCR and economic performance of NH$_3$-SCR technology is the catalyst employed in the process, which should be low in cost, and provide high SCR activity at low temperatures. As a result, manganese-based catalysts have been widely studied over the past few decades owing to their low cost and promisingly high SCR activity at low temperature [5].

However, the SCR activity of pure MnO_x catalysts is profoundly deteriorated by SO_2 and H_2O, and emissions with high concentrations of SO_2 and H_2O can lead to poisoning and the eventual deactivation of pure MnO_x catalysts. These drawbacks can be overcome, and the SCR activity of pure MnO_x catalysts can even be improved by incorporating transition metal or rare earth elements as catalyst promoters [6,7]. For example, Chen et al. reported that Fe-MnO_x mixed-oxide catalysts provided a 100% NO_x conversion in the temperature range from 140 °C to 220 °C [8].

Since SCR activity occurs on the surfaces of catalysts, the SCR activity of catalysts can be greatly enhanced, particularly under low-temperature conditions, by maximizing their surface area via distribution over high surface area supports. Numerous types of supports have been employed for Mn-based catalysts, such as molecular sieves, carbon materials, and metal oxides. Molecular sieves modified with Mn-based catalyst materials have shown excellent SCR activity because of their regular pore structures, high strength, and high specific surface areas [9]. Numerous Mn-based molecular sieve supported catalysts, such as Mn supported on ZSM-5 zeolite (denoted as Mn/ZSM-5) [10] and MnO_x/silicoaluminophosphate zeolite (SAPO-34) [11], have been shown to be highly active for NO_x reduction at low temperatures. Carbon materials have also been widely used as catalyst supports because of their well-developed porosity, high specific surface area, chemical stability, strong adsorption ability, and excellent thermal conductivity [12]. For example, Lu et al. synthesized MnO_2 supported on carbon nanotubes (CNTs), and the catalyst attained an NO conversion of 89.5% at 180 °C [13]. Finally, metal oxide supports provide high catalyst surface areas, high thermal stability, and surface acid-base properties [14,15]. Manganese-based catalysts supported on metal oxides, such as MnO_x/TiO_2 [16–18], MnO_x/Al_2O_3 [19], MnO_x/CeO_2 [20–22], and MnO_x/Ce-ZrO_2 [23], have generated considerable attention, and represent promising catalysts for the SCR of NO_x at low temperatures.

In recent years, many researchers have focused on catalysts employing supports composed of natural ore materials, such as montmorillonite, saponite, attapulgite, chabazite, and vermiculite (VMT), due to their great abundance, which makes them easily obtainable and inexpensive materials. A number of catalysts have incorporated these natural ore materials, such as Zr-Mn/attapulgite (ATP) [24], Ce-MnO_x/ATP [25], porous clay hetero-structures (PCH) modified with NH_3-Cu [26], Cu-chabazite (CHA) [27–29] and crystalline MnO_2 (c-MnO_2)/TiO_2-palygorskite (PAL) [30]. Among the available natural ore materials, the unique layered structure of VMT makes it an excellent candidate as a support for the SCR of NO. In addition, VMT includes metal oxides that can actively assist in catalytic reactions, and its great abundance in Xinjiang, China makes this material of particular interest to Chinese researchers. Chmielarz et al. investigated PCHs based on montmorillonite, saponite, and VMT modified with Fe or Cu as catalysts [31]. Among these catalysts, PCHs involving VMT achieved the best SCR activity at 400 °C. Samojeden et al. developed VMT supports prepared by modifications with nitric acid, and the resulting catalysts formed by impregnation with Cu provided an NO conversion of 94.3% at 350 °C [32]. Moreover, VMT has been widely investigated for use in heterogeneous catalysts, such as in the photocatalyst TiO_2/VMT [33], NiO/VMT for carbon monoxide methanation [34,35], and a novel $HgCl_2$/EML-VMT-C catalyst employing expanded multilayered (EML) VMT mixed with carbon on the surface (EML-VMT-C) for the hydrochlorination of acetylene [36]. However, the application of catalysts employing VMT supports for the SCR of NO_x at low temperatures remains challenging.

In light of the substantial cost benefits associated with the use of natural ore materials as catalyst supports, the present study employs VMT as supports in the fabrication of MnO_x-Fe_2O_3/VMT catalysts by the impregnation method for the SCR of NO by NH_3. The as-prepared MnO_x-Fe_2O_3/VMT catalysts provide an excellent NO conversion of 96.5% at 200 °C with a gas hourly space velocity (GHSV) of 30,000 h^{-1}. To improve the low temperature performance, corresponding MnO_x-Fe_2O_3/VMT monolithic honeycomb catalysts (MHCs) were prepared by an extrusion method, and the MHCs achieve an NO conversion greater than 98.6% at 150 °C and GHSV = 4000 h^{-1}. In particular, the MnO_x-Fe_2O_3/VMT MHCs provided good SCR activity at room temperature (20 °C), with an NO conversion of 62.2% (GHSV = 1000 h^{-1}). Finally, because flue gases still contain small amounts of

SO$_2$ after desulfurization and water removal, we also investigated the influence of SO$_2$ on the catalytic performance of MnO$_x$-Fe$_2$O$_3$/VMT MHCs.

2. Results and Discussion

The results of catalytic activity testing for the as-prepared MnO$_x$/VMT and MnO$_x$-Fe$_2$O$_3$/VMT catalysts are shown in Figure 1. As shown in Figure 1a, the NO conversion of the MnO$_x$/VMT and MnO$_x$-Fe$_2$O$_3$/VMT catalysts increased obviously with increasing temperature in the low temperature region until attaining a maximum value, after which the NO conversion decreased. The highest NO conversion attained for the MnO$_x$-Fe$_2$O$_3$/VMT catalyst was 96.5% at 200 °C, while the highest NO conversion attained for the MnO$_x$/VMT catalyst was 93.1% at 250 °C. Compared with the MnO$_x$/VMT catalyst, the catalytic activity of the MnO$_x$-Fe$_2$O$_3$/VMT catalyst was between 6% and 16% greater than that of the MnO$_x$/VMT catalyst in the low temperature range of 20–200 °C, indicating that the addition of Fe obviously increased the low temperature activity of the catalyst. For both catalysts, the declining NO conversion at relatively high temperatures originated from the oxidation of the ammonia.

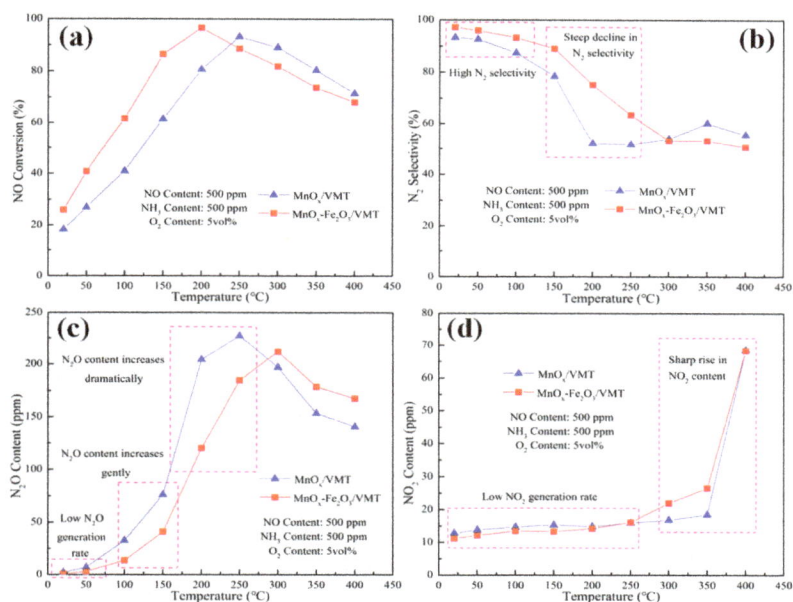

Figure 1. Catalytic activity of MnO$_x$/VMT and MnO$_x$-Fe$_2$O$_3$/VMT catalyst samples: (**a**) NO conversion; (**b**) N$_2$ selectivity; (**c**) N$_2$O content; and (**d**) NO$_2$ content (N$_2$ as balance gas, GHSV = 30,000 h^{-1}).

In addition, Figure 1b shows that the N$_2$ selectivity of the MnO$_x$/VMT and MnO$_x$-Fe$_2$O$_3$/VMT catalysts gradually declined with increasing temperature in the low temperature region (20–100 °C), and sharply declined in the middle temperature region (150–200 °C). The MnO$_x$-Fe$_2$O$_3$/VMT catalyst exhibited excellent N$_2$ selectivities of 97.1% and 95.9% at 20 °C and 50 °C, respectively, and its N$_2$ selectivity was greater than that of the MnO$_x$/VMT catalyst at temperatures less than 250 °C. However, we note that the sharp drop in the N$_2$ selectivity of the MnO$_x$-Fe$_2$O$_3$/VMT catalyst at 150 °C resulted in an N$_2$ selectivity that was less than that of the MnO$_x$/VMT catalyst at temperatures greater than 250 °C. The N$_2$ selectivity results are readily correlated with the measured N$_2$O and NO$_2$ contents shown in Figure 1c,d, respectively. Here, we note that the gradual decline in the N$_2$ selectivity of both catalysts at low temperature corresponds with a gradually increasing generation of N$_2$O and NO$_2$ over a similar temperature range. In addition, the sharply declining N$_2$ selectivity of both catalysts in the middle temperature region is mainly because the N_2O content is increasing rapidly for temperatures greater than 150 °C. We also note that the region over which the N_2 selectivity of the MnO_x/VMT catalyst was greater than that of the MnO_x-Fe_2O_3/VMT catalyst (i.e., at temperatures greater than 250 °C) corresponds with the fact that both the N_2O and the NO_2 contents were less for the MnO_x/VMT catalyst than for the MnO_x-Fe_2O_3/VMT catalyst at temperatures greater than 250 °C. Nevertheless, we note that the MnO_x-Fe_2O_3/VMT catalyst demonstrated both a better NO conversion and a better N_2 selectivity that those of the MnO_x/VMT catalyst in the temperature range of 20–200 °C.

Figure 2a presents X-ray diffractometer (XRD) patterns of the VMT support and as-prepared MnO_x/VMT and MnO_x-Fe_2O_3/VMT powdered catalyst samples. We note that VMT presents several strong peaks that exhibit greatly decreased intensities after impregnation. In addition, several diffraction peaks indicative of Mn_3O_4 (PDF#18-0803) are observed for the MnO_x/VMT catalyst at 18.1°, 29.2°, 32.4°, 36.2°, 51.0°, and 58.5°, while no others crystal phases of MnO_x are evident. These results suggest that the existence of diffraction peaks could be due to large crystals of MnO_x, resulting in weak XRD peaks indicative of MnO_x and VMT [37]. The MnO_x-Fe_2O_3/VMT catalyst sample presents even weaker XRD peaks indicative of MnO_x and VMT, and no additional crystal phases are observed. Here, the coexistence of manganese and iron oxides enhances dispersion, and consequently reduces the crystallinity, indicating the presence of strong interactions between these two metal oxides [38].

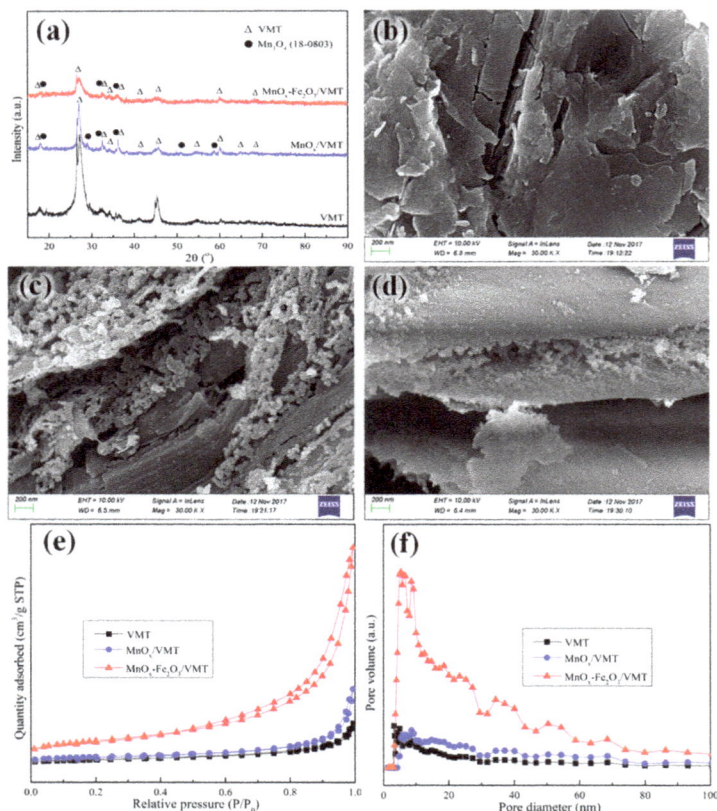

Figure 2. XRD patterns (**a**); N_2 isotherms (**b**) and pore diameter distribution curves (**c**); scanning electron microscopy (SEM) images (**d**–**f**) of the VMT support and MnO_x/VMT and MnO_x-Fe_2O_3/VMT catalyst samples.

Figure 2b–d present SEM micrographs indicative of the morphologies of the VMT support and as-prepared MnO_x/VMT and MnO_x-Fe_2O_3/VMT catalysts, respectively. We note the distinct layered structure of the VMT support, and that the layer surfaces are very smooth, without obvious pores or wrinkles. After impregnation, the MnO_x/VMT catalyst exhibits a distribution of irregular loose particles, and the layer surfaces of the VMT support appear to be very rough (Figure 2c), indicating the formation of many new channels on the VMT surfaces. However, the sizes of the irregular loose particles of the MnO_x-Fe_2O_3/VMT catalyst are substantially decreased. These results explain the reason for the increased surface area and decreased pore diameter of the MnO_x-Fe_2O_3/VMT catalyst relative to those of the MnO_x/VMT catalyst.

The N_2 adsorption-desorption isotherm plots and the corresponding BJH pore size distribution curves of the VMT support and as-prepared MnO_x/VMT and MnO_x-Fe_2O_3/VMT catalyst samples are presented in Figure 2e,f, respectively. From Figure 2e, we note the presence of well-defined type II hysteresis loops with sloping adsorption branches for all samples, which is particularly pronounced for the MnO_x-Fe_2O_3/VMT catalyst. From Figure 2f, we note that both catalyst samples exhibit three narrow peaks in the pore diameter range of 2–10 nm. The textural data for all samples are listed in Table 1. From the table, we find that the impregnation of 20 wt% Mn more than doubled the BET surface area of the MnO_x/VMT catalyst sample relative to that of the VMT support, but, surprisingly, the pore diameter decreased by about 30%. Moreover, the impregnation of 20 wt% Mn and 5 wt% Fe further increased the BET surface area of the MnO_x-Fe_2O_3/VMT catalyst sample by a factor greater than 3 relative to the MnO_x/VMT catalyst sample, while the pore diameter further decreased by about 12%. These results indicate that the doping of second metal can change the surface structure of the MnO_x-Fe_2O_3/VMT catalyst [39], and the calcination subsequent to impregnation may have formed additional channels in the surfaces of the VMT support. A high BET surface area is beneficial toward increasing the number of active sites of a catalyst and, thus, provides an increased NO conversion.

Table 1. Physical properties of VMT support and catalyst samples.

Samples	BET Surface Area (m^2/g)	Pore Volume (cm^3/g)	Pore Diameter (nm)
VMT support	2.9	0.02	26.2
MnO_x/VMT	6.7	0.03	18.4
MnO_x-Fe_2O_3/VMT	21.9	0.09	16.2
MnO_x-Fe_2O_3/VMT MHCs	15.4	0.07	17.9

The XPS spectra of the as-prepared MnO_x/VMT and MnO_x-Fe_2O_3/VMT catalysts are shown in Figure 3, and their principle surface compositions obtained from the fitted spectra in Figure 3b–d are listed in Table 2. Sharp photoelectron peaks are observed in Figure 3a for Fe, Mn, O, and C elements with binding energies of 712.1 eV (Fe $2p_{3/2}$), 642.1 eV (Mn $2p_{3/2}$), 532.1 eV (O 1s), and 284.1 eV (C 1s), respectively. We note that the XPS spectrum obtained for the MnO_x/VMT catalyst also exhibits a small peak indicative of Fe because the VMT support material naturally contains a small concentration of Fe. The Fe 2p, Mn 2p, and O 1s spectra of the catalyst samples are individually discussed in detail below.

Table 2. Surface compositions of representative catalyst samples obtained by XPS analysis.

Samples	Mn^{4+}/Mn_{total} (%)	O_{ads}/O_{total} (%)	Fe^{3+}/Fe_{total} (%)
MnO_x/VMT	26.3	38.5	51.5
MnO_x-Fe_2O_3/VMT	35.9	46.0	59.1

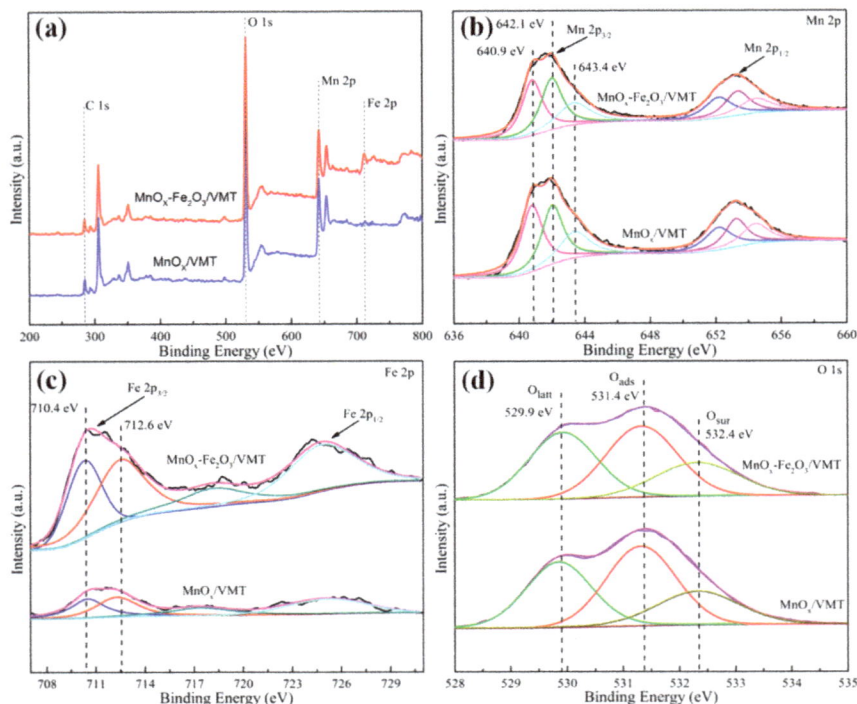

Figure 3. XPS survey spectra (**a**); Mn 2p spectra (**b**); Fe 2p spectra (**c**); O 1s spectra (**d**) of MnO$_x$/VMT and MnO$_x$-Fe$_2$O$_3$/VMT catalyst samples.

As shown in Figure 3b, the Mn 2p spectra include two main peaks at binding energies of 653.8 ± 0.4 eV and 641.9 ± 0.4 eV, which are assigned to Mn 2p$_{3/2}$ and Mn 2p$_{1/2}$ electron states, respectively. To identify the specific Mn species of each sample, the Mn 2p$_{3/2}$ peak was deconvoluted into three peaks, corresponding to Mn^{2+} (641.0 ± 0.4 eV), Mn^{3+} (642.1 ± 0.4 eV), and Mn^{4+} (643.5 ± 0.4 eV), respectively [40]. The percentage of Mn atoms in the Mn^{4+} state listed in Table 2 was then determined as the area under the curve representative of Mn^{4+} relative to the total area under the Mn 2p$_{3/2}$ curve. These values are indicative of the molar concentration of MnO$_2$ relative to all MnO$_x$ on the surfaces of the MnO$_x$-Fe$_2$O$_3$/VMT and MnO$_x$/VMT catalysts. We note that the Mn^{4+} percentage increased by about 37% from the MnO$_x$/VMT catalyst to the MnO$_x$-Fe$_2$O$_3$/VMT catalyst. This indicates that the addition of Fe facilitates the conversion of MnO$_x$ to MnO$_2$ on the catalyst surface. It has been reported [41–43] that the NO conversion capability of pure manganese oxides can be ranked as MnO$_2$ > Mn$_5$O$_8$ > Mn$_2$O$_3$ > Mn$_3$O$_4$. In addition, it has been reported that a greater concentration of MnO$_2$ on the catalyst surface promotes the SCR reaction [44]. Therefore, it can be expected that the MnO$_x$-Fe$_2$O$_3$/VMT catalyst will provide an improved NO conversion relative to that of the MnO$_x$/VMT catalyst.

The Fe 2p spectra presented in Figure 3c exhibit electron binding energy peak values for Fe 2p$_{3/2}$ and Fe 2p$_{1/2}$ states, and a satellite peak of 710.7, 725.0, and 718.6 eV, respectively. The satellite peak energy corresponds well with that reported for Fe$_2$O$_3$ [45]. The Fe 2p$_{3/2}$ peak was deconvolved into two components with peaks at 710.4 eV and 712.6 eV indicative of Fe^{2+} and Fe^{3+} phases [46], respectively. The results indicate that the percentage of Fe atoms in the Fe^{3+} state listed in Table 2 are about 13% less in the VMT support of the MnO$_x$/VMT catalyst than that of the MnO$_x$-Fe$_2$O$_3$/VMT catalyst.

According to a past study [47], Fe^{3+} sites may facilitate the reduction of NO_x at low temperature. Thus, the MnO_x-Fe_2O_3/VMT catalyst can be expected to provide a slightly better low-temperature activity than that of the MnO_x/VMT catalyst.

As shown in Figure 3d the O 1s peaks of the MnO_x/VMT and MnO_x-Fe_2O_3/VMT catalysts at 528–535 eV were deconvolved into three peaks, denoted as O_{latt} (529.9 eV), O_{ads} (531.4 eV), and O_{sur} (532.4 eV), which are attributed to O atoms bonded with metal cations, in adsorbed water, and in surface hydroxyl groups, respectively [48]. It has been widely reported that oxygen in the gas phase can be activated by oxygen vacancies on the surface of SCR catalysts. Therefore, the relative abundance of O_{ads} is of particular interest because an increased percentage of O_{ads} can promote the oxidation of NO to NO_2, and enhance the SCR performance at low temperature through a rapid NH_3-SCR route [49,50]. As shown in Table 2, the ratio of O_{ads}/O_{total} on the surface of the MnO_x-Fe_2O_3/VMT catalyst is about 19% greater than that for the MnO_x/VMT catalyst. As a result, we can expect this factor to further enhance the low-temperature activity of the MnO_x-Fe_2O_3/VMT catalyst relative to that of the MnO_x/VMT catalyst.

The results of H_2-TPR testing are presented in Figure 4 for the VMT support and the as-prepared MnO_x/VMT and MnO_x-Fe_2O_3/VMT catalysts. We note that several peaks are observable for all samples within the temperature range of 200–800 °C. The H_2-TPR curve for the VMT support includes only two peaks, with a reduction peak attributed to $Fe_2O_3 \rightarrow Fe_3O_4$ at 274 °C, and a second peak located at 688 °C that may be attributed to $Fe_3O_4 \rightarrow FeO$ [51]. Compared with the VMT support curve, two new reduction peaks are observed in the H_2-TPR curve for the MnO_x/VMT catalyst due to the addition of Mn. From previous studies [52], the reduction peaks of MnO_x can be assigned to the reduction processes of MnO_2 via Mn_2O_3 to MnO. Here, the first peak observed at 269 °C can be attributed to the $MnO_2 \rightarrow Mn_2O_3$ reduction transition, while the peak at 469 °C can be attributed to $Mn_2O_3 \rightarrow MnO$ [53]. Therefore, because the MnO_x/VMT catalyst includes Fe_2O_3 in the support, the peak centered at 269 °C can be attributed to both $Fe_2O_3 \rightarrow Fe_3O_4$ and $MnO_2 \rightarrow Mn_2O_3$. Meanwhile, the third peak for the MnO_x/VMT sample located at 608 °C can be attributed to $Fe_3O_4 \rightarrow FeO$, and the forth peak located at 742 °C can be attributed to $FeO \rightarrow Fe$ [54]. The H_2-TPR curve for the MnO_x-Fe_2O_3/VMT catalyst exhibited similar reduction peaks above 260 °C, but which were shifted to lower temperatures relative to those for the MnO_x/VMT catalyst, suggesting that the Fe and Mn species in the MnO_x-Fe_2O_3/VMT catalyst were more easily reduced. This can be ascribed to the previously discussed synergetic effect between Mn and Fe, which could effectively promote the redox properties of the MnO_x-Fe_2O_3/VMT catalyst and improve its catalytic activity.

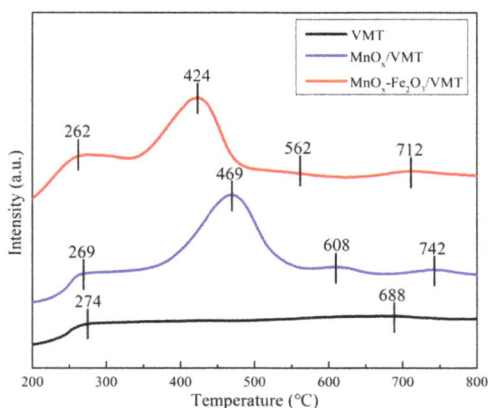

Figure 4. H_2-TPR curves of the VMT support and MnO_x/VMT and MnO_x-Fe_2O_3/VMT catalyst samples.

A photograph of the as-prepared MnO_x-Fe_2O_3/VMT MHCs is shown in Figure 5a. The macrostructure of the MnO_x-Fe_2O_3/VMT MHCs is important for promoting catalytic activity in industrial applications, and their primary physical properties have significant effects on their mechanical and catalytic performances. The SEM micrograph of an as-prepared MnO_x-Fe_2O_3/VMT MHCs in Figure 5b shows that the glass fibers were uniformly distributed throughout the MnO_x-Fe_2O_3/VMT MHCs, which can be expected to provide enhanced mechanical strength. The XRD pattern of a representative MnO_x-Fe_2O_3/VMT MHCs is shown in Figure 5c. In contrast with the XRD results for the MnO_x-Fe_2O_3/VMT powdered catalyst (Figure 2a), the bentonite content in the MHCs, which has a large proportion of SiO_2, yields a sharp peak attributable to SiO_2 (PDF#27-0605) at 21.6°. In addition, two sharp peaks attributable to carbon (PDF#41-1487) are observed at 26.4° and 54.5° owing to the decomposition of the organic additives. Figure 5d presents the N_2 isotherms and corresponding pore size distribution curve (inset) of a MnO_x-Fe_2O_3/VMT MHCs, and the corresponding textural data are listed Table 1. Here, we note that, while the BET surface area and poor volume of the MnO_x-Fe_2O_3/VMT MHCs were less than those of its powdered counterpart, the average pore diameter was increased by around 10%. The porous texture of the MnO_x-Fe_2O_3/VMT MHCs can be expected to play a significant role in the SCR of NO_x owing to an enhanced transportation and adsorption of reactant gases.

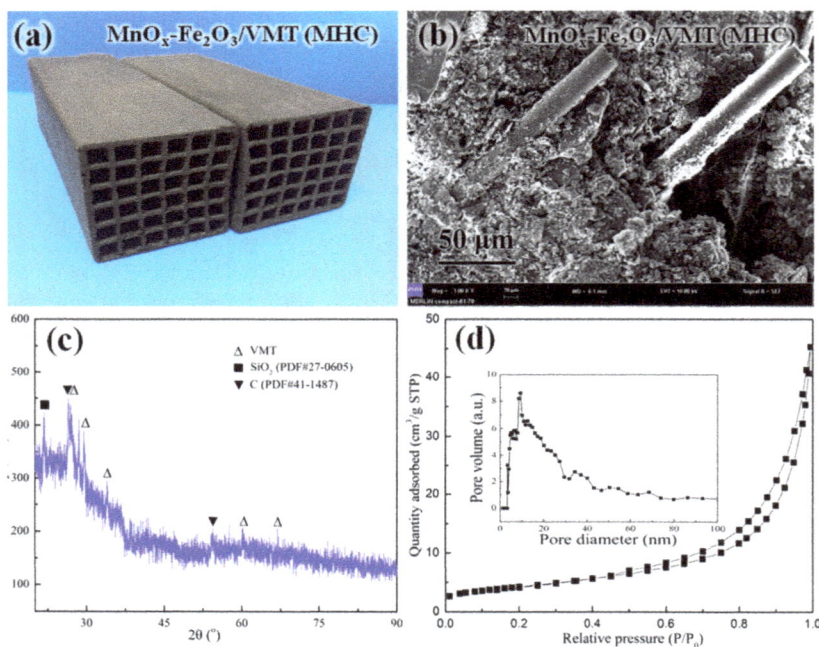

Figure 5. MnO_x-Fe_2O_3/VMT MHCs: (**a**) photograph of as-prepared MHCs; (**b**) SEM image; (**c**) XRD pattern; (**d**) N_2 isotherms and corresponding pore size distribution curve (inset).

The NO conversion of MnO_x-Fe_2O_3/VMT MHCs at different GHSV values is shown as a function of reaction temperature in Figure 6a. We note that the NO conversion decreased with an increasing GHSV from 4000 h^{-1} to 8000 h^{-1}. The temperature region of highest activity was between 150 °C and 200 °C, and the NO conversion attained a maximum value greater than 98% with GHSV = 4000 h^{-1}. Meanwhile, the MnO_x-Fe_2O_3/VMT MHCs exhibited an excellent NO conversion of 39.2% and 53.4% at 20 °C and 50 °C, respectively, with GHSV = 4000 h^{-1}. The cycling stabilities of MnO_x-Fe_2O_3/VMT

MHCs are presented in Figure 6b at various temperatures with different GHSV values. We note that the NO conversion was stable, with no significant changes over a full 10 h of testing. We also find that the MnO_x-Fe_2O_3/VMT MHCs provided excellent NO conversion values of 98.6% and 85.8% at 150 °C and 100 °C, respectively, with GHSV = 4000 h^{-1}. The results obtained are comparable to those reported in the literature (Figure 6c). Thus, a NO conversion of 92% at 500 °C has been reported for WO_3-TiO_2 (GHSV of 11,000 h^{-1}) [55], a conversion of 98% at 400 °C was reached by V_2O_5-MoO_3/TiO_2 (GHSV of 6000 h^{-1}) [56]. A TiO_2 catalyst showed a conversion of 80% at 300 °C (GHSV of 25,000 h^{-1}) [57], 75% at 250 °C for a V_2O_5-WO_3/TiO_2 catalyst (GHSV of 27,000 h^{-1}) [58], and 90% at 160 °C for a Cr-V/TiO_2 catalyst (GHSV of 4000 h^{-1}) [59]. Even at GHSV = 2000 h^{-1}, the MnO_x-Fe_2O_3/VMT MHCs provided an NO conversion of 79.1% at 50 °C. Surprisingly, the MHCs exhibited excellent NO conversion values of 62.2% and 50.2% with GHSV = 1000 h^{-1} and 2000 h^{-1}, respectively, at 20 °C. The excellent SCR performance obtained for the proposed MnO_x-Fe_2O_3/VMT MHCs will strongly contribute to low-temperature De-NO_x processes, and suggests that even room temperature processes are feasible.

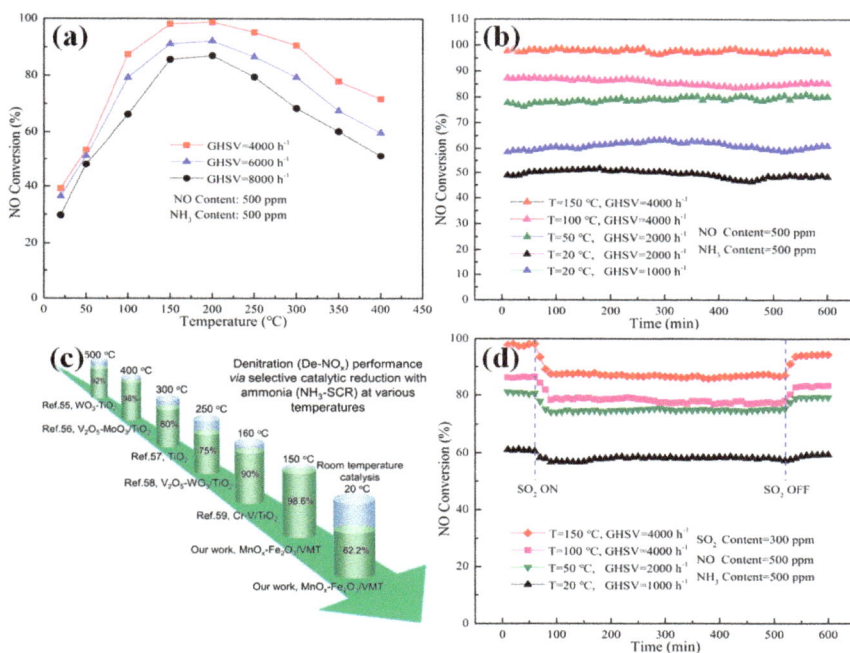

Figure 6. Catalytic activity of MnO_x-Fe_2O_3/VMT MHCs: (**a**) NO conversion; (**b**) cycling stabilities; (**c**) Comparison of pervious reported activity of various MHCs for NH_3-SCR; (**d**) SO_2 resistance at various temperatures with different GHSV.

The impact of SO_2 on the cycling performance of MnO_x-Fe_2O_3/VMT MHCs at different temperatures and GHSV values is shown in Figure 6d. After the first 1 h of testing when 300 ppm SO_2 was introduced into the reaction gas, we note that the extent to which the NO conversion decreased was only slight at 20 °C (GHSV = 1000 h^{-1}), but increased with increasing temperature from 50 °C (GHSV = 2000 h^{-1}) to 150 °C (GHSV = 4000 h^{-1}). This is because the crystallization temperature of sulfate is greater than 40 °C. We also note that the NO conversion of the MnO_x-Fe_2O_3/VMT MHCs was stable over the entire period of SO_2 addition, which indicates good SO_2 resistance. However, the NO conversion values obtained under all conditions did not recover to their original values when the addition of SO_2 was discontinued. These results suggest that the decreased activity of the MHCs was not due to the competitive adsorption of SO_2, but because of the formation of sulfates covering the active sites of the catalysts [60].

3. Materials and Methods

3.1. Catalysts Preparation

3.1.1. Preparation of MnO_x/VMT and MnO_x-Fe_2O_3/VMT Catalysts

Vermiculite supports were prepared by a microwave method as follows. The raw VMT (Xinjiang Yuli Xinlong Vermiculite Co., Ltd., Korla, China) was washed with water until no trace of foreign material was observable under visual inspection, and then dried in an oven at 100 °C. Finally, the washed VMT was placed in a 500 mL beaker and expanded in a microwave. The VMT was collected and placed in a sealed container, and then crushed prior to use. All the catalysts were prepared by the impregnation method. We completely dissolved $Mn(CH_3COO)_2 \cdot 4H_2O$ in water under stirring in an appropriate ratio to the VMT content (i.e., 20 wt% Mn). We added the VMT powder, and continued stirring for 10 h. The sample was dried in air at 100 °C for 12 h, and then crushed and sieved in an 80–100 mesh sieve. Finally, the sample was calcined in air at 500 °C for 5 h. We prepared MnO_x-Fe_2O_3/VMT catalysts by an equivalent method using $Mn(CH_3COO)_2 \cdot 4H_2O$ and $Fe(NO_3)_3 \cdot 9H_2O$ (20 wt% Mn-5 wt% Fe relative to VMT).

3.1.2. Preparation of MnO_x-Fe_2O_3/VMT Monolithic Honeycomb Catalysts

We synthesized MnO_x-Fe_2O_3/VMT MHCs by an extrusion molding method. The powdered MnO_x-Fe_2O_3/VMT catalyst was dry mixed with 10 wt% bentonite and 3 wt% carboxy methyl cellulose (CMC) in a blender mixer. The materials in the blender mixer were then wet mixed with 10 wt% glycerin, followed by a sufficient amount of water to ensure an appropriate viscosity for extrusion. The resulting mixture was kneaded by hand for 30 min until achieving a uniform consistency. The sample was subjected to vacuum de-airing and aged for 24 h to increase the plasticity. The aged sample was molded by an extruding machine to obtain a monolithic honeycomb structure. The honeycomb sample was dried at 70 °C in air for 12 h in a muffle furnace, further dried at 100 °C for 12 h, and then calcined at 500 °C for 5 h. The rate of temperature increase was 1 °C/min for all the above steps. The mold size was 3.3 cm × 3.3 cm with 36 channels (i.e., 6 × 6 cells).

3.2. Catalyst Characterization

The morphologies of the powder catalysts and MHCs were characterized by scanning electron microscopy (SEM; Hitachi S-4300, Hitachi Limited, Tokyo, Japan). A BET apparatus (Micromeritics ASAP 2020, Micromeritics Instrument Ltd., Norcross, GA, USA) was employed to measure the Brunauer-Emmett-Teller (BET) specific surface area and Barrett-Joyner-Halenda (BJH) pore structure of the powder catalysts and MHCs. The samples were degassed in vacuum at 200 °C for 4 h prior to measurement. The total pore volume was calculated from the volume of nitrogen adsorbed at $P/P_0 = 0.99$. Powder X-ray diffraction (XRD) patterns were recorded on a Bruker D8 Advance X-ray diffractometer (Bruker Biosciences Corporation, Billerica, MA, USA) with Cu Kα radiation ($\lambda = 1.5406$ Å) operated at 40 kV and 40 mA. X-ray photoelectron spectroscopy (XPS) data were obtained with an AMICUS/ESCA 3400 electron spectrometer from Kratos Analytical (Manchester, UK) using Mg Kα radiation (20 mA, 12 kV). Binding energies were referenced to the C 1s line at 284.8 eV for adventitious carbon. We conducted H_2 temperature programmed reduction (H_2-TPR) testing to analyze the redox properties of the catalysts using a Micromeritics ChemiSorb 2720TPx system (Micromeritics Instrument Ltd., Norcross, GA, USA) in a temperature range of 20 °C to 800 °C at a rate of 10 °C/min with a gas (10 vol% H_2 relative to Ar) flow rate of 40 mL/min, and then retained for 20 min at 800 °C.

3.3. Activity Measurement

We prepared MnO_x/VMT and MnO_x-Fe_2O_3/VMT catalytic activity studies using a fixed bed microreactor. The reactor was composed of a stainless steel tube with a 10.0 mm inner diameter. Prior to conducting the experiments, quartz sand and quartz wool were placed inside the reaction tube to ensure contact between the powdered catalysts and the thermocouple. The typical composition of the simulated flue gas was 500 ppm NO (denoted as $[NO]_{in}$), 500 ppm NH_3 (denoted as $[NH_3]_{in}$), and 5 vol% O_2 with N_2 as the balance gas. The total volume flow was 100 mL/min, representing a GHSV of 30,000 h^{-1}. The catalytic activity of the powdered catalysts was evaluated in the temperature range of 20 °C to 400 °C during testing according to the exiting concentrations of NO (denoted as $[NO]_{out}$) and NH_3 (denoted as $[NH_3]_{out}$) determined by Fourier transform infrared (FTIR) spectroscopy (Nicolet IS10, Thermo Fisher Scientific, Waltham, MA, USA). The NO conversion ($[NO]_{conversion}$) and N_2 selectivity ($[N_2]_{selectivity}$) were calculated using the following equations:

$$[NO]_{conversion} = \frac{[NO]_{in} - [NO]_{out}}{[NO]_{in}} \times 100\% \tag{1}$$

$$[N_2]_{selectivity} = \left[1 - \frac{[NO_2]_{out} + 2[N_2O]_{out}}{[NO]_{in} - [NO]_{out} + [NH_3]_{in} - [NH_3]_{out}}\right] \times 100\% \tag{2}$$

The catalytic performance of MnO_x-Fe_2O_3/VMT MHCs was evaluated using a similar fixed bed microreactor composed of a quartz tube with a 5.0 cm inner diameter. The simulated flue gas was composed of 500 ppm NO and 500 ppm NH_3 with air as the balance gas. Prior to testing, each MHCs sample was placed in the reaction tube, and then exposed to the simulated flue gas for 1 h to eliminate the influence of adsorption on the catalysts. Then, $[NO]_{out}$ was measured online using a flue gas analyzer (QUINTOX-KM9106, Kane International, New York, NY, USA), and $[NO]_{conversion}$ was calculated using Equation (1). Unless otherwise stated here, all other conditions were equivalent to the conditions employed for powdered catalysts.

4. Conclusions

This paper presented the successful preparation of MnO_x-Fe_2O_3/VMT catalysts with a layered structure by an impregnation method for the first time. The as-prepared MnO_x-Fe_2O_3/VMT catalysts exhibited high NO conversion and N_2 selectivity for the NH_3-SCR of NO in the temperature range of 20–200 °C. The catalysts provided an excellent NO conversion of 96.5% at 200 °C with GHSV = 30,000 h^{-1} and an NO concentration of 500 ppm. Compared with VMT and MnO_x/VMT catalysts, the results of extensive characterization indicated that the high catalytic activity of the MnO_x-Fe_2O_3/VMT catalysts can be attributed to a number of advantageous properties, such as a large specific surface area, high ratios of Mn^{4+}/Mn_{total} and Fe^{3+}/Fe_{total}, and easily reduced Mn species. In addition, the MnO_x-Fe_2O_3/VMT powdered catalyst was successfully employed to form MHCs by an extrusion method. The as-prepared MnO_x-Fe_2O_3/VMT MHCs provided an NO conversion of 98.6% at 150 °C with GHSV = 4000 h^{-1}. Moreover, the MHCs presented excellent De-NO_x performance at low temperature, obtaining an NO conversion of 62.2% at 20 °C with GHSV = 1000 h^{-1}. Furthermore, the MnO_x-Fe_2O_3/VMT MHCs also provided excellent cycling stability, and maintained comparable NO conversion values even after 10 h. Finally, the MnO_x-Fe_2O_3/VMT MHCs demonstrated excellent SO_2 resistance at low temperature (particularly at room temperature). Therefore, the prepared MnO_x-Fe_2O_3/VMT MHCs offer considerable potential for low or even room temperature De-NO_x applications in stationary stack source emissions.

Acknowledgments: The work was supported by National High Technology Research and Development Program of China (863 program) (no. 2015AA03A401), Program for Changjiang Scholars and Innovative Research Team in University (no. IRT_15R46) and the Program of Science and Technology Innovation Team in Bingtuan (no. 2015BD003).

Author Contributions: F.Y. and B.D. designed and administered the experiments. K.Z. performed experiments. M.Z., J.D., X.W., and J.Z. collected and analyzed data. All authors discussed the data and wrote the manuscript.

Conflicts of Interest: The authors declare no conflicts of interests.

References

1. Cai, S.; Liu, J.; Zha, K.; Li, H.; Shi, L.; Zhang, D. A general strategy for the in situ decoration of porous Mn-Co bi-metal oxides on metal mesh/foam for high performance de-NO_x monolith catalysts. *Nanoscale* **2017**, *9*, 5648–5657. [CrossRef] [PubMed]

2. Boningari, T.; Smirniotis, P.G. Impact of nitrogen oxides on the environment and human health: Mn-based materials for the NO_x abatement. *Curr. Opin. Chem. Eng.* **2016**, *13*, 133–141. [CrossRef]

3. Li, J.H.; Chang, H.Z.; Ma, L.; Hao, J.M.; Yang, R.T. Low-temperature selective catalytic reduction of NO_x with NH_3 over metal oxide and zeolite catalysts—A review. *Catal. Today* **2011**, *175*, 147–156. [CrossRef]

4. Zhu, Y.; Chen, B.; Zhao, R.; Zhao, Q.; Gies, H.; Xiao, F.; Vos, D.E.D.; Yokoi, T.; Bao, X.; Kolb, U.; et al. Fe-doped Beta Zeolite from Organotemplate-free Synthesis for NH_3-SCR of NO_x. *Catal. Sci. Technol.* **2016**, *6*, 6581–6592. [CrossRef]

5. Liu, C.; Shi, J.W.; Niu, C. Manganese oxide-based catalysts for low-temperature selective catalytic reduction of NO_x with NH_3: A review. *Appl. Catal. A Gen.* **2016**, *522*, 54–69. [CrossRef]

6. Long, R.Q.; Yang, R.T.; Chang, R. Low temperature selective catalytic reduction (SCR) of NO with NH_3 over Fe-Mn based catalysts. *Chem. Commun.* **2012**, *5*, 452–453.

7. Boningari, T.; Pappas, D.K.; Ettireddy, P.R.; Kotrba, A.; Smirniotis, P.G. Influence of SiO_2 on M/TiO_2 M = Cu, Mn, and Ce) formulations for low-temperature selective catalytic reduction of NO_x with NH_3: Surface properties and key components in relation to the activity of NO_x reduction. *Ind. Eng. Chem. Res.* **2015**, *54*, 2261–2273. [CrossRef]

8. Chen, Z.; Wang, F.; Li, H.; Yang, Q.; Wang, L.F.; Li, X.H. Low-Temperature Selective Catalytic Reduction of NO_x with NH_3 over Fe–Mn Mixed-Oxide Catalysts Containing $Fe_3Mn_3O_8$ Phase. *Ind. Eng. Chem. Res.* **2012**, *51*, 202–212. [CrossRef]

9. Li, R.; Li, Z.B.; Chen, L.Q.; Dong, Y.L.; Ma, S.B.; Yuan, F.L.; Zhu, Y.J. Synthesis of MnNi–SAPO-34 by a one-pot hydrothermal method and its excellent performance for the selective catalytic reduction of NO by NH_3. *Catal. Sci. Technol.* **2017**, *7*, 4989–4995. [CrossRef]

10. Lou, X.; Liu, P.; Li, J.; Li, Z.; He, K. Effects of calcination temperature on Mn species and catalytic activities of Mn/ZSM-5 catalyst for selective catalytic reduction of NO with ammonia. *Appl. Sur. Sci.* **2014**, *307*, 382–387. [CrossRef]

11. Yu, C.; Dong, L.; Feng, C.; Liu, X.; Huang, B. Low-temperature SCR of NO_x by NH_3 over $MnO_x/SAPO-34$ prepared by two different methods: A comparative study. *Environ. Technol.* **2017**, *38*, 1030–1042. [CrossRef] [PubMed]

12. Liu, J.; Zhang, K.; Si, M.; Lian, J.H.; Liu, L.S.; Gou, X. Experimental Research on Catalysts of V_2O_5/AC and $V_2O_5/CNTs$ for Low Temperature SCR Denitrification. *Appl. Mech. Mater.* **2014**, *694*, 478–483. [CrossRef]

13. Lu, X.; Zheng, Y.; Zhang, Y.; Qiu, H. Low-temperature selective catalytic reduction of NO over carbon nanotubes supported MnO_2 fabricated by co-precipitation method. *Micro Nano Lett.* **2015**, *10*, 666–669. [CrossRef]

14. Lin, F.; Wu, X.; Liu, S.; Weng, D.; Huang, Y. Preparation of $MnO_x–CeO_2–Al_2O_3$ mixed oxides for NO_x-assisted soot oxidation: Activity, structure and thermal stability. *Chem. Eng. J.* **2013**, *226*, 105–112. [CrossRef]

15. Jin, R.; Liu, Y.; Wu, Z.; Wang, H.; Gu, T. Low-temperature selective catalytic reduction of NO with NH_3 over MnCe oxides supported on TiO_2 and Al_2O_3: A comparative study. *Chemosphere* **2010**, *78*, 1160–1166. [CrossRef] [PubMed]

16. Pappas, D.K.; Boningari, T.; Boolchand, P.; Smirniotis, P.G. Novel manganese oxide confined interweaved titania nanotubes for the low-temperature selective catalytic reduction (SCR) of NO_x by NH_3. *J. Catal.* **2016**, *334*, 1–13. [CrossRef]

17. Ettireddy, P.R.; Ettireddy, N.; Boningari, T.; Pardemann, R.; Smirniotis, P.G. Investigation of the selective catalytic reduction of nitric oxide with ammonia over Mn/TiO_2 catalysts through transient isotopic labeling and in situ FT-IR studies. *J. Catal.* **2012**, *292*, 53–63. [CrossRef]

18. Smirniotis, P.G.; Peña, D.A.; Uphade, B.S. Low-Temperature Selective Catalytic Reduction (SCR) of NO with NH_3 by Using Mn, Cr, and Cu Oxides Supported on Hombikat TiO_2. *Angew. Chem. Int. Ed. Eng.* **2001**, *40*, 2479–2482. [CrossRef]

19. Singoredjo, L.; Korver, R.; Kapteijn, F.; Moulijn, J. Alumina supported manganese oxides for the low-temperature selective catalytic reduction of nitric oxide with ammonia. *Appl. Catal. B* **1992**, *1*, 297–316. [CrossRef]

20. Ma, K.; Zou, W.X.; Zhang, L.; Li, L.L.; Yu, S.H.; Tang, C.J.; Gao, F.; Dong, L. Construction of hybrid multi-shell hollow structured CeO_2-MnO_x materials for selective catalytic reduction of NO with NH_3. *RSC Adv.* **2017**, *7*, 5989–5999. [CrossRef]

21. Liu, C.; Gao, G.; Shi, J.W.; He, C.; Li, G.D.; Bai, N.; Niu, C. MnO_x-CeO_2 shell-in-shell microspheres for NH_3-SCR de-NO_x at low temperature. *Catal. Commun.* **2016**, *86*, 36–40. [CrossRef]

22. Qi, G.; Yang, R.T.; Chang, R. MnO_x-CeO_2 mixed oxides prepared by co-precipitation for selective catalytic reduction of NO with NH_3 at low temperatures. *Appl. Catal. B Environ.* **2004**, *51*, 93–106. [CrossRef]

23. Hu, H.; Zha, K.; Li, H.; Shi, L.; Zhang, D. In situ DRIFTs investigation of the reaction mechanism over MnO_x-MO_y/$Ce_{0.75}Zr_{0.25}O_2$ (M = Fe, Co, Ni, Cu) for the selective catalytic reduction of NO_x with NH_3. *Appl. Sur. Sci.* **2016**, *387*, 921–928. [CrossRef]

24. Xiao, F.; Gu, Y.; Tang, Z.; Han, F.; Shao, J.; Xu, Q.; Zhu, H. ZrO_2 Modified MnO_x/Attapulgite Catalysts for NH_3-SCR of NO at Low Temperature. *J. Chem. Eng. Jpn.* **2015**, *48*, 481–487. [CrossRef]

25. Xie, A.; Zhou, X.; Huang, X.; Ji, L.; Zhou, W.; Luo, S.; Yao, C. Cerium-loaded MnO_x/attapulgite catalyst for the low-temperature NH_3-selective catalytic reduction. *J. Ind. Eng. Chem.* **2017**, *49*, 230–241. [CrossRef]

26. Chmielarz, L.; Kuśtrowski, P.; Dziembaj, R.; Cool, P.; Vansant, E.F. Selective catalytic reduction of NO with ammonia over porous clay heterostructures modified with copper and iron species. *Catal. Today* **2007**, *119*, 181–186. [CrossRef]

27. Hou, X.X.; Schmieg, S.J.; Wei, L.; Epling, W.S. NH_3 pulsing adsorption and SCR reactions over a Cu-CHA SCR catalyst. *Catal. Today* **2012**, *197*, 9–17. [CrossRef]

28. Schmieg, S.J.; Oh, S.H.; Kim, C.H.; Brown, D.B.; Lee, J.H.; Pedenc, C.H.F.; Kim, D.H. Thermal durability of Cu-CHA NH_3-SCR catalysts for diesel NO_x reduction. *Catal. Today* **2012**, *184*, 252–261. [CrossRef]

29. Brookshear, D.W.; Nam, J.G.; Ke, N.; Toops, T.J.; Binder, A. Impact of sulfation and desulfation on NO_x reduction using Cu-chabazite SCR catalysts. *Catal. Today* **2015**, *258*, 359–366. [CrossRef]

30. Luo, S.; Zhou, W.; Xie, A.; Wu, F.; Yao, C.; Li, X.; Zuo, S.; Liu, T. Effect of MnO_2 polymorphs structure on the selective catalytic reduction of NO_x with NH_3 over TiO_2–Palygorskite. *Chem. Eng. J.* **2016**, *286*, 291–299. [CrossRef]

31. Chmielarz, L.; Kuśtrowski, P.; Piwowarska, Z.; Dudek, B.; Gil, B.; Michalik, M. Montmorillonite, vermiculite and saponite based porous clay heterostructures modified with transition metals as catalysts for the DeNO$_x$ process. *Appl. Catal. B Environ.* **2009**, *88*, 331–340. [CrossRef]

32. Samojeden, B.; Możdżeń, M. The influence of amount of copper of modified vermiculites on catalytic properties in SCR-NH_3. *Energy Fuels* **2017**, *14*, 02020. [CrossRef]

33. Ying, S. Preparation and Properties of Vermiculite Supported TiO_2 Photocatalyst. *Chin. J. Inorg. Chem.* **2011**, *27*, 40–46.

34. Li, P.; Zhu, M.; Dan, J.; Kang, L.; Lai, L.; Cai, X.; Zhang, J.; Yu, F.; Tian, Z.; Dai, B. Two-dimensional porous SiO_2 nanomesh supported high dispersed Ni nanoparticles for CO methanation. *Chem. Eng. J.* **2017**, *326*, 774–780. [CrossRef]

35. Li, P.; Wen, B.; Yu, F.; Zhu, M.; Guo, X.; Han, Y.; Kang, L.; Huang, X.; Dan, J.; Ouyang, F.; et al. High efficient nickel/vermiculite catalyst prepared via microwave irradiation-assisted synthesis for carbon monoxide methanation. *Fuel* **2016**, *171*, 263–269. [CrossRef]

36. Huang, X.; Yu, F.; Zhu, M.; Ouyang, F.; Dan, J. Hydrochlorination of acetylene using expanded multilayered vermiculite (EML-VMT)-supported catalysts. *Chin. Chem. Lett.* **2015**, *26*, 1101–1104. [CrossRef]

37. Shen, B.; Liu, T.; Zhao, N.; Yang, X.; Deng, L. Iron-doped Mn-Ce/TiO_2 catalyst for low temperature selective catalytic reduction of NO with NH_3. *J. Environ. Sci. Chin.* **2010**, *22*, 1447–1454. [CrossRef]

38. Lin, Q.; Li, J.; Ma, L.; Hao, J. Selective catalytic reduction of NO with NH_3 over Mn–Fe/USY under lean burn conditions. *Catal. Today* **2010**, *151*, 251–256. [CrossRef]

39. Thirupathi, B.; Smirniotis, P.G. Co-doping a metal (Cr, Fe, Co, Ni, Cu, Zn, Ce, and Zr) on Mn/TiO_2 catalyst and its effect on the selective reduction of NO with NH_3 at low-temperatures. *Appl. Catal. B Environ.* **2011**, *110*, 195–206. [CrossRef]

40. Fan, Z.; Shi, J.W.; Gao, C.; Gao, G.; Wang, B.; Niu, C. Rationally Designed Porous MnO_x-FeO_x Nanoneedles for Low-Temperature Selective Catalytic Reduction of NO_x by NH_3. *ACS Appl. Mater. Interfaces* **2017**, *9*, 16117. [CrossRef] [PubMed]

41. Thirupathi, B.; Smirniotis, P.G. Nickel-doped Mn/TiO_2 as an efficient catalyst for the low-temperature SCR of NO with NH_3: Catalytic evaluation and characterizations. *J. Catal.* **2012**, *288*, 74–83. [CrossRef]

42. Kapteijn, F.; Singoredjo, L.; Andreini, A.; Moulijn, J.A. Activity and selectivity of pure manganese oxides in the selective catalytic reduction of nitric oxides with ammonia. *Appl. Catal. B Environ.* **1994**, *3*, 173–189. [CrossRef]

43. Reddy, G.K.; He, J.; Thiel, S.W.; Pinto, N.G.; Smirniotis, P.G. Sulfur-tolerant Mn-Ce-Ti sorbents for elemental mercury removal from flue gas: Mechanistic investigation by XPS. *J. Phys. Chem. C* **2015**, *119*, 8634–8644. [CrossRef]

44. Gao, G.; Shi, J.W.; Liu, C.; Gao, C.; Fan, Z.; Niu, C. Mn/CeO_2 catalysts for SCR of NO_x with NH_3: Comparative study on the effect of supports on low-temperature catalytic activity. *Appl. Sur. Sci.* **2017**, *411*, 338–346. [CrossRef]

45. Li, Y.; Wan, Y.; Li, Y.; Zhan, S.; Guan, Q.; Tian, Y. Low-Temperature Selective Catalytic Reduction of NO with NH_3 over Mn_2O_3-doped Fe_2O_3 Hexagonal Microsheets. *ACS Appl. Mater. Interfaces* **2016**, *8*, 5224–5233. [CrossRef] [PubMed]

46. Zhang, J.; Qu, H. Low-temperature selective catalytic reduction of NO_x with NH_3 over Fe–Cu mixed oxide/ZSM-5 catalysts containing Fe_2CuO_4 phase. *Res. Chem. Interfaces* **2015**, *41*, 4961–4975. [CrossRef]

47. Lin, Z.; Lei, Z.; Qu, H.; Qin, Z. A study on chemisorbed oxygen and reaction process of Fe-CuO_x/ZSM-5 via ultrasonic impregnation method for low-temperature NH_3-SCR. *J. Mol. Catal. A Chem.* **2015**, *409*, 207–215.

48. Haidi, X.U.; Fang, Z.; Cao, Y.; Kong, S.; Lin, T.; Gong, M.; Chen, Y. Influence of Mn/(Mn + Ce) Ratio of MnO_x-CeO_2/WO_3-ZrO_2 Monolith Catalyst on Selective Catalytic Reduction of NO_x with Ammonia. *Chin. J. Catal.* **2012**, *33*, 1927–1937.

49. Fang, C.; Zhang, D.; Cai, S.; Zhang, L.; Huang, L.; Li, H.; Maitarad, P.; Shi, L.; Gao, R.; Zhang, J. Low-temperature selective catalytic reduction of NO with NH_3 over nanoflaky MnO_x on carbon nanotubes in situ prepared via a chemical bath deposition route. *Nanoscale* **2013**, *5*, 9199–9207. [CrossRef] [PubMed]

50. Boningari, T.; Ettireddy, P.R.; Somogyvari, A.; Liu, Y.; Vorontsov, A.; Mcdonald, C.A.; Smirniotis, P.G. Influence of elevated surface texture hydrated titania on Ce-doped Mn/TiO_2 catalysts for the low-temperature SCR of NO_x under oxygen-rich conditions. *J. Catal.* **2015**, *325*, 145–155. [CrossRef]

51. Khan, A.; Smirniotis, P.G. Relationship between temperature-programmed reduction profile and activity of modified ferrite-based catalysts for WGS reaction. *J. Mol. Catal. A Chem.* **2008**, *280*, 43–51. [CrossRef]

52. Cai, S.; Hu, H.; Li, H.; Shi, L.; Zhang, D. Design of multi-shell Fe_2O_3@MnO_x@CNTs for the selective catalytic reduction of NO with NH_3: Improvement of catalytic activity and SO_2 tolerance. *Nanoscale* **2016**, *8*, 3588–3598. [CrossRef] [PubMed]

53. Liu, Z.; Zhu, J.; Li, J.; Ma, L.; Woo, S.I. Novel Mn-Ce-Ti mixed-oxide catalyst for the selective catalytic reduction of NO_x with NH_3. *ACS Appl. Mater. Interfaces* **2014**, *6*, 14500–14508. [CrossRef] [PubMed]

54. France, L.J.; Yang, Q.; Li, W.; Chen, Z.; Guang, J.; Guo, D.; Wang, L.F.; Li, X.H. Ceria modified $FeMnO_x$-Enhanced performance and sulphur resistance for low-temperature SCR of NO_x. *Appl. Catal. B Environ.* **2017**, *206*, 203–215. [CrossRef]

55. Kobayashi, M.; Miyoshi, K. WO_3-TiO_2 monolithic catalysts for high temperature SCR of NO by NH_3: Influence of preparation method on structural and physico-chemical properties, activity and durability. *Appl. Catal. B Environ.* **2007**, *72*, 253–261. [CrossRef]

56. Qiu, Y.; Liu, B.; Du, J.; Tang, Q.; Liu, Z.; Liu, R.; Tao, C. The monolithic cordierite supported V_2O_5-MoO_3/TiO_2 catalyst for NH_3-SCR. *Chem. Eng. J.* **2016**, *294*, 264–272. [CrossRef]

57. Hwang, J.; Ha, H.J.; Ryu, J.; Choi, J.J.; Ahn, C.W.; Kim, J.W.; Hahn, B.D.; Yoon, W.H.; Lee, H.; Choi, J.H. Enhancement of washcoat adhesion for SCR catalysts to convert nitrogen oxide using powder spray coating of TiO_2 on metallic honeycomb substrate. *Catal. Commun.* **2017**, *94*, 1–4. [CrossRef]

58. Gan, L.; Lei, S.; Yu, J.; Ma, H.; Yamamoto, Y.; Suzuki, Y.; Xu, G.; Zhang, Z. Development of highly active coated monolith SCR catalyst with strong abrasion resistance for low-temperature application. *Front. Environ. Sci. Eng.* **2015**, *9*, 979–987. [CrossRef]

59. Huang, H.F.; Jin, L.L.; Lu, H.F.; Yu, H.; Chen, Y.J. Monolithic Cr–V/TiO$_2$/cordierite catalysts prepared by in-situ precipitation and impregnation for low-temperature NH$_3$-SCR reactions. *Catal. Commun.* **2013**, *34*, 1–4. [CrossRef]

60. Gao, C.; Shi, J.W.; Fan, Z.; Niu, C. Sulfur and Water Resistance of Mn-Based Catalysts for Low-Temperature Selective Catalytic Reduction of NO$_x$: A Review. *Catalysts* **2018**, *8*, 11. [CrossRef]

catalysts

MDPI

Article

Preparation and Performance of Modified Red Mud-Based Catalysts for Selective Catalytic Reduction of NO$_x$ with NH$_3$

Jingkun Wu, Zhiqiang Gong, Chunmei Lu *, Shengli Niu, Kai Ding, Liting Xu and Kang Zhang

Department of Energy and Power Engineering, Shandong University, Jinan 250061, China;
jkwu@mail.sdu.edu.cn (J.W.); gongzq@gmail.com (Z.G.); nsl@sdu.edu.cn (S.N.); 18818223769@163.com (K.D.);
15165315337@163.com (L.X.); sdnengdongzk@163.com (K.Z.)
* Correspondence: cml@sdu.edu.cn; Tel.: +86-151-6686-2166

Received: 16 December 2017; Accepted: 17 January 2018; Published: 19 January 2018

Abstract: Bayer red mud was selected, and the NH$_3$-SCR activity was tested in a fixed bed in which the typical flue gas atmosphere was simulated. Combined with XRF, XRD, BET, SEM, TG and NH$_3$-Temperature Programmed Desorption (TPD) characterization, the denitration characteristics of Ce-doped red mud catalysts were studied on the basis of alkali-removed red mud. The results showed that typical red mud was a feasible material for denitration catalyst. Acid washing and calcining comprised the best treatment process for raw red mud, which reduced the content of alkaline substances, cleared the catalyst pore and optimized the particle morphology with dispersion. In the temperature range of 300–400 °C, the denitrification efficiency of calcined acid washing of red mud catalyst (ARM) was more than 70%. The doping of Ce significantly enhanced NH$_3$ adsorption from weak, medium and strong acid sites, reduced the crystallinity of α-Fe$_2$O$_3$ in ARM, optimized the specific surface area and broadened the active temperature window, which increased the NO$_x$ conversion rate by an average of nearly 20% points from 250–350 °C. The denitration efficiency of Ce$_{0.3}$/ARM at 300 °C was as high as 88%. The optimum conditions for the denitration reaction of the Ce$_{0.3}$/ARM catalyst were controlled as follows: Gas Hourly Space Velocity (GHSV) of 30,000 h^{-1}, O$_2$ volume fraction of 3.5–4% and the NH$_3$/NO molar ratio ([NH$_3$/NO]) of 1.0. The presence of SO$_2$ in the feed had an irreversible negative effect on the activity of the Ce$_{0.3}$/ARM catalyst.

Keywords: red mud; deNO$_x$ catalyst; NH$_3$-SCR; acid washing and calcining; cerium

---

1. Introduction

Red mud, which is a major solid waste derived from the aluminum industry, has been producing in huge quantities over the decades. At present, most of the red mud is treated for landfilling, which brings a heavy burden to the environment. At the same time, Bayer red mud is considered hazardous due to its strong alkalinity (pH = 10–12). Therefore, utilization of red mud has become the focus of study and an urgent issue all over the world. Since red mud contains Fe$_2$O$_3$ (20–50%) and large amounts of stable materials (SiO$_2$, Al$_2$O$_3$ and TiO$_2$), it is a potential alternative catalyst in wastewater and exhaust cleaning [1–5], which provides a cost-effective route of controlling waste by waste. However, the large-scale catalysis application of red mud is limited because of its alkalinity [6], mainly originating from Na and Ca, and their oxides can cause sintering in the catalyst and reduce the catalytic activity [7]. Recently, Li et al. [8] developed a ball milling and acid-base neutralization method to reuse red mud as an efficient Fe-Ti/Si-Al denitration catalyst free of alkali. Cao et al. [9] reported an approach that used hydrochloric acid to treat red mud followed by alkali precipitation, by which the alkaline reduced and the CO catalytic oxidation performance was improved. Besides, for the first time, Lamonier et al. [10] compared the denitration performance of raw red mud and Cu-doped red mud, improving the activity of red mud. After that, Rajanikanth et al. [11,12] used red mud as a NO_x adsorbent, which was arranged after the plasma purifier in the tail gas treatment module of a biodiesel engine, and the results showed that in this system, most exhausted NO_2 can be converted. Nevertheless, few studies have involved red mud deNO$_x$ catalyst in coal-fired flue gas processing.

Selective Catalytic Reduction of NO_x with NH_3 (NH_3-SCR) is the most efficient method that has been applied in many thermal power plants for flue gas denitrification, while its technical core is the SCR catalyst with high efficiency and stability. So far, the commercial SCR deNO$_x$ catalyst is V_2O_5-WO_3/TiO_2 [13], but it suffers problems such as high manufacturing cost, heavy metal loss and secondary pollution caused by volatility at high temperature. From this, the development of low-cost and environment-friendly catalysts that replace V-Ti-based ones has attracted much attention in the energy and environment engineering field [14,15]. Till now, some progress has been made in iron-based metal oxides catalysts [16]. On the other hand, various metal oxide wastes that include red mud [17,18], fly ash [19] and aluminum dross [20] are inexpensive materials that can be directly used in catalyst preparation, among which the red mud contains abundant effective iron oxides and thereby is valuable in research on SCR deNO$_x$ catalyst. Therefore, a systematic study is needed to assess the feasibility of red mud for SCR deNO$_x$ in coal-fired flue gas processing.

In view of the fact that alkaline substances such as Na, K, Ca and Mg in red mud can be toxic to SCR activity caused by the poisoning of acid sites on the catalyst surface, dealkalization was applied in catalyst preparation. In addition, we noticed that cerium is an important rare earth element that can contribute to various types of catalytic reaction [21,22], and doping of Ce can increase the amount of acid sites on the catalyst surface [23,24]. Recently, Xu et al. [5] prepared cerium-modified red mud catalyst for catalytic ozonation with the result of improved activity. In our present study, Bayer red mud was used to prepare the low-cost SCR deNO$_x$ catalyst, and on the basis of alkali-removed red mud, the denitrification characteristics of Ce-doped red mud catalysts were evaluated with simulated flue gas atmosphere. X-ray Fluorescence (XRF), X-ray Diffraction (XRD), N_2 isotherm adsorption-desorption analysis (BET), Scanning Electron Microscopy (SEM), Thermogravimetric Measurement (TG) and NH_3-Temperature Programmed Desorption (NH_3-TPD) measurements were employed to investigate the internal mechanism between the properties and performance of red mud.

2. Results

2.1. Effect of Dealkalization Method

2.1.1. Components Analysis of RM and Alkali-Removed RM

The chemical composition of red mud catalysts prepared by different dealkalization methods is shown in Table 1. In raw red mud (RM), SiO_2, Al_2O_3 and Fe_2O_3 were the major components (totaling more than 60%); the next ones were Na_2O, TiO_2, CaO and a small amount of MgO and K_2O, as well. Among them, Na and Ca oxides cause sintering in catalyst at high temperatures [7] and reduce the acidity of the catalyst surface, which is seriously toxic to the SCR catalytic activity [25]. After moderate dealkalization, red mud can be activated while most of effective substances are still preserved [26]; hence, the effect of the dealkalization method on the properties and SCR activities of red mud was studied.

Table 1. The composition of RM, WRM1, WRM2 and ARM (wt%). RM: red mud; WRM*n*: Water washing of red mud catalyst, *n* = 1, 2, refers to washing times; ARM: acid washing of red mud catalyst.

Samples	LOSS	SiO_2	Al_2O_3	Fe_2O_3	CaO	MgO	SO_3	TiO_2	K_2O	Na_2O	Others
RM	13.02	20.74	24.26	23.89	2.77	0.56	0.93	2.95	0.57	10.13	0.18
WRM1	13.75	19.75	27.81	22.61	5.52	0.37	0.85	1.55	0.07	7.43	0.29
WRM2	14.14	19.26	27.40	23.23	5.42	0.36	1.01	1.43	0.13	7.35	0.27
ARM	13.06	18.19	35.34	23.12	3.28	0.24	0.25	2.17	-	4.23	0.12

As shown in Table 1, Na and K contents in WRM1 (WRMn, Water washing of red mud catalyst, n = 1, 2, refers to washing times), WRM2 and ARM were found to be reduced significantly towards water washing and acid washing, but minor changes were observed between WRM1 and WRM2 towards adding washing times. It was noticed that Ca content had increased after washing as it was more insoluble in water, and its absolute quality had decreased less or even basically remained unchanged, so that the relative proportion increased. Thus, the water washing method had certain limitations, but it can be seen that the acid washing method can reduce all kinds of alkaline substances to a certain degree, which is considered acceptable for dealkalization of RM.

Figure 1 shows the XRD patterns of RM, ARM and ARM(400) (ARM(t), where t refers to the calcining temperature). The crystalline phases determined in three samples were α-Fe_2O_3 (hematite, PDF-33-0664#), TiO_2 (anatase, PDF-89-4921#), SiO_2 (quartz, PDF-46-1045#), as well as trace phases of carbonates and aluminosilicates. No obvious other diffraction peaks were found, suggesting that other components were well-dispersed in red mud. This result was in accordance with the recent research on phases of red mud produced in China [27]. In ARM, the Fe+3O(OH) (goethite, PDF-29-0713#) diffraction peaks considerably diminished, and in ARM(400), they finally vanished, which affirmed that the crystalline Fe+3O(OH) had transformed due to dehydration after the calcination process.

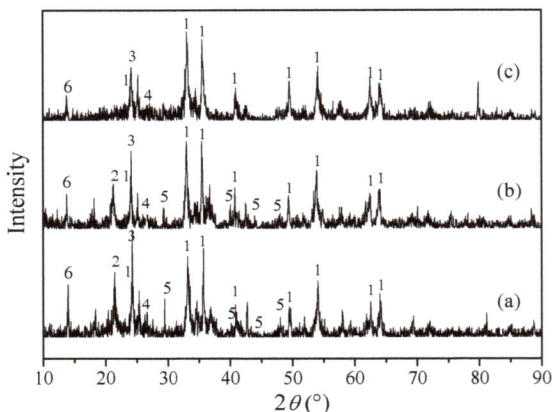

Figure 1. XRD patterns of: (**a**) RM, (**b**) ARM and (**c**) ARM(400). 1: α-Fe_2O_3; 2: Fe+3O(OH); 3: anatase (TiO_2); 4: quartz (SiO_2); 5: calcite ($CaCO_3$); 6: aluminosilicate.

2.1.2. Catalytic Activity of RM and Alkali-Removed RM

Figure 2 shows the NO_x conversion rate of RM, ARM and ARM(400) as a function of reaction temperature. In comparison with RM, the denitration efficiency of ARM increased by about 10% points in the range of 150–375 °C. Unlike the behavior of RM and ARM, better performance was acquired by ARM(400) in the range of 350–400 °C, where the active temperature window had been broadened. Thus, calcining after acid washing can reduce the adverse effect of alkali in RM and effectively improve the catalytic activity at high temperatures, which can be considered as a cost-effective treatment method.

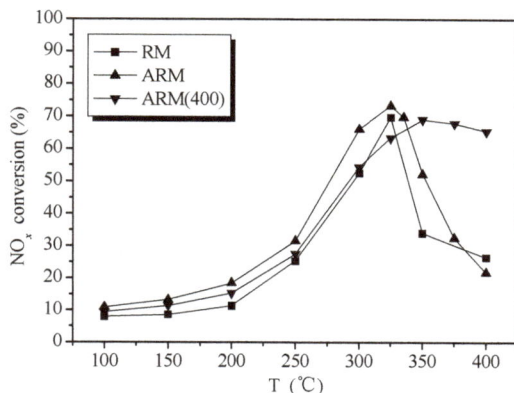

Figure 2. Catalytic activity of RM, ARM and ARM(400). Reaction conditions: reaction gas pressure = 0.1 MPa, N_2 balance, total flow = 2000 mL·min^{-1}, Gas Hourly Space Velocity (GHSV) = 30,000 h^{-1}, [NH_3/NO] = 1.0, initial concentration of NH_3, NO and O_2 = 0.05%, 0.05% and >3.5%, respectively.

2.1.3. Structure Characterization of RM and Alkali-Removed RM

Specific surface areas and pore structure are of significant importance of heterogeneous catalyst. The BET surface areas of samples are shown in Table 2. In general, dealkalization had a positive effect on the BET surface areas, but pore volumes and pore diameter decreased to a certain degree after acid washing treatment. The BET surface areas of ARM and ARM(400) were 50.54 m^2·g^{-1} and 57.19 m^2·g^{-1}, respectively. This suggests that acid washing can remove most of the alkaline substances from red mud and clear the inner channel of red mud minerals [28], so that the specific surface area can be further optimized after calcination, which is beneficial to the diffusion and adsorption of reaction gas on the catalyst surface.

Table 2. BET surface areas, pore volumes and average pore diameter of RM, ARM and ARM(400). ARM(*t*), where *t* refers to the calcining temperature.

Samples	BET Specific Surface Area (m^2·g^{-1})	Pore Volume (cm^3·g^{-1})	Average Pore Diameter (nm)
RM	42.7	0.1919	109.2
ARM	50.5	0.1048	82.9
ARM(400)	57.2	0.1041	72.8

Figure 3 shows the SEM images of samples. As shown in Figure 3a,b, WRM1 showed a similar cementation structure to RM: it displayed a large block, columnar and lamellar structure, surrounded by flocculent small particles of crystallization, where the surface of particles was rough, and the adhesion and pore structure of the catalyst had obviously collapsed and been blocked; while in Figure 3c,d, ARM had larger voids and highly-dispersed particles, owing to the displacement of impurities by H$^+$ occurring in mineral layers, involving K$^+$, Na$^+$, Ca^{2+} and Mg^{2+} ions. After calcination, ARM(400) had further homogeneous morphology with ball-shaped particles, and the surface became smooth and better connected, which provided a larger specific surface area and exposed sufficient active sites for gas diffusion and adsorption in the SCR reaction, and it was the significant reason for the improved activity.

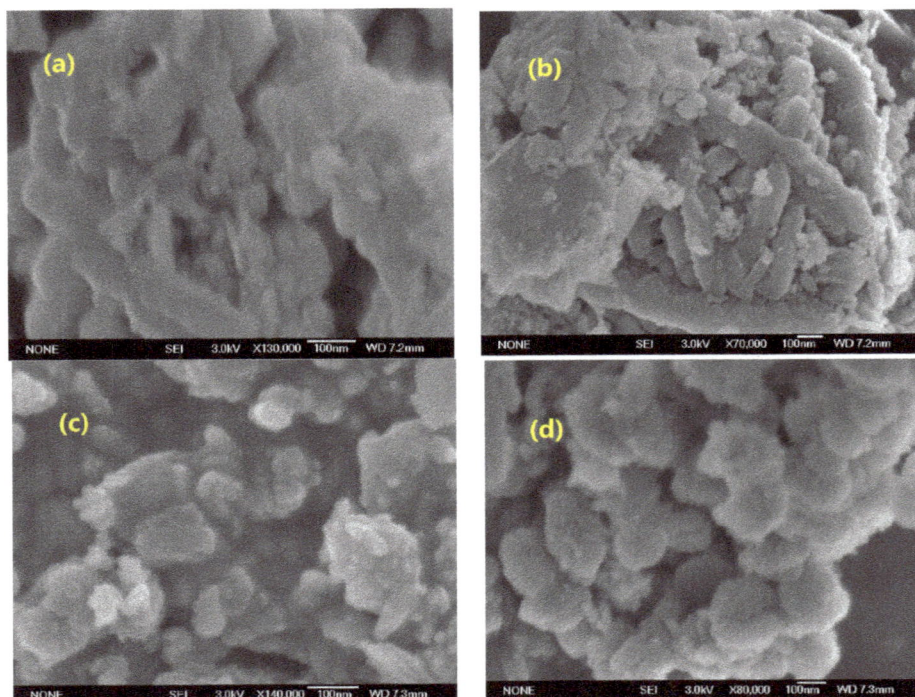

Figure 3. SEM images: (**a**) RM, (**b**) WRM1, (**c**) ARM and (**d**) ARM(400).

2.1.4. Effect of Calcination Temperature on Catalytic Activity of ARM

Calcination temperatures can affect the catalytic activity of red mud. For the TG results in Figure 4a, the fresh ARM sample had a continuous mass loss and finally stabilized at 700 °C. The weight loss peak before 100 and 200 °C corresponds to the removal of adsorbed water and crystalline water in red mud, which is necessary during calcination activation. Moreover, the weight loss stage in the range of 200–300 °C and 300–350 °C can be assigned to the removal of crystalline water from the gibbsite and goethite structure, respectively. To prevent sintering by residual alkaline substance in ARM, the calcination temperature should not exceed 600 °C, so the ARM(400), ARM(450), ARM(500) and ARM(550) catalysts were prepared to study. With the comparison of RM, their catalytic activities are illustrated in Figure 4b. The trends of the conversion-temperature curves of calcined ARM present as the same, but in the range of 200–350 °C reaction temperatures, the catalytic activity of four samples was in the order of ARM(500) > ARM(550) > ARM(400) > ARM(450). Therefore, the calcination temperature of 500 °C was chosen to prepare ARM-supported catalyst with further treatment.

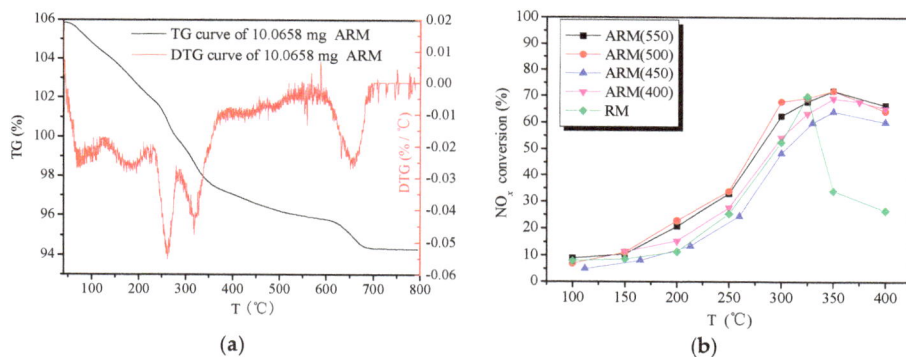

Figure 4. (**a**) The TG-DTG curves of 10.0658 mg ARM; (**b**) catalytic activity of RM, ARM(400), ARM(450), ARM(500) and ARM(550). Reaction conditions: reaction gas pressure = 0.1 MPa, N_2 balance, total flow = 2000 mL·min^{-1}, GHSV = 30,000 h^{-1}, [NH$_3$/NO] = 1.0, initial concentration of NH$_3$, NO and O_2 = 0.05%, 0.05% and >3.5%, respectively.

2.2. Effect of Ce-Doping

2.2.1. Catalytic Activity of Ce/ARM

Though ARM(500) gained a good performance in the experiments above, it still had a certain difference from commercial catalyst. In this section, Ce-doped red mud catalysts were prepared with 500 °C calcination, using ARM as a support. Figure 5 shows the catalytic activity of Ce/ARM with different Ce loading. The amount of Ce loading did not have a linear influence on activities of Ce/ARM in the whole range of reaction temperatures. Compared with ARM(500), the NO$_x$ conversion increased by an average of 5% in Ce$_{0.1}$/ARM and by 10–20% in others. With the increase of Ce loading, the active temperature window of Ce$_{0.5}$/ARM and Ce$_{0.7}$/ARM slightly shifted to high temperatures. The highest NO$_x$ conversion of 88% was gained by Ce$_{0.3}$/ARM at 300 °C and 90% by Ce$_{0.7}$/ARM at 350 °C. It can be viewed that red mud-based Ce/ARM catalysts have good performance for denitration in 275–400 °C, which is comparable with Ce/TiO$_2$ (5% Ce-doped) [29] and V_2O_5/TiO$_2$ catalysts [30].

Based on the above analysis, the SCR activities of Ce-doped red mud were remarkably promoted in both medium and high temperatures, and the active temperature window of catalysts had been apparently broadened. For better understanding the results of differences in activity over Ce/ARM and ARM(500), the NH$_3$-TPD curves are shown in Figure 6 to identify the acidity difference. Like the performance of activity presented in Figure 5, the shapes of the ARM(500) and Ce$_{0.1}$/ARM curves were similar, and a marked increase of desorption intensity was found in Ce$_{0.1}$/ARM, which definitely identified the promoting effect of Ce doping on the amount of surface acid sites and improved activity. For the peaks of the Ce$_{0.3}$/ARM sample, both the peak area of low and high temperatures dramatically increased, which was assigned to enhanced NH$_3$ adsorption from weak, medium and strong acid sites after further Ce doping. Combined with the analysis in Section 2.1, it is concluded that the removal of alkali and NH$_3$ adsorption were of great importance to the catalytic behavior of red mud owing to the changes in structure and acid sites provided by acid treatment and Ce doping, both of which favored the SCR reaction.

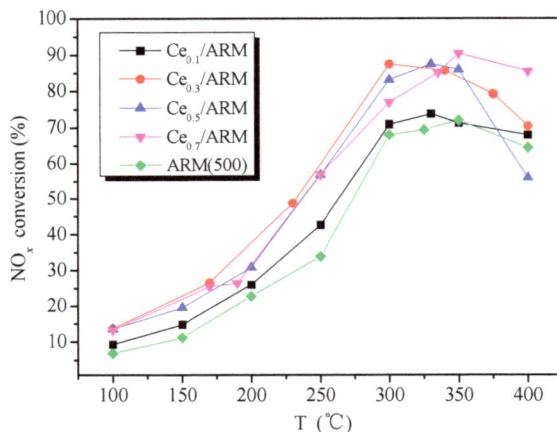

Figure 5. Catalytic activity of ARM(500) and Ce$_x$/ARM catalysts. Reaction conditions: reaction gas pressure = 0.1 MPa, N$_2$ balance, total flow = 2000 mL·min^{-1}, GHSV = 30,000 h^{-1}, [NH$_3$/NO] = 1.0, initial concentration of NH$_3$, NO and O$_2$ = 0.05%, 0.05% and >3.5%, respectively.

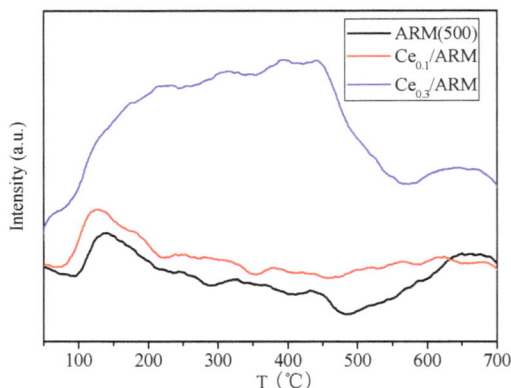

Figure 6. The NH$_3$-Temperature Programmed Desorption (TPD) curves of Ce$_{0.3}$/ARM, Ce$_{0.1}$/ARM and ARM(500).

2.2.2. Structure Characterization of Ce/ARM

Figure 7a shows the XRD patterns of Ce$_{0.1}$/ARM, Ce$_{0.3}$/ARM, Ce$_{0.5}$/ARM and Ce$_{0.7}$/ARM. Local patterns in the range of 40–60° are shown in Figure 7b. Compared with JCPDS standard cards, the main crystalline phase in Ce/ARM samples was α-Fe$_2$O$_3$ (PDF-33-0664#), which was consistent with the result of ARM. Moreover, diffraction peaks of CeO$_2$ (PDF-34-0394#) were observed, except in Ce$_{0.1}$/ARM, which had the least Ce loading. As a result, Ce species had formed the CeO$_2$ crystalline phase in the preparation. In all of four samples, there were no other obvious peaks, indicating that other species were well-dispersed in Ce/ARM after Ce-doping and 500 °C calcination.

According to the Scherrer Equation (1) [31], the peak width at half height (FWHM) of crystal lattice diffraction peaks is inversely proportional to the diameter of crystalline grains, so it can be deduced from the local patterns that in Ce$_{0.3}$/ARM, the average crystalline size of α-Fe$_2$O$_3$ was lower than other samples:

$$D = 0.89\lambda/(\beta\cos\theta) \tag{1}$$

where D is the average thickness of a crystalline grain perpendicular to the crystal surface; β is the FWHM of sample diffraction peaks; θ is the diffraction angle; λ is incident X-ray light wavelength of 0.154 nm.

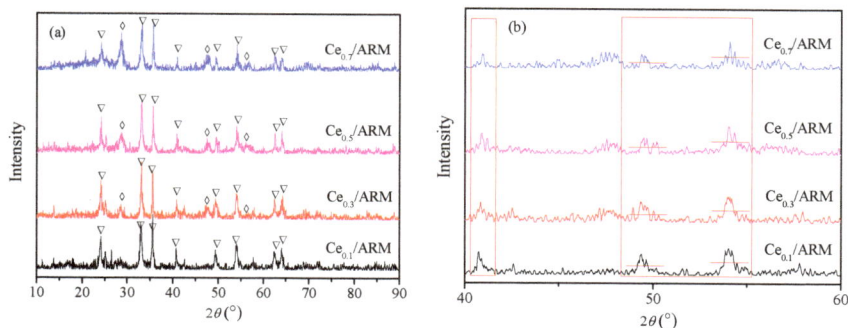

Figure 7. XRD patterns of Ce/ARM catalyst. (**a**) Complete patterns ($\triangledown$: α-Fe$_2$O$_3$ $\lozenge$: CeO$_2$); (**b**) local patterns.

When the composition and macrostructure of catalyst are fixed, the catalytic activity is primarily affected by the specific surface area. Table 3 lists the BET surface areas of Ce/ARM.

Table 3. BET surface areas, pore volumes and average pore diameter of Ce/ARM catalysts.

Samples	BET Specific Surface Area (m^2·g^{-1})	Pore Volume (cm^3·g^{-1})	Average Pore Diameter (nm)
Ce$_{0.1}$/ARM	46.5	0.0922	87.9
Ce$_{0.3}$/ARM	54.4	0.1002	81.6
Ce$_{0.5}$/ARM	45.1	0.1071	98.0
Ce$_{0.7}$/ARM	46.1	0.1181	103.1

Compared to ARM (50.54 m^2·g^{-1}) in Table 2, The BET specific surface areas of Ce/ARM had decreased to some degree, yet it was calculated that the crystalline grain diameter of α-Fe$_2$O$_3$ of Ce/ARM dropped to 15–30 nm compared to that of ARM. In the case of this study, Ce$_{0.3}$/ARM not only had low crystalline grain diameter, but also provided larger BET specific surface area (54.4 m^2·g^{-1}), as well as enhanced the surface acidity, which consequently improved the catalytic activity just under a simple preparation method with a small amount of Ce-doping (less than 5%).

2.2.3. NH$_3$ and O$_2$ Transient Response of Ce$_{0.3}$/ARM and the Effect of Major Operating Parameters

As shown in Figure 8a–e, the fresh Ce$_{0.3}$/ARM catalyst was selected to investigate the effect of major operating parameters on its performance. The experiments were conducted at a reaction temperature of 300 °C, where they could gain the highest denitration efficiency.

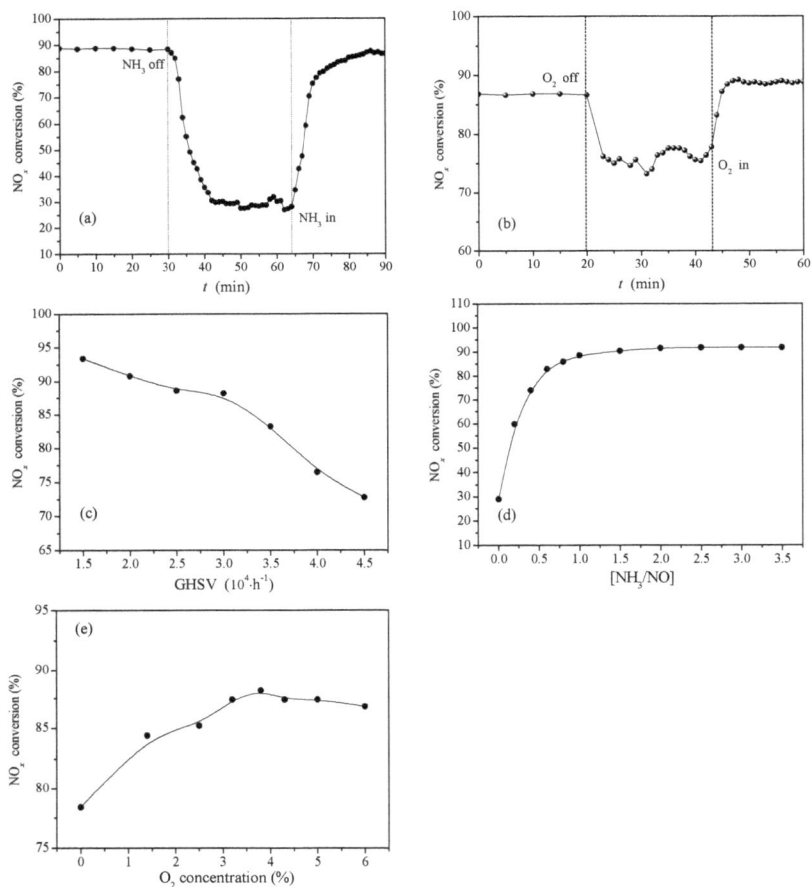

Figure 8. Ce$_{0.3}$/ARM catalyst: (**a**) NH$_3$ transient response; (**b**) O$_2$ transient response. Reaction conditions: reaction temperature = 300 °C, reaction gas pressure = 0.1 MPa, N$_2$ balance, total flow = 2000 mL·min^{-1}, GHSV = 30,000 h^{-1}, [NH$_3$/NO] = 1.0, initial concentration of NH$_3$, NO and O$_2$ = 0.05%, 0.05% and >3.5%, respectively. (**c**) Effect of GHSV. Reaction conditions: reaction temperature = 300 °C, reaction gas pressure = 0.1 MPa, N$_2$ balance, total flow = 2000 mL·min^{-1}, GHSV = 15,000–30,000 h^{-1}, [NH$_3$/NO] = 1.0, initial concentration of NH$_3$, NO and O$_2$ = 0.05%, 0.05% and >3.5%, respectively. (**d**) Effect of [NH$_3$/NO]. Reaction conditions: reaction temperature = 300 °C, reaction gas pressure = 0.1 MPa, N$_2$ balance, total flow = 2000 mL·min^{-1}, GHSV = 30,000 h^{-1}, [NH$_3$/NO] = 0–3.5, initial concentration of NH$_3$, NO and O$_2$ = 0.05%, 0.05% and >3.5%, respectively. (**e**) Effect of O$_2$ volume fraction. Reaction conditions: reaction temperature = 300 °C, reaction gas pressure = 0.1 MPa, N$_2$ balance, total flow = 2000 mL·min^{-1}, GHSV = 30,000 h^{-1}, [NH$_3$/NO] = 0–3.5, initial concentration of NH$_3$, NO and O$_2$ = 0.05%, 0.05% and 0–6%, respectively.

As is known from the NH$_3$-SCR reaction Equations (2)–(5), insufficient NH$_3$ will lead to inadequate SCR reaction and low NO$_x$ conversion. Figure 8a shows the transient response of NH$_3$ on Ce$_{0.3}$/ARM. In the initial half-hour experiment, the whole of the reaction gas was thoroughly mixed and flowed through the catalyst bed for a steady-state reaction. When the NH$_3$ feed was instantly cut off at 30 min, there was hardly a concentration of gaseous NH$_3$ in the flow, but NO$_x$ conversion did not drop immediately till it declined to lower than 30% 12 min later, which was attributed to residual adsorbed NH_3 continuously reacted with NO; it also demonstrated that in this period, the adsorbed NH_3 was quickly consumed in the complete SCR reaction. After the plateau between 42 min and 62 min, the NH_3 feed was turned on. NO_x conversion rapidly raised at the beginning of 5 min, then slowly increased and finally recovered after 20 min. Overall, the adsorption of NH_3 on the catalyst surface played an important role in the SCR reaction on $Ce_{0.3}$/ARM catalyst, which clearly agreed with the requirement of acid sites for improved SCR activity, as discussed above in Figure 6.

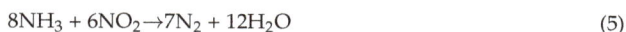

$$4NH_3 + 4NO + O_2 \rightarrow 4N_2 + 6H_2O \qquad (2)$$

$$4NH_3 + 2NO_2 + O_2 \rightarrow 3N_2 + 6H_2O \qquad (3)$$

$$4NH_3 + 6NO \rightarrow 5N_2 + 6H_2O \qquad (4)$$

$$8NH_3 + 6NO_2 \rightarrow 7N_2 + 12H_2O \qquad (5)$$

According to "standard SCR" and "fast SCR" Equations (2) and (3) [32], the SCR reaction occurs more quickly in the presence of O_2. As seen from Figure 8b, NO_x conversion dropped and recovered immediately when the O_2 feed was instantly cut off and turned on, which indicated that gaseous O_2 reacted on the catalyst surface. During the time interval between 22 min and 43 min, the NO_x conversion of $Ce_{0.3}$/ARM remained as high as 78%, which may result from the superior capacity for cyclic oxygen storage and redox of CeO_2 formed in red mud, which ensured that there were sufficient adsorbed oxygen and lattice oxygen participating in the SCR reaction and contributing to the performance of $Ce_{0.3}$/ARM.

Figure 8c shows the effect of Gas Hourly Space Velocity (GHSV) on the catalytic activity of $Ce_{0.3}$/ARM. Different GHSV was generated by changing the filling volume of catalysts. Before GHSV of 30,000 h^{-1}, NO_x conversion remained more than 88% and decreased slightly with the increase of flue gas flow velocity, but it obviously dropped in GHSV of 30,000–45,000 h^{-1}, because when using a small amount of catalyst, the residence time was too short to make the reaction gas fully contact the catalyst surface. The results showed that the $Ce_{0.3}$/ARM catalyst was adapted to GHSV of 15,000–30,000 h^{-1}.

Figure 8d shows the effect of $[NH_3/NO]$. In the presence of NH_3, the NO_x conversion increased sharply and reached 74% with $[NH_3/NO]$ of 1:2 (0.5), and it reached nearly 90% with $[NH_3/NO]$ of 1.0, then stayed at the same level with the increase of $[NH_3/NO]$ before 3.5%. To save the NH_3 consumption and considering the efficiency, it is best to control the $[NH_3/NO]$ at 1.0, which also reveals that the reaction on the catalyst surface of $Ce_{0.3}$/ARM followed the stoichiometric ratio of the "standard SCR" reaction.

Figure 8e shows the effect of O_2 concentration. In accordance with the result in Figure 7b, the NO_x conversion of $Ce_{0.3}$/ARM maintained about 75% in absence of O_2, and it further raised as the O_2 concentration increased. According to "fast SCR" Equation (3), with excessive O_2, NO could be oxidized to NO_2, which accelerates the SCR reaction rate and enhances the NO_x conversion. In this work, the initial O_2 concentration of 3.5–4% in atmosphere is appropriate, where the NO_x conversion of $Ce_{0.3}$/ARM was the best.

2.2.4. Effect of SO_2 on Catalytic Activity of $Ce_{0.3}$/ARM

As shown in Figure 9, a twelve-hour stability experiment of SO_2 resistance was carried out at 300 °C to primarily investigate the effect of SO_2 on the catalytic behaviors of $Ce_{0.3}$/ARM, and another experiment of fresh catalyst without SO_2 was done as a comparison. Starting with 0.03% SO_2, the NO_x conversion decreased obviously from 90–85% in 30 min, then gradually dropped to 73% in the following 2.5 h and maintained a plateau. The inhibition on activity during those 6 h could be explained by competitive adsorption of SO_2 on reaction active sites and continued blocking of the intermediate product. After SO_2 was cut off, the NO_x conversion did not recover and finally stabilized at 64%, which indicated that SO_2 had led to irreversible sulfate deposition or metal sulfate species formation [33] on $Ce_{0.3}$/ARM, so that some active sites were blocked. Thus, the SO_2 resistance features of $Ce_{0.3}$/ARM need further development in the future.

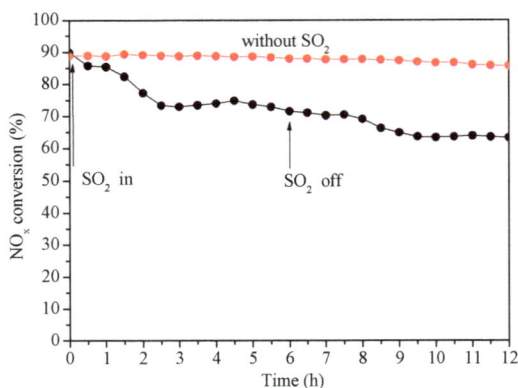

Figure 9. Catalytic activity of $Ce_{0.3}$/ARM in the presence of SO_2 and without SO_2. Reaction conditions: Reaction temperature = 300 °C, reaction gas pressure = 0.1 MPa, N_2 balance, total flow = 2000 mL·min^{-1}, GHSV = 30,000 h^{-1}, [NH_3/NO] = 1.0, initial concentration of NH_3, NO, O_2 and SO_2 = 0.05%, 0.05%, >3.5% and 0.03%, respectively.

3. Discussion

Admittedly, the results in this work demonstrated the advantages of $Ce_{0.3}$/ARM catalyst for simple preparation and promising application in medium and high temperature denitration with typical compositions of flue gas, but the problem of inferior SO_2 resistance should be overcome for the purpose of practical use; further study is still needed on its performance in actual flue gas, where the effects of H_2O and fly ash cannot be ignored. Moreover, the polymetallic property of red mud will increase its uncertainty of activity, which remains to be investigated. In the case of this study, it can be concluded that treated red mud was a feasible support for Ce-doped NH_3-SCR catalyst, and the effects of the dealkalization method and Ce doping on structure and surface acidity had been discussed. In a next step, a point-by-point comparison with previous work should be made.

Recently, in an important progress [8] related to red mud-based (also received from the Shandong Aluminum Industry) SCR catalyst, a variety of samples prepared were investigated, which showed some controversial results at first glance. However, the distinction of work between those samples and this study should be considered in detail, which is shown in Table 4.

Table 4. Comparison of works.

Options	Reference [8]	This Study
Synthesis method	acid-base neutralization	acid washing; impregnation
Activation treatment	washing, drying and calcination; SO_2-activated	washing, drying and calcination
Catalyst composition	Fe_2O_3-TiO_2/(SiO_2-Al_2O_3) (RM-based catalyst)	Ce-doped mixed oxides with inevitable residual alkali (Ce/ARM)
Major active component	α-Fe_2O_3, $Fe_2(SO_4)_3$	α-Fe_2O_3, CeO_2 and others
Active temperature window	350–450 °C	275–375 °C
Operating conditions	GHSV of 60,000 h^{-1}	GHSV of 30,000 h^{-1}

On account of different catalyst precursors, this may explain the reason why the activity of the RM-based catalyst increased with a lower calcination temperature in the range of 500–600 °C [8], but a relatively high temperature was chose from 400–600 °C for ARM, because a relatively lower calcination temperature may be needed for suitably phase structure transformation of the metal hydroxide formed by the neutralization method, while for impregnated catalyst, a low calcination temperature will make it inadequate to establish a strong interaction between the active component and catalyst support. In addition, conflicting performance is also observed between the SO_2 resistance of SO_2–activated RM and $Ce_{0.3}$/ARM. For one thing, increasing the GHSV could indeed reduce the risk of ammonium sulfate deposition. For another, the inferior performance of $Ce_{0.3}$/ARM, by contrast, is possibly due to the fact that some ammonium sulfate could not completely decompose at 300 °C, where the stability experiment was carried out, but the $Ce_{0.3}$/ARM catalyst had the best efficiency at that temperature, so adding a low-temperature sulfur-resistant component to existing $Ce_{0.3}$/ARM catalyst or improving its high-temperature activities will have a good prospect in future work. Nevertheless, SO_2 activation may also have a promoting effect on $Ce_{0.3}$/ARM, though at present, it still lacks the understanding of the interaction between SO_2 and the mixture of metal oxides in red mud.

Furthermore, better understanding of the effect of alkaline substances in red mud on SCR behaviors should be made to help confirm the real reaction sites and even tune the catalyst properties. An interesting work [34] proposed the idea that the surface basicity of K-deposited Ce-supported sulfated zirconia catalyst could also promote the NO conversion especially in a relatively lower temperature range by enhancing the adsorption and oxidation of NO as long as sufficient surface acidity was available, which gives inspiration to improve the activity of existing Ce/ARM catalyst. Herein, the assisted component in red mud could be designed with respect to a certain study object in future work.

4. Materials and Methods

4.1. Preparation of RM and Alkali-Removed RM

Raw red mud catalyst (RM): Bayer red mud was received from Shandong Aluminum Industry Corporation (Zibo, China). The uniform powdered red mud was obtained by crushing and grinding received red mud, then it was screened by 50–120 mesh, after which a small amount of deionized water was added to make powders into strip granules, followed by air drying at 105 °C for 6 h. Finally, the catalyst particles of RM were obtained via screening by 40–60 mesh (0.25–0.38 mm) [35] from those broken strips.

Water washing of red mud catalyst (WRMn, n = 1, 2, refers to washing times): The above-mentioned powdered red mud and deionized water were stirred in a liquid-solid ratio of 10 mL·g^{-1} to form a slurry, with heating in an 80 °C water bath. The resulting solution was washed n times with deionized water by the vacuum filtration method until the pH of filtrate reached 7.0, and then, the filter cake was air dried at 105 °C for 6 h until the weight was constant. Finally, the catalyst particles of WRMn were obtained via grinding and screening by 40–60 mesh from the dried cake.

Acid washing of red mud catalyst (ARM): The prepared red mud slurry was washed once and then titrated with 0.2 mol/L HNO_3 until the pH of the solution reached 7.0. The resulting solution was repeatedly washed to remove residual ions and impurities until the pH of the filtrate reached 7.0, and then, the filter cake was air dried at 105 °C for 6 h until the weight was constant. Finally, the catalyst particles of ARM were obtained via grinding and screening by 40–60 mesh from the dried cake.

In addition, some of the ARM catalyst particles were treated by calcining in air at different temperatures (400 °C, 450 °C, 500 °C and 550 °C) for 4 h. The obtained catalysts were defined as ARM(t), where t refers to the calcining temperature.

4.2. Preparation of Ce-Doped Red Mud Catalyst

Ce-doped red mud catalysts (Ce_x/ARM, x refers to the impregnated solution concentration, for which the unit is $mol \cdot L^{-1}$) were prepared by the impregnation method, using a certain amount of $Ce(NO_3)_3$ solution mixed with ARM powders in a liquid-solid ratio of $1\ mL \cdot g^{-1}$, followed by stirring for 1 h and ultrasonic treatment for 30 min. The resulting slurry was dried rapidly for 20 min by microwaving with a power of 210 W and then air dried at 105 °C for 6 h until the weight was constant. After that, the dried bulk was treated by calcining in air at 500 °C for 4 h. Finally, the Ce/ARM catalyst particles were obtained via grinding and screening by 40–60 mesh from the bulk. Table 5 lists the actual weight percentage of Ce content in red mud catalysts.

Table 5. Ce content in Ce/ARM catalysts.

Samples	Ce Content (wt%)
$Ce_{0.1}$/ARM	1.40
$Ce_{0.3}$/ARM	4.20
$Ce_{0.5}$/ARM	7.01
$Ce_{0.7}$/ARM	9.81

4.3. Catalytic deNO$_x$ Activity Measurements

The NH_3-SCR denitration reaction was carried out on a fixed-bed flow reactor that contained a vertical quartz tube, 0.8 cm inside diameter and 60 cm in length, fitted within a temperature-controlling (room temperature to 400 °C available) electrical heating furnace. Four milliliters of the catalyst sample particles were loaded on a quartz baffle plate fixed on the constant temperature area of the tube, and a K-type thermocouple was inserted in the catalyst bed to measure the accurate catalyst temperature. Simulated coal-fired SCR flue gas consisted of NH_3, NO and O_2, in volume fractions of 0.05% (500 ppm), 0.05% and more than 3.5%, respectively. Industrial nitrogen was used as the carrier gas. Total flow into the reactor was controlled as $2000\ mL \cdot min^{-1}$, and the Gas Hourly Space Velocity (GHSV) was $30,000\ h^{-1}$. Before each experiment point at different reaction temperature (T, °C), reaction gas was thoroughly mixed and flowed through the catalyst bed for 30 min to avoid the measurement error of concentration caused by gas adsorption. Catalytic deNO$_x$ activity of red mud was evaluated by the NO$_x$ conversion rate:

$$NO_x\ \text{Conversion}(\%) = [(NO_x(inlet) - NO_x(outlet))/NO_x(inlet)] \times 100\% \qquad (6)$$

where $NO_x(inlet)$ and $NO_x(outlet)$ signifies the inflow and outflow NO$_x$ (NO and NO_2) concentrations which were measured by the MGA Flue Gas Analyzer (MRU Corporation, Neckarsulm, Obereisesheim, Germany).

4.4. Characterization of Catalysts

The component of samples was analyzed by X-ray Fluorescence (XRF) using the ZSX Primus II Analyzer (Rigaku Corporation, Tokyo, Japan). The crystal structure of samples was examined by X-ray Diffraction (XRD) using the D/max 2500 PC Diffractometer (Rigaku Corporation, Tokyo, Japan) equipped with Cu $K\alpha$ radiation with a wavelength of 0.154 nm. The samples were investigated in the 2θ range of 10–90° at a scanning speed of $1.2° \cdot min^{-1}$, using the MDI Jade software to analyze phases. The Brunauer–Emmett–Teller (BET) specific surface area and pore volume of samples were determined by N_2 isotherm adsorption-desorption using the ASAP2020 Surface Area and Porosity Analyzer (Micromeritics Corporation, Norcross, GA, USA). The microscopic morphology of samples was observed by scanning electron microscope (SEM) using the SUPRA 55 Instrument (ZEISS, Oberkochen, Germany). The calcining parameters were decided by Thermogravimetric Measurement (TG) using the TGA/SDTA851 Analyzer (Mettler Toledo, Zurich, Switzerland) under nitrogen atmosphere and a heating rate of 15 °C·min^{-1} from room temperature to 800 °C.

The NH$_3$-TPD experiments were performed on a ChemAuto 2920 Instrument (Micromeritics, Norcross, GA, USA). First, 90 mg of sample were loaded on a quartz U-tube and heated from room temperature to 300 °C for an hour in argon atmosphere, then cooled to 50 °C. After blowing of argon for 30 min, 10 vol% NH$_3$ (Ar balance) was switched on to complete the pre-absorption for 30 min and heated to 100 °C for 60 min with argon blowing. After the baseline was smooth, the NH$_3$-TPD test was performed by heating the sample to 700 °C at 5 °C·min^{-1}. The desorption amount was detected by the TCD detector. When the test finished, the argon atmosphere was switched on for natural cooling.

5. Conclusions

In this study, Bayer red mud from industrial waste was prepared as low-cost red mud-based catalyst to evaluate the performance of denitration with simulated coal-fired flue gas. It was proven that red mud was a feasible material for the NH$_3$-SCR catalyst.

Compared with RM and alkali-removed RM, acid washing and calcining comprised a cost-effective treatment process for raw red mud, which reduced the alkali content and improved the catalytic activity of ARM at high temperatures. The increase of BET surface area was attributed to the unobstructed catalyst pore and the homogeneous particle morphology with dispersion.

In Ce/ARM, doping of Ce significantly enhanced NH$_3$ adsorption from weak, medium and strong acid sites, reduced the crystallinity of α-Fe$_2$O$_3$ in ARM, optimized the specific surface area and broadened the active temperature window. The optimum doping amount was acquired by Ce$_{0.3}$/ARM, in which the NO$_x$ conversion rate increased by an average of nearly 20% points between 250 and 350 °C, and the highest denitration efficiency reached 88% at 300 °C. The optimum conditions for the denitration reaction on Ce$_{0.3}$/ARM catalyst were controlled as follows: GHSV of 30,000 h^{-1}, O$_2$ volume fraction of 3.5–4% and [NH$_3$/NO] of 1.0.

Acknowledgments: This research was supported by the National Natural Science Foundation of China (NSFC: 51576117) and the Interdisciplinary Development Program of Shandong University (2015JC024).

Author Contributions: Jingkun Wu conceived of and designed the experiments. Jingkun Wu performed the experiments. Zhiqiang Gong, Chunmei Lu and Shengli Niu gave technical support and conceptual advice. Jingkun Wu wrote the paper. Jingkun Wu, Chunmei Lu and Zhiqiang Gong revised the paper. Kai Ding, Liting Xu and Kang Zhang contributed reagents/materials/analysis tools.

Conflicts of Interest: The authors declare no conflict of interest.

References

1. Hu, Z.P.; Zhao, H.; Gao, Z.M.; Yuan, Z.Y. High-surface-area activated red mud supported Co$_3$O$_4$ catalysts for efficient catalytic oxidation of CO. *RSC Adv.* **2016**, *6*, 94748–94755. [CrossRef]
2. Hu, Z.P.; Zhu, Y.P.; Gao, Z.M.; Wang, G.X.; Liu, Y.P.; Liu, X.Y.; Yuan, Z.Y. CuO catalysts supported on activated red mud for efficient catalytic carbon monoxide oxidation. *Chem. Eng. J.* **2016**, *302*, 23–32. [CrossRef]
3. Paredes, J.R.; Ordonez, S.; Vega, A.; Diez, F.V. Catalytic combustion of methane over red mud-based catalysts. *Appl. Catal. B* **2004**, *47*, 37–45. [CrossRef]
4. Bento, N.I.; Santos, P.S.C.; de Souza, T.E.; Oliveira, L.C.A.; Castro, C.S. Composites based on PET and red mud residues as catalyst for organic removal from water. *J. Harzard. Mater.* **2016**, *314*, 304–311. [CrossRef] [PubMed]
5. Xu, B.B.; Qi, F.; Sun, D.Z.; Chen, Z.L.; Robert, D. Cerium doped red mud catalytic ozonation for bezafibrate degradation in wastewater: Efficiency, intermediates, and toxicity. *Chemosphere* **2016**, *146*, 22–31. [CrossRef] [PubMed]
6. Koumanova, B.; Drame, M.; Popangelova, M. Phosphate removal from aqueous solutions using red mud wasted in bauxite Bayer's process. *Resour. Conserv. Recycl.* **1997**, *19*, 11–20. [CrossRef]
7. Ordonez, S.; Sastre, H.; Diez, F.V. Characterisation and deactivation studies of sulfided red mud used as catalyst for the hydrodechlorination of tetrachloroethylene. *Appl. Catal. B* **2001**, *29*, 263–273. [CrossRef]

8. Li, C.M.; Zeng, H.; Liu, P.L.; Yu, J.; Guo, F.; Xu, G.W.; Zhang, Z.G. The recycle of red mud as excellent SCR catalyst for removal of NO_x. *RSC Adv.* **2017**, *7*, 53622–53630. [CrossRef]

9. Cao, J.L.; Wang, Y.; Li, G.J.; Li, K.; Wang, Y.; Ma, M. Mesoporous modified red mud supported CuO nanocatalysts for carbon monoxide oxidation. *Curr. Nanosci.* **2015**, *11*, 413–418. [CrossRef]

10. Lamonier, J.F.; Leclerco, G.; Dufour, M.; Leclercq, L. Utilization of red mud. Catalytic properties in selective reduction of nitric oxide by ammonia. Récents Progrès en Génie des Procédés, Boues industrielles: Traitements. *Recents Prog. Genie Procedes* **1995**, *43*, 31–36.

11. Bhattacharyya, A.; Rajanikanth, B.S. Discharge plasma combined with bauxite residue for biodiesel exhaust cleaning: A case study on NO_x removal. *IEEE Trans. Plasma Sci.* **2015**, *43*, 1974–1982. [CrossRef]

12. Bhattacharyya, A.; Rajanikanth, B.S. Biodiesel exhaust treatment with HFAC plasma supported by red mud: Study on $DeNO_x$ and power consumption. *Energy Procedia* **2015**, *75*, 2371–2378. [CrossRef]

13. Liu, X.; Li, J.H.; Li, X.; Peng, Y.; Wang, H.; Jiang, X.M.; Wang, L.W. NH_3 selective catalytic reduction of NO: A large surface TiO_2 support and its promotion of V_2O_5 dispersion on the prepared catalyst. *Chin. J. Catal.* **2016**, *37*, 878–887. [CrossRef]

14. Li, J.H.; Chang, H.Z.; Ma, L.; Hao, J.M.; Yang, R.T. Low-temperature selective catalytic reduction of NO_x with NH_3 over metal oxide and zeolite catalysts—A review. *Catal. Today* **2011**, *175*, 147–156. [CrossRef]

15. Liu, C.; Shi, J.W.; Gao, C.; Niu, C.M. Manganese oxide-based catalysts for low-temperature selective catalytic reduction of NO_x with NH_3: A review. *Appl. Catal. A* **2016**, *522*, 54–69. [CrossRef]

16. Liang, H.; Gui, K.T.; Zha, X.B. DRIFTS study of γ-Fe_2O_3 nano-catalyst for low-temperature selective catalytic reduction of NO_x with NH_3. *Can. J. Chem. Eng.* **2016**, *94*, 1668–1675. [CrossRef]

17. Sushil, S.; Batra, V.S. Modification of red mud by acid treatment and its application for CO removal. *J. Harzard. Mater.* **2012**, *203–204*, 264–273. [CrossRef] [PubMed]

18. Wang, S.B.; Ang, H.M.; Tade, M.O. Novel applications of red mud as coagulant, adsorbent and catalyst for environmentally benign processes. *Chemosphere* **2008**, *72*, 1621–1635. [CrossRef] [PubMed]

19. Xuan, X.P.; Yue, C.T.; Li, S.Y.; Yao, Q. Selective catalytic reduction of NO by ammonia with fly ash catalyst. *Fuel* **2003**, *82*, 575–579. [CrossRef]

20. Das, B.R.; Dash, B.; Tripathy, B.C.; Bhattacharya, I.N.; Das, S.C. Production of η-alumina from waste aluminium dross. *Miner. Eng.* **2007**, *20*, 252–258. [CrossRef]

21. Zhao, H.; Zhang, D.X.; Wang, F.F.; Wu, T.T.; Gao, J.S. Modification of ferrite-manganese oxide sorbent by doping with cerium oxide. *Process Saf. Environ.* **2008**, *86*, 448–454. [CrossRef]

22. Gao, X.; Jiang, Y.; Zhong, Y.; Luo, Z.Y.; Cen, K.F. The activity and characterization of CeO_2-TiO_2 catalysts prepared by the sol-gel method for selective catalytic reduction of NO with NH_3. *J. Harzard. Mater.* **2010**, *174*, 734–739. [CrossRef] [PubMed]

23. Liu, X.; Li, J.H.; Li, X.; Peng, Y.; Wang, H.; Jiang, X.M.; Wang, L.W. Mechanism of selective catalytic reduction of NO_x with NH_3 over CeO_2-WO_3 catalysts. *Chin. J. Catal.* **2011**, *32*, 836–841.

24. Wang, X.B.; Zhang, L.; Wu, S.G.; Zou, W.X.; Yu, S.H.; Shao, Y.; Dong, L. Promotional effect of Ce on iron-based catalysts for selective catalytic reduction of NO with NH_3. *Catalysts* **2016**, *6*, 112. [CrossRef]

25. Xu, C.H.F.; Chen, K.H.; Gu, Z.F.; Ma, D.D.; Rao, G.H. Effect of operation conditions on activity of de-NO_x SCR catalysts before and after poisoned by alkali metal. *Adv. Mater. Res.* **2014**, *955–959*, 702–705.

26. Seo, P.W.; Cho, S.P.; Hong, S.H.; Hong, S.C. The influence of lattice oxygen in titania on selective catalytic reduction in the low temperature region. *Appl. Catal. A* **2010**, *380*, 21–27. [CrossRef]

27. Cao, S.T.; Ma, H.J.; Zhang, Y.; Chen, X.F.; Zhang, Y.F.; Zhang, Y. The phase transition in bayer red mud from China in high caustic sodium aluminate solutions. *Hydrometallurgy* **2013**, *140*, 111–119. [CrossRef]

28. Cengeloglu, Y.; Tor, A.; Ersoz, M.; Arslan, G. Removal of nitrate from aqueous solution by using red mud. *Sep. Purif. Technol.* **2006**, *51*, 374–378. [CrossRef]

29. Xu, W.Q.; Yu, Y.B.; Zhang, C.B.; He, H. Selective catalytic reduction of NO by NH_3 over a Ce/TiO_2 catalyst. *Catal. Commun.* **2008**, *9*, 1453–1457. [CrossRef]

30. Georgiadou, I.; Papadopoulou, C.; Matralis, H.K.; Voyiatzis, G.A.; Lycourghiotis, A.; Kordulis, C. Preparation, characterization, and catalytic properties for the SCR of NO by NH_3 of V_2O_5/TiO_2 catalysts prepared by equilibrium deposition filtration. *J. Phys. Chem. B* **1998**, *102*, 8459–8468. [CrossRef]

31. Mamulova, K.K.; Tokarsky, J.; Kovar, P.; Vojteskova, S.; Kovarova, A.; Smetana, B.; Kukutschova, J.; Capkova, P.; Matejka, V. Preparation and characterization of photoactive composite kaolinite/TiO_2. *J. Harzard. Mater.* **2011**, *188*, 212–220. [CrossRef] [PubMed]

32. Iwasaki, M.; Shinjoh, H. A comparative study of "standard", "fast" and "NO$_2$" SCR reactions over Fe/zeolite catalyst. *Appl. Catal. A* **2010**, *390*, 71–77. [CrossRef]
33. Jiang, B.Q.; Wu, Z.B.; Liu, Y.; Lee, S.C.; Ho, W.K. DRIFT study of the SO$_2$ effect on low-temperature SCR reaction over Fe-Mn/TiO$_2$. *J. Phys. Chem. C* **2010**, *114*, 4961–4965. [CrossRef]
34. Wang, H.Q.; Gao, S.; Yu, F.X.; Liu, Y.; Weng, X.L.; Wu, Z.B. An effective way to control the performance of a ceria-based deNO$_x$ catalyst with improved alkali resistance: Acid-base adjusting. *J. Phys. Chem. C* **2015**, *119*, 15077–15084. [CrossRef]
35. Xiong, Z.B.; Lu, C.M.; Guo, D.X.; Zhang, X.L.; Han, K.H. Selective catalytic reduction of NO$_x$ with NH$_3$ over iron-cerium mixed oxide catalyst: Catalytic performance and characterization. *J. Chem. Technol. Biotechnol.* **2013**, *88*, 1258–1265.

catalysts

MDPI

Review

Recent Progress in Atomic-Level Understanding of Cu/SSZ-13 Selective Catalytic Reduction Catalysts

Feng Gao * and Charles H. F. Peden *

Institute for Integrated Catalysis, Pacific Northwest National Laboratory, Richland, WA 99352, USA
* Correspondence: feng.gao@pnnl.gov (F.G.); chuck.peden@pnnl.gov (C.H.F.P.)

Received: 2 March 2018; Accepted: 28 March 2018; Published: 31 March 2018

Abstract: Cu/SSZ-13 Selective Catalytic Reduction (SCR) catalysts have been extensively studied for the past five-plus years. New and exciting fundamental and applied science has appeared in the literature quite frequently over this time. In this short review, a few topics specifically focused on a molecular-level understanding of this catalyst are summarized: (1) The nature of the active sites and, in particular, their transformations under varying reaction conditions that include dehydration, the presence of the various SCR reactants and hydrothermal aging; (2) Discussions of standard and fast SCR reaction mechanisms. Considerable progress has been made, especially in the last couple of years, on standard SCR mechanisms. In contrast, mechanisms for fast SCR are much less understood. Possible reaction paths are hypothesized for this latter case to stimulate further investigations; (3) Discussions of rational catalyst design based on new knowledge obtained regarding catalyst stability, overall catalytic performance and mechanistic catalytic chemistry.

Keywords: selective catalytic reduction (SCR); zeolite; Cu/SSZ-13; reaction mechanisms; hydrothermal stability; standard SCR; fast SCR

1. Introduction

The discovery [1–4] and the rapid commercialization shortly thereafter, of Cu/SSZ-13 as a selective catalytic reduction (SCR) catalyst for NO_x abatement for the transportation industry, has been one of the most exciting milestones in environmental catalysis in recent years [5–9]. Inspired by this and also by the fact that SSZ-13 itself [10], a Chabazite silica-alumina zeolite, offers great structural simplicity for basic research, there has been an exponential growth in the number of published studies of Cu/SSZ-13 appearing in the open literature in recent years. In particular, Figure 1 depicts the numbers of papers published each year for the past 8 years containing "Cu", "SSZ-13" and "SCR" in their contents, using the "Web of Science" search engine. Notably, ~180 papers, with total citations greater than 4000, have been published in this short period of time. While likely an incomplete search particularly for 2017, these numbers clearly reflect the importance, popularity and enthusiasm for research of the now commercialized Cu/SSZ-13 catalyst. For comparison, from 1970 to 1980, during which time the three-way catalyst was developed and commercialized, fewer than 30 publications appeared in the open literature addressing "emission control catalyst," searched using the same engine.

In 2015, we published the first comprehensive review of the early understandings of Cu/SSZ-13 catalyst regarding its synthesis, structural details from spectroscopy and theory, SCR kinetics, as well as key aspects of the catalyst for practical applications [6]. Now after 3 years, new and exciting results have appeared in literature, especially with respect to the developing molecular-level understanding of the nature of the active Cu species under realistic operating (i.e., situ/operando) conditions, as well as the SCR reaction mechanisms [11–26]. As such, it seems timely to review these new fundamental understandings. We note, however, that this is not a comprehensive review but a summary of selected topics; in particular, topics relevant for the molecular-level understanding about the nature of Cu and the SCR reaction mechanisms. More specifically, we summarize very recent results on the mechanistic relevance of a binuclear Cu complex formed from mobile mononuclear complexes and how identification of several catalytically active Cu species led to an understanding of critical rate-limiting steps of reaction (Section 3). To do so, we also need to first briefly review what is now known about the multiple forms of Cu species present under various conditions of temperature and gas-phase composition (Section 2). Because the forms of Cu present also effect another critical aspect of these catalysts for practical applications, notably their stability, we also summarize some recent studies aimed at identifying optimum catalysts for both reactivity and stability in Section 4.

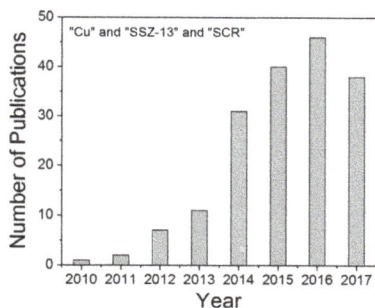

Figure 1. Number of recent publications on Cu/SSZ-13 catalysts for NH_3 Selective Catalytic Reduction (SCR).

2. Transformations of Cu Active Species

Cu/SSZ-13 is typically prepared via solution ion-exchange of NH_4/SSZ-13 with Cu salts at low pHs to avoid Cu agglomeration, or by so-called one-pot synthesis methods [27–31]. Unless Cu is present at levels exceeding the ion-exchange capacities of the particular SSZ-13 composition used, the current general consensus is that, in the freshly prepared specimens, Cu is present as isolated ions, including Cu^{2+} balanced by two nearby framework negative charges (abbreviated as Cu^{2+}-2Z, where Z represents a charged zeolite framework site) and a $[Cu(OH)]^+$ species balanced by one framework negative charge (as $[Cu(OH)]^+$-Z) [17,31–33]. It is also generally agreed, based on almost identical X-ray absorption near edge structure (XANES) spectra with model Cu salt solutions, that these ions are fully solvated as $[Cu(H_2O)_6]^{2+}$ and $[Cu(OH)(H_2O)_5]^+$ complexes in hydrated ambient samples [20,34,35]. These two ions are, for practical purposes, spectroscopically indistinguishable and they reside in the Chabazite (CHA) cages where they have relatively small (i.e., longer range) interactions with the CHA framework. Starting from this hydrated state, three scenarios are described next regarding the transformations of these Cu species due to the following changing conditions: (1) dehydration in vacuum or an inert gas; (2) hydrothermal aging (HTA); or (3) exposure to NH_3, including that occurring during low-temperature SCR reaction conditions.

2.1. Dehydration

During dehydration, complete H_2O desorption occurs between ~250–300 °C depending somewhat on the specific dehydration conditions. With the removal of H_2O ligands, Cu ions now migrate to cationic exchange positions and bond to lattice O (O_L) of the zeolite framework. Such a change is well reflected by changes in unit cell parameters of the CHA substrate from X-ray diffraction (XRD) [27,35], -OH and H_2O vibrations from FTIR [35], X-ray absorption and emission spectra (XAS and XES) [20] and hyperfine interactions between the unpaired electron and the nuclear spin of Cu(II) (I = 3/2) from electron paramagnetic resonance (EPR) [36]. It is now well-documented that in dehydrated Cu/SSZ-13, Cu^{2+}-2Z with Cu ions located in windows of 6-membered rings (6 MR) are the energetically most favorable configuration. As such, Schneider and coworkers predicted that exchanged Cu^{2+} ions first saturate these 2Al sites before populating unpaired, or 1Al, sites as $Cu(OH)]^+$ [17]. Note that the relative populations of Cu^{2+}-2Z and $[Cu(OH)]^+$-Z are dependent both on Si/Al and Cu/Al ratios of the Cu/SSZ-13 material, as shown in Figure 2.

This prediction, originated primarily from thermodynamics grounds, does not rigorously describe actual Cu/SSZ-13 catalyst compositions. In particular, a number of studies discovered that $[Cu(OH)]^+$-Z can populate before Cu^{2+}-2Z saturation [18,20,37]. A likely explanation is that, in fully hydrated samples, dynamic equilibrium exists between $[Cu(H_2O)_6]^{2+}$ and $[Cu(OH)(H_2O)_5]^+$ ions that can be described using Reaction (1):

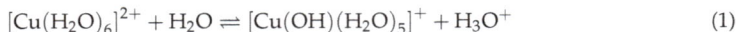

$$[Cu(H_2O)_6]^{2+} + H_2O \rightleftharpoons [Cu(OH)(H_2O)_5]^+ + H_3O^+ \tag{1}$$

Then, depending critically on the composition (Si/Al and Cu/Al ratios), specific Cu distributions and how dehydration is carried out, $[Cu(OH)]^+$-Z can very well be kinetically stabilized in the dehydrated form. For example, Martini et al. very recently demonstrated that a Cu/SSZ-13 sample with Si/Al = 14 and Cu/Al = 0.1 can contain substantial amounts of $[Cu(OH)]^+$-Z as determined with XANES [20], even though all of the Cu ions can, in principle, be more thermodynamically stabilized as Cu^{2+}-2Z (Figure 2).

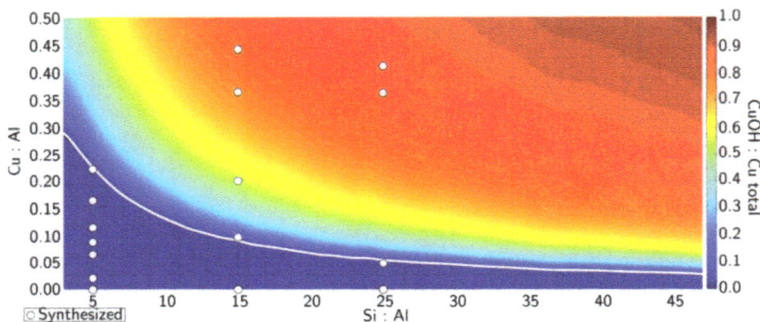

Figure 2. The predicted Cu site compositional phase diagram versus Si:Al and Cu:Al ratios, with the color scale indicating predicted fractions of CuOH. The white line demarcates the transition from a [Z2-CuII]-only region to a mixed [Z2-CuII]/[Z-CuII-OH] region. White circles indicate compositions of synthesized Cu-SSZ-13 samples used to verify the computed predictions. Reprinted with permission from *J. Am. Chem. Soc.*, **2016**, *138*, 6028–6048. Copyright 2016, American Chemical Society.

Even though $[Cu(H_2O)_6]^{2+}$ and $[Cu(OH)(H_2O)_5]^+$ ions are likely spectroscopically indistinguishable by XANES, indirect evidence does exist for the reversible reaction described in Reaction (1) from electron paramagnetic resonance (EPR) measurements. Figure 3 presents EPR spectra collected during dehydration of a Cu/SSZ-13 sample with Si/Al = 6 and Cu/Al = 0.032 [38]. At such Al abundance and low Cu loadings, it is expected that $[Cu(H_2O)_6]^{2+}$ is the only Cu species in the hydrated form and it will be converted exclusively to Cu^{2+}-2Z upon dehydration. However, at intermediate dehydration temperatures (e.g., 150 °C), the clear signal intensity loss suggests formation of EPR silent $[Cu(OH)]^+$ [36], even though it eventually converts to EPR active Cu^{2+}-2Z as evidenced by the signal intensity recovery at higher temperatures. For catalysts with higher Si/Al and Cu/Al ratios, $[Cu(OH)]^+$-Z can indeed be thermodynamically stabilized during dehydration.

Another interesting Cu ion transformation during dehydration is the so-called "auto-reduction"; that is, formation of Cu(I) moieties in vacuum or in an inert gas flow without addition of reductants [36]. Although not conclusive, a likely mechanism can be described as follows, involving formation of an OH radical:

$$[Cu(OH)]^+ \rightarrow Cu^+ + HO^{\cdot} \qquad (2)$$

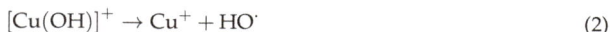

It is also notable that auto-reduction even occurs in the presence of O_2, as long as the temperature is high enough [25,31]. Applying multivariate curve resolution (MCR) techniques to XANES linear combination fit (LCF) analysis, Martini et al. [20] recently studied dehydration of a series of Cu/SSZ-13 materials with varying compositions. Using 5 model structures, detailed Cu transformations as a function of temperature were quantified and the results are shown in Figure 4. Besides the expected result that lower Si/Al ratios favor Cu^{2+}-2Z population, the authors also discovered that the extents of autoreduction are the highest at intermediate Si/Al ratios, which they rationalized as due to autoreduction proceeding though a cooperative multi-step process involving proximal acid sites. However, this conclusion is in contrast to H_2 temperature-programmed reduction (H_2-TPR) results by Gao et al. [31], who demonstrated, based on H_2 consumption differences between ambient and autoreduced samples, that the extent of autoreduction increases with increasing Si/Al ratios at similar Cu/Al ratios. Besides the well-known positioning for Cu^{2+}-2Z and $[Cu(OH)]^+$-Z in dehydrated Cu/SSZ-13, N_2 adsorption FTIR measurements by Martini et al. [20] suggest that autoreduced Cu ions (i.e., Cu^+-Z) occupy both windows of 6 MRs and 8 MRs, with relative populations that depend on Cu/Al and Si/Al ratios.

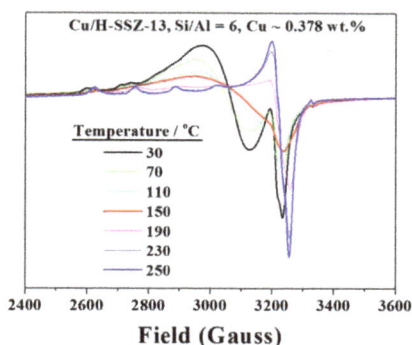

Figure 3. Electron Paramagnetic Resonance (EPR) spectra of a Cu/SSZ-13 sample (Si/Al = 6, Cu/Al = 0.032) during dehydration. Reprinted from *J. Catal.*, **2014**, *319*, 1–14. Copyright 2014, with permission from Elsevier.

2.2. Hydrothermal Aging

Hydrothermal stability of Cu/SSZ-13 is a critical criterion for its practical applications; as such, hydrothermal aging effects have been extensively addressed [29,37,39–41]. Hydrothermal stability depends on various parameters including catalyst composition, aging temperature and duration. Typically, aging at ~700 °C and lower causes limited structural degradation of Cu/SSZ-13 (e.g., dealumination, surface area loss), while higher aging temperatures induce partial to even complete structural damage [37,41].

Recently, Luo et al. [18] hydrothermally aged a state-of-the-art commercial Cu/SSZ-13 catalyst with Si/Al = 9.5 and Cu/Al = 0.3 at 600 °C for varying durations from 1 to 100 h. Subsequently, NH_3 temperature-programmed desorption (NH_3-TPD, following NH_3 adsorption at 150 °C) was performed and the results are shown in Figure 5. The TPD feature above 400 °C is due to NH_3 desorption from Brønsted acid sites, while the feature below 300 °C is attributed to NH_3 desorption from Cu. Note that the aging temperature of 600 °C causes essentially no dealumination as determined from ^{27}Al NMR; as such, Brønsted acidity loss during aging is presumed to be caused by the reverse of Reaction (1):

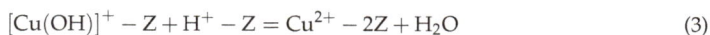

$$[Cu(OH)]^+ - Z + H^+ - Z = Cu^{2+} - 2Z + H_2O \qquad (3)$$

This notion is fully consistent with NH_3/Cu ratio quantifications using the TPD data, where it was found that, at an NH_3 adsorption temperature 150 °C, ~2 molecules of NH_3 adsorbs on Cu^{2+}-2Z and ~1 molecule of NH_3 adsorbs on $[Cu(OH)]^+$-Z. The fact that 1 molecule of NH_3 adsorbs on 1 Brønsted acid site nicely explains the largely invariant NH_3 yields as a function of aging time that are evident in Figure 5. Two key additional points are worth noting for this study. First, even though $[Cu(OH)]^+$-Z can be kinetically stabilized prior to Cu^{2+}-2Z saturation in freshly prepared catalysts, given enough time and thermal energy during hydrothermal aging, $[Cu(OH)]^+$-Z will convert to the thermodynamically more stable Cu^{2+}-2Z species. Second, although NH_3 can fully solvate Cu^{2+}-2Z, forming a detached $[Cu(NH_3)_4]^{2+}$ complex at ambient temperatures, at higher temperatures, Cu(II) sites interacting with both lattice O (O_L) and NH_3 (e.g., as partially solvated $[Cu(O_L)_2(NH_3)_2]^{2+}$) appear to better describe these species. More details on Cu and NH_3 interactions will be given below.

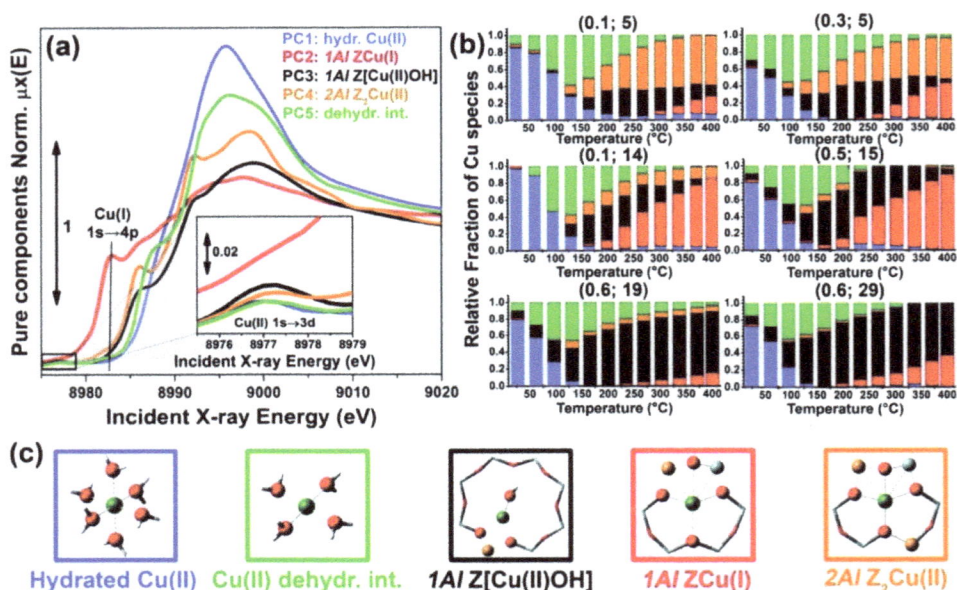

Figure 4. Multivariate curve resolution (MCR) coupled with the alternating least square (ALS) analysis of global temperature-dependent XANES dataset collected for six Cu-CHA samples with different compositions during He-activation from 25 to 400 °C, assuming the presence of 5 pure components. (**a**) XANES spectra of pure components $\mu_i(E)$ derived from MCR-ALS. The inset reports a magnification of the Cu(II) 1s → 3d transition region in the theoretical spectra; (**b**) Temperature-dependent abundance of pure species, w_i^p(Cu/Al; Si/Al, T), in each of the catalysts (bars have the same colors as the corresponding spectra in panel (**a**)); (**c**) Proposed assignment of the five pure components (PC) to specific Cu-species/sites formed in the Cu-CHA catalyst as a function of composition and activation temperature, using the same color code as in parts (**a**) and (**b**); Blue (PC1): mobile Cu(II)-aquo-complexes $[Cu(II)(H_2O)_n]^{2+}/[Cu(II)(H_2O)_{n-1}(OH)]^+$ with $n = 6$; green (PC5): Cu(II) dehydration intermediate, possibly represented by mobile $[Cu(II)(H_2O)_n]^{2+}/[Cu(II)(H_2O)_{n-1}(OH)]^+$ complexes with $n = 4$; black (PC3): 1Al Z[Cu(II)-OH] sites in their oxidized form; red (PC2): 1Al Z-Cu(I) sites in their reduced form, resulting from self-reduction of 1Al Z[Cu(II)OH] species; orange (PC4): 2Al Z2-Cu(II) sites. Atom color code: Cu: green; H: white; O: red; Si: grey; Al: yellow. *Chem. Sci.*, **2017**, *8*, 6836–6851—Published by The Royal Society of Chemistry.

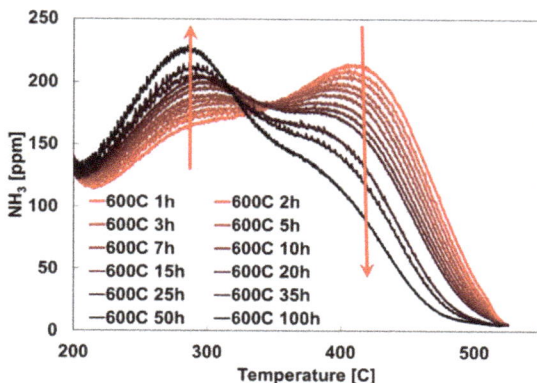

Figure 5. NH$_3$-TPD profiles for Cu/SSZ-13 upon progressive aging at 600 °C; arrows indicate increasing aging times. Reprinted from *J. Catal.*, **2017**, *348*, 291–299. Copyright 2017, with permission from Elsevier.

Hydrothermal aging under more harsh conditions can cause additional types of Cu transformations, including especially formation of copper oxide clusters (CuO$_x$). Song et al. [37] recently conducted detailed aging studies of a Cu/SSZ-13 catalyst with Si/Al = 12 and Cu loading 2.1 wt % (corresponding to Cu/Al = 0.25), where aging temperatures systematically varied from 550 to 900 °C. The fresh and aged samples were examined with EPR in both hydrated and dehydrated forms at cryogenic temperatures in order to quantify Cu^{2+}-2Z, [Cu(OH)]$^+$-Z and CuO$_x$ clusters in these samples, based on the facts that Cu^{2+}-2Z is EPR active in both hydrated (i.e., as [Cu(H$_2$O)$_6$]$^{2+}$) and dehydrated forms, CuO$_x$ clusters are always EPR silent due to antiferromagnetic effects and [Cu(OH)]$^+$-Z is EPR active when hydrated (as [Cu(OH)(H$_2$O)$_5$]$^+$) but becomes EPR silent when dehydrated due to pseudo Jahn-Teller effects [36]; that is, absorbed microwave energy by [Cu(OH)]$^+$-Z dissipates too rapidly for the time scale of EPR measurements.

From the results shown in Figure 6, [Cu(OH)]$^+$-Z content monotonically decreases while CuO$_x$ content increases with increasing aging temperature. Cu^{2+}-2Z content first increases, reaching its highest amount of ~1.5 wt % in the sample aged at 700 °C and then declines slightly before stabilizing at ~1.2 wt % at higher aging temperatures. From such quantifications, it is evident that [Cu(OH)]$^+$-Z species primarily convert to Cu^{2+}-2Z and perhaps a portion to CuO$_x$ clusters at aging temperatures ≤700 °C. The former transformation, as described by Reaction (3), is also found for the current state-of-the-art commercial Cu/SSZ-13 catalyst (Figure 5) [18]. At aging temperatures of 750 °C and above, Cu^{2+}-2Z stabilizes at ~1.2 wt % indicating that this portion of Cu^{2+} must be located in windows of 6MRs with paired Al sites; that is, at the energetically most stable locations for Cu^{2+}-2Z [33]. The remaining Cu^{2+} species are less stable, ultimately converting to CuO$_x$ at higher aging temperatures. An important outcome of this result is the remarkable hydrothermal stability evident for Cu^{2+}-2Z located in windows of SSZ-13 6 MRs, where they survive even under extremely harsh hydrothermal aging temperatures of 900 °C. On the other hand, [Cu(OH)]$^+$-Z is hydrothermally much less stable; some of which converts to more stable Cu^{2+}-2Z and the rest to CuOx clusters during aging.

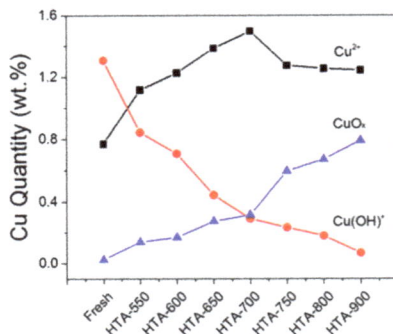

Figure 6. Estimation of Cu^{2+}, $Cu(OH)^+$ and CuO_x in fresh and hydrothermally aged Cu/SSZ-13 samples with Si/Al = 12 and Cu/Al = 0.25. HTA represents hydrothermal aging. Reprinted with permission from *ACS Catal.*, **2017**, *7*, 8214–8227. Copyright 2017, American Chemical Society.

The processes by which $[Cu(OH)]^+$-Z converts to CuO_x clusters are not apparent at present. Song et al. [37] proposed recently that $[Cu(OH)]^+$-Z first undergoes a hydrolysis reaction to form mobile $Cu(OH)_2$ (Reaction (4)) during hydrothermal aging, with the latter then migrating and agglomerating, to eventually form CuOx clusters.

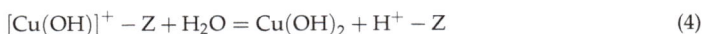

$$[Cu(OH)]^+ - Z + H_2O = Cu(OH)_2 + H^+ - Z \tag{4}$$

Hydrolysis activation barriers simulated from DFT are fully consistent with the hydrothermal stability difference between $[Cu(OH)]^+$-Z and Cu^{2+}-2Z, where it was found that the former hydrolyzes more readily than the latter.

The nature of the CuO_x clusters formed during hydrothermal aging, for example, their sizes, locations and whether they incorporate any aluminum (from zeolite substrate dealumination), is also essentially unknown at present due, in part, to a lack of techniques for their unambiguous characterization. For example, using scanning TEM, Song et al. [37] failed to identify clusters that can be assigned to CuO_x in aged Cu/SSZ-13. Notable also is that a lack of surface Al and Cu segregation evident from XPS analyses indicates that these moieties may still be located in pores and channels of SSZ-13. Weckhuysen and coworkers [22] recently utilized atom probe tomography (APT) to compare aged Cu/ZSM-5 and Cu/SSZ-13. In Cu/ZSM-5, severe Cu and Al aggregation was demonstrated and mapped in 3D and a Cu/Al atomic ratio of ~2 for the aggregates indicates formation of copper aluminate. Similarly, irreversible formation of a Cu-Al phase was confirmed via XAFS measurements in a study by Vennestrøm et al. [42] for aged Cu/ZSM-5. In both studies, a phase similar to $CuAl_2O_4$ spinel was suggested. However, formation of a crystalline spinel phase is likely to be difficult in the confined spaces of zeolite channels and cages. Instead, formation of distorted amorphous phases with stoichiometry perhaps close to that of $CuAl_2O_4$ spinel may better describe the Cu-Al phases formed in aged Cu/ZSM-5. In contrast, APT analysis of aged Cu/SSZ-13 shows some Cu and Al clustering but to a limited degree (relative to Cu/ZSM-5) with undetermined stoichiometries [22]. It can be concluded, therefore, that CuO_x formed in aged Cu/SSZ 13 can indeed interact with Al from substrate dealumination but stoichiometry and structures of the thus formed species are still undetermined.

2.3. Interactions with NH_3

In comparison to NO_x and H_2O, interactions between Cu ions and NH_3 are much stronger; this renders a generally agreed scenario that, under low-temperature NH_3-SCR reaction conditions, Cu-NH_3 complexes are abundant and some of these are key intermediates for reaction [15,17,20]. Therefore, it is of considerable relevance to study species formed by adsorption of NH_3 on Cu/SSZ-13 and identify their mechanistic involvement in NH_3-SCR. Cu-NH_3 interactions are dependent on nature of Cu ions (Cu^{2+}, $[Cu(OH)]^+$ or Cu^+), local environments of these Cu ions, gas-phase compositions and temperature.

Schneider and coworkers [17] utilized DFT thermodynamic screening to predict species that are likely relevant to NH_3-SCR; that is, in the presence of 300 ppm NH_3, 2% H_2O and varying O_2 pressures as a function of temperature. From the results shown in Figure 7, Cu^{2+}-2Z and $[Cu(OH)]^+$-Z display completely different reactivity toward NH_3, where the latter is much more readily reducible to Cu^+-Z by NH_3. Under low-temperature (e.g., 200 °C) SCR conditions, this simulation predicts a dominance of $[Cu(NH_3)_4]^{2+}$, $[Cu(OH)(NH_3)_3]^+$ and $[Cu(NH_3)_2]^+$ species, all of which detach from the SSZ-13 framework as mobile species [17]. Among these, $[Cu(NH_3)_2]^+$ has the highest mobility and can readily pass through 8MR openings that are shared by neighboring unit cells. Note that, while the studies discussed above by Luo et al. [18] demonstrate via NH_3-TPD that likely Cu-NH_3 complexes are $[Cu(O_L)_2(NH_3)_2]^{2+}$ and $[Cu(O_L)_2(OH)(NH_3)]^+$ following NH_3 adsorption and purging at 150 °C, the apparent discrepancy can be understood from the gas-phase NH_3 pressure differences, where the presence of gas phase NH_3 allows for higher coordination with NH_3 ligands.

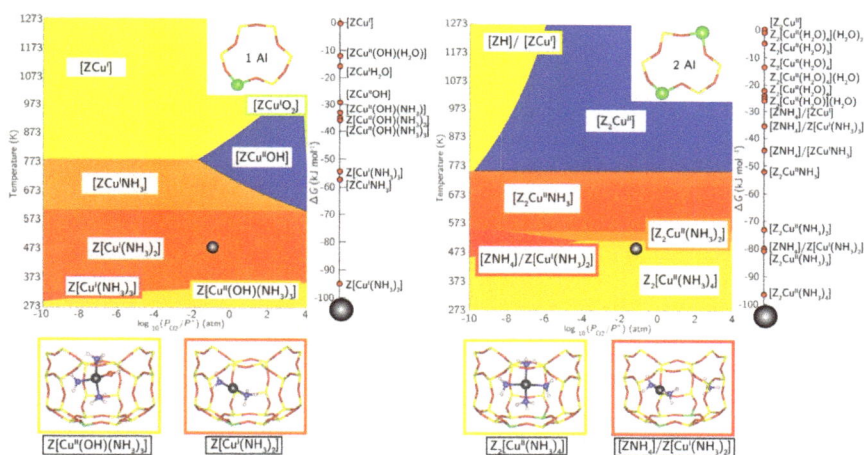

Figure 7. Phase diagrams for 1Al (**left**) and 2Al (**right**) sites with varying T and PO_2 at 300 ppm of NH_3 and 2% H_2O. Relative rankings for all stable species—$\Delta G_{form} < 0$ at 473 K and 10% O_2 (chrome spheres on the phase diagrams)—are given to the right of each phase diagram. The structures shown on the bottom are the most stable CuI (red outline) and CuII (golden outline) species under these conditions. Reprinted with permission from *J. Am. Chem. Soc.*, **2016**, *138*, 6028–6048. Copyright 2016, American Chemical Society.

EPR can be used to characterize Cu(II)-NH_3 complexes in a quantitative way. An added benefit of this technique is that Cu(I)-NH_3 complexes, being EPR silent, do not complicate the measurements. As shown by the spectra in Figure 8, obtained subsequent to NH_3 and $^{15}NH_3$ adsorption on a one-pot synthesized Cu/SSZ-13 followed by degassing at 25 °C, second derivatives of the EPR signals display super-hyperfine structures that can be used to determine Cu(II)-NH_3 coordination [14]. The super-hyperfine structures come from interactions of the Cu(II) unpaired electron with the nuclear spin of ^{14}N (I = 1) or ^{15}N (I = 1/2), which splits the EPR lines into $2nI + 1$ components, where I is the nuclear spin and n is the number of nuclei. Based on this, species B in Figure 8 can be assigned to $[Cu(NH_3)_5]^{2+}$ and species C to $[Cu(O_L)_2(NH_3)_2]^{2+}$ [14]. The former complex is not stable at higher temperatures, where it loses NH_3 ligands sequentially with increasing temperature. Possible structures that form in this process were simulated with DFT [14].

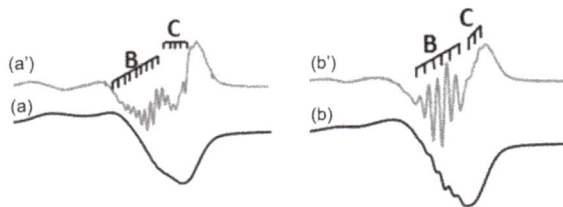

Figure 8. EPR spectra recorded at 25 °C after the adsorption of 6 NH$_3$/Cu atom (**a**) or 6 ^{15}NH$_3$ per Cu (**b**) on Cu-SSZ-13 followed by degassing at 25 °C. (**a'** and **b'**) are second derivatives of the spectra, used to determine the super-hyperfine structure of EPR signals B and C. Reprinted with permission from *J. Phys. Chem. Lett.*, **2015**, *6*, 1011–1017. Copyright 2015, American Chemical Society.

It is now well established that NH$_3$ itself can reduce Cu(II) to Cu(I) [17,43]. In particular, the theoretical phase diagrams shown in Figure 7 indicate that [Cu(OH)]$^+$ is readily reduced at low temperatures irrespective of the O$_2$ pressure. In the absence of NH$_3$ complexation, Cu^{2+} reduction is more difficult, requiring higher temperatures and low O$_2$ pressures. Experimentally, it has been shown that EPR signals disappear almost completely by heating NH$_3$ saturated Cu/SSZ-13 to 250 °C, demonstrating formation of EPR silent Cu(I) moieties [14]. A sharp Cu K-edge feature at 8982.5 eV, characteristic for Cu(I) upon NH$_3$ introduction as probed with XANES also clearly indicates Cu(II) reduction to Cu(I) [20,44]. Based on XAS and XES experimental findings, prior literature assignments and DFT simulations, Giordanino et al. first proposed that these Cu(I) moieties are O$_L$-Cu$^+$-NH$_3$ and NH$_3$-Cu$^+$-NH$_3$ species and emphasized the linear nature of the latter [43]. Their initial simulated structures are shown in Figure 9, although it is now known that the NH$_3$-Cu$^+$-NH$_3$ species are much more delocalized in the zeolite than that indicated in the figure.

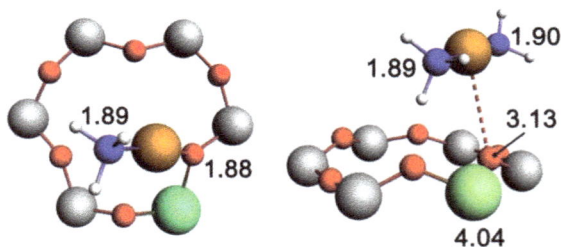

Figure 9. Local Cu environments after adsorption of one (**left**) and two (**right**) NH$_3$ molecules. Color code: orange, Cu; green, Al; gray, Si; red, O; blue, N; white, H. Distances between Cu and neighboring atoms are indicated in angstroms. Reprinted with permission from *J. Phys. Chem. Lett.*, **2014**, *5*, 1552–1559. Copyright 2014, American Chemical Society.

Detailed reaction mechanisms for Cu(II) reduction to Cu(I) by NH$_3$ are not understood. Possible reaction pathways are shown as follows:

$$Cu^{2+} + NH_3 = Cu^+ - NH_2 + H^+ \tag{5}$$

$$[Cu(OH)]^+ + NH_3 = Cu^+ - NH_2 + H_2O \tag{6}$$

Reaction (5) has been initially suggested to be energetically unfavorable due to a high activation barrier of +119 kJ/mol from DFT simulations [11]. However, in the presence of surplus NH$_3$ molecules to provide sufficient solvation effects, energy barriers can drop substantially to 30–34 kJ/mol, making the process much more feasible [14]. Therefore, a more likely Reaction (5) should be modified to, for example, Reaction (7).

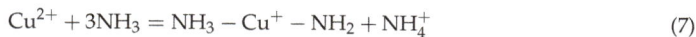

$$Cu^{2+} + 3NH_3 = NH_3 - Cu^+ - NH_2 + NH_4^+ \tag{7}$$

Reaction (6), on the other hand, is expected to require much less NH_3 solvation assistance since H_2O formation is a strong driving force energetically. This may well explain why $[Cu(OH)]^+$ is much more readily reducible than uncomplexed Cu^{2+} ions [17]. Note, however, that Reactions (5–7) have not been proven experimentally. In particular, the presence of an $-NH_2$ species has not been demonstrated. Possible reasons are that this species is overwhelmed by the more prevalent NH_3 ligands, or its instability leads to rapid consumption, for example, via formation and decomposition of N_2H_4 [43]. Clearly, more work is needed to gain further details on 'direct' Cu(II) ion reduction by NH_3.

The high mobility for $[Cu(NH_3)_2]^+$, as will be shown in the next section, plays a critical role in low-temperature SCR. Of note is that this characteristic is also used to prepare Cu/SSZ-13 using NH_3-assisted, reaction-driven ion exchange at low temperatures [44,45]. For example, treating Cu_2O and zeolite physical mixtures in NH_3 or $NH_3 + NO$ at 250 °C effectively converts Cu_2O into Cu ions that are anchored at zeolite ion-exchange sites due to formation, migration and reaction of mobile $[Cu(NH_3)_2]^+$ complexes [44].

3. NH_3-SCR Mechanisms

3.1. Standard NH_3-SCR

The "Standard" NH_3-SCR reaction is stoichiometrically shown in Reaction (8):

$$4NO + 4NH_3 + O_2 = 4N_2 + 6H_2O \tag{8}$$

Complexity of this reaction is clearly evident in the stoichiometry: the reaction is redox in nature and there are 3 reactants to activate. Furthermore, complexity is also evident in the nature of the catalytically active Cu species, as first recognized from SCR kinetics. For example, Figure 10 displays steady state NO_x light-off curve for a Cu/SSZ-13 catalyst with relatively low Cu loading, allowing low- and high-temperature reaction regimes to be clearly resolved [19,38]. Importantly, the NO_x conversion decrease with increasing temperature between ~250 and 350 °C indicates a temperature-dependent change in the nature of active Cu sites, which has now been repeatedly demonstrated via in situ/operando XAS. In particular, below this temperature range, Cu ions are fully solvated by NH_3 as mobile mono-Cu complexes in SSZ-13 cages, while above 350 °C, Cu ions anchor at (complex to) cationic sites of the zeolite framework [15].

In the low-temperature regime, strong Cu loading dependence is found in SCR kinetics. For example, as shown in Figure 11, a linear correlation between SCR rates and Cu loading (i.e., invariant turnover frequencies, TOFs), expected for isolated Cu-ions being the active sites, is only found at intermediate Cu loadings for a series of catalyst samples with essentially only isolated Cu ions [38]. The decrease in TOFs at high Cu loadings can, at least in part, be explained by mass transfer limitations for the reactants. In particular, when Cu loadings are sufficiently high, not all of the Cu ions will be readily accessible to the reactants and, in this way, the ones that are deep in the bulk of the zeolite will be underutilized, causing apparent TOFs to drop. Still, measured reaction activation energies will not decrease in this case, as usually expected from reactant transfer limitations; this is because SCR kinetics will essentially only measure reaction carried out by Cu ions in the outer layers of the catalysts that do not sense such limitations. More significantly, in the low Cu-loading regime, SCR rates also deviate from the linear SCR rate versus Cu loading correlation for completely different reasons. In probing this low Cu-loading regime, critical new insights have been provided for understanding the standard SCR mechanism.

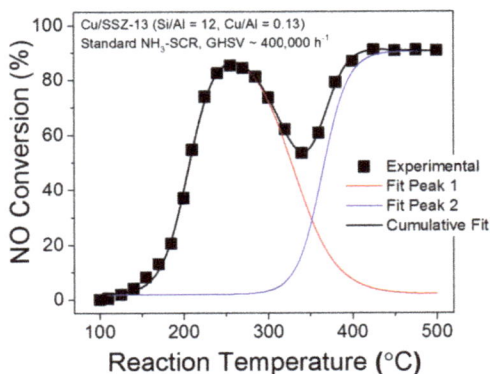

Figure 10. NO conversion versus temperature data (■) for standard SCR over a Cu/SSZ-13 catalyst with Si/Al = 12 and Cu/Al = 0.13. The reactant feed contains 350 ppm NH$_3$, 350 ppm NO, 14% O$_2$, 2.5% H$_2$O balanced with N$_2$ at a gas hourly space velocity (GHSV) of 400,000 h^{-1}. Also included are simulated curves assuming low- and high-temperature reaction routes. Reprinted with permission from *J. Am. Chem. Soc.*, **2017**, *139*, 4935–4942. Copyright 2017, American Chemical Society.

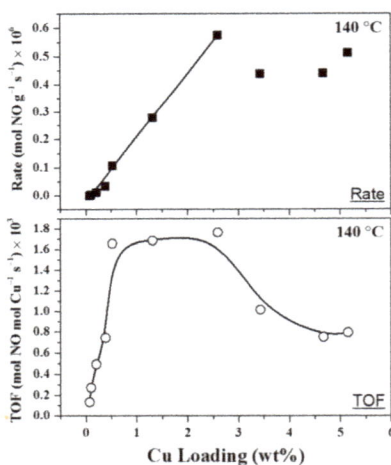

Figure 11. SCR rates (upper panel) and SCR turnover frequencies (lower panel) as a function of Cu loading at a reaction temperature of 140 °C (a temperature used to readily obtain intrinsic, low-conversion reaction kinetics). The reactant feed contains 350 ppm NO, 350 ppm NH$_3$, 14% O$_2$, 2.5% H$_2$O balanced with N$_2$ at a GHSV of 400,000 h^{-1}. Reprinted from *J. Catal.*, **2014**, *319*, 1–14. Copyright 2014, with permission from Elsevier.

Of special note is that reaction rates at low temperatures are limited by O$_2$ activation; notably, for the process by which Cu(I) is re-oxidized back to Cu(II) to finish redox cycling of the active sites [19,21]. In particular, a few important experimental findings led to the surprising proposal that rate-limiting, low-temperature O$_2$ activation requires participation of two Cu(I) sites. First, as shown in Figure 12, for reaction temperatures up to 250 °C, instead of a linear SCR rate versus Cu loading correlation, an interesting linear SCR rate versus (Cu loading)^2 correlation was discovered by Gao et al. [38], strongly suggesting that the reaction rate limiting step involves participation of two Cu ions. This surprising experimental result was later reproduced by Gounder, Ribeiro and coworkers [21]. Secondly, in a study on dry NO oxidation to NO_2 by Cu/SSZ-13, Ribeiro and coworkers discovered that isolated Cu ions do not catalyze this reaction, whereas CuxOy ($x \geq 2$) clusters contribute linearly to the rate of NO oxidation per mole Cu at 300 °C [46]. This again suggests that only multinuclear Cu species are capable of activating O_2 at low to moderate temperatures. Thirdly, a linear rate versus (Cu loading)^2 correlation was also found for low-temperature NH_3 oxidation ($4NH_3 + 3O_2 = 2N_2 + 6H_2O$) [38], suggesting that key intermediates that are associated with the rate-limiting step must contain NH_3 ligands, the common species in both SCR and NH_3 oxidation that, as discussed above, binds with Cu stronger than any other relevant reactants.

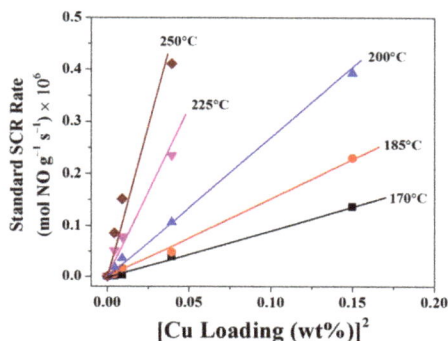

Figure 12. Standard SCR rates (moles NO g^{-1} s^{-1}) as a function of the square of the Cu loading for reaction on Cu/SSZ-13 catalysts with low Cu loadings. The reactant feed contains 350 ppm NO, 350 ppm NH_3, 14% O_2, 2.5% H_2O balanced with N_2 at a GHSV of 400,000 h^{-1}. Reprinted from *J. Catal.*, **2014**, *319*, 1–14. Copyright 2014, with permission from Elsevier.

The experimental findings just discussed strongly implicate two NH_3-ligated Cu(I) species collectively in the activation of O_2 as an essential step for standard NH_3-SCR. Assuming that no more than one Cu ion is present in any Chabazite cage (which is indeed the prevalent scenario), migration of such Cu(I) moieties across shared 8 MR windows between neighboring unit cells therefore becomes essential. The realization that linear $[Cu(NH_3)_2]^+$ is the most stable and most mobile Cu(I) ion complex makes this species the best candidate for O_2 activation. Based on this, Gao et al. [19] used DFT computer simulations to first propose the following pathway as an essential elemental step for the O_2 activation, that is, the oxidation half-cycle of SCR:

$$2[Cu(NH_3)_2]^+ + O_2 \rightarrow [Cu(NH_3)_2]^+ - O_2 - [Cu(NH_3)_2]^+ \qquad (9)$$

Concurrently, Paolucci, et al. identified this same process as the rate-limiting step for SCR at low temperatures. This group's subsequent Science paper [21] provided many additional important details on the rates of complexed Cu-ion transport and the formation of the binuclear Cu complex. Significantly, Reaction (9) and especially the $[Cu(NH_3)_2]^+$ diffusion and Cu-O_2-Cu binding steps, have been proven to be thermodynamically feasible from theoretical simulations [19,21]. At the lowest Cu loadings, formation of the Cu-O_2-Cu intermediate is the rate-limiting, essentially controlled by the entropy costly diffusion of $[Cu(NH_3)_2]^+$. Furthermore, low-temperature SCR kinetics measured under such conditions show strong characteristics of mass-transfer limited reaction, with low pre-exponential factors and apparent activation energies [19]. At somewhat higher Cu loadings where 'encounters' of two complexes become more facile and frequent, diffusion of $[Cu(NH_3)_2]^+$ and the corresponding activation of O_2, is no longer rate limiting. As a result, all isolated Cu ions become catalytically equivalent, as reflected by the onset of the TOF invariant regime at intermediate Cu loadings (Figure 11).

It is important to note from Reaction (9) that activation of one O_2 molecule (with eventual formation of $2O^{2-}$ ions) is a four-electron reduction, while oxidation of two Cu(I) centers to Cu(II) is only a two-electron process. Therefore, this chemistry is only realized in the presence of another electron provider. Gao et al. [19] proposed that NO serves as another electron provider, which is oxidized to NO_2 to facilitate Cu(I) oxidation according to the following reaction:

$$[Cu(NH_3)_2]^+ - O_2 - [Cu(NH_3)_2]^+ + NO \rightarrow [Cu(NH_3)_2]^{2+} - O^{2-} - [Cu(NH_3)_2]^{2+} + NO_2 \quad (10)$$

This proposed Cu(I) → Cu(II) pathway has yet to be proven experimentally, even though computed activation barriers should be readily overcome under low-temperature SCR conditions [19].

Using $[Cu(OH)]^+$-Z as the active site, a full standard SCR cycle has been simulated with DFT and the results are shown in Figure 13. Besides key intermediate formation pathways described by Reactions (9) and (10), this mechanism also involves energetically feasible HONO (which can then lead to NH_4NO_2 formation) and Cu-O-Cu hydrolysis steps to complete catalytic turnover [19]. A similar mechanism has been proposed by others without specifying how a charge balance can be maintained during the reduction of O_2 to $2O^{2-}$ and corresponding oxidation of two Cu(I) ions to Cu(II), nor how dinuclear intermediates split into two isolated Cu ions to complete a catalytic cycle [21]. These are some of the remaining questions for the low-temperature standard SCR mechanisms that require additional research.

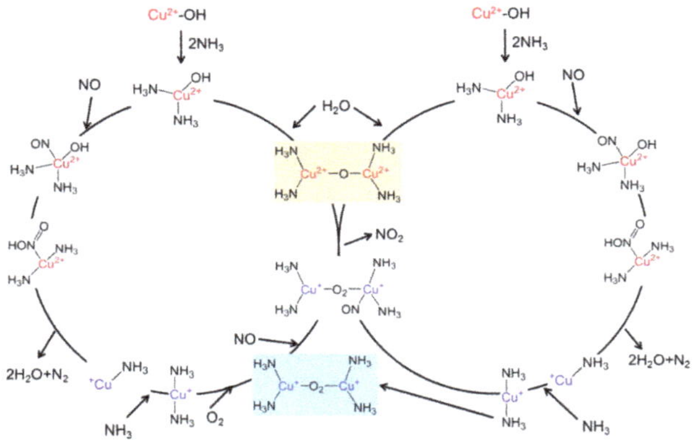

Figure 13. Complete redox cycling mechanism for low-temperature standard NH_3-SCR derived from DFT calculations that involves two Cu(I) centers in the initiation of the oxidation half-cycle. Adapted with permission from *J. Am. Chem. Soc.*, **2017**, *139*, 4935–4942. Copyright 2017, American Chemical Society.

Much less is known about standard SCR mechanisms for the high-temperature (>350 °C) regime shown in Figure 10. Under practical SCR reaction operations, this regime is characterized by high NO_x conversions where the light-off curves contain little, if any, kinetic information. By using low Cu loadings and very high space velocities to maintain differential NO_x conversions and also by carefully subtracting reactivity from the SSZ-13 substrate, Gao et al. [38] were able to collect kinetically meaningful data shown in Figure 14 in the form of Arrhenius plots. This regime is characterized by kinetic features and parameters that are markedly different from the low-temperature regime. First, SCR is apparently carried out on single Cu ions for all Cu loadings studied, indicating that O_2 activation no longer requires paired Cu ion complexes. Secondly, apparent reaction activation energies of ~140 kJ/mol are much higher than those typically found in the low-temperature regime (60–90 kJ/mol), suggesting rather demanding reaction rate-limiting steps.

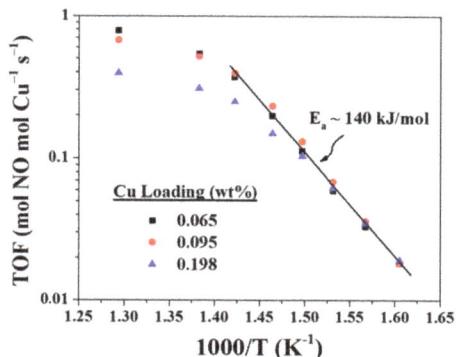

Figure 14. Turnover frequencies (TOFs, mol NO mol Cu^{-1} s^{-1}) as a function of temperature in Arrhenius plots for low Cu loaded Cu/SSZ-13 (Si/Al = 6). The reactant feed contains 350 ppm NO, 350 ppm NH_3, 14% O_2, 2.5% H_2O balanced with N_2 at a GHSV of 1,200,000 h^{-1}. Different symbols represent samples with different Cu loadings. Reprinted from *J. Catal.*, **2014**, *319*, 1–14. Copyright 2014, with permission from Elsevier.

Recent operando XAS studies by Lomachenko et al. [15] demonstrate that, above 300 °C, mobile Cu components quickly diminish and four coordinated Cu(II) complexes convert to 3 coordinated ones that spectroscopically resemble a [Cu(II)-NO_3^-]-Z species. This finding may be used to justify an SCR mechanism proposed by Janssens et al. [13], who postulated that Cu(I) oxidation to Cu(II) is achieved by nitrate intermediate formation from NO+O_2 as the rate-limiting step:

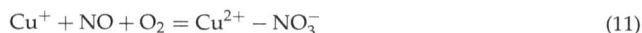

$$Cu^+ + NO + O_2 = Cu^{2+} - NO_3^- \tag{11}$$

Even though these authors did not specify a temperature window for this mechanism proposal, from our current understanding as discussed above, such a mechanism is clearly not suitable to describe the SCR reaction in the low-temperature regime for the following reasons: (1) it is inconsistent with the Cu loading dependent kinetics found at low temperatures; and (2) Cu ions are heavily ligated with NH_3 in the low temperature regime. According to a very recent operando XAS study by Marberger et al. [26], Cu nitrate species appear only when NH_3 is not readily available; that is, away from SCR conditions. However, at elevated (>300 °C) reaction temperatures, there are currently no experimental findings that disprove this proposal. Nitrate, however, does not interact with NH_3 to form intermediates that directly leads to N_2 + H_2O formation due to a mismatch in charge transfer; that is, the N atom in NO_3^- has an oxidation state of +5 while the N atom in NH_3 has an oxidation state of -3. Thus, it is necessary to generate intermediates with an N oxidation state of +3 for a viable mechanism. In the proposal by Janssens et al. [13], this is achieved by the following reaction:

$$Cu^{2+} - NO_3^- + NO \rightleftharpoons Cu^{2+} - NO_2^- + NO_2 \tag{12}$$

Upon nitrite formation, NH_3 can then interact with it to form an NH_4NO_2 intermediate, which is highly unstable and decomposes selectively to N_2 and H_2O.

Alternatively, high-temperature SCR may follow an Eley-Rideal type of mechanism involving activation of chemisorbed NH_3 as the rate-limiting step. In this case, NH_3 is activated first via an N-H bond cleavage:

$$Cu^{2+} - NH_3 = Cu^+ - NH_2 + H^+ \tag{13}$$

Following Reaction (13), gas-phase NO could interact with the resulting species to form a highly unstable nitrosoamide intermediate (NH_2NO), which also decomposes selectively to N_2 and H_2O.

Note that in these two possible mechanisms, both the nitrate formation and N–H cleavage steps are energetically rather demanding, with DFT-calculated activation energies of ~104 kJ/mol [13] and ~119 kJ/mol [11], respectively. These values are in reasonable agreement with the apparent activation energy of ~140 kJ/mol found experimentally (Figure 14), suggesting that the rate-limiting steps described above are rather plausible. However, more experimental and theoretical studies are certainly needed for a more detailed understanding of the standard SCR mechanism at elevated temperatures.

3.2. Fast NH$_3$-SCR

The so-called "fast" SCR reaction, occurring when NOx is composed of both NO and NO$_2$ ideally in equal amounts, is described by the following stoichiometry [47]:

$$NO + NO_2 + 2NH_3 = 2N_2 + 3H_2O \tag{14}$$

Despite extensive recent research efforts on Cu/SSZ-13 catalysts, detailed kinetics studies about the fast SCR (Reaction (14)) on this catalyst are surprisingly scarce. There are obvious experimental difficulties: (1) the accumulation of NH$_4$NO$_3$ at low reaction temperatures poisons the catalyst and prevents steady-state measurements; (2) high SCR reaction rates at elevated temperatures generally preclude low conversion kinetic measurements; and (3) background reactivity of the SSZ-13 support itself. For example, at reaction temperatures above ~350 °C, fast SCR can proceed so readily on H/SSZ-13 that the presence of Cu has little influence on NO$_x$ conversions [6]. While recent studies by Li et al. [48] of fast SCR mechanisms for H/SSZ-13, using both experimental and theoretical approaches are outside the scope of this review, clearly it is the understanding of lower temperature fast SCR mechanisms on Cu that is of higher significance.

In the low temperature regime, fast SCR most certainly occurs on NH$_3$-ligated Cu sites. In situ XAS measurements demonstrate that, unlike standard SCR where Cu(II) and Cu(I) coexist, Cu(II) is the only observed Cu oxidation state under fast SCR reaction conditions [34]. This does not at all necessarily mean that Cu(I)/Cu(II) redox cycling is not involved in fast SCR; instead, this result likely indicates that Cu(I) species are extremely short-lived under fast SCR conditions. Indeed, recent studies postulate that NO$_2$ oxidants accelerate SCR rates both by enhancing Cu(I) oxidation kinetics and by engaging a larger fraction of Cu sites in the catalyst [13,21]. However, to our knowledge, there has not been strong experimental evidence to suggest that fast SCR proceeds at higher rates than standard SCR over Cu/SSZ-13. Indeed, for the isostructural Cu/SAPO-34 catalyst, NO$_2$ was even found to inhibit low-temperature activity due to NH$_4$NO$_3$ poisoning [49]. It can be postulated that this poisoning effect is particularly severe for small pore Cu/Chabazite since pore plugging is clearly more facile in comparison to zeolites with larger pore openings.

With the rather limited kinetics and spectroscopic data available, fast SCR mechanisms for Cu/SSZ-13 can only be speculated about. We note, therefore, that the discussions that follow should not be treated as affirmative but rather suggestive, aiming to encourage more research to fully understand this chemistry. In constructing plausible reaction paths, the following considerations are either established or assumed: (1) active Cu(II) sites at lower reaction temperatures are isolated and they are ligated with NH$_3$; (2) active NO$_x$ species must have an N oxidation state of +3; (3) intermediates that lead to N$_2$ and H$_2$O formation are either nitrosoamide (NH$_2$NO) or ammonium nitrite (NH$_4$NO$_2$), which are known to be highly unstable and decompose readily to the target products; (4) Cu(I) sites are re-oxidized back to the active Cu(II) form by NO$_2$; and (5) NH$_4$NO$_3$ acts as a major side product, poisoning the fast SCR reaction at lower temperatures.

Depending on whether or not the formation of the active NO$_x$ species with an N oxidation state of +3 requires participation of Cu(II), two scenarios are possible. Without Cu(II) participation for charge transfer, NO can be activated by NO$_2$ to form N$_2$O$_3$ non-catalytically, with subsequent interaction of N^{+3} cation species with NH$_3$-ligated Cu(II) to form intermediates that lead to N$_2$ + H$_2$O. The following reaction pathways summarize this first scenario:

$$NO + NO_2 \rightleftharpoons N_2O_3 \tag{15}$$

$$N_2O_3 + H_2O = 2HONO \tag{16}$$

$$Cu^{2+} - NH_3 + HONO = Cu^{2+} - NH_4NO_2 \tag{17}$$

In these reactions, since no charge transfer occurs between Cu and NO$_x$, the role of Cu(II) is to provide a reactive form of NH$_3$ and to provide a location for the formation of reaction intermediates. Note that Reactions (15–17) are, again, only hypothetical and likely incomplete. Other processes, for example direct NH$_3$ + N$_2$O$_3$ reactions, are also possible:

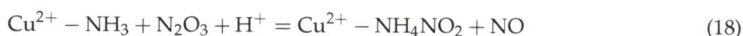

$$Cu^{2+} - NH_3 + N_2O_3 + H^+ = Cu^{2+} - NH_4NO_2 + NO \tag{18}$$

In contrast to the above, if charge transfer between Cu(II) and NOx does occur (i.e., fast SCR follows a redox mechanism as does standard SCR), then a reduction half-cycle similar to that for standard SCR is also applicable here [11,13]. In particular, a possible second fast SCR mechanism is as follows:

$$Cu^{2+} - NH_3 + NO = Cu^+ - NH_2NO + H^+ \tag{19}$$

$$Cu^+ + NO_2 = Cu^{2+} - NO_2^- \tag{20}$$

$$Cu^{2+} - NO_2^- + NH_3 + H^+ = Cu^{2+} - NH_4NO_2 \tag{21}$$

In this case, NO is activated by Cu(II) via formation of a NH$_2$NO intermediate and Cu(I) is re-oxidized by NO$_2$ to complete redox cycling.

As noted above, it is not yet clear whether low-temperature fast SCR follows a redox mechanism or not. The lack of observable Cu(I) species under fast SCR conditions may either be due to their non-existence or transient nature. As such, more research is required to elucidate further details of this chemistry. It is clear, though, that side reactions do occur from NO$_2$ disproportionation that render fast SCR on Cu/SSZ-13 not any faster than standard SCR. In particular, these reactions are described as follows:

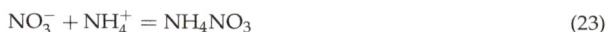

$$2NO_2 \rightleftharpoons N_2O_4 \rightleftharpoons NO_3^- + NO^+ \tag{22}$$

$$NO_3^- + NH_4^+ = NH_4NO_3 \tag{23}$$

Note that H$_2$O can participate in these latter reactions, in which case HNO$_3$ and HONO will form as intermediates [50]. However, this will not alter the final product formed, NH$_4$NO$_3$. In fact, NH$_4$NO$_3$ has been found to be particularly stable on Cu/SSZ-13, thus displaying a considerable low-temperature inhibition of SCR by blocking pores and channels of the small-pore support [51].

To summarize, in comparison to higher temperature reaction with H/SSZ-13, low-temperature fast SCR is certainly enhanced by the presence of Cu [6]. However, the precise roles that Cu ions play in this reaction is not clear at present. Whether Cu ions actively participate in NO activation (partial oxidation), or they simply provide locations for certain reaction steps to occur, still awaits further study.

4. Towards Rational Design of Cu/SSZ-13

In the above discussions, we have described the current understanding of various transformations of Cu in Cu/SSZ-13 catalysts during synthesis and during SCR reaction. In particular, we have summarized how the roles of these various Cu species in the mechanistic processes that are central to the catalytic chemistry. With this understanding, it becomes possible to suggest catalyst compositions, catalyst processing procedures and reaction conditions for optimum SCR performance. Furthermore, as will be discussed below, Cu speciation can have a profound effect on the stability of Cu/SSZ-13 catalysts. With this in mind, this last section summarizes recent studies that have: (1) determined the known routes for the degradation of Cu/SSZ-13 catalysts; (2) identified undesirable forms of Cu for long-term catalyst stability and possible mechanistic routes for degrading catalyst performance; and (3) suggested 'design rules' for selecting and synthesizing optimum Cu/SSZ-13 catalysts.

Two major degradation mechanisms have recently been identified for Cu/SSZ-13 SCR catalysts: hydrothermal aging and sulfur poisoning [16]. Sulfur poisoning is typically reversible; adsorbed sulfur (mainly in the form of sulfate) can be removed during high-temperature regeneration treatments. In contrast, hydrothermal degradation is typically irreversible due to permanent structural changes in the zeolite that result in, for example, loss of Brønsted acid sites and changes to the chemical or physical structure of Cu active centers. This explains why hydrothermal aging effects have been so extensively studied [39–41,52–54]. Prior studies have suggested that Cu/SSZ-13 durability enhancement can be achieved in a number of ways: (1) composition optimization, that is, choosing optimized Si/Al and Cu/Al ratios [31]; (2) zeolite particle size optimization [55]; (3) new synthesis methods development [30,56–59]; and (4) introduction of stability enhancement additives. Among these, some appear not to be as useful as others. For example, although it is generally believed that larger zeolite particle size leads to better hydrothermal stability, the study by Prodinger et al. [55] demonstrates that Cu/SSZ-13 with sub-micrometer particle size does not display inferior hydrothermal stability than Cu/SSZ-13 with particle sizes ~10 times higher. From our own experience, too large of a particle size can even be detrimental due to poor heat dissipation during structure directing agent (SDA) combustion that induces rather serious thermal degradation of the SSZ-13 substrate. Also, new Cu/SSZ-13 synthesis methods, including one-pot ones, do not necessarily lead to catalysts with better stability than catalysts prepared with more traditional solution ion-exchange methods [30,56–59].

In terms of Si/Al ratios, we note that this parameter has changed from ~17.5 for the first generation commercial (BASF) catalyst [40], to ~9.0 for the current state-of-the-art material [16,18]. This change is nontrivial; as suggested by Figure 2, a lower Si/Al ratio allows for a higher percentage of Cu^{2+}-2Z sites at similar Cu loadings, thus reducing the formation of undesired CuO_x clusters during hydrothermal aging. As judged from our Cu/SSZ-13 materials synthesized in-house, lowering Si/Al ratios further to ~6 does not lead to catalysts with inferior hydrothermal stability in comparison to catalysts with higher Si/Al ratios, as long as Cu/Al ratios are properly adjusted (more details will be given below). However, very high Si/Al ratios (e.g., 36) are clearly detrimental to hydrothermal stabilities. Therefore, the suggested Si/Al ratios should fall in the range of 6–10 based on our current understanding.

Once Si/Al ratios for an SSZ-13 material is fixed, the most important parameter that influences stability of the final Cu/SSZ-13 product is Cu content, that is, the Cu/Al ratio. Intuitively, one might expect that a full Cu exchange, where all zeolite ion-exchange sites are occupied by Cu ions, best prevents zeolite dealumination and structural degradation. However, even in the very first hydrothermal stability testing of this catalyst, Lobo and coworkers already realized that high Cu/Al ratios are actually detrimental to hydrothermal stability [29]. In particular, as shown in Figure 15 for both Cu/SSZ-13 and Cu/SSZ-16, increasing Cu loading from intermediate to high levels invariably leads to decreased catalytic performance for aged catalysts. These initial results were later fully reproduced by Nam and coworkers, with their XRD measurements demonstrating severe structural damage for highly Cu-loaded catalysts following hydrothermal aging [41]. Two explanations seem possible: (1) somewhat non-intuitively, high Cu loadings induce a high degree of dealumination that destabilizes the SSZ-13 substrate; or (2) as suggested by Lobo and coworkers, copper aluminate and/or copper oxide species are formed during hydrothermal aging that perhaps catalyze the zeolite structure collapse [29].

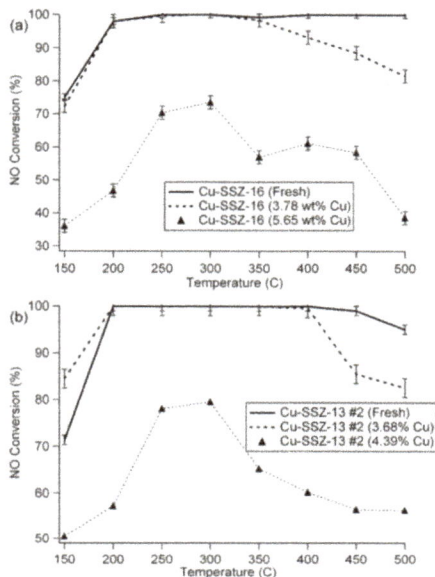

Figure 15. Comparing the effect of copper loading on the NH_3-SCR activity and hydrothermal stability for (**a**) Cu-SSZ-16 and (**b**) Cu-SSZ-13. Reprinted from *Appl. Catal. B-Environ.*, **2011**, *102*, 441–448. Copyright 2011, with permission from Elsevier.

In a recent study of Fe/SSZ-13 catalysts, Kovarik et al. demonstrated that, even with ~80% of dealumination during hydrothermal aging as determined from ^{27}Al NMR quantification, XRD patterns and surface area/pore volume measurements do not show measurable differences in comparison to a fresh catalyst [60]. This strongly suggests that dealumination is not necessarily the primary cause of structural degradation for SSZ-13. A likely explanation is that, unlike many zeolite materials that are only stable within a small range of Si/Al ratios, SSZ-13 is stable at Si/Al from 1 to infinity. In contrast, it is demonstrated below that CuOx clusters formed during hydrothermal aging are more likely to be responsible for structural damage.

Figure 16 compares micro and mesopore size distributions of a Cu/SSZ-13 catalyst with Si/Al = 12 and Cu/Al = 0.25 (the same catalyst used for EPR studies shown in Figure 6) in fresh and two hydrothermally aged (at 700 and 800 °C) forms. These results demonstrate that hydrothermal aging induces no change in micropore size; however, it does cause mesopore formation. For the sample aged at 800 °C in particular, the amount of mesopores that are ≤4 nm increases significantly, largely from micropore damage. It is important to note that both hydrothermally aged catalysts have similar extents of dealumination, yet contain increasing amounts of CuOx species relative to the fresh catalyst (Figure 6), likely rationalizing enhanced mesopore formation. Accumulation of such mesopores eventually leads to structural instability and even collapse. It can be expected that such a process depends on temperature, aging time and Cu loading, where higher aging temperatures, longer aging durations and higher Cu loadings promote more damage. Prior literature is fully in line with this structure damage mechanism [39–41,53,61] On the other hand, dealumination alone is apparently much less destructive for this zeolite catalyst.

Identification of CuOx formation as the main cause of structural instability discussed immediately above and the fact that [Cu(OH)]$^+$-Z selectively converts to CuOx during hydrothermal aging (as discussed in Section 2.2 above), allows for an estimation of Cu/Al ratios that would provide optimum hydrothermal stabilities. For catalysts that contain both kinetically stabilized

[Cu(OH)]$^+$-Z and available 2Al "empty" sites, a hydrothermal treatment at relatively low temperatures (i.e., degreening) promotes Reaction (3), which leads to more stable catalysts. An extra benefit for this treatment, is that a mildly aged Cu/SSZ-13 catalyst will be more resistant to sulfur poisoning. Notably, studies by Luo et al. [16] and Epling and coworkers [23] provide strong evidence to demonstrate that [Cu(OH)]$^+$-Z is much more vulnerable than Cu^{2+}-2Z toward sulfur poisoning.

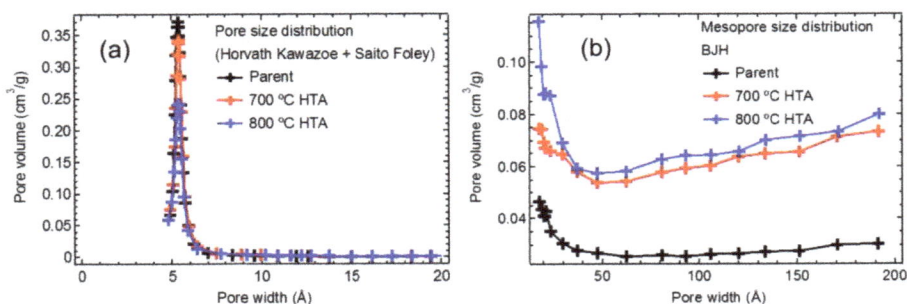

Figure 16. (**a**) Micropore size distributions of the fresh, HTA-700 and HTA-800 samples determined with the Horvath-Kawazoe and Saito-Foley methods; (**b**) Mesopore size distributions of the fresh, HTA-700 and HTA-800 samples determined with the BJH method. Reprinted with permission from *ACS Catal.*, **2017**, *7*, 8214–8227. Copyright 2017, American Chemical Society.

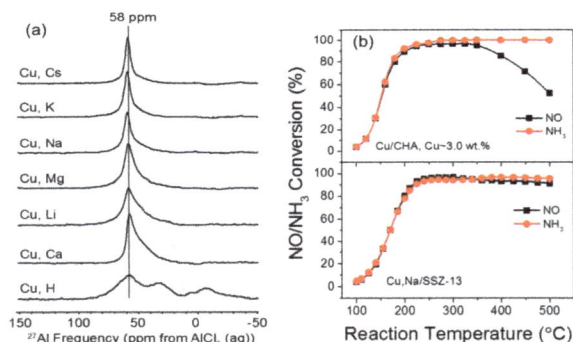

Figure 17. (**a**) ^{27}Al NMR spectra of hydrothermally aged (HTA) Cu,H and Cu,M/SSZ-13 catalysts, where M represents alkali and alkaline earth coactions; (**b**) NO and NH$_3$ light-off curves for standard SCR over an HTA Cu/SSZ-13 catalyst with Si/Al = 12.5 and 3.0% Cu and an HTA Cu,Na/SSZ-13 catalyst with Si/Al = 6, 1.0% Cu and 1.8% Na. Adapted with permission from *ACS Catal.*, **2015**, *5*, 6780–6791. Copyright 2015, American Chemical Society.

CuO$_x$ formed during hydrothermal aging also causes SCR selectivity to decrease because this species does not efficiently catalyze SCR but is instead active in catalyzing the undesirable NH$_3$ oxidation reaction at temperatures above ~300 °C. This problem may be circumvented with low Cu loaded catalysts where the dominance of Cu^{2+}-2Z prevents CuOx formation during hydrothermal aging. However, for SSZ-13 substrates that are not synthesized with optimized stability themselves, a low Cu loaded composition does not alone satisfactorily improve hydrothermal stabilities. In this case, coaction additives can be added to reduce structural degradation from -Si-O(H)-Al-hydrolysis and even dealumination. In a study by Gao et al. [62], a number of alkali and alkaline earth coactions were tested. As shown in Figure 17a, to varying degrees, all of the coaction additives show positive effects in preventing structural degradation of the SSZ-13 substrate in comparison to the Cu,H/SSZ-13 counterpart. In the latter undoped case, significant framework Al distortion, partial -Si-O(H)-Al-hydrolysis and dealumination are much more severe than other samples as evidenced by the NMR spectral differences. Figure 17b compares SCR selectivities (i.e., differences between NO and NH$_3$ conversion) for a hydrothermally aged Cu/SSZ-13 with 3% Cu loading and a hydrothermally aged Cu,Na/SSZ-13 with 1% Cu and 1.8% Na loadings. Clearly, above ~350 °C SCR selectivities for the high Cu loaded sample markedly decrease due to the presence of CuO$_x$, whereas high SCR selectivities maintain at all reaction temperatures for the low Cu-loaded Cu,Na/SSZ-13 sample due to the absence of CuO$_x$ in this hydrothermally aged catalyst. We note that the Cu and Na loadings used here are not optimized; the purpose of this study was to show that the concept of using additives to enhance Cu/SSZ-13 catalyst stability and performance is indeed a valid one.

5. Concluding Remarks

This short review has focused on Cu/SSZ-13 catalysts that are currently commercialized for diesel vehicle NO$_x$ emission control. In particular, our aims have been to summarize exciting recent insights into the nature of the catalytically active Cu species, some of the critical rate-liming reactions involved in the complex redox-based standard SCR mechanism at typical reaction temperatures and the effects of Cu speciation on catalyst stability with respect to hydrothermal aging and sulfur poisoning. Of note has been the identification of mobile NH$_3$-complexed, monomeric Cu ions that are active for the NO$_x$ reduction cycle of the redox mechanism. Significantly and unusually, the thus formed Cu$^+$-ion complexes subsequently migrate between zeolite cages to form dimeric species that are the critical catalytic structures for O$_2$ activation, with the latter reaction being an essential rate-limiting process for the re-oxidation of Cu that completes the catalytic cycle. These conclusions were based on highly unusual reaction kinetics observed experimentally several years ago, yet only recently explained via the application of computational simulations. Studies of Cu speciation under various reaction conditions have also provided clear evidence for the undesirable properties of CuO$_x$ clusters for both SCR selectivity (due to promotion of the NH$_3$ oxidation reaction) and catalyst stability, where recent results demonstrate a possible 'catalytic' role for CuO$_x$ clusters in degrading the zeolite structure. Processes leading to formation of CuO$_x$ and the 'precursor' forms of Cu ions most susceptible to CuO$_x$ formation were described above. Finally, based on these new molecular-level insights, we discussed how such fundamental information can be used to 'design' Cu/SSZ-13 catalysts for optimized performance and stability.

Acknowledgments: The authors gratefully acknowledge the US Department of Energy (DOE), Energy Efficiency and Renewable Energy, Vehicle Technologies Office for the support of this work at Pacific Northwest National Laboratory (PNNL). PNNL is operated for the US DOE by Battelle.

Author Contributions: Both authors contributed equally to the preparation of the review paper.

Conflicts of Interest: The authors declare no conflict of interest.

References

1. Andersen, P.J.; Collier, J.E.; Casci, J.L.; Chen, H.-Y.; Fedeyko, J.M.; Foo, R.K.S.; Rajaram, R.R. Transition metal/zeolite SCR catalysts. International Patent WO/2008/132452, 11 June 2008.
2. Bull, I.; Xue, W.M.; Burk, P.; Boorse, R.S.; Jaglowski, W.M.; Koermer, G.S.; Moini, A.; Patchett, J.A.; Dettling, J.C.; Caudle, M.T. Copper CHA zeolite catalysts. U.S. Patent 7,601,662 B2, 13 October 2009.
3. Bull, I.; Moini, A.; Koermer, G.S.; Patchett, J.A.; Jaglowski, W.M.; Roth, S. Zeolite catalyst with improved NOx reduction in SCR. U.S. Patent 7,704,475 B2, 27 April 2010.
4. Andersen, P.J.; Chen, H.-Y.; Fedeyko, J.M.; Weigert, E. Small pore molecular sieve supported copper catalysts durable against lean/rich aging for the reduction of nitrogen oxides. U.S. Patent 7,998,443 B2, 16 August 2011.
5. Gao, F.; Kwak, J.H.; Szanyi, J.; Peden, C.H.F. Current understanding of Cu-exchanged Chabazite molecular sieves for use as commercial diesel engine DeNO$_x$ catalysts. *Top. Catal.* **2013**, *56*, 1441–1459. [CrossRef]

6. Beale, A.M.; Gao, F.; Lezcano-Gonzalez, I.; Peden, C.H.F.; Szanyi, J. Recent advances in automotive catalysis for NO$_x$ emission control by small-pore microporous materials. *Chem. Soc. Rev.* **2015**, *44*, 7371–7405. [CrossRef] [PubMed]

7. Paolucci, C.; di Iorio, J.R.; Ribeiro, F.H.; Gounder, R.; Schneider, W.F. Catalysis science of NO$_x$ selective catalytic reduction with ammonia over Cu-SSZ-13 and Cu-SAPO-34. *Adv. Catal.* **2016**, *59*, 1–107.

8. Wang, J.H.; Zhao, H.W.; Haller, G.; Li, Y.D. Recent advances in the selective catalytic reduction of NO$_x$ with NH$_3$ on Cu-Chabazite catalysts. *Appl. Catal. B Environ.* **2017**, *202*, 346–354. [CrossRef]

9. Nova, I.; Tronconi, E. *Urea-SCR Technology for deNO$_x$ after Treatment of Diesel Exhausts*; Springer Science & Business Media: New York, NY, USA, 2014.

10. Zones, S.I. Zeolite SSZ-13 and its method of preparation. U.S. Patent 4,544,538, 1 October 1985.

11. Paolucci, C.; Verma, A.A.; Bates, S.A.; Kispersky, V.F.; Miller, J.T.; Gounder, R.; Delgass, W.N.; Ribeiro, F.H.; Schneider, W.F. Isolation of the Copper Redox Steps in the Standard Selective Catalytic Reduction on Cu-SSZ-13. *Angew. Chem. Int. Ed.* **2014**, *53*, 11828–11833. [CrossRef] [PubMed]

12. Di Iorio, J.R.; Bates, S.A.; Verma, A.A.; Delgass, W.N.; Ribeiro, F.H.; Miller, J.T.; Gounder, R. The dynamic nature of Brønsted acid sites in Cu–Zeolites during NO$_x$ selective catalytic reduction: Quantification by gas-phase ammonia titration. *Top. Catal.* **2015**, *58*, 424–434. [CrossRef]

13. Janssens, T.V.W.; Falsig, H.; Lundegaard, L.F.; Vennestrom, P.N.R.; Rasmussen, S.B.; Moses, P.G.; Giordanino, F.; Borfecchia, E.; Lomachenko, K.A.; Lamberti, C.; et al. A consistent reaction scheme for the selective catalytic reduction of nitrogen oxides with ammonia. *ACS Catal.* **2015**, *5*, 2832–2845. [CrossRef]

14. Moreno-Gonzalez, M.; Hueso, B.; Boronat, M.; Blasco, T.; Corma, A. Ammonia-containing species formed in Cu-Chabazite as per in situ EPR, solid-state NMR and DFT calculations. *J. Phys. Chem. Lett.* **2015**, *6*, 1011–1017. [CrossRef] [PubMed]

15. Lomachenko, K.A.; Borfecchia, E.; Negri, C.; Berlier, G.; Lamberti, C.; Beato, P.; Falsig, H.; Bordiga, S. The Cu-CHA deNO$_x$ Catalyst in Action: Temperature-Dependent NH$_3$-Assisted Selective Catalytic Reduction Monitored by Operando XAS and XES. *J. Am. Chem. Soc.* **2016**, *138*, 12025–12028. [CrossRef] [PubMed]

16. Luo, J.Y.; Wang, D.; Kumar, A.; Li, J.H.; Kamasamudram, K.; Currier, N.; Yezerets, A. Identification of two types of Cu sites in Cu/SSZ-13 and their unique responses to hydrothermal aging and sulfur poisoning. *Catal. Today* **2016**, *267*, 3–9. [CrossRef]

17. Paolucci, C.; Parekh, A.A.; Khurana, I.; Di Iorio, J.R.; Li, H.; Albarracin-Caballero, J.D.; Shih, A.J.; Anggara, T.; Delgass, W.N.; Miller, J.T.; et al. Catalysis in a cage: Condition-dependent speciation and dynamics of exchanged Cu cations in SSZ-13 zeolites. *J. Am. Chem. Soc.* **2016**, *138*, 6028–6048. [CrossRef] [PubMed]

18. Luo, J.Y.; Gao, F.; Kamasamudram, K.; Currier, N.; Peden, C.H.F.; Yezerets, A. New insights into Cu/SSZ-13 SCR catalyst acidity. Part I: Nature of acidic sites probed by NH$_3$ titration. *J. Catal.* **2017**, *348*, 291–299. [CrossRef]

19. Gao, F.; Mei, D.H.; Wang, Y.L.; Szanyi, J.; Peden, C.H.F. Selective catalytic reduction over Cu/SSZ-13: Linking homo-and heterogeneous catalysis. *J. Am. Chem. Soc.* **2017**, *139*, 4935–4942. [CrossRef] [PubMed]

20. Martini, A.; Borfecchia, E.; Lomachenko, K.A.; Pankin, I.A.; Negri, C.; Berlier, G.; Beato, P.; Falsig, H.; Bordiga, S.; Lamberti, C. Composition-driven Cu-speciation and reducibility in Cu-CHA zeolite catalysts: A multivariate XAS/FTIR approach to complexity. *Chem. Sci.* **2017**, *8*, 6836–6851. [CrossRef] [PubMed]

21. Paolucci, C.; Khurana, I.; Parekh, A.A.; Li, S.C.; Shih, A.J.; Li, H.; Di Iorio, J.R.; Albarracin-Caballero, J.D.; Yezerets, A.; Miller, J.T.; et al. Dynamic multinuclear sites formed by mobilized copper ions in NO$_x$ selective catalytic reduction. *Science* **2017**, *357*, 898–903. [CrossRef] [PubMed]

22. Schmidt, J.E.; Oord, R.; Guo, W.; Poplawsky, J.D.; Weckhuysen, B.M. Nanoscale tomography reveals the deactivation of automotive copper-exchanged zeolite catalysts. *Nat. Commun.* **2017**, *8*. [CrossRef]

23. Jangjou, Y.; Do, Q.; Gu, Y.T.; Lim, L.-G.; Sun, H.; Wang, D.; Kumar, A.; Li, J.H.; Grabow, L.C.; Epling, W.S. On the nature of Cu active centers in Cu-SSZ-13 and their responses to SO$_2$ exposure. *ACS Catal.* **2018**, *8*, 1325–1337. [CrossRef]

24. Godiksen, A.; Isaksen, O.L.; Rasmussen, S.B.; Vennestrom, P.N.R.; Mossin, S. Site-Specific Reactivity of Copper Chabazite Zeolites with Nitric Oxide, Ammonia and Oxygen. *ChemCatChem* **2018**, *10*, 366–370. [CrossRef]

25. Andersen, C.W.; Borfecchia, E.; Bremholm, M.; Jorgensen, M.R.V.; Vennestrom, P.N.R.; Lamberti, C.; Lundegaard, L.F.; Iversen, B.B. Redox-Driven Migration of Copper Ions in the Cu-CHA Zeolite as Shown by the In Situ PXRD/XANES Technique. *Angew. Chem. Int. Ed.* **2017**, *56*, 10367–10372. [CrossRef] [PubMed]

26. Marberger, A.; Petrov, A.W.; Steiger, P.; Elsener, M.; Kröcher, O.; Nachtegaal, M.; Ferri, D. Time-resolved copper speciation during selective catalytic reduction of NO on Cu-SSZ-13. *Nat. Catal.* **2018**, *1*, 221–227. [CrossRef]

27. Fickel, D.W.; Lobo, R.F. Copper coordination in Cu-SSZ-13 and Cu-SSZ-16 investigated by variable-temperature XRD. *J. Phys. Chem. C* **2010**, *114*, 1633–1640. [CrossRef]

28. Kwak, J.H.; Tonkyn, R.G.; Kim, D.H.; Szanyi, J.; Peden, C.H.F. Excellent activity and selectivity of Cu-SSZ-13 in the selective catalytic reduction of NO$_x$ with NH$_3$. *J. Catal.* **2010**, *275*, 187–190. [CrossRef]

29. Fickel, D.W.; D'Addio, E.; Lauterbach, J.A.; Lobo, R.F. The ammonia selective catalytic reduction activity of copper-exchanged small-pore zeolites. *Appl. Catal. B Environ.* **2011**, *102*, 441–448. [CrossRef]

30. Ren, L.M.; Zhu, L.F.; Yang, C.G.; Chen, Y.M.; Sun, Q.; Zhang, H.Y.; Li, C.J.; Nawaz, F.; Meng, X.J.; Xiao, F.S. Designed copper–amine complex as an efficient template for one-pot synthesis of Cu-SSZ-13 zeolite with excellent activity for selective catalytic reduction of NO$_x$ by NH$_3$. *Chem. Commun.* **2011**, *47*, 9789–9791. [CrossRef] [PubMed]

31. Gao, F.; Washton, N.M.; Wang, Y.L.; Kollar, M.; Szanyi, J.; Peden, C.H.F. Effects of Si/Al ratio on Cu/SSZ-13 NH$_3$-SCR catalysts: Implications for the active Cu species and the roles of Brønsted acidity. *J. Catal.* **2015**, *331*, 25–38. [CrossRef]

32. Kwak, J.H.; Zhu, H.Y.; Lee, J.H.; Peden, C.H.F.; Szanyi, J. Two different cationic positions in Cu-SSZ-13? *Chem. Commun.* **2012**, *48*, 4758–4760. [CrossRef] [PubMed]

33. Andersen, C.W.; Bremholm, M.; Vennestrøm, P.N.R.; Blichfeld, A.B.; Lundegaard, L.F.; Iversen, B.B. Location of Cu^{2+} in CHA zeolite investigated by X-ray diffraction using the Rietveld/maximum entropy method. *IUCrJ* **2014**, *1*, 382–386. [CrossRef] [PubMed]

34. McEwen, J.S.; Anggara, T.; Schneider, W.F.; Kispersky, V.F.; Miller, J.T.; Delgass, W.N.; Ribeiro, F.H. Integrated operando X-ray absorption and DFT characterization of Cu–SSZ-13 exchange sites during the selective catalytic reduction of NO$_x$ with NH$_3$. *Catal. Today* **2012**, *184*, 129–144. [CrossRef]

35. Kwak, J.H.; Varga, T.; Peden, C.H.F.; Gao, F.; Hanson, J.C.; Szanyi, J. Following the movement of Cu ions in a SSZ-13 zeolite during dehydration, reduction and adsorption: A combined in situ TP-XRD, XANES/DRIFTS study. *J. Catal.* **2014**, *314*, 83–93. [CrossRef]

36. Godiksen, A.; Stappen, F.N.; Vennestrom, P.N.R.; Giordanino, F.; Rasmussen, S.B.; Lundegaard, L.F.; Mossin, S. Coordination environment of copper sites in Cu-CHA zeolite investigated by electron paramagnetic resonance. *J. Phys. Chem. C* **2014**, *118*, 23126–23138. [CrossRef]

37. Song, J.; Wang, Y.L.; Walter, E.D.; Washton, N.M.; Mei, D.H.; Kovarik, L.; Engelhard, M.H.; Prodinger, S.; Wang, Y.; Peden, C.H.F.; Gao, F. Toward Rational Design of Cu/SSZ-13 Selective Catalytic Reduction Catalysts: Implications from Atomic-Level Understanding of Hydrothermal Stability. *ACS Catal.* **2017**, *7*, 8214–8227. [CrossRef]

38. Gao, F.; Walter, E.D.; Kollar, M.; Wang, Y.L.; Szanyi, J.; Peden, C.H.F. Understanding ammonia selective catalytic reduction kinetics over Cu/SSZ-13 from motion of the Cu ions. *J. Catal.* **2014**, *319*, 1–14. [CrossRef]

39. Kwak, J.H.; Tran, D.; Burton, S.D.; Szanyi, J.; Lee, J.H.; Peden, C.H.F. Effects of hydrothermal aging on NH$_3$-SCR reaction over Cu/zeolites. *J. Catal.* **2012**, *287*, 203–209. [CrossRef]

40. Schmieg, S.J.; Oh, S.H.; Kim, C.H.; Brown, D.B.; Lee, J.H.; Peden, C.H.F.; Kim, D.H. Thermal durability of Cu-CHA NH$_3$-SCR catalysts for diesel NO$_x$ reduction. *Catal. Today* **2012**, *184*, 252–261. [CrossRef]

41. Kim, Y.J.; Lee, J.K.; Min, K.M.; Hong, S.B.; Nam, I.S.; Cho, B.K. Hydrothermal stability of CuSSZ13 for reducing NOx by NH$_3$. *J. Catal.* **2014**, *311*, 447–457. [CrossRef]

42. Vennestrom, P.N.R.; Janssens, T.V.W.; Kustov, A.; Grill, M.; Puig-Molina, A.; Lundegaard, L.F.; Tiruvalam, R.R.; Concepcion, P.; Corma, A. Influence of lattice stability on hydrothermal deactivation of Cu-ZSM-5 and Cu-IM-5 zeolites for selective catalytic reduction of NO$_x$ by NH$_3$. *J. Catal.* **2014**, *309*, 477–490. [CrossRef]

43. Giordanino, F.; Borfecchia, E.; Lomachenko, K.A.; Lazzarini, A.; Agostini, G.; Gallo, E.; Soldatov, A.V.; Beato, P.; Bordiga, S.; Lamberti, C. Interaction of NH$_3$ with Cu-SSZ-13 catalyst: A complementary FTIR, XANES and XES study. *J. Phys. Chem. Lett.* **2014**, *5*, 1552–1559. [CrossRef] [PubMed]

44. Shwan, S.; Skoglundh, M.; Lundegaard, L.F.; Tiruvalam, R.R.; Janssens, T.V.W.; Carlsson, A.; Vennestrom, P.N.R. Solid-state ion-exchange of copper into zeolites facilitated by ammonia at low temperature. *ACS Catal.* **2015**, *5*, 16–19. [CrossRef]

45. Clemens, A.K.S.; Shishkin, A.; Carlsson, P.A.; Skoglundh, M.; Martinez-Casado, F.J.; Matej, Z.; Balmes, O.; Harelind, H. Reaction-driven ion exchange of copper into zeolite SSZ-13. *ACS Catal.* **2015**, *5*, 6209–6218. [CrossRef]

46. Verma, A.A.; Bates, S.A.; Anggara, T.; Paolucci, C.; Parekh, A.A.; Kamasamudram, K.; Yezerets, A.; Miller, J.T.; Delgass, W.N.; Schneider, W.F.; Ribeiro, F.H. NO oxidation: A probe reaction on Cu-SSZ-13. *J. Catal.* **2014**, *312*, 179–190. [CrossRef]

47. Brandenberger, S.; Krocher, O.; Tissler, A.; Althoff, R. The state of the art in selective catalytic reduction of NO$_x$ by Ammonia using metal-exchanged zeolite catalysts. *Catal. Rev. Sci. Eng.* **2008**, *50*, 492–531. [CrossRef]

48. Li, S.C.; Zheng, Y.; Gao, F.; Szanyi, J.; Schneider, W.F. Experimental and Computational Interrogation of Fast SCR Mechanism and Active Sites on H-Form SSZ-13. *ACS Catal.* **2017**, *7*, 5087–5096. [CrossRef]

49. Hao, T.; Wang, J.; Yu, T.; Wang, J.Q.; Shen, M.Q. Effect of NO$_2$ on the Selective Catalytic Reduction of NO with NH$_3$ over Cu/SAPO-34 Molecular Sieve Catalyst. *Acta Phys. Chim. Sin.* **2014**, *30*, 1567–1574.

50. Gao, F.; Wang, Y.L.; Kollar, M.; Washton, N.M.; Szanyi, J.; Peden, C.H.F. A comparative kinetics study between Cu/SSZ-13 and Fe/SSZ-13 SCR catalysts. *Catal. Today* **2015**, *258*, 347–358. [CrossRef]

51. Chen, H.-Y.; Wei, Z.H.; Kollar, M.; Gao, F.; Wang, Y.L.; Szanyi, J.; Peden, C.H.F. A comparative study of N$_2$O formation during the selective catalytic reduction of NO$_x$ with NH$_3$ on zeolite supported Cu catalysts. *J. Catal.* **2015**, *329*, 490–498. [CrossRef]

52. Ye, Q.; Wang, L.F.; Yang, R.T. Activity, propene poisoning resistance and hydrothermal stability of copper exchanged Chabazite-like zeolite catalysts for SCR of NO with ammonia in comparison to Cu/ZSM-5. *Appl. Catal. A Gen.* **2012**, *427–428*, 24–34. [CrossRef]

53. Blakeman, P.G.; Burkholder, E.M.; Chen, H.Y.; Collier, J.E.; Fedeyko, J.M.; Jobson, H.; Rajaram, R.R. The role of pore size on the thermal stability of zeolite supported Cu SCR catalysts. *Catal. Today* **2014**, *231*, 56–63. [CrossRef]

54. Lezcano-Gonzalez, I.; Deka, U.; van der Bij, H.E.; Paalanen, P.; Arstad, B.; Weckhuysen, B.M.; Beale, A.M. Chemical deactivation of Cu-SSZ-13 ammonia selective catalytic reduction (NH$_3$-SCR) systems. *Appl. Catal. B Environ.* **2014**, *154–155*, 339–349. [CrossRef]

55. Prodinger, S.; Derewinski, M.A.; Wang, Y.L.; Washton, N.M.; Walter, E.D.; Szanyi, J.; Gao, F.; Wang, Y.; Peden, C.H.F. Sub-micron Cu/SSZ-13: Synthesis and application as selective catalytic reduction (SCR) catalysts. *Appl. Catal. B Environ.* **2017**, *201*, 461–469. [CrossRef]

56. Martinez-Franco, R.; Moliner, M.; Thogersen, J.R.; Corma, A. Efficient One-Pot Preparation of Cu-SSZ-13 Materials using Cooperative OSDAs for their Catalytic Application in the SCR of NO$_x$. *ChemCatChem* **2013**, *5*, 3316–3323. [CrossRef]

57. Martin, N.; Moliner, M.; Corma, A. High yield synthesis of high-silica Chabazite by combining the role of zeolite precursors and tetraethylammonium: SCR of NO$_x$. *Chem. Commun.* **2015**, *51*, 9965–9968. [CrossRef] [PubMed]

58. Wang, X.; Wu, Q.M.; Chen, C.Y.; Pan, S.X.; Zhang, W.P.; Meng, X.J.; Maurer, S.; Feyen, M.; Muller, U.; Xiao, F.-S. Atom-economical synthesis of a high silica CHA zeolite using a solvent-free route. *Chem. Commun.* **2015**, *51*, 16920–16923. [CrossRef] [PubMed]

59. Zhao, Z.C.; Yu, R.; Zhao, R.R.; Shi, C.; Gies, H.; Xiao, F.S.; de Vos, D.; Yokoi, T.; Bao, X.H.; Kolb, U.; et al. Cu-exchanged Al-rich SSZ-13 zeolite from organotemplate-free synthesis as NH$_3$-SCR catalyst: Effects of Na+ ions on the activity and hydrothermal stability. *Appl. Catal. B Environ.* **2017**, *217*, 421–428. [CrossRef]

60. Kovarik, L.; Washton, N.M.; Kukadapu, R.; Devaraj, A.; Wang, A.Y.; Wang, Y.L.; Szanyi, J.; Peden, C.H.F.; Gao, F. Transformation of active sites in Fe/SSZ-13 SCR catalysts during hydrothermal aging: A spectroscopic, microscopic and kinetics study. *ACS Catal.* **2017**, *7*, 2458–2470. [CrossRef]

61. Wang, D.; Jangjou, Y.; Liu, Y.; Sharma, M.K.; Luo, J.Y.; Li, J.H.; Kamasamudram, K.; Epling, W.S. A comparison of hydrothermal aging effects on NH$_3$-SCR of NO$_x$ over Cu-SSZ-13 and Cu-SAPO-34 catalysts. *Appl. Catal. B Environ.* **2015**, *165*, 438–445. [CrossRef]

62. Gao, F.; Wang, Y.L.; Washton, N.M.; Kollar, M.; Szanyi, J.; Peden, C.H.F. Effects of alkali and alkaline earth cocations on the activity and hydrothermal stability of Cu/SSZ-13 NH$_3$–SCR catalysts. *ACS Catal.* **2015**, *5*, 6780–6791. [CrossRef]

catalysts

MDPI

Article

Mobility of NH_3-Solvated Cu^{II} Ions in Cu-SSZ-13 and Cu-ZSM-5 NH_3-SCR Catalysts: A Comparative Impedance Spectroscopy Study

Valentina Rizzotto [1], Peirong Chen [1,2,*] and Ulrich Simon [1,*]

[1] Institute of Inorganic Chemistry (IAC), RWTH Aachen University, Landoltweg 1, 52074 Aachen, Germany; valentina.rizzotto@ac.rwth-aachen.de

[2] Guangdong Provincial Key Laboratory of Atmospheric Environment and Pollution Control, School of Environment and Energy, South China University of Technology, 510006 Guangzhou, China

* Correspondence: chenpr@scut.edu.cn (P.C.); ulrich.simon@ac.rwth-aachen.de (U.S.);
  Tel.: +86-20-39380508 (P.C.); +49-241-80-94644 (U.S.)

Received: 14 March 2018; Accepted: 9 April 2018; Published: 18 April 2018

Abstract: The mobility of NH_3-solvated Cu ions within the zeolite framework has been recently identified as a key factor for the kinetics of the selective catalytic reduction of NO_x with NH_3 (NH_3-SCR) over Cu-zeolite catalysts at low temperatures. Here, we utilize in situ impedance spectroscopy to explore the mobility of NH_3-solvated Cu^{II} ions, i.e., $Cu^{II}(NH_3)_n$, in Cu-SSZ-13 and Cu-ZSM-5 zeolites with varied Cu ion exchange levels, and observed that both the zeolite framework (CHA or MFI) and the Cu exchange level influence the high-frequency dielectric relaxation processes that are associated with the short-range (local) motion of $Cu^{II}(NH_3)_n$. Our results suggest that the local motion of $Cu^{II}(NH_3)_n$ species is favored within the CHA framework due to the unique cage structure, and thereby contribute to the overall ion conductivity at high frequencies, which, on the contrary, is not observed for ZSM-5, where NH_3-solvated Cu^{2+} ions do not experience a comparable constrained space for local motion. This study sheds new light on the mobility of Cu active sites under NH_3-SCR related reaction conditions and may contribute to an advanced understanding of the underlying mechanism.

Keywords: selective catalytic reduction of NO_x; impedance spectroscopy; Cu-SSZ-13 zeolite; Cu-ZSM-5 zeolite

1. Introduction

Nitric oxide and nitrogen dioxide (NO and NO_2, commonly indicated as "NO_x") are harmful atmospheric pollutants and are byproducts of anthropogenic high-temperature combustions [1,2]. "Three-way" catalysts, although being successfully employed for NO_x abatement in gasoline-powered vehicles, are not sufficiently effective in the O_2-rich exhaust environment typically in "lean-burn" engines (e.g., diesel engines). In this case, one of the leading strategies for NO_x emission control ($DeNO_x$) is the selective catalytic reduction with ammonia as reducing agent (NH_3-SCR) [3]. In NH_3-SCR systems, an aqueous urea solution (AdBlue™, VDA, Berlin, Germany) is injected into the hot exhaust gas mixture and it subsequently decomposes to form gaseous NH_3 [3–5]. Depending on the NO to NO_2 ratio, different SCR reactions may occur and the most important one is the so-called "standard SCR" [2,4,6]:

$$4NH_3 + 4NO + O_2 \rightarrow 4N_2 + 6H_2O \qquad (1)$$

Cu- and Fe-exchanged zeolites are among the most common catalysts for the SCR reaction in diesel-powered automobiles because of their high activity and stability under the respective reaction conditions [7]. Particularly, Cu-zeolites show outstanding low-temperature performance and low sensitivity to NO_2 concentration in the exhaust flow [4,5]. Due to their better performance (in terms of both activity and hydrothermal stability) when compared to large- and medium-pore zeolite materials, Cu-exchanged small-pore chabazite (CHA) zeolites (e.g., Cu-SSZ-13 and Cu-SAPO-34) are now the most common choice for NH_3-SCR [2–6,8,9]. Nevertheless, a deeper mechanistic understanding of Cu-zeolite catalyzed NH_3-SCR is still needed in order to develop even more efficient catalysts for meeting the increasingly stringent NO_x emission legislation [1,3,4].

Zeolites are microporous aluminosilicates, in which the ordered porous structure is generated from the crystalline structure of the material itself: the tetrahedral TO_4 primary building units (T being usually Si or Al), sharing corners, form channels and cages of molecular sizes. The inequivalent substitution of a Si T-atom with a similar-size Al T-atom results in a net negative framework charge, which requires being balanced by extra-framework cations, e.g., H^+ in proton-form zeolites or metallic cations in metal-exchanged zeolites. For this reason, even if zeolites are electrically non-conductive, they are ionically conductive due to the motion of extra-framework cations [10].

In Cu-zeolites, Cu^{2+} cations are introduced into the zeolite framework acting as active sites for the catalytic conversion of NO_x. Under SCR conditions, the Cu^{2+} centers are solvated by NH_3, leading to the formation of Cu-NH_3 complexes (hereon indicated as "$Cu^{II}(NH_3)_n$ species"). Due to the solvation effect of NH_3, the electrostatic interaction between Cu^{2+} and the negatively charged zeolite framework is weakened, increasing the mobility of the Cu ions within the framework [11]. Further interaction of $Cu^{II}(NH_3)_n$ with NO leads to the reduction of Cu^{II} to Cu^{I}, together with the conversion of NO and one of NH_3 molecules to N_2 and H_2O [5]. Recently, several research groups [12,13] identified that the transient formation of $[Cu^{I}(NH_3)_2]^+$-O_2-$[Cu^{I}(NH_3)_2]^+$ intermediates, which are necessary to activate O_2, is a rate-limiting step for the $Cu^{I} \rightarrow Cu^{II}$ re-oxidation half-cycle. In particular, for Cu-SSZ-13 with low Cu density, the mobility of $[Cu^{I}(NH_3)_2]^+$ complexes was found to determine the pairing of two $[Cu^{I}(NH_3)_2]^+$ in neighboring CHA cages, and thus plays a vital role in the Cu redox cycle, and consequently, the overall NH_3-SCR reaction. Due to a weaker electrostatic interaction, Cu^{I} species are considered to be more mobile than Cu^{II} species [11].

Despite its importance in SCR catalysis, direct experimental investigation of the Cu ion mobility remains a challenge. In this context, in situ impedance spectroscopy (IS) becomes a reliable technique, since it allows for studying the frequency-dependent conduction and the polarization processes of zeolites, in which the solvated Cu species are involved. By performing in situ IS under NH_3-SCR related conditions, we have already investigated the electrical properties of a series of commercially relevant zeolite catalysts (e.g., H-ZSM-5, Cu-ZSM-5, Fe-ZSM-5, Cu-SAPO-34) [10,14–16]. Moreover, a novel analytical approach combining IS with diffuse reflectance infrared Fourier transform spectroscopy (DRIFTS) was developed and allows for understanding the NH_3-induced electrical properties (e.g., proton transport) of zeolite catalysts at a molecular level [17–22]. A very recent study of our group has shown how IS combined with density-functional theory calculations enables to analyze the local movement of $[Cu^{I}(NH_3)_2]^+$ under SCR conditions [23].

In view of these findings, here, we designed a series of in situ IS experiments to probe the mobility of $Cu^{II}(NH_3)_n$ species, which are formed after NH_3-loading on the Cu-exchanged zeolite catalysts. Cu-SSZ-13 and Cu-ZSM-5 catalysts with varied Cu loadings were compared, in order to probe framework-related effects. Even though further experiments and theoretical support are needed, this proof-of-concept study confirms that in situ IS is capable of probing the mobility of Cu^{II} intermediates within the zeolite framework under NH_3-SCR related reaction conditions.

2. Results and Discussion

2.1. Catalyst Characterization

For each zeolite framework, we synthesized three samples with different Cu/Al ratios via aqueous ion-exchange, following previously established protocols [19]. The Cu-exchanged zeolites will be hereon named as "Cu(*x*)-SSZ-13" and "Cu(*x*)-ZSM-5", respectively, where *x* refers to the Cu/Al ratio determined by inductively coupled plasma optical-emission spectroscopy (ICP-OES) (Table 1).

Table 1. Cu loadings and Cu/Al ratios determined by ICP-OES analysis.

Sample	Cu (wt. %)	Cu/Al
H-SSZ-13	-	-
Cu(0.14)-SSZ-13	0.58	0.14
Cu(0.18)-SSZ-13	0.76	0.18
Cu(0.24)-SSZ-13	1.11	0.24
H-ZSM-5	-	-
Cu(0.09)-ZSM-5	0.64	0.09
Cu(0.14)-ZSM-5	0.99	0.14
Cu(0.16)-ZSM-5	1.17	0.16

The obtained SSZ-13 and ZSM-5 samples show typical X-ray diffraction (XRD) patterns, which are associated to CHA and MFI crystal structures, respectively (Figure 1a,b) [24]. No visible copper oxide (CuO$_x$) reflection was noticed in the XRD patterns for the Cu-exchange zeolite samples, implying the absence of large CuO$_x$ agglomerates. Nevertheless, the presence of highly dispersed CuO$_x$ clusters or small particles, which may be below the detection limit of XRD, cannot be fully excluded, especially in samples with high Cu/Al ratios, i.e., Cu(0.24)-SSZ-13 and Cu(0.16)-ZSM-5. Scanning electron microscopy (SEM) was applied to study the surface morphology of the samples and the homogeneity of the zeolite film on the inter-digital electrodes (IDE) that were used in IS measurements. No significant morphologic change was observed after ion exchange and the zeolite particles in the thick film on IDE are densely packed and homogeneously dispersed (Figure 2a–d).

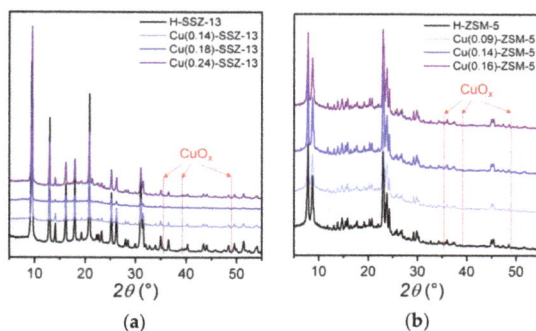

Figure 1. X-ray diffraction (XRD) patterns of SSZ-13 (**a**) and ZSM-5 (**b**) samples. No visible reflections from crystalline CuO$_x$ (ICCD 5-661; see dotted lines in red) were noted in the patterns.

The surface acidity and identity of NH$_3$ adsorption sites were investigated by temperature-programmed desorption using NH$_3$ as a probe molecule (NH$_3$-TPD). Figure 3a,b compare the NH$_3$-TPD profiles of H- and Cu(0.14)-SSZ-13 and of H- and Cu(0.14)-ZSM-5 catalysts, respectively. The diagrams show two main desorption peaks for all of the samples: the low-temperature peak (ca. 140 and 160 °C for SSZ-13 and ZSM-5, respectively) is associated to weakly bound NH$_3$, namely physisorbed NH$_3$ or NH$_3$ molecules adsorbed on extra-framework Al sites; the high-temperature peak (ca. 380 and 345 °C for SSZ-13 and ZSM-5, respectively), corresponds to NH$_4^+$ strongly bounded to Brønsted sites. The Cu-exchanged zeolite samples present an additional peak (ca. 240 and 230 °C for SSZ-13 and ZSM-5, respectively) that is overlapping partially with the high- and low-temperature ones, which is attributed to NH$_3$ desorption from Cu sites [5,18,25]. As compared to the respective

H-form zeolite, the Cu-form zeolite shows a reduced desorption of NH_3 from Brønsted sites, as a result of the replacement of H^+ on Brønsted sites with Cu^{2+} after ion exchange [18,26].

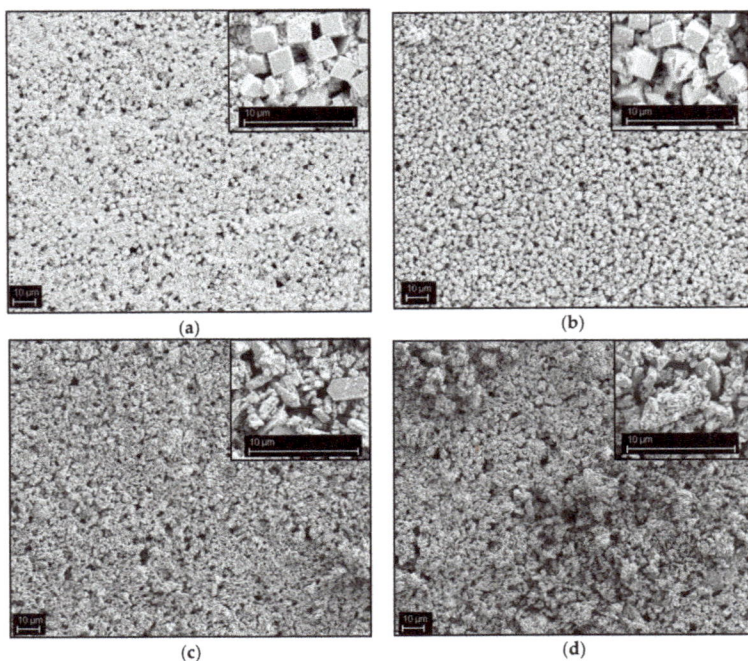

Figure 2. Representative scanning electron microscopy (SEM) images of H-SSZ-13 (**a**), Cu(0.18)-SSZ-13 (**b**), H-ZSM-5 (**c**), and Cu(0.14)-ZSM-5 (**d**) zeolite films on inter-digital electrode (IDE) chips.

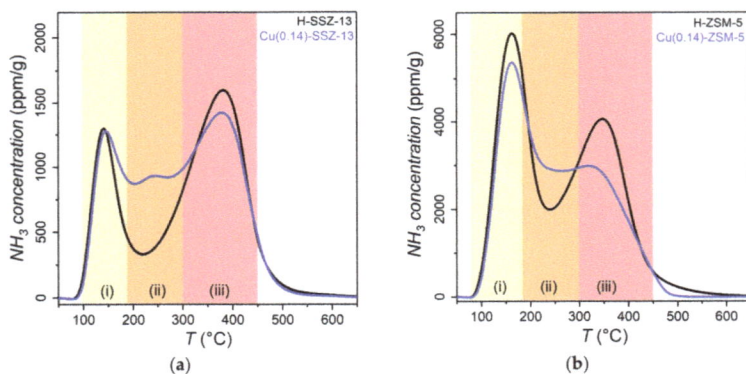

Figure 3. NH_3 desorption profiles obtained for SSZ-13 (**a**) and ZSM-5 (**b**) catalysts by NH_3-TPD, with a ramping rate of 2 °C/min. Before TPD measurements, the NH_3-loaded samples were flushed with N_2 for 3 h at 50 °C to remove physically adsorbed surface NH_3 species. The three desorption ranges are indicated by the colored areas: (i) weakly-bound NH_3 (yellow); (ii) NH_3 adsorbed on Cu sites (orange); and, (iii) NH_3 adsorbed on Brønsted sites (red).

2.2. In Situ IS: Modulus Spectra

Impedance spectroscopy (IS) is an electric perturbation technique, in which an alternating voltage U, with angular frequency ω (=$2\pi\nu$) and amplitude U_0, is applied to a system in thermodynamic equilibrium. As for zeolites, the as-such generated electric field induces motion of the extra-framework cations, which is macroscopically measured as a current $I(\omega)$, inversely proportional to the complex impedance $Z(\omega)$:

$$Z(\omega) = \frac{U(\omega)}{I(\omega)}. \tag{2}$$

Complex impedance can be expressed by its real (Z') and imaginary (Z'') parts as:

$$Z(\omega) = Z'(\omega) + jZ''(\omega). \tag{3}$$

The so-called Argand plot (also known as Nyquist plot), which displays Z'' against Z', is dominated by the low-frequency tail that is related to phenomena such as sample/electrode interface polarization. For the study of high-frequency processes, modulus plot (M'' vs. *log* ν) has been proved to be more suitable. In M'' plots, the experimentally collected impedance values (Z) are transformed according to the equation:

$$M'' = 2\pi\nu C_0 Z' \tag{4}$$

where ν is the perturbing frequency, C_0 is the capacitance of the empty capacitor, and Z' the real part of the complex impedance [10,17,27]. In the modulus plot, the position of a local maximum corresponds to a resonance frequency (ν_{res}) of the system [17,18]. The relaxation time (τ) of the involved ion movement phenomenon can be thus derived according to

$$\tau = 1/\nu_{res} \tag{5}$$

The modulus spectra obtained at 200 °C in the presence of NH_3 are reported in Figure 4a,b for the whole sets of SSZ-13 and ZSM-5 zeolites, respectively. At this temperature, a minimum amount of weakly bound NH_3 is adsorbed on the zeolites, while the adsorption of NH_3 on Cu sites is widely unaffected (see NH_3-TPD profiles in Figure 3a,b). Therefore, the effect of $Cu^{II}(NH_3)_n$ species on the ion movement within zeolites is expected to be most pronounced at ca. 200 °C. Each modulus spectrum presents two resonance peaks, corresponding to two dielectric relaxation modes. Previous studies proved that the low-frequency maximum (LF, i.e., 1–100 Hz) is related to long-range ion transport, i.e., translational ion motion, while at high frequencies (HF, i.e., 10^4–10^6 Hz), short-range (local) ion motion is displayed [10,17,27]. The LF ion transport processes have been extensively studied in our previous investigations [19,28]. The main focus of this study will be the HF resonance peak, to which the short-range movement of $Cu^{II}(NH_3)_n$ species is expected to contribute [12]. As displayed in Figure 4a, for SSZ-13 the HF resonance peak undergoes a significant shift to higher frequencies with increasing Cu loading (Cu/Al ratio) in the samples, from the starting value *log* ν_{res} = 4.13 (ca. 1.36×10^4 Hz) in H-SSZ-13 to the value *log* ν_{res} = 5.33 (ca. 2.15×10^5 Hz) for Cu(0.24)-SSZ-13. On the contrary, in the case of ZSM-5 zeolites, a slight shift of HF to lower frequencies was observed. The H-ZSM-5 presents the HF peak at *log* ν_{res} = 5.27 (ca. 1.84×10^5 Hz), while the HF peak for the sample with the highest Cu-loading was shifted to *log* ν_{res} = 4.93 (ca. 8.58×10^4 Hz).

Notably, even for Cu-exchange zeolites, due to the fact that only a fraction of Brønsted sites is occupied with Cu ions, there is still the presence of protons in the zeolite framework, and NH_4^+ ions will be formed in the respective temperature range and undergo local motion [15,29]. Therefore, the HF resonance peak is supposed to result from a superposition of the local movement of both NH_3-solvated H^+ and Cu^{2+} ions. The shift to higher frequencies of the HF peak in SSZ-13 with increasing Cu/Al implies that the formed $Cu^{II}(NH_3)_n$ species leads to a shorter relaxation time (Table 2), and thus becomes the frequency-determining species for the HF process. The fact that an opposite trend was observed for ZSM-5, i.e., an increase of the HF relaxation time with increasing Cu loading (Table 2), indicates that the crystalline framework plays a decisive role in the local motion of $Cu^{II}(NH_3)_n$. It is however ruled out that non-solvated Cu^{2+} ions can get similarly mobile, since a shift to higher frequencies was not observed under pure N_2 (Figure S1). On the contrary, the HF ν_{res} for pristine SSZ-13 zeolite in N_2 was significantly shifted to lower frequencies by the presence of non-solvated Cu^{2+}. This suggests that the occupancy of Brønsted sites by the bivalent metal ions decreases the amount of H^+ that is available for local and translational charge transport.

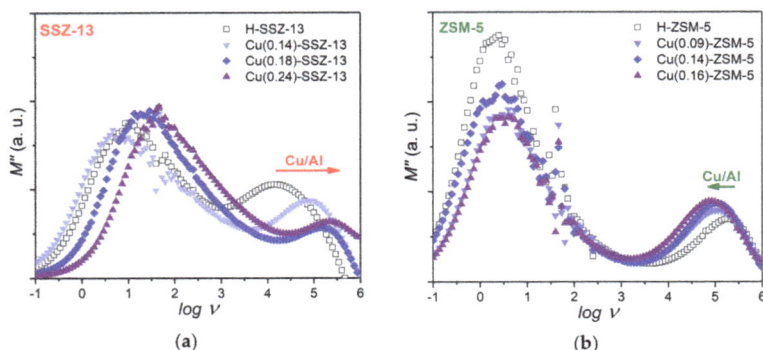

Figure 4. Modulus plots of in situ IS results collected in NH_3 (100 ppm in N_2) at 200 °C over H-form and Cu-exchanged (**a**) SSZ-13 and (**b**) ZSM-5 zeolites.

Table 2. Resonance frequencies (ν_{res}) and corresponding relaxation times (τ) for local ion motion within the studied zeolites. The ν_{res} values were obtained from the modulus plots of the in situ IS experiments under NH_3 (100 ppm in N_2) at 200 °C.

Sample	HF ν_{res}	log ν_{res}	HF τ
H-SSZ-13	1.36×10^4	4.13	7.36×10^{-5}
Cu(0.14)-SSZ-13	7.36×10^4	4.87	1.36×10^{-5}
Cu(0.19)-SSZ-13	1.58×10^5	5.20	6.31×10^{-6}
Cu(0.24)-SSZ-13	2.15×10^5	5.33	5.41×10^{-6}
H-ZSM-5	1.84×10^5	5.27	5.41×10^{-6}
Cu(0.09)-ZSM-5	1.17×10^5	5.07	8.58×10^{-6}
Cu(0.14)-ZSM-5	1.00×10^5	5.00	1.00×10^{-5}
Cu(0.16)-ZSM-5	8.58×10^4	4.93	1.17×10^{-5}

2.3. In Situ IS: HF Arrhenius Plots

The observed ion conducting processes are temperature-dependent, and their activation energy (E_a) can be derived from the equation

$$ln\left(Y'_{res}T\right) \sim ln(\sigma T) = A + \frac{E_a}{k_B T} \tag{6}$$

where Y' is the real part of the admittance (i.e., $Y' = 1/Z'$) at the temperature-dependent ν_{res}, σ is the specific conductivity of the zeolite, A is the pre-exponential factor (which depends on the charge and number of mobile species, their on-site oscillation frequencies and the hopping distances), T is temperature in Kelvin and k_B is the Boltzmann constant [17,18,27]. Here, ν_{res} is always intended as the resonance frequency associated to the HF maximum (i.e., for the short-range ion motion) in the corresponding modulus plot at the selected temperature. Arrhenius-like plots (i.e., $ln\ Y'T$ vs. $1/T$) are therefore helpful in the study of the temperature-dependent ion motion processes in zeolites [10,17,27]. Figures 5a–d and 6a–d illustrate the Arrhenius plots for the SSZ-13 and ZSM-5 catalysts, respectively.

A linear increase of ln $(Y'T)$ with decreasing $1/T$ was observed in measurements that were performed over pristine H-SSZ-13 and H-ZSM-5 in N_2 (Figures 5a and 6a, black squares). In both cases, the measured conductivity can only be associated to proton mobility, since no other cations are present and no solvent molecule (e.g., H_2O, NH_3) can be supportive [15]. The observed mechanism is characterized by a higher E_a in H-SSZ-13 ($E_a = 94.1 \pm 0.7$ kJ·mol^{-1}), as compared with H-ZSM-5 ($E_a = 58.7 \pm 0.6$ kJ·mol^{-1}). We assume that this difference is related to the structural differences between the two zeolites, which is consistent with the finding that the local mobility of protons is lower in CHA zeolites than in other zeolites [30]. The NH_3-loaded samples show, in general, higher conductivity than the pristine counterparts, thanks to the contribution of NH_3 to form NH_4^+ that exhibits a higher mobility in zeolites. The non-linearity of the profiles below 350 °C reflects a combined effect of NH_3 solvation and thermal activation, as revealed in our previous studies on proton transport in zeolites [15,29]. Above 350 °C, NH_3 is completely desorbed and the ion mobility follows the same mechanism as observed in the pristine zeolites.

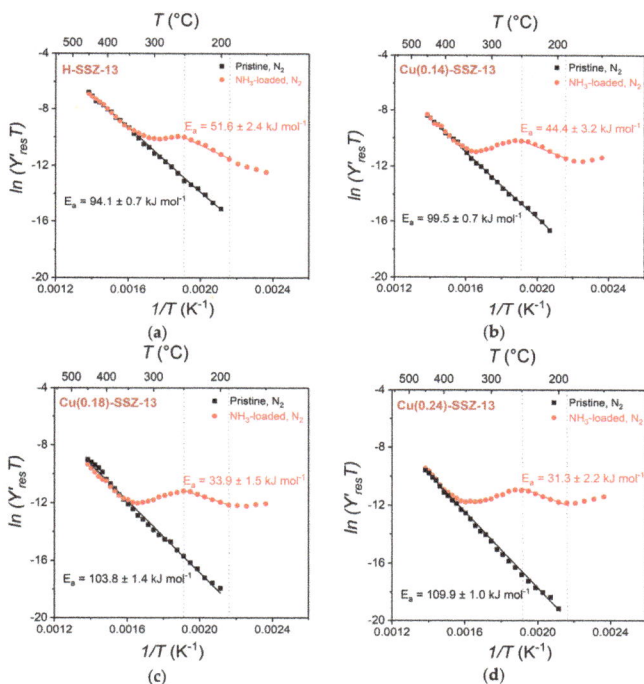

Figure 5. Arrhenius plots of the in situ IS results collected under different conditions (black squares: pristine zeolite in N_2; red circles: NH_3-loaded zeolite in N_2) for (**a**) H-SSZ-13, (**b**) Cu(0.14)-SSZ-13, (**c**) Cu(0.19)-SSZ-13, and (**d**) Cu(0.24)-SSZ-13. The indicated E_a value was derived from the corresponding linear fitting of temperature-dependent ionic conductivities [i.e., ln $(Y'_{res}T)$]. The dotted lines highlight the temperature range (190 and 250 °C) from which the E_a value for the NH_3-loaded SSZ-13 zeolite was derived.

A linear increase of conductivity with temperature under N_2 is also observed for the pristine Cu-exchanged zeolites, even though the effect of the Cu exchange level is different for the two crystalline frameworks (Figures 5b–d and 6b–d). In both catalyst systems, the increased Cu loading leads to a progressive increase of the E_a for the short-range H$^+$ motion (Figure 7). We therefore assume in a first approximation that the occupation of Brønsted sites by Cu^{2+}, independently whether single- or double-bounded, affect the total background charge to the zeolite lattice in a way that the proton motion gets less likely with increasing Cu content and thus requires a higher activation energy.

The analysis of NH_3-loaded zeolites is even more complex, since the temperature ranges for desorption of solvent molecules from Cu^{2+} sites and for the removal of weakly bound NH_3 species are not clearly separated. Nevertheless, it is possible to individuate a "temperature window", between 190 and 250 °C, in which the impedance measurements are supposed to be most sensitive to $Cu^{II}(NH_3)_n$ species, whereas still the contribution of NH_4^+ has to be taken into account because of its co-existence in the respective temperature range (see NH_3-TPD profiles in Figure 3). The E_a associated to the local ion motion in the respective zeolite was calculated and is displayed in Figure 7. After NH_3 adsorption, a significant decrease of the E_a value was observed for both Cu-SSZ-13 and Cu-ZSM-5, as compared to that for the corresponding pristine zeolite. Cu-ZSM-5 shows an increase of E_a with increasing Cu/Al, which is similar to the trend that was observed for the pristine zeolites in the NH_3-free state. On the contrary, for Cu-SSZ-13, the E_a is progressively lowered by the increase of the Cu amount, suggesting a favorable short range motion of $Cu^{II}(NH_3)_n$ at a higher Cu loading.

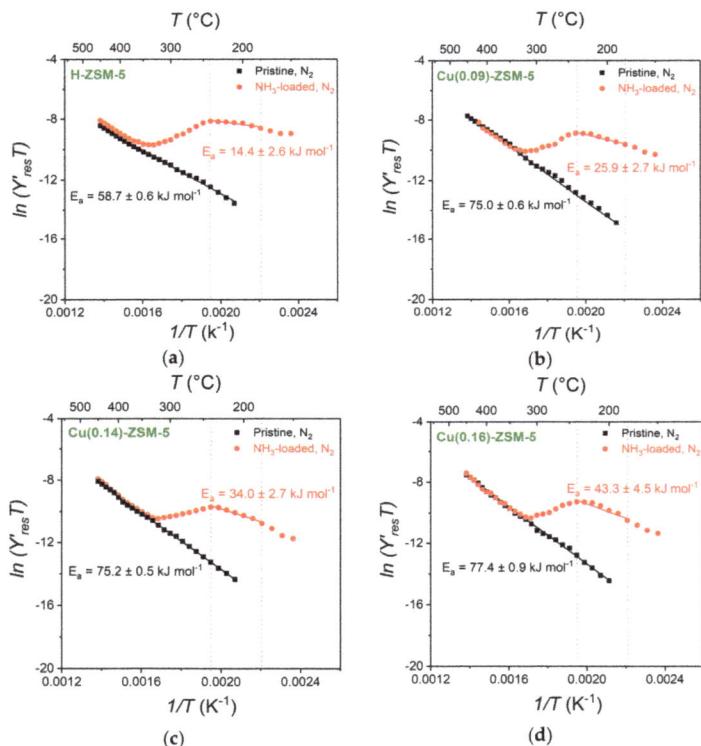

Figure 6. Arrhenius plots of the in situ IS results under different conditions (black squares: pristine zeolite in N_2; red circles: NH_3-loaded zeolite in N_2) for (**a**) H-ZSM-5, (**b**) Cu(0.09)-ZSM-5, (**c**) Cu(0.14)-ZSM-5, and (**d**) Cu(0.16)-ZSM-5. The indicated E_a value was derived from the corresponding linear fitting of temperature-dependent ionic conductivities [i.e., $ln\,(Y'_{res}T)$]. The dotted lines highlight the temperature range (180 and 240 °C) from which the E_a value for the NH_3-loaded ZSM-5 zeolite was derived.

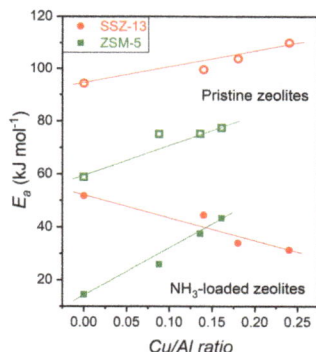

Figure 7. Activation energy E_a for the local motion of ions in the pristine zeolite (empty symbols) and in the NH$_3$-loaded zeolites (full symbols) as a function of Cu/Al ratio for SSZ-13 (red symbols) and ZSM-5 (green symbols) catalysts.

Combining the information that was obtained from the modulus plot analysis and the temperature-dependent measurements, it emerges that the presence of Cu^{II}(NH$_3$)$_n$ is affecting the dipolar relaxation processes that take place in zeolites at high frequencies. In both CHA and MFI frameworks, while the ion motion properties changed with the availability of Cu^{II}(NH$_3$)$_n$ species, a substantial difference was observed. For ZSM-5, the formation of Cu^{II}(NH$_3$)$_n$ species led to an increase in the relaxation time and in the overall activation energy that was associated to the ion motion processes in the selected temperature window (180–240 °C). SSZ-13 shows different properties: the increase of Cu amount favors an ion motion process with a shorter relaxation time and lower activation energy, pointing to enhanced local ion motion at a higher Cu loading.

The different local environments of Cu ion in ZSM-5 and SSZ-13 may be responsible for their distinct ion motion properties. Theoretical calculations [31] proved that Cu^{2+} in ZSM-5 is usually coordinated to an Al site in a six-member ring (6-MR) that is connected directly with a 10-MR channel, while a second Al required for charge balance is located in an adjacent 5-MR, which is part of a smaller channel. The larger dimensions of the MFI principal channels do not allow for Cu^{II}(NH$_3$)$_n$ to find intermediate position of minimized potential energy. As a result, the transport of Cu^{II}(NH$_3$)$_n$ to another couple of Al atoms facilitates long-range rather than local motion of the species within ZSM-5 zeolites. In this way, we assume that Cu^{II}(NH$_3$)$_n$ species do not sensibly contribute to the HF conductivity. Besides, due to the Cu ion exchange, the amount of H$^+$/NH$_4^+$ that could contribute to the HF mobility decreased with the increase of Cu/Al ratio.

On the contrary, for SSZ-13 zeolite, the movement of Cu^{II}(NH$_3$)$_n$ is confined inside the CHA cage. When considering this, the shift of HF maximum to higher frequencies, as well as the lowering of E_a with increasing Cu-loading is consistent with an increasing influence of the intra-cage mobility of the Cu^{II}(NH$_3$)$_n$. It has been repeatedly reported that NH$_3$-solvated and -mobilized Cu ions are the active sites for low-temperature NH$_3$-SCR reactions (below 250 °C) over Cu-SSZ-13 [11–13]. Our observations now document that the unique CHA cage plays a crucial role to sustain high local mobility of the NH$_3$-solvated Cu ions. It is thus of high interest to understand further the potential correlation between the NH$_3$-SCR reactivity of Cu species and their local mobility in zeolite catalysts, as experimentally accessed in this work.

3. Materials and Methods

3.1. Zeolite Synthesis

H-SSZ-13 (CHA-type framework) was synthesized following a procedure reported in literature [32]. The H-form ZSM-5 (MFI-type framework) is a commercially available zeolite from Clariant (Muttenz, Switzerland). According to ICP-OES, the two H-form zeolites present similar Si/Al ratios (i.e., 12.5 and 13.5 for H-SSZ-13, and H-ZSM-5, respectively).

The H-form zeolites were converted to Cu-zeolites via wet ion-exchange for 24 h in aqueous solutions of $Cu(NO_3)_2 \cdot 3H_2O$ ($\geq$99.5%, Merck, Darmstadt, Germany). In the case of SSZ-13, the temperature for the Cu ion exchange was set to 80 °C, in order to facilitate the ion diffusion in the small-pore zeolite framework [32]. After ion exchange, the zeolite sample was recovered, washed three times with ultra-pure water, and subsequently, dried overnight at 100 °C. The Cu-exchange process was repeated up to three times, to obtain samples with increasing Cu^{2+} contents. The dried products were then ground and calcined at 500 °C for 2 h. For comparison, the same calcination treatment was carried out over the H-form zeolites as well.

3.2. Zeolites Characterization

The crystalline phase of the samples was determined by XRD using a STOE STADI P diffractometer (STOE, Darmstadt, Germany) with a Cu Kα radiation source (λ = 1.54059 Å). The morphology of the catalysts was investigated by SEM with a Zeiss DSM 982 Gemini microscope (Zeiss, Oberkochen, Germany). The Si, Al and Cu contents of each sample were calculated from the elemental analysis obtained via ICP-OES. Before measuring, the zeolite samples were dissolved in HF.

NH_3-TPD was used to study the acidity of the zeolites and the identity of the NH_3 adsorption sites. An amount of ca. 50 mg of catalyst powder was introduced in a tubular quartz reactor, which was blocked with quartz wool and heated in a tube furnace (Carbolite, TZF, Sheffield, UK). Before each measurement, the sample was preheated for 1 h at 500 °C (ramping rate = 7.5 °C/min) in O_2 (flow rate = 50 sccm). The sample was then cooled to 50 °C in O_2 and it was hold for another 5 h at the same temperature in the N_2 atmosphere. The adsorption of NH_3 was carried out at 50 °C in flowing NH_3 atmosphere (1000 ppm in N_2, 50 sccm) for 2 h. In order to remove the physisorbed NH_3, the reactor was flushed with N_2 for 3 h at 50 °C. The temperature was subsequently increased from 50 to 700 °C (ramping rate = 2 °C/min) and the desorption of NH_3 was measured by means of an UV gas analyzer (LIMAS11 UV, ABB, Zürich, Switzerland).

3.3. In Situ IS Measurements

3.3.1. IS Set-Up

To perform in situ IS measurements, alumina chips that were equipped with screen-printed gold inter-digital electrodes (IDEs) and backside integrated heaters were employed (Figure 8). A thick film of zeolite catalyst was deposited on the IDE and was subsequently calcined at 480 °C for 10 h. The homogeneity of the film was examined by means of SEM (Figure 2a–d). The chip was inserted in a stainless-steel chamber (inner volume = 30 cm^3) and was connected by electrical contact to an impedance analyzer (SI 1260, Solartron, Bognor Regis, UK) and a digital multimeter (Keithley 2400, Solon, OH, USA) for the power supply of the integrated heater. A ZnSe-window was used to cap the chamber and allows for calibrating the chip temperature by means of a pyrometer (KT 19.82, Heitronics, Wiesbaden, Germany). The gas atmosphere in the chamber is regulated by mass flow controllers (MFCs; MKS 1179A, MKS 1259C and MKS-647C, MKS Instruments, Andover, MA, USA).

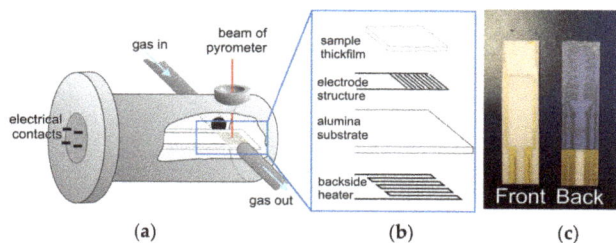

Figure 8. Schematic representations of the IS measurement chamber (**a**) and the IDE chip (**b**) (adapted with permission from [16]); and, a picture of a typical IDE chip with zeolite thick film (**c**).

3.3.2. Multi-Frequency In Situ IS

The in situ IS measurements were performed by applying an alternating voltage of 0.1 V rms, in frequency range between 0.1 Hz and 1 MHz. For each modulus spectrum at a selected temperature, 106 measuring points at equidistant logarithmic steps were recorded, with a total measurement duration of ca. 30 min.

Impedance values for modulus plots were collected at 200 °C in pure N_2, and subsequently, in 100 ppm NH_3 in N_2 (flow rate = 100 sccm). In both cases, the IS measurement was initialized after the chip had been flushed for 40 min in the desired atmosphere. Before each measurement, the sample was pretreated for 1.5 h in pure O_2 at 300 °C.

To obtain the Arrhenius plots, a set of modulus spectra were registered in the temperature range between 150 and 450 °C, at a step size of 10 °C. In each step, the temperature was stabilized for 10 min before collecting IS data. Prior to each experiment, the measured sample was pretreated for 1.5 h in pure O_2 at 450 °C.

4. Conclusions

In the present study, in situ impedance spectroscopy was applied to investigate the formation and local mobility of $Cu^{II}(NH_3)_n$ species resulting from the solvation and mobilization of Cu cations in Cu-exchanged zeolite catalysts during NH_3-SCR. A combination of the local movement of H^+/NH_4^+ and $Cu^{II}(NH_3)_n$ species led to a resonance peak at high frequencies (i.e., 10^4–10^6 Hz) in the modulus plots of the IS data. With the increasing of Cu loading, the HF peak of SSZ-13 shifted significantly to higher frequencies, while a gradual shift to lower frequencies was observed for ZSM-5. This contrary effect of Cu loading on the local ion mobility is believed to result from a structural effect of zeolite: the CHA framework, because of its unique cage structure, seems to facilitate the local motion of the $Cu^{II}(NH_3)_n$ species, so that they can contribute to the overall ion conductivity at high frequencies; in contrast, after NH_3 solvation, Cu^{2+} ions in ZSM-5 appear to move across the zeolite lattice rather than locally. This assumption was corroborated by temperature-dependent IS measurements, which evidenced a decreasing activation energy for the high-frequency ion conduction process with increasing Cu loading, specifically for SSZ-13. These results show that Cu^{II} species experience enhanced mobility in SSZ-13 under SCR-like conditions. This sheds new light on the participation of Cu^{II} in the full reaction cycle and the specific properties of CHA zeolites in NH_3-SCR. As the mobility of Cu^{II} is a new parameter to be considered in the further refinement of the reaction mechanism, the impedance data reported here may be a valuable complementation to the established experimental and theoretical methods that generated the existing knowledge of the nature of NH_3-SCR.

Supplementary Materials: The following results are available online at http://www.mdpi.com/2073-4344/8/4/162/s1, Figure S1: Modulus plots collected in N_2 at 200 °C.

Acknowledgments: This work was supported by the German Research Foundation (DFG) under grant SI 609/14-1, the Federal Ministry of Education and Research in the context of the DeNOx project (13XP5042A) and the Exploratory Research Space of RWTH Aachen University financed by the Excellence Initiative of the German federal and state governments to promote science and research at German universities. We thank Dieter Rauch and Ralf Moos for fruitful discussions and for providing IDE chips, Anna Clemens for providing the H-SSZ-13, and Martin Tabak for Cu-SSZ-13 synthesis.

Author Contributions: Peirong Chen and Valentina Rizzotto conceived and designed the experiments; Valentina Rizzotto performed the experiments; Valentina Rizzotto, Peirong Chen and Ulrich Simon analyzed the data; Peirong Chen and Ulrich Simon supervised the project and prepared the paper after the first draft made by Valentina Rizzotto.

Conflicts of Interest: The authors declare no conflict of interest. The founding sponsors had no role in the design of the study; in the collection, analyses, or interpretation of data; in the writing of the manuscript and in the decision to publish the results.

References

1. European Environment Agency. *Air Quality in Europe—2016 Report*; EEA report, No 28/2016; Publication Office of the European Union: Luxembourg, 2016; ISBN 978-92-9213-847-9. [CrossRef]

2. Zhang, R.; Liu, N.; Lei, Z.; Chen, B. Selective transformation of various nitrogen-containing exhaust gases toward N_2 over zeolites catalysts. *Chem. Rev.* **2016**, *116*, 3658–3721. [CrossRef] [PubMed]

3. Johnson, T.; Joshi, A. *Review of Vehicle Engine Efficiency and Emissions*; SAE Technical Paper 2017-01-0907; SAE International: Warrendale, PA, USA, 2017. [CrossRef]

4. Nova, I.; Tronconi, E. *Urea-SCR Technology for DeNOx after Treatment of Diesel Exhausts*, 1st ed.; Springer: Berlin/Heidelberg, Germany, 2014; ISBN 978-1-4899-8071-7.

5. Beale, A.M.; Gao, F.; Lezcano-Gonzalez, I.; Peden, C.H.F.; Szanyi, J. Recent advances in automotive catalysis for NO_x emission control by small-pore microporous materials. *Chem. Soc. Rev.* **2015**, *44*, 7371–7405. [CrossRef] [PubMed]

6. Guan, B.; Zhan, R.; Lin, H.; Huang, Z. Review of state of the art technologies of selective catalytic reduction of NO_x from diesel engine exhaust. *Appl. Therm. Eng.* **2014**, *66*, 395–414. [CrossRef]

7. Brandenberger, S.; Kröcher, O.; Tissler, A.; Althoff, R. The state of the art in selective catalytic reduction of NO_x by ammonia using metal-exchanged zeolite catalysts. *Catal. Rev. Sci. Eng.* **2008**, *50*, 492–531. [CrossRef]

8. Kwak, J.H.; Tonkyn, G.R.; Kim, D.H.; Szanyi, J.; Peden, C.H.F. Excellent activity and selectivity of Cu-SSZ-13 in the selective catalytic reduction of NO_x with NH_3. *J. Catal.* **2010**, *275*, 187–190. [CrossRef]

9. Paolucci, C.; Di Iorio, J.R.; Ribeiro, F.H.; Gounder, R.; Schneider, W.F. Catalysis science of NO_x selective catalytic reduction with ammonia over Cu-SSZ-13 and Cu-SAPO-34. *Adv. Catal.* **2016**, *59*, 1–107. [CrossRef]

10. Simon, U.; Franke, M.E. Electrical properties of nanoscaled host/guest compounds. *Microporous Mesoporous Mater.* **2000**, *41*, 1–36. [CrossRef]

11. Paolucci, C.; Parekh, A.A.; Khurana, I.; Di Iorio, J.R.; Li, H.; Albarracin-Caballero, J.D.; Shih, A.J.; Angarra, T.; Delgass, W.N.; Miller, J.T.; et al. Catalysis in a cage: Condition-dependent Speciation and dynamics of exchanged Cu cations in SSZ-13 zeolites. *J. Am. Chem. Soc.* **2016**, *138*, 6028–6048. [CrossRef] [PubMed]

12. Paolucci, C.; Khurana, I.; Parekh, A.A.; Li, S.; Shih, A.J.; Li, H.; Di Iorio, J.R.; Albarracin Caballero, J.D.; Yezerets, A.; Miller, J.T.; et al. Dynamic multinuclear sites formed by moblized copper ions in NO_x selective catalytic reduction. *Science* **2017**, *357*, 898–903. [CrossRef] [PubMed]

13. Gao, F.; Mei, D.; Wang, Y.; Szanyi, J.; Peden, C.H.F. Selective catalytic reduction over Cu/SSZ-13: Linking homo- and heterogeneous catalysis. *J. Am. Chem. Soc.* **2017**, *139*, 4935–4942. [CrossRef] [PubMed]

14. Franke, M.E.; Simon, U. Proton mobility in H-ZSM5 studied by impedance spectroscopy. *Solid State Ion.* **1999**, *118*, 311–316. [CrossRef]

15. Franke, M.E.; Simon, U. Solvate supported proton transport in zeolites. *Chem. Phys. Chem.* **2004**, *5*, 465–472. [CrossRef] [PubMed]

16. Simons, T.; Simon, U. Zeolites as nanoporous, gas-sensitive materials for in situ monitoring of $DeNO_x$-SCR. *Beilstein J. Nanotechnol.* **2012**, *3*, 667–673. [CrossRef] [PubMed]

17. Chen, P.; Schönebaum, S.; Simons, T.; Rauch, D.; Dietrich, M.; Moos, R.; Simon, U. Correlating the integral sensing properties of zeolites with molecular processes by combining broadband impedance and DRIFT spectroscopy—A new approach for bridging the scales. *Sensors* **2015**, *15*, 28915–28941. [CrossRef] [PubMed]

18. Chen, P.; Simon, U. In Situ spectroscopic studies of proton transport in zeolite catalysts for NH_3-SCR. *Catalysts* **2016**, *6*, 204. [CrossRef]

19. Chen, P.; Rauch, D.; Weide, P.; Schönebaum, S.; Simons, T.; Muhler, M.; Moos, R.; Simon, U. The effect of Cu and Fe cations on NH_3-supported proton transport in $DeNO_x$-SCR zeolite catalysts. *Catal. Sci. Technol.* **2016**, *6*, 3362–3366. [CrossRef]

20. Chen, P.; Simböck, J.; Schönebaum, S.; Rauch, D.; Simons, T.; Palkovits, R.; Moos, R.; Simon, U. Monitoring NH_3 storage and conversion in Cu-ZSM-5 and Cu-SAPO-34 catalysts for NH_3-SCR by simultaneous impedance and DRIFTS spectroscopy. *Sens. Actuators B* **2016**, *236*, 1075–1082. [CrossRef]

21. Chen, P.; Jabłońska, M.; Weide, P.; Caumanns, T.; Weirich, T.; Muhler, M.; Moos, R.; Palkovits, R.; Simon, U. Formation and effect of NH_4^+ intermediates in NH_3-SCR over Fe-ZSM-5 zeolite catalysts. *ACS Catal.* **2016**, *6*, 7696–7700. [CrossRef]

22. Simons, T.; Chen, P.; Rauch, D.; Moos, R.; Simon, U. Sensing catalytic conversion: Simultaneous DRIFT and impedance spectroscopy for in situ monitoring of $DeNO_x$-SCR on zeolites. *Sens. Actuators B Chem.* **2016**, *224*, 492–499. [CrossRef]

23. Chen, P.; Khetan, A.; Jabłońska, M.; Simböck, J.; Muhler, M.; Palkovits, R.; Pitsch, H.; Simon, U. Local dynamics of copper active sites in zeolite catalysts for selective catalytic reduction of NO_x with NH_3. Unpublished manuscript.

24. Treacy, M.M.J.; Higgins, J.B. *Collection of Simulated XRD Powder Patterns for Zeolites*, 4th ed.; Elsevier: Amsterdam, The Netherlands, 2001.

25. Lezcano-Gonzalez, I.; Deka, U.; Arstad, B.; Van Yperen-De Deyne, A.; Hemelsoet, K.; Waroquier, M.; Van Speybroeck, V.; Weckhuysen, B.M.; Beale, A.M. Determining the storage, availability and reactivity of NH_3 within Cu-Chabazite-based ammonia selective catalytic reduction systems. *Phys. Chem. Chem. Phys.* **2014**, *16*, 1639–1650. [CrossRef] [PubMed]

26. Brandenberger, S.; Kröcher, O.; Wokaun, A.; Tissler, A.; Althoff, R. The role of Brønsted acidity in the selective catalytic reduction of NO with ammonia over Fe-ZSM-5. *J. Catal.* **2009**, *268*, 297–306. [CrossRef]

27. Simon, U.; Flesch, U. Cation-Cation interaction in dehydrated zeolites X and Y monitored by modulus spectroscopy. *J. Porous Mater.* **1999**, *6*, 33–40. [CrossRef]

28. Chen, P.; Moos, R.; Simon, U. Metal loading affects the proton transport properties and the reation monitoring performance of Fe-ZSM-5 and Cu-ZSM-5 in NH_3-SCR. *J. Phys. Chem. C* **2016**, *120*, 25361–25370. [CrossRef]

29. Rodriguez-Gonzalez, L.; Rodriguez-Castellon, E.; Jimenez-Lopez, A.; Simon, U. Correlation of TPD and impedance measurements on the desorption of NH_3 from zeolites H-ZSM-5. *Solid State Ion.* **2008**, *179*, 1968–1973. [CrossRef]

30. Sierka, M.; Sauer, J. Proton mobility in Chabazite, Faujasite, and ZSM-5 zeolite catalysts. Comparison bases on ab initio calculations. *J. Phys. Chem. B* **2001**, *105*, 1603–1613. [CrossRef]

31. Deka, U.; Lezcano-Gonzalez, I.; Weckhuysen, B.M.; Beale, A.M. Local environment and nature of Cu active sites in zeolite-based catalysts for the selective catalytic reduction of NO_x. *ACS Catal.* **2013**, *3*, 413–427. [CrossRef]

32. Shishkin, A.; Kannisto, H.; Carlsson, P.; Härelind, H.; Skoglundh, M. Synthesis and functionalization of SSZ-13 as an NH_3-SCR catalyst. *Catal. Sci. Technol.* **2014**, *4*, 3917–3926. [CrossRef]

catalysts

Article

Influence of the Sodium Impregnation Solvent on the Deactivation of Cu/FER-Exchanged Zeolites Dedicated to the SCR of NO$_x$ with NH$_3$

Marie-Laure Tarot [1], Mathias Barreau [1], Daniel Duprez [1], Vincent Lauga [2], Eduard Emil Iojoiu [2], Xavier Courtois [1,*] and Fabien Can [1,*]

[1] CNRS, UMR 7285 Institut de Chimie des Milieux et Matériaux de Poitiers (IC2MP), Université de Poitiers, 4 rue Michel Brunet—TSA 51106—86073 Poitiers CEDEX 9, France; marie.laure.tarot@univ-poitiers.fr (M.-L.T.); mathias.barreau@univ-poitiers.fr (M.B.); daniel.duprez@univ-poitiers.fr (D.D.)

[2] Renault Trucks—Volvo Group Trucks Technology—Powertrain Engineering Lyon, 99 route de Lyon—69806 Saint-Priest CEDEX, France; vincent.lauga@volvo.com (V.L.); eduard.emil.iojoiu@volvo.com (E.E.I.)

* Correspondence: xavier.courtois@univ-poitiers.fr (X.C.); fabien.can@univ-poitiers.fr (F.C.); Tel.: +33-519453994 (X.C.); +33-549453997 (F.C.)

Received: 24 November 2017; Accepted: 19 December 2017; Published: 23 December 2017

Abstract: The effect of the sodium addition mode was investigated on model Cu/FER selective catalytic reduction (SCR) catalysts with two copper loadings (2.8 wt. % and 6.1 wt. %) in order to compare samples with or without over-exchanged copper. Na was added by wet-impregnation using two solvents: water or ethanol. Catalysts were evaluated in Standard and Fast-SCR conditions, as well as in NO and NH$_3$ oxidation. They were characterized by H$_2$-TPR, NO and NH$_3$ adsorption monitored by FT-IR. As expected, whatever the copper loading, ammonia adsorption capacity was decreased by Na additions. Interestingly, characterizations also showed that Na impregnation in water favors the migration of the Cu-exchanged species, leading to the formation of CuO extra-framework compounds. Consequently, for both copper loadings, Na impregnation in water led to a stronger catalyst deactivation than impregnation in ethanol. Finally, the NO$_x$ conversion at low temperature (250 °C) appeared mainly affected by the loss in NH$_3$ adsorption capacity whereas the deNO$_x$ deactivation at high temperature (500 °C) was rather governed by the decrease in the exchanged copper ratio, which also induced a partial inhibition of NO and NH$_3$ oxidation behaviors.

Keywords: NH$_3$-SCR; sodium poisoning; copper exchanged zeolite; acidity

---

1. Introduction

Selective catalytic reduction of NO$_x$ by ammonia (NH$_3$-SCR) is an effective method to control nitrogen oxides emissions from fixed-source exhaust gases. The main involved reactions are the "Standard-SCR" reaction (Equation (1)) and the "Fast-SCR" reaction (Equation (2)).

$$2NH_3 + 2NO + \frac{1}{2}O_2 \rightarrow 2N_2 + 3H_2O \tag{1}$$

$$2NH_3 + NO + NO_2 \rightarrow 2N_2 + 3H_2O \tag{2}$$

In fact, the activity obtained in Standard-SCR condition is largely improved by favoring the oxidation of NO to NO$_2$ to provide the most favorable NO$_2$/NO$_x$ ratio of 0.5, leading to the "Fast-SCR" stoichiometry.

More recently, this NH$_3$-SCR technology was adapted to mobile sources to reach the environmental standards dedicated to passenger cars and heavy duty vehicles. Ammonia is then usually obtained by the decomposition of an aqueous solution of urea, which is injected in the exhaust pipe. The first generations of NH_3-SCR catalysts were based on V_2O_5-WO_3-TiO_2 materials, as usually implemented for NO_x treatment from power plants and stationary sources. Subsequently, new systems have emerged to expand the operating temperature window and to solve the high temperature deactivation drawback due to the anatase-rutile transition of TiO_2 and the possible V_2O_5 sublimation. Metal-exchanged zeolite-based catalysts were reported to be relevant materials for this purpose. A variety of zeolites have been proposed (ZSM-5, mordenite, beta, ferrierite, Y-zeolite, chabazite, …) and in the last few years, new generations of exchanged catalysts based on small pore zeolites have emerged. For instance, SAPO-34 or SZZ-13-based zeolites showed very good hydrothermal stabilities and also a resistance to hydrocarbon (HC) poisoning [1–4]. Zeolites are usually promoted by transition metals such as iron or copper [5]. Compared with iron-exchanged zeolites, copper-zeolite catalysts were reported to be more active at low temperature (<300 °C), less sensitive to the inlet NO_2/NO_x ratio, but produced more N_2O [6].

Besides, CO_2 emissions from common fossil sources induce a worrying situation that requires to increase the role of renewable resources for energy production. Biomass can be used as feedstock for the production of biodiesel which can be then utilized in the transport sector to replace fossil fuels. According to the recent European legislation (Directive 2003/30/EC), 10% share of biofuels in the transport sector must be achieved by 2020. However, biodiesel contains minerals traces such as sodium, potassium or phosphorus, which are authorized in limited amounts (maximum of 5 and 4 ppm for Na + K and P, respectively in European Union; EN 14214). Considering that (i) the Euro VI standard for heavy duty vehicles requires a durability of 700,000 km and (ii) some truck fleets use biodiesel exclusively, catalysts in the exhaust pipe may be subjected to kilograms of these poisons. Consequently, this study is focused on the influence of Na deposit on the NH_3-SCR process over copper-exchanged zeolite.

The effects of alkali on vanadia-based SCR catalysts are well described in the literature. For instance, exchange of alkali metal cation from the acids groups of V-OH species decreases the acidity, leading to a loss of ammonia-sorption properties [7]. The deactivation of the $V^{5+} = O$ redox sites was also highlighted [8]. In addition, potassium was reported to induce a stronger deactivation than sodium, for four similar weight contents of sodium and potassium (0.1, 0.5, 1, 2 wt. %) in the same catalyst [9]. Similar conclusion was reported by Kern et al. [10] for iron-exchanged catalysts with K and Na contents from 0 to 0.5 mmol/g_{cata}. However, only few studies concerning the poisoning of zeolites by alkali are available. Sodium and potassium are supposed to lead to the same type of deactivation. Shwan et al. studied the poisoning of a Fe-BEA zeolite by exposition to vapors from an aqueous solution of KNO_3 [11]. This poisoning led to a decrease in the Brønsted acidity and to a loss of active isolated iron species, but also increased both the NO_x storage behavior and the NO oxidation activity. Ma et al. [12] observed the same tendencies on the Cu/SAPO-34 poisoned by incipient wetness impregnation. They also noticed that the Standard-SCR activity is not impacted with 0.5 wt. % of potassium. Adding potassium firstly affected the low temperature activity (<300 °C) together with an increase in the N_2O formation. In addition, Brookshear et al. [13,14] showed the importance of the diesel exhaust pipe design toward the sodium impact on the SCR catalyst. The heavy duty configuration (Diesel Oxidation Catalyst (DOC)—Diesel Particulate Filter (DPF)—SCR) was less impacted by sodium than in the light-duty configuration (DOC-SCR-DPF), because sodium was preferentially trapped by the DPF.

The aim of this study was to investigate the deactivation of Cu-FER catalysts by sodium, with a special focus on the influence of the sodium impregnation solvent. The ferrierite (FER) structure was selected as host support because of its medium pore size allowing a high thermal stability [15]. FER structure was also claimed as appropriate structure for iron or copper-exchanged SCR catalysts [16]. Impregnation of sodium in water was used especially to simulate cold start/engine switch off periods, for which water condensation occurs in the exhaust pipe [11,12,17]. However, the use of such polar solvent should not be representative of Na-catalyst interactions in the exhaust pipe at usual operating temperature. Then, sodium impregnations were also performed in ethanol, which is less polar than water, to avoid interactions between water and the exchanged zeolite. This is assumed to be more representative of a gas phase poisoning. In order to compare the different mechanisms of deactivation depending on the Na impregnation solvent, two catalysts with or without extra-framework copper were used, containing 6.1 wt. % and 2.8 wt. % of copper, respectively.

2. Results

2.1. Structural/Textural Characterisation of the Catalysts, Elemental Analysis

2.1.1. Sodium Free Catalysts

XRD patterns of the zeolite as received and after the hydrothermal treatment at 600 °C are reported in Supplementary Materials (Figure S1). No evolution of the diffraction peaks were evidenced, which is also consistent with the stability of the specific surface area at 395–405 $m^2 \cdot g^{-1}$. However, the infrared structural band (T–O–T vibrations) were shifted from 1096 to 1103 cm^{-1} after aging, indicating a small dealumination of the zeolite (Supplementary Materials, Figure S2).

After copper exchange, no evolution of the zeolite structure was observed by XRD analysis. No new diffraction peak appeared over $Cu_{2.8}$/FER, whereas broad diffraction peaks appeared on the diffractogram of $Cu_{6.1}$/FER near $2\theta = 35.5°$ and $38.7°$, which are attributed to CuO (Supplementary Materials, Figures S3 and S4). For both copper loadings, metal exchange and subsequent hydrothermal treatment induced a decrease of about 10% of the specific surface area, at around 352 $m^2\ g^{-1}$ for $Cu_{2.8}$/FER and $Cu_{6.1}$/FER (Table 1).

Table 1. Catalysts characterizations.

[Cu] (wt. %)	n_{Cu}/n_{Al}	Na Impregn Solvent	[Na] (ppm/µmol g⁻¹)	n_{Na}/n_{Al}	S_{BET} (m² g⁻¹)	$V\mu$ (i) (cm³ g⁻¹)	V_{total} (ii) (cm³ g⁻¹)	H_2-TPR (n_{H2}/n_{Cu})	L:B (iii)
2.8 (Cu₂.₈/FER)	0.30	Na free sample	0	0	352	0.12	0.22	1.02	83:17
		Water (W)	4500/195	0.13				0.98	91:9
			5400/235	0.16				1.00	96:4
			12,800/560	0.37				1.03	100:0
			19,000/830	0.56	270	0.11	0.17	1.04	100:0
		Ethanol (Et)	3900/170	0.11				1.05	91:9
			9600/420	0.28				1.02	89:11
			14,200/620	0.41	238	0.06	0.17	1.10	92:8
6.1 (Cu₆.₁/FER)	0.67	Na free sample	0	0	352	0.12	0.23	1.03	96:4
		Water (W)	4300/190	0.12				1.03	97:3
			11,800/515	0.34				1.02	100:0
			13,800/600	0.40				1.08	100:0
			20,300/885	0.59	265	0.09	0.16	1.02	100:0
		Ethanol Et)	8700/380	0.25				1.06	98:2
			15,200/665	0.44	251	0.09	0.15	0.96	99:1

(i): microporous volume deduced from N_2 adsorption/desorption at 77 K; (ii): total porous volume deduced from N_2 adsorption/desorption at 77 K; (iii): Lewis acid sites (L); Brønsted acid sites (B) ratio determined by NH_3 adsorption at 50 °C.

2.1.2. Na Impregnated Catalysts

All the sodium loading from Inductively Coupled Plasma (ICP) elemental analysis are reported in Table 1 depending on the copper content and the Na impregnation solvent.

For each $Cu_{2.8}$/FER and $Cu_{6.1}$/FER catalyst, four Na loadings from 4300–4500 ppm to 19,000–20,300 ppm were obtained by sodium impregnation in water. The specific surface areas of the highest loaded samples were decreased to 265–270 $m^2\ g^{-1}$, compared with 352 $m^2\ g^{-1}$ for the fresh catalysts (Table 1). In addition, the XRD patterns of the corresponding samples point out the appearance of CuO for $Cu_{2.8}$/FER whereas the intensity of the CuO peaks significantly increased for $Cu_{6.1}$/FER (Supplementary Materials Figures S3 and S4).

Sodium loadings after impregnation in ethanol were ranked between 3900–14,200 ppm and 8700–15,200 ppm for $Cu_{2.8}$/FER and $Cu_{6.1}$/FER, respectively. No evolution of the XRD patterns was observed compared with the sodium free catalyst (patterns not shown), unlike the impregnation of Na in water. This first result already predicts an effect of the sodium solvent toward physico-chemical properties of the materials. The specific surface areas for the higher Na amounts decreased to 240–250 $m^2 g^{-1}$ (Table 1).

2.2. Catalytic Activity

NO_x conversions obtained over stabilized $Cu_{2.8}$/FER and $Cu_{6.1}$/FER samples (without sodium) are reported in Figure 1 for the Standard-SCR and Fast-SCR conditions. Please remind that the catalytic tests in Fast-SCR condition were performed with lower amount of catalyst (15 mg versus 50 mg for the Standard-SCR condition) to limit the $deNO_x$ efficiency and to possibly distinguish a deactivation by sodium addition.

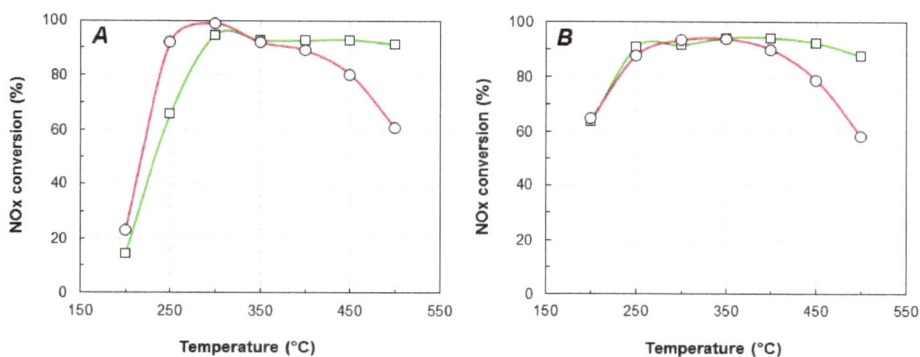

Figure 1. NO_x conversion by NH_3 over Cu/FER-exchanged zeolite. (**A**) standard-SCR; (**B**) Fast-SCR. (—, □): $Cu_{2.8}$/FER; (—, ○): $Cu_{6.1}$/FER.

In Standard-SCR condition (Figure 1A), the best activity at low temperature was achieved by the $Cu_{6.1}$/FER catalyst with a full NO_x conversion at around 300 °C. However, the NO_x conversion showed a net drop in the 350–500 °C temperature range. It is assigned to a competition between the NO_x reduction reaction and the ammonia oxidation reaction, leading to a lack in ammonia to reduce NO_x. To the opposite, $Cu_{2.8}$/FER sample presented an expanded operating temperature window, with around 90–95% of NO_x conversion from 300 to 500 °C. The maximum outlet N_2O concentration was recorded at 500 °C, at 1.5 and 5 ppm for $Cu_{2.8}$/FER and $Cu_{6.1}$/FER, respectively.

Compared with the Standard-SCR condition, higher NO_x conversions were observed at low temperature (200 °C) in Fast-SCR condition (Figure 1B), especially taking into account that the implemented catalyst weight was divided by more than 3. Again, a decrease in NO_x conversion was denoted at temperature higher than 350–400 °C, especially for $Cu_{6.1}$/FER samples, attributed to the ammonia oxidation reaction. The maximum outlet N_2O concentration (500 °C) was then a little increased, at 4 and 7 ppm for $Cu_{2.8}$/FER and $Cu_{6.1}$/FER, respectively.

Standard-SCR and Fast-SCR catalytic tests were also performed on the Cu/FER zeolite containing various sodium loadings. Preliminary tests were performed with catalysts submitted to the impregnation procedures in water or ethanol, but without sodium salt. Results (not shown) indicated no change in the catalytic behavior.

In order to illustrate the influence of both sodium content and sodium impregnation solvent (water or ethanol) on the $deNO_x$ efficiency of both $Cu_{2.8}$/FER and $Cu_{6.1}$/FER catalysts, two representative temperatures were selected: 250 °C was chosen to compare the $deNO_x$ behaviors at "low" temperature, while 500 °C was selected for the "high" temperature comparisons. Figure 2 compares the loss of the NO$_x$ conversion (expressed as a percentage of loss compared with the Na free samples) at "low" and "high" selected temperatures, depending on the sodium content (water and ethanol solvents are presented in blue and red lines, respectively). Note that the N$_2$O outlet concentration was not affected by sodium additions.

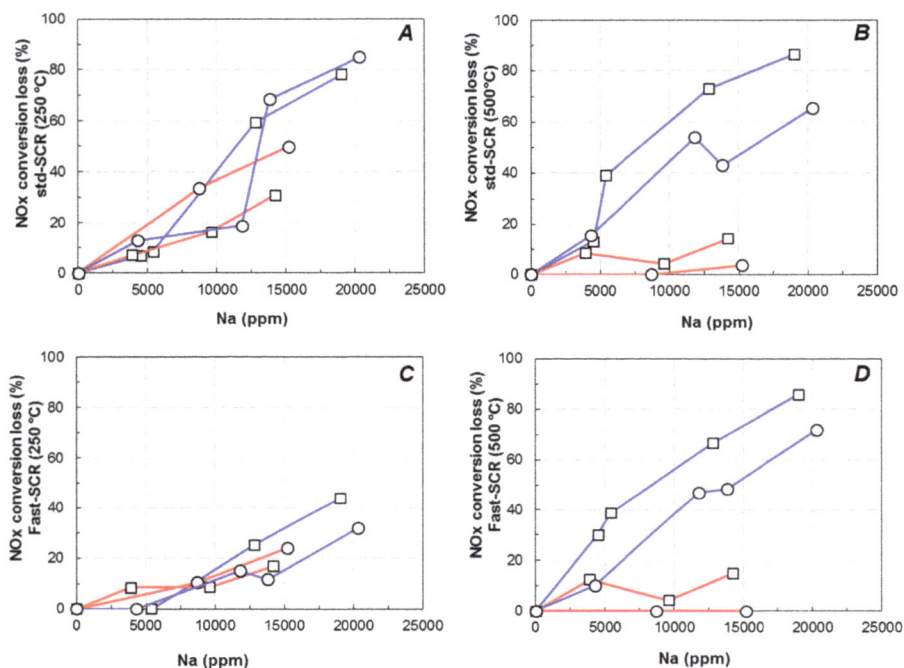

Figure 2. NO$_x$ conversion loss in function of the sodium loading on Cu/FER zeolites in standard (**A,B**) and Fast-SCR (**C,D**) at 250 °C (**A,C**) and 500 °C (**B,D**). □: Cu$_{2.8}$/FER; ○: Cu$_{6.1}$/FER. (Na impregnation in water: blue line; sodium impregnation in ethanol: red line).

It appears that the NO$_x$ conversion was strongly affected by the sodium loading. However, the deactivation acted differently depending on both the Na solvent (i.e., water or ethanol) and the SCR inlet condition (Standard or Fast-SCR experiments).

In Standard-SCR condition, when sodium was impregnated on Cu$_{6.1}$/FER via the water pathway (blue line), the deactivation at 250 °C (Figure 2A) appeared rather limited until 12,500 ppm of Na content, with a loss of NO$_x$ conversion of 20%. Beyond, the deNO$_x$ efficiency rapidly dropped to achieve deactivation over 80% for the higher Na content (around 2% of Na deposit). Interestingly, Na deposition using ethanol as solvent (red line) led to different observations. For approximately the same Na content (15,000 ppm), the loss in NO$_x$ conversion was around 50% when sodium was impregnated in ethanol versus 70% when sodium was impregnated in water.

Over Cu$_{2.8}$/FER, same trends were obtained: the most pronounced decrease was observed for sodium impregnation in water (blue line). However, for this catalyst with the low copper loading, the influence of the sodium solvent appeared a little less marked than for Cu$_{6.1}$/FER.

Finally, for Na contents higher than 14,200 ppm impregnated in water, similar NO$_x$ conversions were observed in Standard-SCR condition whatever the catalyst (i.e., Cu$_{6.1}$/FER or Cu$_{2.8}$/FER).

This suggests that beyond a threshold value of Na deposit, both catalysts tend to exhibit similar remaining active sites.

At high temperature, it was previously described that the competitive reaction between ammonia oxidation and NO_x reduction was enhanced over the most loaded copper catalyst ($Cu_{6.1}$/FER). Nevertheless, Figure 2B clearly shows that at 500 °C, for both studied copper loading, sodium impregnation in ethanol (red lines) had few impact on the NO_x conversion up to 14,200 ppm Na content. On the opposite, strong deactivations were denoted when Na was impregnated in water (blue line). The NO_x conversion loss for approximately 2% Na then reached 65% and 86% for $Cu_{6.1}$/FER and $Cu_{2.8}$/FER, respectively.

In Fast-SCR condition, no influence of Na deposits on the NO_2 availability (from NO oxidation into NO_2) is expected since the optimal inlet NO_2/NO_x ratio is already fixed to 0.5. Only other parameters should be involved in the deactivation mechanism. In fact, Figure 2C reports that the NO_x conversion at 250 °C was decreased with similar tendencies with sodium contents, whatever the Na impregnation route (water or ethanol) and regardless the copper loading ($Cu_{2.8}$/FER or $Cu_{6.1}$/FER). The decrease rate appeared less marked than in Standard-SCR condition.

At 500 °C (Figure 2D), similar behaviors to those denoted in Figure 2B (Standard-SCR) are observed: the ethanol impregnation solvent did not induce a strong deactivation of the catalysts whereas impregnation with water induced the same NO_x conversion loss than in Standard-SCR condition.

Finally, for the two studied catalysts with different copper loadings, the sodium impregnation in water strongly impacted the catalytic activity compare with the impregnation in ethanol. Moreover, when Na was added in water, almost the same catalytic activities were reached for the highest sodium loadings for both catalysts with different copper contents, whereas the initial catalytic activities were different. Considering the needed functions for the NH_3-SCR activity, the decrease in activity of the catalysts could be attributed to alteration of the redox properties and/or to a poisoning of the acidic sites. Results described in this section, indicate that sodium addition in water or ethanol did not led to the same effects on the catalysts behaviors. Consequently, dedicated characterizations were performed to gain information about the Na poisoning and the influence of the Na impregnation solvent.

2.3. Redox Properties and Copper Characterization

In this section, the influence of the sodium impregnation solvent was examined through redox properties of $Cu_{2.8}$/FER and $Cu_{6.1}$/FER catalysts. Both NH_3 and NO oxidation catalytic tests and H_2-TPR experiments were carried out. Additionally, NO adsorption monitored by infrared spectroscopy was also performed to assess to the copper state.

2.3.1. NH_3 and NO Oxidation

As described in the previous section, the oxidation performances of $Cu_{2.8}$/FER and $Cu_{6.1}$/FER are suspected to be differently affected by the sodium loading and impregnation solvent. Catalytic oxidation of NH_3 and NO experiments were then performed over all the studied samples. The ammonia oxidation by oxygen is not expected since it induces an undesired reductant consumption, but it also accounts for ammonia activation. NO oxidation is mainly involved in Standard-SCR condition to promote the Fast-SCR stoichiometry. Experiments were performed from 300 °C up to 500 °C for ammonia oxidation and from 350 °C up to 500 °C for NO oxidation, but only data recorded at 450 °C are reported in Figure 3 as representative results.

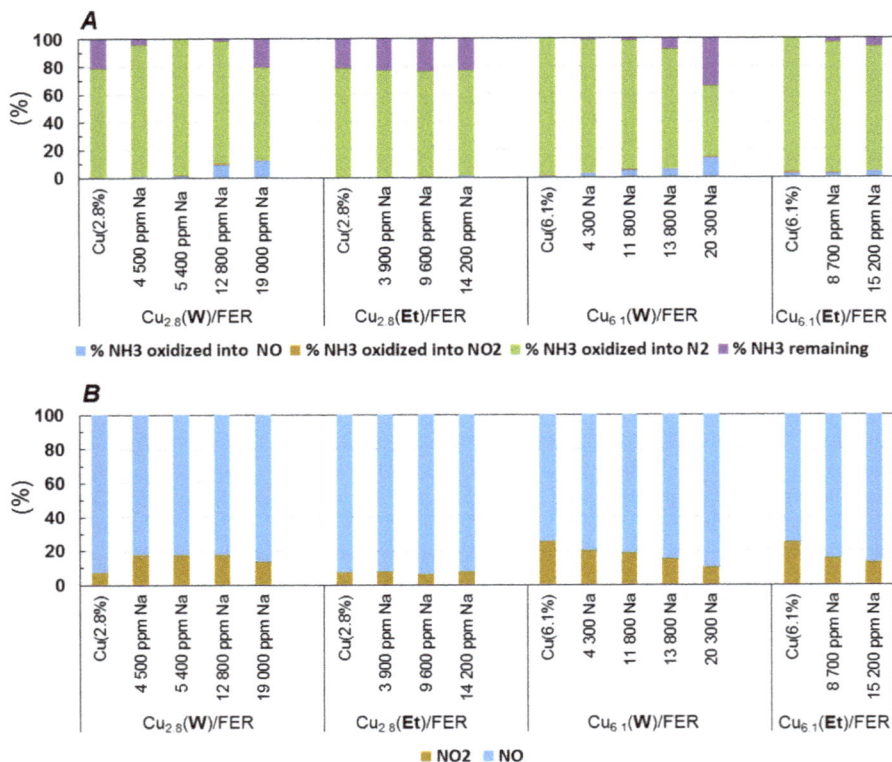

Figure 3. Effect of Na loading on NH$_3$ (**A**) and NO (**B**) oxidation activities at 450 °C over Cu$_{2.8}$/FER and Cu$_{6.1}$/FER, depending on the Na loading impregnated in water (W) or ethanol (Et).

With sodium free samples, full ammonia oxidation was observed with the highest loaded copper catalyst (Cu$_{6.1}$/FER, Figure 3A). The ammonia oxidation then appeared fully selective in N$_2$ since neither NO$_x$ nor N$_2$O were recorded (detection limits close to 1 ppm for these N-products). To the opposite, 78% of NH$_3$ was converted over Cu$_{2.8}$/FER sample. These results are in accordance with SCR tests presented Figure 1 which illustrate that the competitive reactivity between NH$_3$ oxidation by O$_2$ and NO$_x$ reduction by NH$_3$ at high temperature is more significant over the Cu$_{6.1}$/FER catalyst. Both NH$_3$ oxidation activity and selectivity were impacted by Na poisoning. Concerning the Cu$_{2.8}$/FER catalyst, Na deposits in water (Cu$_{2.8}$(W)/FER) firstly enhanced the NH$_3$ conversion into N$_2$. Increasing the sodium content at 12,800 ppm led to NO formation which was also associated with a decrease in the NH$_3$ conversion, especially for 19,000 ppm Na. To the opposite, sodium impregnation in ethanol (Cu$_{2.8}$(Et)/FER) did not alter the oxidation of NH$_3$ by O$_2$. For Cu$_{6.1}$/FER sample, both Na impregnations in water or ethanol led to a deactivation of the catalyst, but in a less extent with ethanol, as illustrated in Figure 3A. Interestingly, at 450 °C, the deactivation of NH$_3$ oxidation properties was again associated with the formation of oxidized compounds, mainly NO (no N$_2$O was recorded). NO was emitted in a lower extent at 400 °C, but similar evolutions of the ammonia conversion were observed depending on the Na loading, as illustrated in Figure S5 (Supplementary Materials).

The NO oxidation rates recorded at 450 °C depending on the sodium loading are depicted Figure 3B. Whatever the test, the NO oxidation rate remained low, with a maximum of 30% recorded with sodium free Cu$_{6.1}$/FER. Indeed, as expected, the highest NO oxidation activities were observed for the highest loaded copper catalyst [18,19]. With the sodium free $Cu_{2.8}$/FER sample, the NO oxidation rate was limited at 7%. Addition of sodium by impregnation in water led first to an improvement of the NO oxidation rate, until 18% for 5400 ppm Na, but the rate then decreased down to 14% for approximately 2% Na. On the contrary, no influence of the Na content was observed when the alkali element was added by impregnation in ethanol. Results differed on $Cu_{6.1}$/FER catalyst: Na addition led to a similar decrease in the oxidation behavior whatever the impregnation solvent. Note that the same tendencies were observed at lower temperatures (data not shown). Finally, the sodium impregnation in water or ethanol resulted in similar trends than those observed for NH_3 oxidation. Especially, the redox function appeared not affected on the $Cu_{2.8}$/FER sample when Na was added in ethanol.

Concerning the general influence of Na addition in water, in copper-exchanged zeolite, various assumption can be advanced to explain why NH_3 and NO oxidation properties were enhanced over $Cu_{2.8}$/FER, whereas opposite results emerged over higher copper content material ($Cu_{6.1}$/FER). Indeed, the increase in NO oxidation behavior of copper-exchanged zeolite is commonly attributed to the formation of bulk copper (CuO), whether by thermal aging or copper loading increase [18,19].

In fact, when sodium was added by impregnation in water, the XRD patterns pointed out the appearance of CuO for $Cu_{2.8}$/FER while the intensity of the CuO peaks significantly increased for $Cu_{6.1}$/FER. On the contrary, no evolution of the XRD patterns was observed when sodium was added in ethanol. Consequently, the increase in oxidation properties after Na addition in water over $Cu_{2.8}$/FER could be attributed to the formation of CuO while the observed decrease in oxidation behavior for $Cu_{6.1}$/FER could be attributed to CuO sintering. A copper migration from exchange location to extra-framework, leading to CuO formation can be then suspected. After impregnation in ethanol, only copper sintering of extra-framework copper species can be postulated (after Na addition, samples were hydrothermally aged at 700 °C for 16 h).

In order to gain more information about of the sodium poisoning effects toward the copper state, especially depending on the Na solvent, samples were thereafter characterized by H_2-TPR and NO adsorption monitored by FT-IR.

2.3.2. Temperature Programmed Reduction by Hydrogen (H_2-TPR)

Generally, for copper-exchanged-based zeolites, two reductions peaks are expected. The first one is attributed to the reduction of Cu^{2+} species to Cu^{+} and the second one at higher temperature corresponds to the reduction of Cu^{+} to Cu^0 [20]. However, for some zeolites, such as Cu/Faujasite (FAU), two H_2 consumption peaks related to the reduction of Cu^{2+} to Cu^{+} can be detected, depending of the location of the copper in the zeolite [21]. In case of extra-framework copper, the direct reduction of CuO species into metallic copper can be observed in the 200–400 °C temperature range [20,22]. In addition, coper aluminate species ($CuAl_2O_4$) may be formed if traces of extra-framework aluminum (EFAL) remains in the zeolite samples, due for instance to dealumination caused by hydrothermal treatments. $CuAl_2O_4$ reduction by H_2 should then occurs around 500 °C [23]. Besides, it was reported that the reduction temperature of exchanged copper species are depending on the zeolite Si/Al ratio, with a decrease in the reduction peak temperature with the increase of the Si/Al ratio [24]. Moreover, the temperatures of copper reduction also decrease with the increase the sodium content as cocation [20,24].

Obviously, no reduction peak was observed for the host Ferrierite zeolite (results not shown). Figure 4 reports the TPR profiles of $Cu_{2.8}$/FER and $Cu_{6.1}$/FER TPR samples depending on the Na loading for Na addition performed in water. For all catalysts, the measured H_2/Cu ratio based on hydrogen consumption until 1000 °C was always very close to 1 (Table 1). It indicates a total Cu^{2+} to Cu^0 reduction, whatever the copper and sodium loadings. Consequently, there is neither evidence of copper stabilization in Cu^{I} state nor reduction of other species than Cu^{2+}, such as carbonates associated with Na addition for instance.

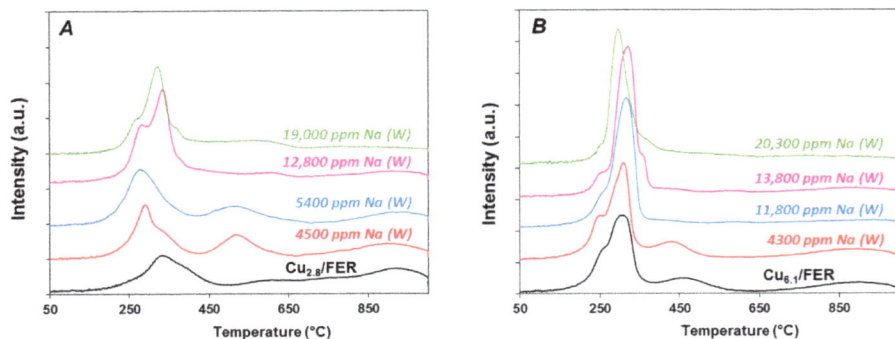

Figure 4. H_2-TPR profiles: Effect of Na loading impregnated in water (W) on (**A**) copper-exchanged zeolite $Cu_{2.8}$/FER and (**B**) on over-exchanged zeolite $Cu_{6.1}$/FER.

The $Cu_{2.8}$/FER reduction profile (Figure 4A) presents two main reduction peaks centered at 330 °C and 950 °C, and a broad reduction peak in the 500–800 °C temperature range. Varying copper loading and calcinations temperature (results not shown), the two main peaks at 330 °C and 950 °C are attributed to the two steps reduction process of exchanged Cu^{2+} via intermediate formation of Cu^+. The species reduced between 500 °C and 800 °C are assigned to $CuAl_2O_4$ (trace of dealumination leading to alumina was suggested by IR skeleton T-O-T characterization, Figure S2 in Supplementary Materials). However, different species of Cu^{2+} and Cu^+ may be also reduced in this temperature range [23,25].

After addition of 4500 ppm Na by impregnation in water ($Cu_{2.8}$/FER), the first peak corresponding to the reduction of exchanged Cu^{2+} to Cu^+ was shifted to lower temperature, at 290 °C. The reduction peak assigned to the second-step of Cu^+ reduction to Cu^0 is still observed at higher temperature (T = 950 °C), but the corresponding H_2 consumption was decreased compared with the sodium free sample. It became lower than for the reduction peak associated to the first reduction-step. Additionally, reduction peak centered at 520 °C was then well defined in the intermediate temperature range which corresponds to the reduction of $CuAl_2O_4$ species. Same evolutions were observed with the increase in the sodium contents, with a continuous decrease in the high temperature peak consumption (950 °C) in favor of the 250–450 °C region hydrogen consumption. The intermediate reduction peak was more or less well defined depending on the Na loading.

The $Cu_{6.1}$/FER sample exhibited TPR profiles with a first reduction-step with two components (T = 250 °C (shoulder) and 300 °C, Figure 4B). Extra-framework copper was expected for these sample since for 100% exchange (Cu/Al = 0.5), the copper loading should be around 4.5 wt. % copper. Then, the first reduction peak contained Cu^{2+} and CuO reduction. The second peak around 460 °C can be associate with copper aluminate and the last peak at 900 °C corresponds to the second step of exchanged copper reduction, from Cu^+ to Cu^0. Again, addition of sodium by impregnation in water led to the disappearance of the high temperature reduction peak while the H_2 consumption in the 200–300 °C temperature range increased.

Based on this results and literature data, H_2-TPR profiles were deconvoluted to assess to the copper distribution (Figure S6): the H_2 consumption related to the peak centered at around 950 °C (attributed to Cu^+ in exchanged position) was deduced to the integration of the broad low reduction peak(s). The remaining H_2 consumption was attributed to the one step bulk CuO reduction. In accordance with the XRD characterizations, H_2 consumptions attributed to CuO species increased with the sodium loadings. The intermediate temperature range reduction was attributed to copper aluminates reduction. However, this peak was not always well defined and some H_2 consumptions were then difficult to address and not attributed. Finally, the copper distribution deduced from these deconvolutions are reported in Figure 5.

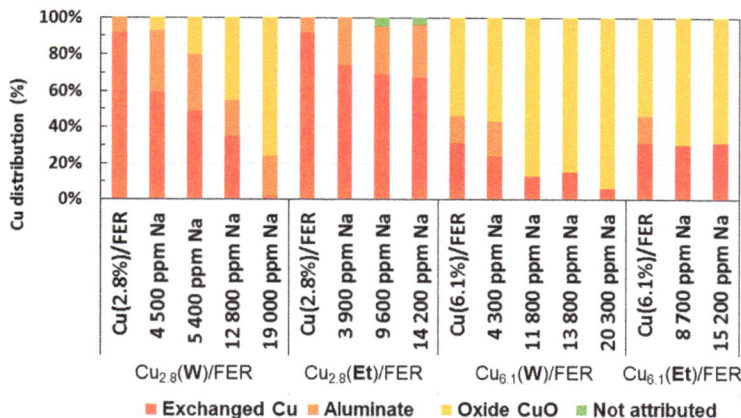

Figure 5. Copper distribution deduced from H_2-TPR experiments in function of Na impregnation solvent (water (W) or ethanol (Et)) over $Cu_{2.8}$/FER and $Cu_{6.1}$/FER catalysts. (■): exchanged Cu; (■): aluminate Cu; (■): oxide Cu; (■): no attributed.

For the sodium free $Cu_{2.8}$/FER sample, more than 90% of copper appeared to be in exchanged position. No CuO was evidenced, only few percent of copper aluminate were deduced from the deconvolution of the TPR profile. After addition of 4500 ppm Na (in water), only 60% of copper remained in exchange position, whereas approximately 33% corresponded to aluminate and 7% to CuO. Increasing the sodium loadings did not significantly affect the ratio of copper aluminate anymore, but it led to the disappearance on exchanged Cu^{2+} in favor of CuO.

Before sodium addition, the $Cu_{6.1}$/FER catalyst exhibited around 31% of exchange copper, 15% of aluminate and 54% of CuO. Again, it clearly appears that the amount of exchanged copper decreased with a remarkable continuous drop with Na loading (water as solvent), in favor of CuO formation. For instance, the amount of exchanged copper in $Cu_{6.1(W)}$/FER sample dropped to 24% with 4300 ppm Na and only 7% with 20,300 ppm of sodium.

These results are in accordance with the assumption previously postulated about the oxidation behavior of the corresponding catalysts (Section 2.3.1). After Na impregnation with water as solvent, the highest NO and NH_3 oxidation activities is correlated with the presence of copper oxide (CuO). During this impregnation procedure, sodium and copper probably compete to take place in the exchange positions. Consequently, part of the exchanged copper was moving out the zeolite framework, leading to the formation of extra-framework CuO and possibly $CuAl_2O_4$ due to the presence of traces of alumina.

When sodium amounts were added using ethanol as impregnation solvent, results significantly differ.

On $Cu_{2.8}$/FER sample (H_2-TPR profiles not shown), the Na addition in ethanol did not led to a shift the reduction temperatures of Cu^{2+} to Cu^+ and Cu^+ to Cu^0. However, a reductions peak appeared at 520 °C. It represents around 25% of hydrogen consumed for each sample containing sodium and can be attributed to $CuAl_2O_4$. Then, a quarter of the initially exchanged copper migrated out the framework and react with residual alumina. A weak peak was also observed near 740 °C (excepted for 3900 ppm of Na). The corresponding hydrogen consumption was only 2% of the total hydrogen consumption and its attribution is not clear. It is reported as "not attributed" in Figure 5. Interestingly, the hydrogen consumption of the first peak ($Cu^{2+} \rightarrow Cu^+$) was very close to the hydrogen consumption of the highest temperature reduction peak (Cu$^+$ → Cu^0), indicating that there is no clear evidence of CuO formation, as illustrated in Figure 5.

Same trends were observed for Cu$_{6.1}$/FER catalyst. Figure 5 show that sodium impregnation in ethanol led only to small changes in copper distribution. Especially, no supplementary CuO formation was observed.

Finally, it appears that sodium impregnation in ethanol led to significantly lower interaction with exchanged copper than using water as sodium salt solvent. Using water, the amount of exchanged copper continuously decreased with the Na loading, leading to CuO formation. On the opposite, the sodium impregnation in ethanol did not significantly affect the copper distribution

In the next section, the copper state was also examined by NO adsorption monitored by FT-IR spectroscopy.

2.3.3. NO Adsorption Monitored by FT-IR Spectroscopy

Copper species in Cu-exchanged zeolite were largely studied in the literature by FT-IR adsorption experiments of probe molecules [26]. NO molecule is commonly used to investigate the oxidation state of copper cations in zeolites, due to its ability to form stable nitrosyl adducts with both Cu^{2+} and Cu$^+$ cations. Several types of Cu species can co-exist, as solvated (Cu^{2+}–(OH)$^-$)$^+$ ions [27] or Cu^{2+} monomers or (Cu^{2+}–O^{2-}–Cu^{2+})$^{2+}$ dimmers [28].

IR spectra of adsorbed NO on the Na-free catalysts containing 2.8% and 6.1% Cu are reported in Figure 6A. The interaction between NO and Cu/FER resulted in the formation of two nitrosyl complexes [Cu^{2+}–NO and Cu$^+$–NO]. The main IR bands at 1905–1910 cm^{-1} are characteristic of NO adsorbed on Cu^{2+} ions. NO adsorbed on Cu$^+$ ions led to nitrosyl species at around 1810 cm^{-1} [29]. The presence of Cu$^+$ sites may result from the high vacuum applied for the FT-IR experiments. However, similar copper species distribution, namely Cu^{2+} and Cu$^+$ sites, were observed over both samples with 2.8% and 6.1% copper. Dinitrosyl species were not observed.

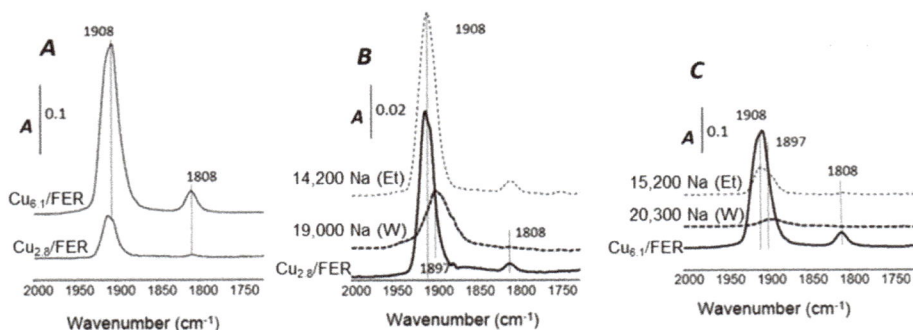

Figure 6. FT-IR spectra of NO adsorption. (**A**) comparison of Na-free copper-exchanged zeolite; (**B**) effect of Na deposit depending on the solvent over Cu$_{2.8}$/FER; (**C**) effect of Na deposit depending on the solvent over Cu$_{6.1}$/FER.

The effect of Na impregnation solvent over Cu species of Cu$_{2.8}$/FER and Cu$_{6.1}$/FER is presented in Figure 6B,C, respectively. Note that only the highest soda-content materials were examined.

For the Cu$_{2.8}$/FER catalyst (Figure 6B), similar IR bands were observed after impregnation of Na in ethanol. Notably, characteristic IR bands of NO adsorbed on Cu^{2+} ions were still observed, in similar intensity. These results are in good agreement with H$_2$-TPR experiments which revealed that Na impregnation in ethanol did not significantly impacted the copper distribution. Interestingly, Na addition in water led to opposite results since Cu^{2+} species were largely affected. More precisely, the band at 1908 cm^{-1} was shifted to lower wavenumbers, at 1897 cm^{-1}. The band at 1908 cm^{-1}

is assigned to the interaction between NO and isolated Cu^{2+} ions having a square pyramidal configuration [30] whereas the component at lower wavenumbers, at around 1897 cm^{-1}, is attributed to Cu^{2+} sites with a square planar configuration and adjacent to a single framework aluminum atom. It only appeared after sodium addition in water which highlights interactions between Na and exchanged copper using this solvent. The band at 1808 cm^{-1}, assigned to nitrosyl species adsorbed on Cu^+ ion, was not observed.

Over $Cu_{6.1}$/FER samples (Figure 6C), similar tendencies were denoted: the impregnation of Na in water affected copper species more than Na addition in ethanol. Again, the band at 1897 cm^{-1} was observed denoting an alteration of copper configuration.

Finally, NO adsorption monitored by FT-IR spectroscopy confirmed that Na deposit by water pathway impacted more severely the copper species than with the ethanol route. These results globally converge with H_2-TPR experiments. However, NH_3-SCR activity also depends on NH_3 adsorption, mainly linked with acidic properties. This point is investigated in the next section.

2.4. Effect of the Na Additions on the Surface OH Groups and NH_3 Adsorption Capacity

The influence on the Na deposits on surface OH groups and ammonia adsorption ability were studied by infrared spectroscopy.

2.4.1. OH Groups

Figure 7A reports the IR spectra in OH stretching vibration region of $Cu_{2.8}$/FER and $Cu_{6.1}$/FER samples before sodium addition and for the highest sodium contents depending on the Na impregnation solvent. After activation (450 °C), IR spectra recorded at RT were characterized by IR band at 3747 cm^{-1} assigned to a single terminal Si–OH groups. The component at 3724 cm^{-1} is assigned to the O–H stretchings of geminal silanols and/or weakly perturbed vicinal pairs of terminal silanols [31]. One tailed band for bridged Al–(OH)–Si hydroxyls at 3602 cm^{-1} was also denoted. This is characteristic of the Ferrerite zeolite Brønsted acid sites, associated with the 8 and 10-membered ring channels. The band at 3642 cm^{-1} is assigned to Al–OH extra-framework groups. Thermal treatments procedures can partially remove aluminum from the crystal framework. It is assumed that the aluminum removed from the framework (EFAL) induces Lewis acidity and/or basic properties [32,33].

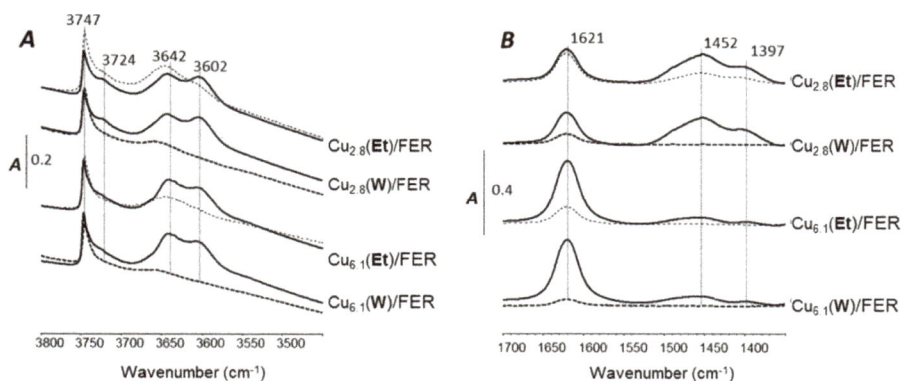

Figure 7. IR spectra for sodium free zeolites (full line) and for the highest Na contents depending on the Na solvent (water, bold dotted line; ethanol, thin dotted line). (**A**) OH stretching vibration recorded at RT after activation and evacuation at 450 °C; (**B**) IR spectra of adsorbed NH_3 after evacuation at 150 °C.

For both catalysts, impregnation of around 2% Na in water (bold dotted line) led to a significant decrease in the OH bands intensity (3602 and 3642 cm^{-1}), especially for bridged Al–(OH)–Si hydroxyls. Therefore, it could be supposed that a large part of the H$^+$ ions in the zeolite framework was exchanged by Na$^+$ ions during the impregnation. Additionally, the intensity of the band at 3602 cm^{-1} preferentially decreased with the Na loading (spectra not shown). To the opposite, Na impregnation in ethanol (thin dotted line) affected in a lesser extent the intensity of the band at 3602 cm^{-1} attributed to acidic Brønsted sites in the zeolite lattice. Consequently, it clearly appears that Na deposit in water poisoned more strongly the Brønsted acidity of the zeolite, whatever the Cu content (Cu$_{2.8}$/FER or Cu$_{6.1}$/FER samples), than for samples poisoned with ethanol as solvent for Na impregnation. Consequently Na deposited in ethanol is suspected to interact mainly with extra-framework aluminum.

Note that no supplementary bands were observed after Na impregnation, to the opposite of Fritz et al. who observed a new band at 3695 cm^{-1}, attributed to Na$^+$ interaction with water hydrogen-bonded to a framework oxygen atom [34].

2.4.2. NH$_3$ Adsorption

NH$_3$ adsorption was monitored by FT-IR. In the 3500–3000 cm^{-1} spectral range (spectra not shown), IR spectra were distinguished by bands at 3356, 3273, 3220 and 3189 cm^{-1} assigned to N–H stretching vibration modes [1,35–37]. More precisely, the band at 3189 cm^{-1} is assigned to Cu$^+$–NH$_3$ [38,39]. The spectral region of N–H stretching vibrations was relatively complex, because the distortion of the NH$_3$ molecule upon coordination led to a splitting of the $\nu_{N-H,asym}$ mode and to the activation of the $\nu_{N-H,sym}$ mode. In fact, the symmetric N–H vibration (ν_{sym}) of NH$_4^+$ is inactive toward IR vibration modes. Below 1800 cm^{-1} (Figure 7B), resolved N–H bending modes of ammonia adsorbed on Brønsted and Lewis sites appeared. This spectral range is dominated by absorption bands at 1452 and 1397 cm^{-1} (δ_{asym}), which are usually assigned to ammonium ions formed by the protonation of ammonia molecules located at Brønsted acid sites. These two components are tentatively assigned to different types of hydrogen-bonded ammonium, including monodentate and bidentate complexes for instance. The sharp band at 1621 cm^{-1} is assigned to bending modes of ammonia coordinatively bound to Lewis acid sites (δ_{asym}). The nature of these sites could be either tricoordinated Al atoms or Al species attached to defective sites, related to framework and extra-framework positions and also copper Lewis sites such as Cu^{2+} [1].

The bands at 1452–1397 cm^{-1} and 1621 cm^{-1} were used to determine the Brønsted and Lewis ammonia adsorption sites, based on the molar extinction coefficient values ($\varepsilon_{Brønsted}$ = 4.6 cm·µmol^{-1}, ε_{Lewis} = 1.1 cm·µmol^{-1} both determined at the laboratory by successive addition of controlled amount of NH$_3$). The calculated amounts depending on the catalyst and the sodium loading and impregnation procedure are reported in Figure 8 (Lewis to Brønsted acid sites ratio is provided in Table 1). This graph shows a decrease in the adsorption of ammonia with increasing the Na content whatever the considered catalyst and the Na introduction mode. Considering the linear trend, one Na atom led to the disappearance of about two sites of ammonia adsorption (including Lewis and Brønsted), as previously mentioned for other materials [40]. However, Na addition led to some small different behavior in function of copper loading. For Cu$_{2.8}$/FER catalyst, the total amount of acid sites was globally not affected until 300–400 µmol Na/g$_{catalyst}$ (7000–9000 ppm Na). To the opposite, adsorbed ammonia of the catalyst containing over-exchanged copper (Cu$_{6.1}$/FER sample) is globally linearly poisoned by Na. Interestingly, the discrimination of the Brønsted and Lewis acid sites concentration also indicates that impregnation in water induced the total disappearance of the Brønsted sites for approximately 12,000 ppm Na, while the Lewis/Brønsted ratio appeared not affected when Na was added using ethanol (Table 1).

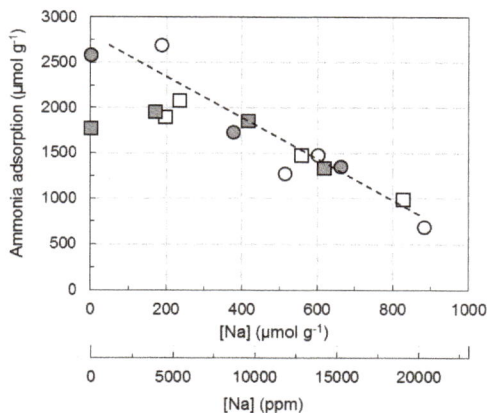

Figure 8. Relationship between the ammonia adsorption amount (determined by NH_3 adsorption evacuated at 50 °C) and the Na loading. □: $Cu_{2.8}(W)/FER$; ■: $Cu_{2.8}(Et)/FER$; ○: $Cu_{6.1}(W)/FER$; ●: $Cu_{6.1}(Et)/FER$.

Finally, FT-IR characterizations indicate a decrease in the ammonia adsorption capacity with increasing the Na content whatever the considered Na solvent. However, addition of around 2% Na in water resulted in a significant decrease in the OH bands intensity, whereas Na impregnation in ethanol affected in a lesser extent acidic Brønsted sites in the zeolite lattice. Water appeared to favor an ion exchange between the H^+ species and Na^+ ions during the impregnation step, as suspected between Na^+ and Cu^{2+} from TPR and XRD experiments.

3. Discussion

3.1. Ammonia Adsorption and Copper State

The previous sections showed that sodium additions impacted both the copper state and the acidity/ammonia adsorption behaviors.

The influence of Na addition on the ammonia adsorption capacity is rather similar depending on the impregnation procedure, and differed a little in regard on the copper loading (Figure 8).

On the opposite, significant differences were obtained concerning the copper distribution depending on the Na loading and impregnation solvent. Water solvent appeared to favor CuO formation at the expense of exchange Cu^{2+} (Figure 5).

The amount of exchanged copper determined from H_2-TPR results is reported in function of ammonia adsorption behavior in Figure 9. This figure reveals that roughly, lower the number of ammonia adsorption sites on Na poisoned sample by impregnation in water, lower the copper in exchange position for both $Cu_{2.8}/FER$ (square white symbol) and $Cu_{6.1}/FER$ samples (circle white symbol). No clear trend is denoted for Na impregnation in ethanol (grey symbol) because the amount of copper in exchanged position varied in a significantly lower extent. As a result, it is assumed that ammonia adsorption sites poisoned by Na deposits by ethanol route did not directly modify the nature of copper sites, whatever the materials ($Cu_{2.8}/FER$ or $Cu_{6.1}/FER$). Additionally, the catalyst containing lower copper content ($Cu_{2.8}/FER$) is more sensitive to Na poisoning than $Cu_{6.1}/FER$ sample which already exhibited large ratio of CuO before Na addition.

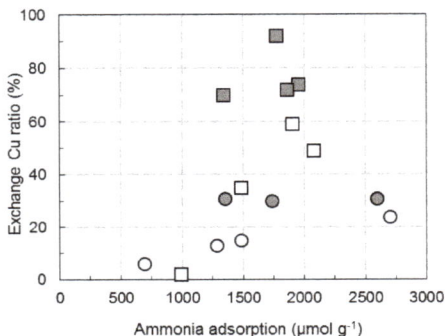

Figure 9. Relationship between the ratio of copper in exchanged position and the amount of adsorbed ammonia. □: $Cu_{2.8}(W)/FER$; ■: $Cu_{2.8}(Et)/FER$; ○: $Cu_{6.1}(W)/FER$; ●: $Cu_{6.1}(Et)/FER$.

3.2. Effect of Na Deposit over NH₃-SCR Behaviour

As presented in the introduction section, NH_3-SCR activity generally depends on both redox and acidity behaviors. Results presented in this study showed that both are potentially affected by sodium addition. In order to present a general overview of the influence of the Na addition, the $deNO_x$ activity recorded in Standard and Fast-SCR at selected temperatures (250 °C and 500 °C) for both $Cu_{2.8}/FER$ and $Cu_{6.1}/FER$ are plotted in function of (i) the ratio of exchanged Cu (assessed from H_2-TPR experiments) (Figure 10A,C) or (ii) the amount of adsorbed NH_3 determined by FT-IR analysis (Figure 10B,D).

At low temperature (250 °C, Figure 10A,B), the $deNO_x$ activity appeared to be sensitive to both copper state and amount of ammonia adsorption sites. More specifically, NO_x conversion increased with the ratio of copper in exchanged position until approximately 50% of exchanged copper, and became relatively constant for higher values for both Standard and Fast-SCR condition. In parallel, a linear trend was obtained between the number of ammonia adsorption sites and the $deNO_x$ efficiency in the whole studied domain, whatever the considered $Cu_{2.8}/FER$ or $Cu_{6.1}/FER$ samples. Additionally, as expected, Fast-SCR condition allowed higher NO_x conversion than Standard-SCR condition, but the same trends were observed for both reaction mixtures. Thus, at low temperature, it appeared that the number of ammonia adsorption sites has potentially more influence than the copper state (redox behavior). These results are consistent with the NO and NH_3 oxidation properties previously discussed in Section 2.3.1 who exhibited very low activities at 250 °C.

$DeNO_x$ activity at high temperature (500 °C) are reported in Figure 10C,D depending on the copper state and the number of ammonia adsorption sites, respectively. Obviously, obtained trends at 500 °C differ compared to those observed at low temperature. The $deNO_x$ efficiency was then strongly dependent of the ratio of exchanged copper, whatever the considered catalyst (Figure 10C). Indeed, large CuO particles (mainly on $Cu_{6.1}/FER$ catalyst) favors the ammonia oxidation at high temperature, then inhibiting the NO_x reduction.

Concerning the role of the ammonia adsorption ability, trends are not clearly evidenced. Plots are largely dispersed for $Cu_{2.8}/FER$. For $Cu_{6.1}/FER$ sample, a trend seems to appear with a decrease in the $deNO_x$ activity with the decrease in the number of ammonia adsorption sites, but to a limited extend. Such results could be expected since the ammonia interaction with adsorption sites (mainly acidic sites) should be very weak at this temperature. As a consequence, the activity is mostly driven by copper state at 500 °C.

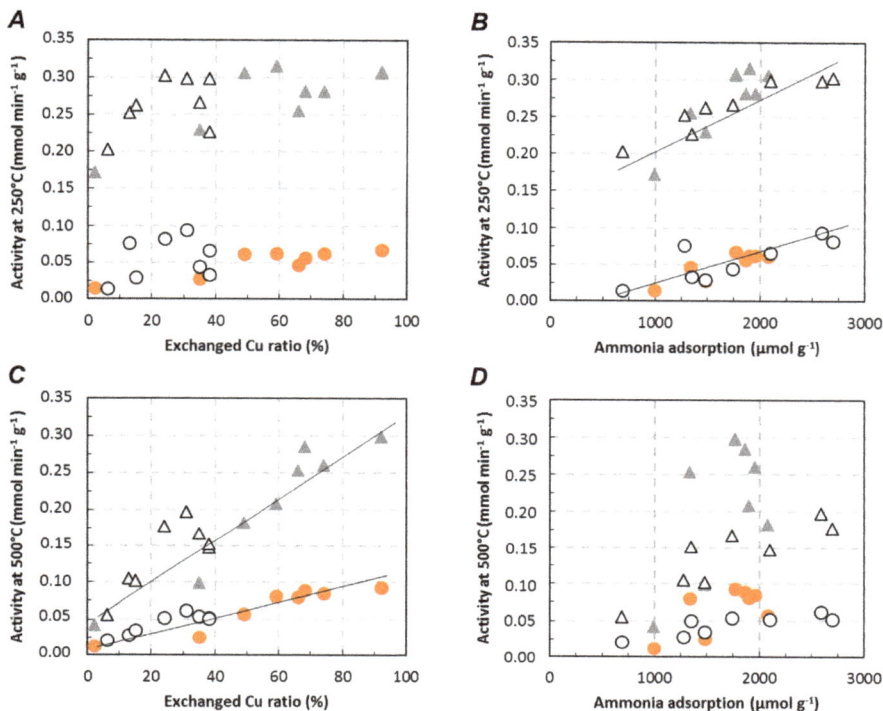

Figure 10. Effect of copper content of Cu$_{2.8}$/FER ($\bullet$, $\blacktriangle$) and Cu$_{6.1}$/FER (O, $\triangle$) catalysts in the relationship between deNOx activity in Standard-SCR (O) and Fast-SCR ($\triangle$) and exchanged Cu ((**A**) 250 °C; (**C**) 500 °C) or ammonia adsorption ((**B**) 250 °C, (**D**) 500 °C).

4. Materials and Methods

4.1. Catalyst Preparation

A ferrierite structure zeolite (SiO$_2$/Al$_2$O$_3$ = 20, Alfa Aesar, Haverhill, MA, USA), which exhibits a channel system of 10 and 8 member rings, was selected to ensure a high hydrothermal stability. Two copper loadings were prepared (2.8 wt. % and 6.1 wt. %, measured by ICP elemental analysis after preparation of the samples) in order to obtain samples with or without extra-framework copper. Copper was added using the ion-exchange method. The ion exchanges were realized with 10 g of zeolite in suspension in Cu(CH$_3$COO)$_2$ (Sigma-Aldrich, L'Isle d'Abeau, France) diluted in ultra-pure water. The pH was adjusted at 5 with HNO$_3$ to ensure the ionic copper state, based on the diagram of copper speciation from Ajmal et al. [41]. The solution was stirred under reflux at 80 °C for 23 h. Solution was then filtered and thoroughly washed with ultra-pure water and dried at 120 °C for 15 h. The resulting powder was firstly treated at 600 °C under 10% O$_2$-90% N$_2$ for 30 min (heating rate: 2 °C min^{-1}). Vapor water (10%) was then added and the catalyst was maintained at 600 °C for 16 h before cooling to room temperature. The corresponding stabilized samples are denoted Cu$_{2.8}$/FER and Cu$_{6.1}$/FER depending on the copper content.

4.2. Sodium Addition

Sodium addition was made by impregnation, using water or ethanol as solvent. 1.5 g of the catalyst was suspended in 4 mL of water or ethanol with the desire amount of sodium (NaNO$_3$, from

Fisher Scientific Labosi, Elancourt, France). The solution was stirred at room temperature during 4 h. After solvent evaporation, the sample was dried one night at 80 °C. Finally, the resulting powder, as the catalyst without sodium, was submitted again to the thermal treatment: 30 min at 600 °C under 10% O_2-90% N_2 and 16 h at 600 °C under 10% O_2, 10% H_2O and N_2. The effective Na contents were measured by ICP, and the amount of sodium deposit was up to 20,300 ppm.

The resulting impregnated catalysts were denoted (W) or (Et) depending on the sodium impregnation solvent (water or ethanol).

4.3. Characterization Techniques and Catalytic Tests

4.3.1. Structural/Textural Characterizations, Elemental Analysis

Elemental analysis was carried out to assess the copper and sodium loadings. After mineralization of the samples, analysis were performed with an ICP-OES apparatus (Perkin Elmer Optima 2000 DV, Waltham, MA, USA).

Specific surface areas were determined by BET method (Micromeritics ASAP 2000, Norcross, GA, USA) using nitrogen adsorption at 77 K. The total pore volume was evaluated at $P/P^0 = 0.97$, and the microporous volume was calculated with t-plot. Prior to the N_2 physisorption, the samples were degassed under vacuum 1 h at 90 °C and 4 h at 350 °C.

The powder XRD patterns were collected using an Empyrean (PANalytical, Almelo, The Netherlands) diffractometer. Data acquisition was recorded from 5° to 70° (2θ) with a scanning step of 0.017. Crystalline phases were identified by comparison with ICDD database files.

4.3.2. Catalytic Activity

The SCR tests were performed in a tubular quartz reactor. For the Standard-SCR catalytic tests, 50 mg of zeolite (sieved in the 0.2–0.4 mm range) diluted in 50 mg of SiC (Prolabo, 0.35 mm) were used. The reactant gas composition was 500 ppm NO, 500 ppm NH_3, 10% O_2, 10% CO_2, 9% H_2O balanced in N_2. The total flow rate was fixed at 250 mL min^{-1}. For the Fast-SCR catalytic tests, 500 ppm NO_x were made of 250 ppm NO and 250 ppm NO_2, and only 15 mg of zeolite diluted in 15 mg of SiC were used in order to obtain conversions allowing the discrimination of catalysts. Gaseous NO, NO_2, NH_3, CO_2, O_2 and N_2 flows were adjusted by mass-flow controllers, while H_2O was added via a saturator. The composition of the feed gas and effluent stream were analyzed online with a MKS 2030 Multigas infrared analyzer.

4.3.3. Redox Properties

Temperature Programmed Reduction (TPR) with Hydrogen

Temperature programmed reduction experiments were performed on a Micromeritics Autochem 2920. Sample (100 mg) was placed in a U-shaped quartz reactor. Prior the reduction, the catalyst was in situ calcined at 450 °C for 30 min under O_2 (heating rate: 5 °C·min^{-1}). After cooling to room temperature and purge under argon flow for 45 min, the reduction was carried out under 1% H_2/Ar up to 1000 °C (rate: 10 °C·min^{-1}). The reduction was followed by a thermal conductivity detector (TCD). Because this TCD is sensitive to water, a H_2O trap was added downstream the reactor, allowing the quantification of the H_2 consumption.

NO and NH_3 Oxidation

The NO and NH_3 oxidation tests were performed in similar condition than the Standard-SCR tests. 50 mg of catalyst (sieved in the 0.2–0.4 mm range), diluted in 50 mg of SiC (Prolabo, 0.35 mm) were used. The gas compositions were 500 ppm NO or 500 ppm NH_3, 10% O_2, 10% CO_2, 9% H_2O balanced with N_2, with a total flow rate fixed at 250 mL·min^{-1}.

4.3.4. Adsorption of Probe Molecules Monitored by FT-IR

Adsorption of probe molecules monitored by infrared spectroscopy was used to characterize (i) ammonia adsorption behavior (the usual pyridine sensor to characterize acidic sites is not appropriate since it is not able to diffuse into the small porosity of the considered materials [31,42]) and (ii) copper state of exchanged zeolites by NO adsorption. IR spectra were collected using a Nexus Nicolet spectrometer equipped with a DTGS detector (deuterium triglyceride sulfur) and a KBr beam splitter. Spectra were normalized to a disc of 10 mg·cm^{-2}. Before adsorption, the catalyst was in situ pre-treated under vacuum at 450 °C. Ammonia was then adsorbed at 25 °C and desorption was performed up to 450 °C. NO adsorption was performed at room temperature, IR spectra are compared with the same n_{NO}/n_{Cu} ($\approx$30%) by adding a known amount of NO on the cell.

5. Conclusions

The impact of sodium, as alkali poison from biodiesel, was investigated on two model NH$_3$-SCR catalysts (Cu$_{2.8\%}$/FER and Cu$_{6.1\%}$/FER), depending on the Na introduction method (impregnation in water or in ethanol). Impregnation in ethanol led mainly to a decrease in ammonia adsorption capacity, which affected the deNO$_x$ efficiency at low temperature whereas NO and NH$_3$ oxidation behaviors were not really affected. Impregnation of Na in water (representative of interaction of catalyst with condensed water, which occurs during engine stop and cold start) furthermore caused a migration of copper species from the exchange sites to extra-framework CuO. Consequently, NH$_3$ and NO oxidation behaviors were decreased. Finally, whatever the catalyst and the reaction mixture (Fast or Standard-SCR), the NO$_x$ conversion decrease with Na addition appeared linked to the decrease in ammonia adsorption at low temperature (250 °C), whereas the activity at high temperature (500 °C) mainly depended on the ratio of exchanged copper.

Supplementary Materials: The following are available online at www.mdpi.com/2073-4344/8/1/3/s1, Figure S1: XRD patterns of fresh and hydrotreated zeolite for 16 h at 600 °C and 750 °C. Figure S2: T–O–T bands obtained with zeolite before and after hydrotreatment at 600 °C and 750 °C during 16 h (samples diluted in KBr). Figure S3: XRD patterns of fresh FER, sodium free Cu$_{2.8}$/FER and Cu$_{2.8}$(w)/FER with 19,000 ppm Na impregnated in water. Figure S4: XRD patterns of fresh FER, sodium free Cu$_{6.1}$/FER and Cu$_{6.1}$(w)/FER with 20,300 ppm Na impregnated in water. Figure S5: Effect of Na loading on NH$_3$ oxidation at 400 °C over Cu$_{2.8}$/FER and Cu$_{6.1}$/FER samples depending on the Na loading impregnated in water (W) or ethanol (Et). Figure S6: H$_2$-TPR profile deconvolution over Cu$_{2.8}$/FER with 19,000 ppm of Na content impregnated in water.

Acknowledgments: The authors gratefully acknowledge the French National Agency for Research (ANR) for its financial support (Appibio Project, Reference ANR-14-CE22-0003).

Author Contributions: Xavier Courtois, Fabien Can, Vincent Lauga and Eduard Emil Iojoiu conceived and designed the experiments, Marie-Laure Tarot and Mathias Barreau performed the experiments; Marie-Laure Tarot analyzed the data, Marie-Laure Tarot, Xavier Courtois and Fabien Can wrote the paper, Daniel Duprez, Vincent Lauga and Eduard Emil Iojoiu also contributed to scientific discussions.

Conflicts of Interest: The authors declare no conflict of interest.

References

1. Wang, D.; Zhang, L.; Li, J.; Kamasamudram, K.; Epling, W.S. NH$_3$-SCR over Cu/SAPO-34—Zeolite acidity and Cu structure changes as a function of Cu loading. *Catal. Today* **2014**, *231*, 64–74. [CrossRef]
2. Ma, L.; Cheng, Y.; Cavataio, G.; McCabe, R.W.; Fu, L.; Li, J. Characterization of commercial Cu-SSZ-13 and Cu-SAPO-34 catalysts with hydrothermal treatment for NH$_3$-SCR of NO$_x$ in diesel exhaust. *Chem. Eng. J.* **2013**, *225*, 323–330. [CrossRef]
3. Kwak, J.H.; Tonkyn, R.G.; Kim, D.H.; Szanyi, J.; Peden, C.H.F. Excellent activity and selectivity of Cu-SSZ-13 in the selective catalytic reduction of NO$_x$ with NH$_3$. *J. Catal.* **2010**, *275*, 187–190. [CrossRef]
4. Kim, Y.J.; Lee, J.K.; Min, K.M.; Hong, S.B.; Nam, I.-S.; Cho, B.K. Hydrothermal stability of CuSSZ13 for reducing NO$_x$ by NH$_3$. *J. Catal.* **2014**, *311*, 447–457. [CrossRef]
5. Brandenberger, S.; Kröcher, O.; Tissler, A.; Althoff, R. The state of the art in selective catalytic reduction of NO$_x$ by ammonia using metal-exchanged zeolite catalysts. *Catal. Rev.* **2008**, *50*, 492–531. [CrossRef]

6. Guan, B.; Zhan, R.; Lin, H.; Huang, Z. Review of state of the art technologies of selective catalytic reduction of NO$_x$ from diesel engine exhaust. *Appl. Therm. Eng.* **2014**, *66*, 395–414. [CrossRef]

7. Nicosia, D.; Czekaj, I.; Kröcher, O. Chemical deactivation of V$_2$O$_5$/WO$_3$–TiO$_2$ SCR catalysts by additives and impurities from fuels, lubrication oils and urea solution: Part II. Characterization study of the effect of alkali and alkaline earth metals. *Appl. Catal. B* **2008**, *77*, 228–236. [CrossRef]

8. Klimczak, M.; Kern, P.; Heinzelmann, T.; Lucas, M.; Claus, P. High-throughput study of the effects of inorganic additives and poisons on NH$_3$-SCR catalysts—Part I: V$_2$O$_5$–WO$_3$/TiO$_2$ catalysts. *Appl. Catal. B* **2010**, *95*, 39–47. [CrossRef]

9. Chen, L.; Li, J.; Ge, M. The poisoning effect of alkali metals doping over nano V$_2$O$_5$–WO$_3$/TiO$_2$ catalysts on selective catalytic reduction of NO$_x$ by NH$_3$. *Chem. Eng. J.* **2011**, *170*, 531–537. [CrossRef]

10. Kern, P.; Klimczak, M.; Heinzelmann, T.; Lucas, M.; Claus, P. High-throughput study of the effects of inorganic additives and poisons on NH$_3$-SCR catalysts. Part II: Fe–zeolite catalysts. *Appl. Catal. B* **2010**, *95*, 48–56. [CrossRef]

11. Shwan, S.; Jansson, J.; Olsson, L.; Skoglundh, M. Chemical deactivation of H-BEA and Fe-BEA as NH$_3$-SCR catalysts—Effect of potassium. *Appl. Catal. B* **2015**, *166–167*, 277–286. [CrossRef]

12. Ma, J.; Si, Z.; Weng, D.; Wu, X.; Ma, Y. Potassium poisoning on Cu-SAPO-34 catalyst for selective catalytic reduction of NO$_x$ with ammonia. *Chem. Eng. J.* **2015**, *267*, 191–200. [CrossRef]

13. Brookshear, D.W.; Nguyen, K.; Toops, T.J.; Bunting, B.G.; Rohr, W.F. Impact of Biodiesel-Based Na on the Selective Catalytic Reduction of NO$_x$ by NH$_3$ over Cu–Zeolite Catalysts. *Top. Catal.* **2013**, *56*, 62–67. [CrossRef]

14. Brookshear, D.W.; Nguyen, K.; Toops, T.J.; Bunting, B.G.; Rohr, W.F.; Howe, J. Investigation of the effects of biodiesel-based Na on emissions control components. *Catal. Today* **2012**, *184*, 205–218. [CrossRef]

15. Rahkamaa-Tolonen, K.; Maunula, T.; Lomma, M.; Huuhtanen, M.; Keiski, R.L. The effect of NO$_2$ on the activity of fresh and aged zeolite catalysts in the NH$_3$-SCR reaction. *Catal. Today* **2005**, *100*, 217–222. [CrossRef]

16. Hamon, C.; Blanchard, G. Device for Treating Exhaust Gases. European Patent EP 2857084 A1, 8 April 2015.

17. Lezcano-Gonzalez, I.; Deka, U.; van der Bij, H.E.; Paalanen, P.; Arstad, B.; Weckhuysen, B.M.; Beale, A.M. Chemical deactivation of Cu-SSZ-13 ammonia selective catalytic reduction (NH$_3$-SCR) systems. *Appl. Catal. B* **2014**, *154–155*, 339–349. [CrossRef]

18. Sultana, A.; Nanba, T.; Haneda, M.; Sasaki, M.; Hamada, H. Influence of co-cations on the formation of Cu$^+$ species in Cu/ZSM-5 and its effect on selective catalytic reduction of NO$_x$ with NH$_3$. *Appl. Catal. B* **2010**, *101*, 61–67. [CrossRef]

19. Mihai, O.; Widyastuti, C.R.; Andonova, S.; Kamasamudram, K.; Li, J.; Joshi, S.Y.; Currier, N.W.; Yezerets, A.; Olsson, L. The effect of Cu-loading on different reactions involved in NH$_3$-SCR over Cu-BEA catalysts. *J. Catal.* **2014**, *311*, 170–181. [CrossRef]

20. Bulánek, R.; Wichterlová, B.; Sobalík, Z.; Tichý, J. Reducibility and oxidation activity of Cu ions in zeolites: Effect of Cu ion coordination and zeolite framework composition. *Appl. Catal. B* **2001**, *31*, 13–25. [CrossRef]

21. Delahay, G.; Valade, D.; Guzmán-Vargas, A.; Coq, B. Selective catalytic reduction of nitric oxide with ammonia on Fe-ZSM-5 catalysts prepared by different methods. *Appl. Catal. B* **2005**, *55*, 149–155. [CrossRef]

22. Delahay, G.; Coq, B.; Broussous, L. Selective catalytic reduction of nitrogen monoxide by decane on copper-exchanged beta zeolites. *Appl. Catal. B* **1997**, *12*, 49–59. [CrossRef]

23. Dumas, J.M.; Geron, C.; Kribii, A.; Barbier, J. Preparation of supported copper catalysts. *Appl. Catal. B* **1989**, *47*, L9–L15. [CrossRef]

24. Torre-Abreu, C.; Ribeiro, M.F.; Henriques, C.; Delahay, G. NO TPD and H$_2$-TPR studies for characterisation of CuMOR catalysts the role of Si/Al ratio, copper content and cocation. *Appl. Catal. B* **1997**, *14*, 261–272. [CrossRef]

25. Severino, F.; Brito, J.L.; Laine, J.; Fierro, J.L.G.; Agudo, A.L. Nature of copper active sites in the carbon monoxide oxidation on CuAl$_2$O$_4$ and cucr$_2$o$_4$ spinel type catalysts. *J. Catal.* **1998**, *177*, 82–95. [CrossRef]

26. Deka, U.; Lezcano-Gonzalez, I.; Weckhuysen, B.M.; Beale, A.M. Local environment and nature of Cu active sites in zeolite-based catalysts for the selective catalytic reduction of NO$_x$. *ACS Catal.* **2013**, *3*, 413–427. [CrossRef]

27. Dědeček, J.; Wichterlová, B. Role of hydrated Cu ion complexes and aluminum distribution in the framework on the Cu ion siting in ZSM-5. *J. Phys. Chem. B* **1997**, *101*, 10233–10240. [CrossRef]

28. Costa, P.D.; Modén, B.; Meitzner, G.D.; Lee, D.K.; Iglesia, E. Spectroscopic and chemical characterization of active and inactive Cu species in NO decomposition catalysts based on Cu-ZSM5. *Phys. Chem. Chem. Phys.* **2002**, *4*, 4590–4601. [CrossRef]

29. Henriques, C.; Ribeiro, M.F.; Abreu, C.; Murphy, D.M.; Poignant, F.; Saussey, J.; Lavalley, J.C. An FT-IR study of NO adsorption over Cu-exchanged MFI catalysts: Effect of Si/Al ratio, copper loading and catalyst pre-treatment. *Appl. Catal. B* **1998**, *16*, 79–95. [CrossRef]

30. Hadjiivanov, K.I. Identification of neutral and charged $N_x O_y$ surface species by IR spectroscopy. *Catal. Rev.* **2000**, *42*, 71–144. [CrossRef]

31. Trombetta, M.; Busca, G.; Rossini, S.; Piccoli, V.; Cornaro, U.; Guercio, A.; Catani, R.; Willey, R.J. FT-IR studies on light olefin skeletal isomerization catalysis: III surface acidity and activity of amorphous and crystalline catalysts belonging to the $SiO_2–Al_2O_3$ system. *J. Catal.* **1998**, *179*, 581–596. [CrossRef]

32. Chakarova, K.; Hadjiivanov, K. FTIR study of N_2 and CO adsorption on H-D-FER. *Microporous Mesoporous Mater.* **2013**, *177*, 59–65. [CrossRef]

33. De Ménorval, B.; Ayrault, P.; Gnep, N.S.; Guisnet, M. Mechanism of n-butene skeletal isomerization over HFER zeolites: A new proposal. *J. Catal.* **2005**, *230*, 38–51. [CrossRef]

34. Fritz, P.O.; Lunsford, J.H. The effect of sodium poisoning on dealuminated Y-type zeolites. *J. Catal.* **1989**, *118*, 85–98. [CrossRef]

35. Datka, J.; Gil, B.; Kubacka, A. Acid properties of NaH-mordenites: Infrared spectroscopic studies of ammonia sorption. *Zeolites* **1995**, *15*, 501–506. [CrossRef]

36. Kosslick, H.; Berndt, H.; Lanh, H.D.; Martin, A.; Miessner, H.; Tuan, V.A.; Jänchen, J. Acid properties of ZSM-20-type zeolite. *J. Chem. Soc. Faraday Trans.* **1994**, *90*, 2837. [CrossRef]

37. Elzey, S.; Mubayi, A.; Larsen, S.C.; Grassian, V.H. FTIR study of the selective catalytic reduction of NO_2 with ammonia on nanocrystalline NaY and CuY. *J. Mol. Catal. A* **2008**, *285*, 48–57. [CrossRef]

38. Poignant, F.; Saussey, J.; Lavalley, J.-C.; Mabilon, G. In situ FT-IR study of NH_3 formation during the reduction of NO_x with propane on H/Cu-ZSM-5 in excess oxygen. *Catal. Today* **1996**, *29*, 93–97. [CrossRef]

39. Poignant, F.; Saussey, J.; Lavalley, J.-C.; Mabilon, G. NH_3 formation during the reduction of nitrogen monoxide by propane on H–Cu–ZSM-5 in excess oxygen. *J. Chem. Soc. Chem. Commun.* **1995**, 89–90. [CrossRef]

40. Can, F.; Travert, A.; Ruaux, V.; Gilson, J.-P.; Maugé, F.; Hu, R.; Wormsbecher, R.F. FCC gasoline sulfur reduction additives: Mechanism and active sites. *J. Catal.* **2007**, *249*, 79–92. [CrossRef]

41. Ajmal, M.; Hussain Khan, A.; Ahmad, S.; Ahmad, A. Role of sawdust in the removal of copper(II) from industrial wastes. *Water Res.* **1998**, *32*, 3085–3091. [CrossRef]

42. Wichterlová, B.; Tvarůžková, Z.; Sobalík, Z.; Sarv, P. Determination and properties of acid sites in H-ferrierite: A comparison of ferrierite and MFI structures. *Microporous Mesoporous Mater.* **1998**, *24*, 223–233. [CrossRef]

catalysts

MDPI

Article

Surface Species and Metal Oxidation State during H$_2$-Assisted NH$_3$-SCR of NO$_x$ over Alumina-Supported Silver and Indium

Linda Ström *, Per-Anders Carlsson, Magnus Skoglundh and Hanna Härelind

Competence Centre for Catalysis, Department of Chemistry and Chemical Engineering,
Chalmers University of Technology, SE-412 96 Göteborg, Sweden; per-anders.carlsson@chalmers.se (P.-A.C.);
skoglund@chalmers.se (M.S.); hanna.harelind@chalmers.se (H.H.)
* Correspondence: linda.strom@chalmers.se; Tel.: +46-31-772-1000

Received: 1 January 2018; Accepted: 17 January 2018; Published: 19 January 2018

Abstract: Alumina-supported silver and indium catalysts are investigated for the hydrogen-assisted selective catalytic reduction (SCR) of NO$_x$ with ammonia. Particularly, we focus on the active phase of the catalyst and the formation of surface species, as a function of the gas environment. Diffuse reflectance ultraviolet-visible (UV-vis) spectroscopy was used to follow the oxidation state of the silver and indium phases, and in situ diffuse reflectance infrared Fourier transform spectroscopy (DRIFTS) was used to elucidate the formation of surface species during SCR conditions. In addition, the NO$_x$ reduction efficiency of the materials was evaluated using H$_2$-assisted NH$_3$-SCR. The DRIFTS results show that the Ag/Al$_2$O$_3$ sample forms NO-containing surface species during SCR conditions to a higher extent compared to the In/Al$_2$O$_3$ sample. The silver sample also appears to be more reduced by H$_2$ than the indium sample, as revealed by UV-vis spectroscopic experiments. Addition of H$_2$, however, may promote the formation of highly dispersed In$_2$O$_3$ clusters, which previously have been suggested to be important for the SCR reaction. The affinity to adsorb NH$_3$ is confirmed by both temperature programmed desorption (NH$_3$-TPD) and in situ DRIFTS to be higher for the In/Al$_2$O$_3$ sample compared to Ag/Al$_2$O$_3$. The strong adsorption of NH$_3$ may inhibit (self-poison) the NH$_3$ activation, thereby hindering further reaction over this catalyst, which is also shown by the lower SCR activity compared to Ag/Al$_2$O$_3$.

Keywords: lean NO$_x$ reduction; silver-alumina; indium-alumina; ammonia-SCR; hydrogen effect

1. Introduction

The development of fuel-efficient engines, operating under lean conditions, is motivated by fluctuating oil prices, more stringent emission legislations, and climate changes. Among the most attractive exhaust aftertreatment techniques for lean NO$_x$ reduction is the selective catalytic reduction with either ammonia (NH$_3$-SCR) or hydrocarbons (HC-SCR). For example, Cu-based zeolites have recently been shown to exhibit high NO$_x$ removal activity in a wide temperature range [1–3]. In contrast to NH$_3$-SCR, the HC-SCR catalysts need to be further improved as to be competitive. This puts pressure on building new understanding of the materials and mechanisms for HC-SCR. It seems, though, that NH$_3$ is a key intermediate for both techniques [4,5]. Due to its excellent thermal and mechanical stability, alumina is the most widely used catalyst support material [6]. The silver-alumina system has been studied for HC-SCR applications and with the pioneering work by Satokawa [7] in 2000, it was shown that the catalytic activity for NO$_x$ reduction can be further improved by the addition of small amounts of hydrogen. This widely studied phenomenon [8–11] is denoted the 'hydrogen effect' and has previously been regarded as limited to silver-based catalysts only [12]. However, recently In/Al$_2$O$_3$, which also has been studied for SCR applications [13–19], was found to exhibit a hydrogen effect, albeit to a lower extent compared to Ag/Al$_2$O$_3$ [20]. The hydrogen effect over Ag/Al$_2$O$_3$ has been suggested to originate from the reduction of adsorbed nitrogen species [10,21,22], changes in the type of Ag species [23–25], and/or enhanced activation of the reductant [9,11]. However, in contrast to other precious metal catalysts, alumina-supported silver is not active for H$_2$-SCR [26]. From a practical point of view, H$_2$ can be provided to the vehicle's exhaust after treatment by on-board reforming of for example solid amine salts or fuel [19,27].

Several studies have focused on the role of silver phases in the Ag/Al$_2$O$_3$ catalyst [28–33], suggesting the active phase for the selective reduction of NO$_x$ to be Ag$_n^{\delta+}$-clusters [32,33] and Ag$^+$ ions [28,29], or a combination of these. Except for these species, metallic silver (Ag^0) is recognized as responsible for complete combustion of the reductant [34]. In the In/Al$_2$O$_3$ catalyst, highly dispersed indium cluster sites (In^{3+}) have been identified as the active component for hydrocarbon activation during HC-SCR [17].

A deeper understanding of the underlying mechanisms of the hydrogen effect is useful, not only to improve the SCR catalysts, but also for facilitating development of systems that reduce NO$_x$ efficiently without the addition of hydrogen. In this study, we compare Ag/Al$_2$O$_3$ and In/Al$_2$O$_3$ catalysts, for the SCR of NO$_x$ with ammonia as model reductant. In particular, we are focusing on how the catalytically active silver and indium phases are affected by the SCR environment, and how the formed surface species interact with these phases, using diffuse reflectance ultraviolet-visible (UV-vis) spectroscopy, in situ diffuse reflectance infrared Fourier transform spectroscopy (DRIFTS), and ammonia-temperature programmed desorption (NH$_3$-TPD).

2. Results

In this study, we compared H$_2$-assisted NH$_3$-SCR of NO$_x$ over Ag/Al$_2$O$_3$ and In/Al$_2$O$_3$. In Section 2.1, we show the NO$_x$ conversion and formed species during the SCR experiments. In addition, we compare the density of acidic sites by NH$_3$-TPD (Section 2.2). We have also focused on the possible changes in the active phase and surface species as a function of gas phase components. These results are presented in Sections 2.3 and 2.4

2.1. Catalytic Activity

The Ag/Al$_2$O$_3$ and In/Al$_2$O$_3$ samples were evaluated as NH$_3$-SCR catalysts using a flow reactor with a feed composition of 500 ppm NO, 500 ppm NH$_3$, 10% O$_2$, 5% H$_2$O, and Ar as carrier gas. The experiments were subsequently repeated with the addition of 1000 ppm H$_2$.

The NO$_x$ reduction and NH$_3$ conversion, as well as the formation of NO$_2$ and N$_2$O over Ag/Al$_2$O$_3$, are shown in Figure 1. Without the addition of H$_2$, the catalyst is inactive for NO$_x$ reduction. However, H$_2$-assisted NH$_3$-SCR reduces NO$_x$ in a broad temperature window and with a maximum reduction just above 80%. The selectivity towards N$_2$ is high, in accordance with previous reports [35,36], 80% when H$_2$ is present in the feed. It can also be seen that the low N$_2$O formation decreases even further when H$_2$ is present, confirming the results of Kondratenko et al., that H$_2$ suppresses the total N$_2$O production [25]. This is assigned to H$_2$-induced Ag^0 formation, which likely is responsible for the N$_2$O decomposition [37]. The negative values of NO$_x$ reduction above 350 °C observed when H$_2$ is absent (black line in Figure 1a) is likely due to the oxidation of NH$_3$ to NO, which also explains the significant amount of NH$_3$ converted in Figure 1b [35]. Shimizu and Satsuma [32] suggest that the addition of H$_2$ to NH$_3$ + O$_2$ enhances the oxidative activation of NH$_3$ by decreasing the activation energy of the rate-determining step, (i.e., formation of NH$_x$).

The In/Al$_2$O$_3$ catalyst is, like Ag/Al$_2$O$_3$, inactive for NO$_x$ reduction with solely NH$_3$, as seen in Figure 2. With the addition of H$_2$, the NO$_x$ reduction clearly increases, however, to a considerably lower level compared to Ag/Al$_2$O$_3$. In order to separate between the effect of the active phase and the Al$_2$O$_3$ support, the equivalent experiments were executed for γ-Al$_2$O$_3$, which is totally inactive for both the NO$_x$ reduction and NH$_3$ conversion, even with addition of H$_2$ (results not shown), in accordance with the findings of Doronkin et al. [19].

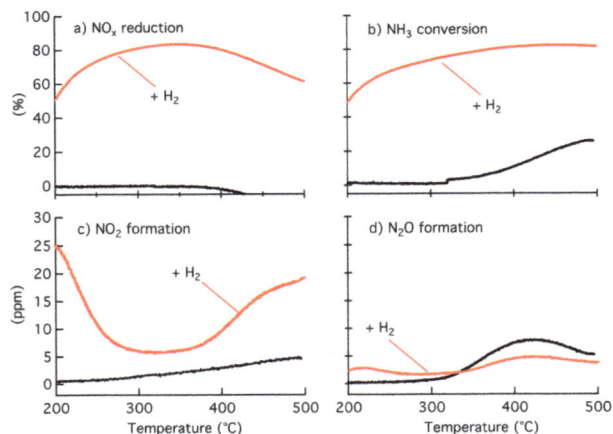

Figure 1. NH$_3$-SCR over Ag/Al$_2$O$_3$. (**a**) NO$_x$ reduction; (**b**) NH$_3$ conversion; (**c**) NO$_2$ formation and (**d**) N$_2$O formation over the Ag/Al$_2$O$_3$ catalyst as a function of temperature. Inlet gas composition: 500 ppm NO, 500 ppm NH$_3$, 10% O$_2$ and 5% H$_2$O, Ar-bal. (red lines represents the addition of 1000 ppm H$_2$).

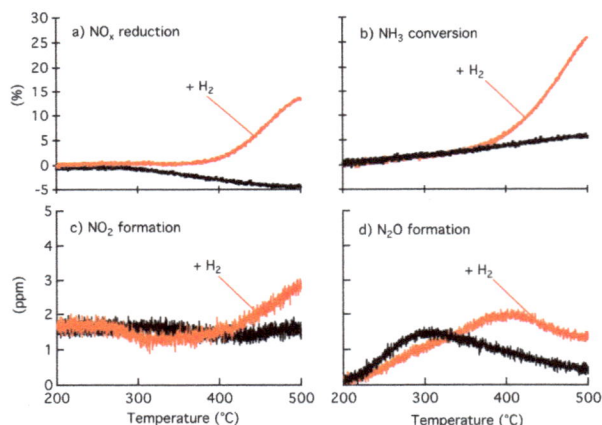

Figure 2. NH$_3$-SCR over In/Al$_2$O$_3$. (**a**) NO$_x$ reduction; (**b**) NH$_3$ conversion; (**c**) NO$_2$ formation and (**d**) N$_2$O formation over the In/Al$_2$O$_3$ sample as a function of temperature. Inlet gas composition: 500 ppm NO, 500 ppm NH$_3$, 10% O$_2$ and 5% H$_2$O, Ar-bal. (red lines represents the addition of 1000 ppm H$_2$).

2.2. Surface Acidity

The surface of the alumina support consists of a combination of aluminum and oxygen ions, which may exhibit lower coordination numbers compared to ions of the bulk. The surface ions hold vacant sites, which, at ambient temperature, are always occupied by either dissociatively adsorbed water in the form of surface hydroxyl (OH) groups, or by coordinated water molecules [38]. Twelve different configurations of OH can be present at the surface, bearing slightly different net charges, consequently possessing different properties, such as variations in acidity [39,40].

Ammonia-TPD experiments were performed in order to investigate the surface acidity of the samples. The desorption profiles of NH$_3$ during the NH$_3$-TPD measurements over the γ-Al$_2$O$_3$, Ag/Al$_2$O$_3$ and In/Al$_2$O$_3$ samples, are shown in Figure 3. The total amount of desorbed ammonia, summarized in Table 1, shows that the highest concentration of acidic sites is found for the γ-Al$_2$O$_3$ sample, followed by In/Al$_2$O$_3$ and Ag/Al$_2$O$_3$, which possessed the lowest concentration of acidic sites. Comparing the NH$_3$ desorption peaks for the samples; the weakest type of acidic site (i.e., the peak with lowest desorption temperature) is found at a higher temperature for the γ-Al$_2$O$_3$ sample, corresponding to 19% of the total desorbed amount of NH$_3$, compared to the Ag/Al$_2$O$_3$ (49%) and In/Al$_2$O$_3$ (16%) samples, respectively. The peak representing the strongest type of acidic site, i.e., the peak at the highest temperature, is also found at a higher temperature for the γ-Al$_2$O$_3$ sample. In addition, the highest desorption-temperature peak of the impregnated samples corresponds to the middle-temperature desorption peak of γ-Al$_2$O$_3$, indicating that the impregnation with silver and indium, respectively, results in less strong acidity for these samples. For the γ-Al$_2$O$_3$ sample, 33% of the total desorbed amount of NH$_3$ is adsorbed on this (strongest acidic) type of site, compared to 6% in the Ag/Al$_2$O$_3$ and 52% in the In/Al$_2$O$_3$ sample. However, the Gaussian peak representing the strongest acidic sites of the γ-Al$_2$O$_3$ sample is centered around almost 100 °C higher temperature compared to the peak holding the most acidic site of In/Al$_2$O$_3$. This implies that the impregnation procedure of γ-Al$_2$O$_3$ leads to an electronical modification and physical blockage of acidic sites at the catalyst support.

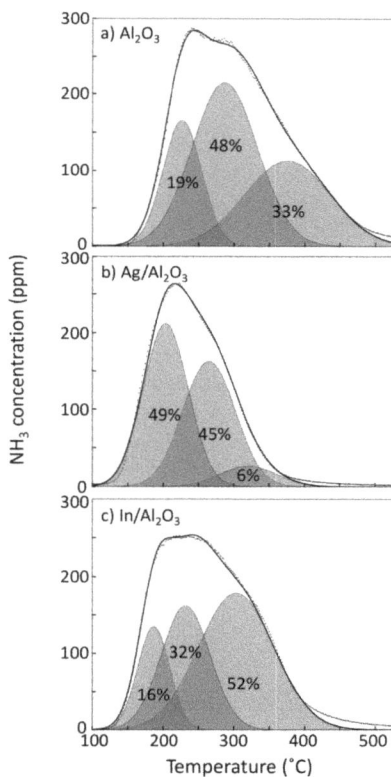

Figure 3. NH$_3$-temperature programmed desorption (TPD) profiles for (**a**) γ-Al$_2$O$_3$; (**b**) Ag/Al$_2$O$_3$ and (**c**) In/Al$_2$O$_3$, with the desorbed NH$_3$ concentration as a function of the temperature. The desorbed amount of NH$_3$ in the deconvoluted peaks are denoted in percentage of the measured total desorbed amount of NH$_3$.

Table 1. Total desorbed amount of NH$_3$ during the NH$_3$-TPD experiments.

Sample	Desorbed NH$_3$ (mmol/cm^2)
γ-Al$_2$O$_3$	11.2
Ag/Al$_2$O$_3$	7.4
In/Al$_2$O$_3$	9.5

2.3. Characterization of the Oxidation State of the Active Phase

Silver and indium species present in the catalyst samples were characterized using diffuse reflectance UV-vis spectroscopy. In order to investigate the influence of the NH$_3$-SCR reaction components, the samples were pretreated with NO, NH$_3$ and H$_2$, respectively. The UV-vis spectrum of the fresh (i.e., non-pretreated) Ag/Al$_2$O$_3$ sample (alumina subtracted) is shown in Figure 4, with absorbance peaks assigned according to Table 2. The spectrum shows that the sample contains a mixture of isolated Ag$^+$-ions, Ag$_n{}^{\delta+}$-clusters and Ag^0. Note that some isolated Ag$^+$-ions, which exhibited peaks below 200 nm, may have been present in the samples without being detected, since the spectrum only contained signals above 200 nm due to instrument limitations [41].

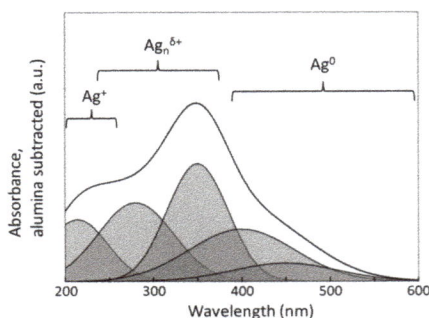

Figure 4. Ultra-violet (UV)-vis spectrum of the fresh (i.e., non-pretreated) Ag/Al$_2$O$_3$ sample, with the absorbance for the alumina sample subtracted, as a function of the wavelength. The peak ranges assigned to isolated Ag$^+$-ions, Ag$_n{}^{\delta+}$-clusters and Ag^0 are denoted.

Table 2. Assignment of absorption peaks in the UV-vis spectra of Ag/Al$_2$O$_3$.

Species	Wavelength [nm]	Reference
Isolated Ag$^+$-ions	192–250	[32]
	196, 212, 224	[42]
	220	[43]
	215–240	[44]
	212, 260	[44]
Ag$_n{}^{\delta+}$-clusters	260–370	[37]
	238–272	[45,46]
	322	[42]
	350, 285	[32]
Ag^0	>390	[28,29,32,42,46–48]

Figure 5 shows, in the same scale as Figure 4, the UV-vis spectra of the Ag/Al$_2$O$_3$ sample after pretreatments in H$_2$, NO and NH$_3$ at 300 °C. The H$_2$ pretreatment results in increased intensity of the bands at wavelengths corresponding to Ag$_n{}^{\delta+}$-clusters and metallic silver, as shown in Figure 5a. In contrast, the absorption spectrum recorded after pretreatment in NO (Figure 5b) shows decreased intensity, compared to the fresh sample, at wavelengths corresponding to $Ag_n^{\delta+}$-clusters and metallic silver, indicating that NO slightly oxidizes the catalyst surface at 300 °C. In Figure 5c, the spectrum recorded after NH$_3$-pretreatment shows that the surface is reduced by NH$_3$. Compared to the fresh sample, this pretreatment shifts peaks from clusters to more completely reduced Ag phases.

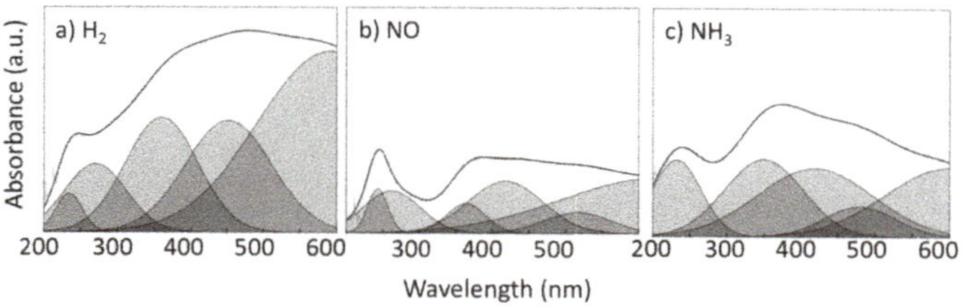

Figure 5. UV-vis spectra for the Ag/Al$_2$O$_3$ sample, with the absorbance for the alumina sample subtracted, plotted as a function of the wavelength. The graphs represent the spectra after pretreatment in (**a**) H$_2$, (**b**) NO, and (**c**) NH$_3$ at 300 °C. The scale of these figures is the same as of Figure 4.

The UV-vis spectrum of the fresh (i.e., non-pretreated) In/Al$_2$O$_3$ sample is shown in Figure 6. Absorption peaks in the range 200–450 nm are assigned to In$_2$O$_3$ [49–54]. Lv et al. [49] experienced that increased concentration of In$_2$O$_3$ results in a broadening and a redshift of the absorbance edge. The spectrum recorded after the H$_2$-pretreatment shows broadened absorption peaks with a slight redshift (see Figure 7a). However, peaks at wavelengths above 450 nm increases somewhat after the H$_2$-pretreatment, indicating the presence of more reduced indium species. Increased and broadened absorption peaks in the range 200–450 nm are detected after pretreatments also with NO and NH$_3$ at 300 °C (see Figure 7b,c), which could indicate increased In$_2$O$_3$ concentration in the sample.

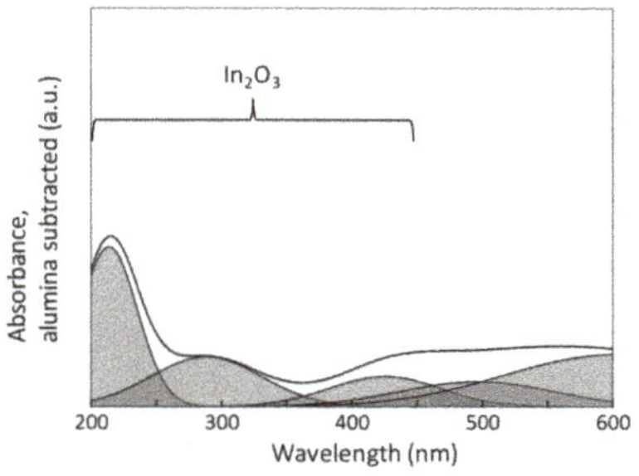

Figure 6. UV-vis spectrum of the fresh (i.e., non-pretreated) In/Al$_2$O$_3$ sample, with the absorbance for the alumina sample subtracted, as a function of the wavelength. The peak range assigned to In$_2$O$_3$ is denoted.

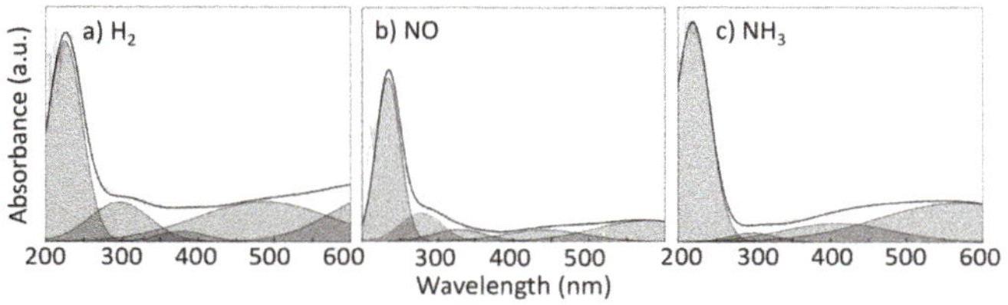

Figure 7. UV-vis spectra for the In/Al$_2$O$_3$ sample, with the absorbance for the alumina sample subtracted, plotted as a function of the wavelength. The graphs represent the spectra after pretreatment in (**a**) H$_2$, (**b**) NO, and (**c**) NH$_3$ at 300 °C. The scale of these figures is the same as of Figure 6.

2.4. Evaluation of Surface Species

The gas environment influences the surface of the catalyst, and in order to study the active phase and connect that to the reaction itself, DRIFTS was used to follow the formation of surface species at reaction conditions. Figure 8 shows the formation of surface species for the Ag/Al_2O_3, In/Al_2O_3 and $\gamma\text{-}Al_2O_3$ samples during exposure to NO, NH_3, H_2 and O_2 at 300 °C. The measurements were performed after 10 min exposure to the gas mixture in the last measurement sequence (see Table 3). All observed peaks remained after the specific gas component that gave rise to the corresponding absorption band, was switched off. This indicates that the surface species are chemisorbed to the samples. Below, absorption bands assigned to the adsorption of NO and NH_3 are presented separately. Also, absorption bands assigned to hydroxyl groups (i.e., bands at 3500–3800 cm^{-1} [55]) were observed (Figure 8). Comparing the relative peak intensities within the OH-band area of the infra-red (IR) patterns, it can be shown that the spectrum of the $\gamma\text{-}Al_2O_3$ sample resembled the one of In/Al_2O_3 more closely, compared to Ag/Al_2O_3.

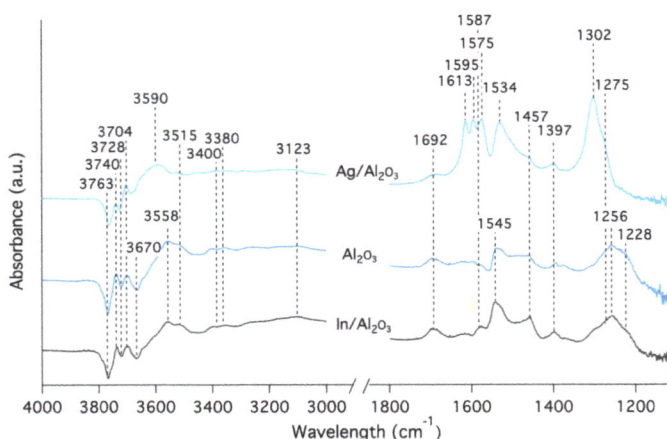

Figure 8. Diffuse reflectance infrared Fourier transform spectroscopy (DRIFTS) spectra showing formation of surface species for the Ag/Al_2O_3, In/Al_2O_3 and $\gamma\text{-}Al_2O_3$ samples during the exposure to selective catalytic reduction (SCR) reaction conditions (NO, NH_3, H_2, O_2, Ar-bal.) at 300 °C.

Table 3. The seven DRIFTS sequences. Gas condition: 500 ppm NO, 1000 ppm H_2 and 500 ppm NH_3. All sequences included 10% O_2 and Ar as the carrier gas.

Sequence	NO	H_2	NH_3
1	NO		
2	NO	H_2	
3	NO	H_2	NH_3
4		H_2	NH_3
5			NH_3
6	NO		NH_3
7	NO	H_2	NH_3

2.4.1. Assignment of NO Absorption Bands

Exposing Al_2O_3-based catalysts to NO and O_2 (or to NO_2) leads to the formation of surface nitrate and nitrite species. Nitrites have been recognized as an intermediate state in the formation of nitrates over Ag/Al_2O_3 [21], which may be a reason for the nitrite band to appear at 1228 cm^{-1} in the $\gamma\text{-}Al_2O_3$ and In/Al_2O_3 spectra, but not in the spectrum of Ag/Al_2O_3. Absorption bands related to the symmetric N=O stretching vibrations are located in the region between 1650 and 1500 cm^{-1}, while the asymmetrical stretching of the O–N–O group can be detected between 1200 and 1350 cm^{-1} [21,34,56]. The spectra of In/Al$_2$O$_3$ and γ-Al$_2$O$_3$ exhibit broad peaks centered around 1256 cm^{-1}, as shown in Figure 8. This band is assigned to (bidentate) nitrate [57]. In the same region, at 1302 cm^{-1}, Ag/Al$_2$O$_3$ exhibits a sharp peak, which is assigned to monodentate nitrate [21,57]. The three peaks located at 1575, 1595 and 1613 cm^{-1} for the Ag/Al$_2$O$_3$ sample are assigned to bridged-, bidentate- and mono-dentate nitrate, respectively [21,41]. A summary of all absorption band assignments upon NO exposure is found in Table 4.

Table 4. Assignments of infra-red (IR) peaks associated with nitrite and nitrate species.

Wavenumber (cm^{-1})	Appears in Sample	Surface Species	References
1228	Al$_2$O$_3$, In/Al$_2$O$_3$	Nitrite	[21]
1256	Al$_2$O$_3$, In/Al$_2$O$_3$	Bidentate nitrate	[21,58]
1302	Ag/Al$_2$O$_3$	Monodentate nitrate	[21]
1534	Ag/Al$_2$O$_3$	Monodentate nitrate	[21]
1545	Al$_2$O$_3$, In/Al$_2$O$_3$	Monodentate nitrate	[21]
1575	Ag/Al$_2$O$_3$	Monodentate nitrate	[21]
1595	Ag/Al$_2$O$_3$	Bidentate nitrate	[21]
1613	Ag/Al$_2$O$_3$	Bridged nitrate	[21,59,60]

2.4.2. Assignment of NH$_3$ Absorption Bands

The symmetric and asymmetric vibration of surface coordinated NH$_3$ results in absorption bands at 1275 and 1587 cm^{-1}, respectively [61]. Moreover, the bands at 1397 and 1692 cm^{-1} are likely due to the adsorption of NH$_4$$^+$ ions at Brønstedt acidic sites [61,62]. The less pronounced absorption bands at 3380 and 3400 cm^{-1} are assigned to the symmetric and asymmetric N–H stretching vibrations of NH$_3$ hydrogen bonded to surface OH [62,63]. A summary of the absorption bands associated with ammonia is found in Table 5.

Table 5. Assignments of IR peaks associated with ammonia surface species.

Wavenumber (cm^{-1})	Vibration	Reference
1275	Symmetric bending of surface coordinated NH$_3$	[61]
1397	NH$_4$$^+$ ions at Brønstedt acidic site	[62,63]
1587	Asymmetric bending of surface coordinated NH$_3$	[61]
1692	NH$_4$$^+$ ions at Brønstedt acidic site	[62,63]
3380	Symmetric NH stretching vibrations of NH$_3$ hydrogen bonded to surface OH	[62,63]
3400	Asymmetric NH stretching vibrations of NH$_3$ hydrogen bonded to surface OH	[63]

3. Discussion

The NH$_3$-SCR results (Figures 1 and 2) show that the H$_2$-assisted reduction of NO at 300 °C is high over the Ag/Al$_2$O$_3$ sample and low over the In/Al$_2$O$_3$ sample. In situ DRIFTS results (Figure 8) show that the formation of NO-containing species is higher over the Ag/Al$_2$O$_3$ sample compared to the In/Al$_2$O$_3$ and γ-Al$_2$O$_3$ samples at this temperature. An efficient adsorption of the gas phase species is crucial for achieving an effective catalytic conversion. However, all absorption bands assigned to NH$_3$ adsorption are more pronounced for the γ-Al$_2$O$_3$ and In/Al$_2$O$_3$ samples, compared to the Ag/Al$_2$O$_3$ sample. This indicates that the adsorption of NH$_3$-surface species is more efficient over the former two samples compared to the latter, which is supported by the NH$_3$-TPD (Figure 3), showing that the γ-Al$_2$O$_3$ and In/Al$_2$O$_3$ samples provide higher density of acidic sites, compared to the Ag/Al$_2$O$_3$ sample. Furthermore, the DRIFTS spectra show differences at the wavelengths corresponding to OH-groups. The In/Al$_2$O$_3$ pattern resembles the one of γ-Al$_2$O$_3$ to a higher degree compared to Ag/Al$_2$O$_3$. Since alumina is impregnated with Ag and In in equivalent molar amounts, this implies that Ag affects the acidic properties of the OH-rich alumina surface to a higher degree compared to In, resulting in a lower concentration of acidic sites for Ag/Al_2O_3.

Silver clusters ($Ag_n^{\delta+}$) have previously been identified as the prime species for the activity in H_2-assisted NH_3-SCR [64], and it has been shown that the activity is linearly proportional to the relative amount of these clusters [32]. Shimizu and Satsuma [32] suggest the following reaction mechanism for H_2-NH_3-SCR over Ag/Al_2O_3: (i) dissociation of H_2 on the Ag site, (ii) spillover of H^+ to form a proton on Al_2O_3, (iii) aggregation of isolated Ag^+ ions to $Ag_n^{\delta+}$-clusters ($n \leq 8$), (iv) reduction of O_2 promoted by $Ag_n^{\delta+}$-clusters and H^+ to O_2^-, H_2O and $Ag_n^{(\delta+x)+}$ or Ag^+, (v) N–H activation by O_2^- to yield NH_x ($x \leq 2$) (vi) oxidation of NO by O_2^- forming NO_2, (vii) reaction between NH_x and NO to yield N_2 and H_2O. The study by Tamm et al. [65] confirms that silver is needed for the dissociation of H_2, which directly participates in the reaction mechanism and also that the NO to NO_2 oxidation is part of this mechanism.

The UV-vis spectra of the In/Al_2O_3 sample show increased peak intensity for bands assigned to In_2O_3 but also increased absorbance at higher wavelengths (>450 nm) after pretreatment in H_2. The reaction mechanism of H_2-assisted NH_3 over In/Al_2O_3 could therefore resemble what Shimizu et al. suggested for Ag/Al_2O_3. However, the NH_3-TPD profiles of the catalysts elucidate the difference in acidity between the catalysts, where the In/Al_2O_3 sample contains significantly stronger acidic sites compared to the Ag/Al_2O_3 sample. An important difference in the mechanisms between the two catalysts could therefore be the stronger affinity between NH_3 and In/Al_2O_3, possibly hindering the NH_3 activation and thereby hindering further reaction over this catalyst to a higher extent compared to Ag/Al_2O_3. Another possible issue that may restrain the NO_x conversion over In/Al_2O_3 is the lower adsorption of NO species at the catalyst surface.

4. Materials and Methods

4.1. Catalyst Preparation and Basic Characterization

Two catalyst samples, 2.0 wt % Ag/Al_2O_3 and 2.1 wt % In/Al_2O_3 (which corresponds to equivalent molar amount of Ag and In, respectively), were prepared by incipient wetness impregnation of γ-Al_2O_3 (PURALOX®SBa 200, Sasol, Hamburg, Germany) using freeze-drying, according to the procedure described in detail previously [20], and briefly below. Silver nitrate (≥99.0% Sigma-Aldrich/Merck, Darmstadt, Germany) and indium nitrate hydrate (99.99% Sigma-Aldrich) were used as the active phase precursor for Ag and In, respectively. After impregnation, the samples were frozen in liquid nitrogen, subsequently freeze-dried, and finally calcined in air at 600 °C for four hours. The as prepared powder samples were characterized with respect to surface area by N_2 sorption (BET) and crystal structure by X-ray diffraction (XRD), as described in details elsewhere [20].

For the evaluation of the catalytic activity for NH_3-SCR, monolith samples with 188 channels (400 CPSI, Ø = 20 mm, L = 20 mm) were cut from a commercial cordierite honeycomb structure (Corning, Corning, NY, USA) and calcined in air at 600 °C for one hour. Binder agent (DISPERAL® P2, Sasol) and one of the powder catalysts (ratio 1:4) in 1:1-ratio ethanol-water solutions were mixed to form washcoat slurries. The monoliths were dipped into the slurries, gently shaken for removal of excess slurry, dried (90 °C in air), and subsequently calcined (500 °C, 3 min). The coating procedure was repeated until the washcoat mass corresponded to 20% of the coated monolith mass. Finally, the monolith samples were calcined in air (600 °C, 1 h).

The catalyst samples have previously been characterized with respect to surface area and crystal structure [20]. The specific surface areas are 197, 185, and 188 m^2/g, for γ-Al_2O_3, Ag/Al_2O_3 and In/Al_2O_3, respectively. X-ray diffractograms indicate that the main crystalline phase in all samples is γ-Al_2O_3, and other crystalline phases (if any) are only present as particles smaller than 3–5 nm [66], in line with previous studies [17,18].

4.2. Lean NO$_x$ Reduction Experiments

The catalytic activity for NH$_3$-SCR was evaluated in extinction experiments (500–100 °C, 10 °C/min) at a flow rate of 3500 mL/min (GHSV of 33,400 h^{-1}) in a flow reactor setup previously described by Kannisto et al. [67]. Briefly, the reactor consisted of a horizontal quarts tube (L = 80 cm, $\varnothing_i$ = 22 mm) heated by a metal coil. The catalyst sample was positioned close to the tube outlet, surrounded by bare monoliths for shielding of heat radiation to the thermocouples [68], which were placed inside and just before the coated monolith. Prior to each measurement, the sample was pretreated in 10% O$_2$ (Ar-balance) at 500 °C for 30 min. During the experiments, the gas composition was 500 ppm NO, 500 ppm NH$_3$, 10% O$_2$, and 5% H$_2$O, in the presence or absence of 1000 ppm H$_2$ (Ar-balance). The outlet gas composition was analyzed by a gas phase Fourier transform infrared (FTIR) spectrometer (MKS 2030, MKS Instruments, Telford, UK). The reduction of NO$_x$ and conversion of NH$_3$ were obtained from the ratios of the differences between the inlet and outlet concentrations to the corresponding inlet concentration.

4.3. UV-Vis Spectroscopy

The samples, pretreated in varying SCR gas components, were characterized by diffuse reflectance UV-vis spectroscopy. Reflectance spectra in the wavelength range 200–1200 nm were recorded using a Varian Cary 5000 UV-vis near-IR (NIR) spectrophotometer, equipped with an external DRA-2500 unit (Agilent, Santa Clara, CA, USA). The same flow reactor equipment as in the NO$_x$ reduction experiments was used for the pretreatments, where the samples were exposed to NO (1500 ppm, Ar balance), NH$_3$ (1500 ppm, Ar balance), and H$_2$ (3000 ppm, Ar balance), respectively. The pretreatments were carried out in a gas flow of 100 mL/min for 20 min at 300 °C. During data processing, the spectrum of the alumina support (which was pretreated in the same way as the Ag/Al$_2$O$_3$ and In/Al$_2$O$_3$ samples) was subtracted, and each spectrum was deconvoluted using Gaussian peaks for evaluation purposes.

4.4. In Situ DRIFTS

The surface species on the samples were characterized during SCR reaction conditions by in situ DRIFTS. The instrument used was a Bruker Vertex70 spectrometer equipped with a high-temperature reaction cell (Harrick Scientific, Pleasantville, NY, USA) with KBr windows. Prior to the measurements, the samples were pressed into tablets and then crushed by a mortar, in order to enlarge the powder particles to avoid channeling during the measurement. Subsequently, the powder fraction of the crushed tablets was sieved to a size range of 38 to 75 μm.

After placing the sample in the reaction cell, the Ag/Al$_2$O$_3$ and In/Al$_2$O$_3$ samples were pretreated at 500 °C in NO (2000 ppm, Ar-bal.) for 30 min, followed by O$_2$ (10%, Ar bal.) for 45 min, and finally H$_2$ (1000 ppm, Ar bal.) for 15 min. The γ-Al$_2$O$_3$ sample was pretreated in O$_2$ (10%, Ar bal.) for 45 min and then H$_2$ (1000 ppm, Ar bal.) for 15 min. The gas conditions during the measurement sequence is listed in Table 3, and corresponded to 500 ppm NO, 500 ppm NH$_3$, 1000 ppm H$_2$, and 10% O$_2$ (Ar balance). Each sequence lasted for 10 min and data were collected a few seconds after starting the experiment, after 5 min, and finally after 10 min. A resolution of 1 cm^{-1} was adapted, and 128 scans were recorded for the background spectra (recorded in Ar at 300 °C), and 64 scans for the measurements. All experiments were carried out at 300 °C in a flow rate of 100 mL/min and the data are presented as absorbance ($\log I/R$).

4.5. NH$_3$-TPD

The concentration and strength of acidic sites of the coated monolith samples were measured by NH$_3$-TPD. The flow reactor presented in Section 2.2 was used for this purpose, where the sample first was pretreated in 10% O$_2$ (20 min) in order to remove carbonaceous matter, flushed with argon (5 min), and then exposed to 1000 ppm H$_2$ (20 min) at 550 °C. During the following NH$_3$-TPD experiment, the catalyst surface was exposed to NH$_3$ (1000 ppm) at 100 °C until saturation, followed by Ar flushing until the NH$_3$ signal vanished. The temperature was subsequently increased linearly to 550 °C (20 °C/min) while the desorbed NH$_3$ was measured continuously. For peak identification purposes, the NH$_3$-TPD profiles obtained are deconvoluted into Gaussian peaks.

5. Conclusions

This work shows that NO-species are formed to a higher extent over Ag/Al$_2$O$_3$ during SCR conditions, compared to In/Al$_2$O$_3$. The latter provide higher density of acidic sites, quantified by NH$_3$-TPD, and also exhibit higher NH$_3$ adsorption, as shown by DRIFTS. Moreover, the Ag/Al$_2$O$_3$ sample is more clearly reduced by H$_2$, compared to In/Al$_2$O$_3$. However, H$_2$ seems to promote the formation of highly dispersed In$_2$O$_3$ clusters, which previously have been suggested to be crucial for HC-SCR and could be important for the activation of the reducing agent also during H$_2$-assisted NH$_3$-SCR. Since adsorption of reactants in suitable proportions is crucial for high catalytic activity, an important difference between the two catalysts could be the stronger affinity for NH$_3$ over In/Al$_2$O$_3$, compared to Ag/Al$_2$O$_3$. This can possibly inhibit the NH$_3$ activation over the former, and thereby hindering further reaction over this catalyst, which is also shown by the lower SCR activity, compared to Ag/Al$_2$O$_3$.

Acknowledgments: This work has been financially supported by the Swedish Research Council and was performed within the Competence Centre for Catalysis, which is hosted by Chalmers University of Technology and financially supported by the Swedish Energy Agency and the member companies: AB Volvo, ECAPS AB, Haldor Topsøe A/S, Scania CV AB, Volvo Car Corporation AB and Wärtsilä Finland Oy.

Author Contributions: L.S., P.-A.C., M.S. and H.H. conceived and designed the experiments; L.S performed the experiments; L.S., P.-A.C., M.S. and H.H. analyzed the data; L.S., P.-A.C., M.S. and H.H. wrote the paper.

Conflicts of Interest: The authors declare no conflict of interest. The founding sponsors had no role in the design of the study; in the collection, analyses, or interpretation of data; in the writing of the manuscript, and in the decision to publish the results.

References

1. Wang, J.H.; Zhao, H.W.; Haller, G.; Li, Y.D. Recent advances in the selective catalytic reduction of NO$_x$ with NH$_3$ on Cu-Chabazite catalysts. *Appl. Catal. B Environ.* **2017**, *202*, 346–354. [CrossRef]

2. Shishkin, A.; Shwan, S.; Pingel, T.N.; Olsson, E.; Clemens, A.; Carlsson, P.A.; Härelind, H.; Skoglundh, M. Functionalization of SSZ-13 and Fe-Beta with copper by NH$_3$ and NO facilitated solid-state ion-exchange. *Catalysts* **2017**, *7*, 232. [CrossRef]

3. Clemens, A.K.S.; Shishkin, A.; Carlsson, P.A.; Skoglundh, M.; Martinez-Casado, F.J.; Matej, Z.; Balmes, O.; Härelind, H. Reaction-driven ion exchange of copper into zeolite SSZ-13. *ACS Catal.* **2015**, *5*, 6209–6218. [CrossRef]

4. Gunnarsson, F.; Pihl, J.A.; Toops, T.J.; Skoglundh, M.; Härelind, H. Lean NO$_x$ reduction over Ag/alumina catalysts via ethanol-SCR using ethanol/gasoline blends. *Appl. Catal. B Environ.* **2017**, *202*, 42–50. [CrossRef]

5. Matarrese, R.; Ingelsten, H.H.; Skoglundh, M. Aspects of reducing agent and role of amine species in the reduction of NO over H-ZSM-5 in oxygen excess. *J. Catal.* **2008**, *258*, 386–392. [CrossRef]

6. Chorkendorff, I.; Niemantsverdriet, J.W. *Concepts of Modern Catalysis and Kinetics*; WILEY-VCH Verlag GmbH & Co. KGaA: Weinheim, Germany, 2007.

7. Satokawa, S. Enhancing the NO/C$_3$H$_8$/O$_2$ reaction by using H$_2$ over Ag/Al$_2$O$_3$ catalysts under lean-exhaust conditions. *Chem. Lett.* **2000**, *29*, 294–295. [CrossRef]

8. Breen, J.P.; Burch, R.; Hardacre, C.; Hill, C.J. Structural investigation of the promotional effect of hydrogen during the selective catalytic reduction of NO$_x$ with hydrocarbons over Ag/Al$_2$O$_3$ catalysts. *J. Phys. Chem. B* **2005**, *109*, 4805–4807. [CrossRef] [PubMed]

9. Satokawa, S.; Shibata, J.; Shimizu, K.; Atsushi, S.; Hattori, T. Promotion effect of H$_2$ on the low temperature activity of the selective reduction of NO by light hydrocarbons over Ag/Al$_2$O$_3$. *Appl. Catal. B Environ.* **2003**, *42*, 179–186. [CrossRef]

10. Kannisto, H.; Ingelsten, H.H.; Skoglundh, M. Aspects of the role of hydrogen in H$_2$-assisted HC-SCR over Ag-Al$_2$O$_3$. *Top. Catal.* **2009**, *52*, 1817. [CrossRef]

11. Shibata, J.; Takada, Y.; Shichi, A.; Satokawa, S.; Satsuma, A.; Hattori, T. Ag cluster as active species for SCR of NO by propane in the presence of hydrogen over Ag-MFI. *J. Catal.* **2004**, *222*, 368–376. [CrossRef]

12. Breen, J.P.; Burch, R. A review of the effect of the addition of hydrogen in the selective catalytic reduction of NO_x with hydrocarbons on silver catalysts. *Top. Catal.* **2006**, *39*, 53–58. [CrossRef]

13. Boutros, M.; Starck, J.; de Tymowski, B.; Trichard, J.-M.; Da Costa, P. On the effect of poor metals (Al, Ga, In) on the NO_x conversion in ethanol selective catalytic reduction. *Top. Catal.* **2009**, *52*, 1786. [CrossRef]

14. Li, J.H.; Hao, J.M.; Cui, X.Y.; Fu, L.X. Influence of preparation methods of In_2O_3/Al_2O_3 catalyst on selective catalytic reduction of NO by propene in the presence of oxygen. *Catal. Lett.* **2005**, *103*, 75–82. [CrossRef]

15. Maunula, T.; Kintaichi, Y.; Inaba, M.; Haneda, M.; Sato, K.; Hamada, H. Enhanced activity of in and Ga-supported sol-gel alumina catalysts for NO reduction by hydrocarbons in lean conditions. *Appl. Catal. B Environ.* **1998**, *15*, 291–304. [CrossRef]

16. Maunula, T.; Kintaichi, Y.; Haneda, M.; Hamada, H. Preparation and reaction mechanistic characterization of sol-gel indium/alumina catalysts developed for NO_x reduction by propene in lean conditions. *Catal. Lett.* **1999**, *61*, 121–130. [CrossRef]

17. Park, P.W.; Ragle, C.S.; Boyer, C.L.; Balmer, M.L.; Engelhard, M.; McCready, D. In_2O_3/Al_2O_3 catalysts for NO_x reduction in lean condition. *J. Catal.* **2002**, *210*, 97–105. [CrossRef]

18. Erkfeldt, S.; Petersson, M.; Palmqvist, A. Alumina-supported In_2O_3, Ga_2O_3 and B_2O_3 catalysts for lean NO_x reduction with dimethyl ether. *Appl. Catal. B Environ.* **2012**, *117*, 369–383. [CrossRef]

19. Doronkin, D.E.; Fogel, S.; Tamm, S.; Olsson, L.; Khan, T.S.; Bligaard, T.; Gabrielsson, P.; Dahl, S. Study of the "Fast SCR"-like mechanism of H_2-assisted SCR of NO_x with ammonia over Ag/Al_2O_3. *Appl. Catal. B Environ.* **2012**, *113*, 228–236. [CrossRef]

20. Ström, L.; Carlsson, P.-A.; Skoglundh, M.; Härelind, H. Hydrogen-assisted SCR of NO_x over alumina-supported silver and indium catalysts using C_2-hydrocarbons and oxygenates. *Appl. Catal. B Environ.* **2016**, *181*, 403–412. [CrossRef]

21. Tamm, S.; Vallim, N.; Skoglundh, M.; Olsson, L. The influence of hydrogen on the stability of nitrates during H_2-assisted SCR over Ag/Al_2O_3 catalysts—A drift study. *J. Catal.* **2013**, *307*, 153–161. [CrossRef]

22. Sadokhina, N.A.; Doronkin, D.E.; Baeva, G.N.; Dahl, S.; Stakheev, A.Y. Reactivity of surface nitrates in H_2-assisted SCR of NO_x over Ag/Al_2O_3 catalyst. *Top. Catal.* **2013**, *56*, 737–744. [CrossRef]

23. Kim, P.S.; Kim, M.K.; Cho, B.K.; Nam, I.-S.; Oh, S.H. Effect of H_2 on denox performance of HC-SCR over Ag/Al_2O_3: Morphological, chemical and kinetic changes. *J. Catal.* **2013**, *301*, 65–76. [CrossRef]

24. Thomas, C. On an additional promoting role of hydrogen in the H_2-assisted C_3H_6-SCR of NO_x on Ag/Al_2O_3: A lowering of the temperature of formation-decomposition of the organo-NO_x intermediates? *Appl. Catal. B Environ.* **2015**, *162*, 454–462. [CrossRef]

25. Kondratenko, E.V.; Kondratenko, V.A.; Richter, M.; Fricke, R. Influence of O_2 and H_2 on NO reduction by NH_3 over Ag/Al_2O_3: A transient isotopic approach. *J. Catal.* **2006**, *239*, 23–33. [CrossRef]

26. Burch, R.; Breen, J.P.; Meunier, F.C. A review of the selective reduction of NO_x, with hydrocarbons under lean-burn conditions with non-zeolitic oxide and platinum group metal catalysts. *Appl. Catal. B Environ.* **2002**, *39*, 283–303. [CrossRef]

27. Gunnarsson, F.; Granlund, M.Z.; Englund, M.; Dawody, J.; Pettersson, L.J.; Härelind, H. Combining HC-SCR over Ag/Al_2O_3 and hydrogen generation over Rh/CeO_2-ZrO_2 using biofuels: An integrated system approach for real applications. *Appl. Catal. B Environ.* **2015**, *162*, 583–592. [CrossRef]

28. Shimizu, K.; Shibata, J.; Yoshida, H.; Satsuma, A.; Hattori, T. Silver-alumina catalysts for selective reduction of NO by higher hydrocarbons: Structure of active sites and reaction mechanism. *Appl. Catal. B Environ.* **2001**, *30*, 151–162. [CrossRef]

29. Bogdanchikova, N.; Meunier, F.C.; Avalos-Borja, M.; Breen, J.P.; Pestryakov, A. On the nature of the silver phases of Ag/Al_2O_3 catalysts for reactions involving nitric oxide. *Appl. Catal. B Environ.* **2002**, *36*, 287–297. [CrossRef]

30. Iglesias-Juez, A.; Hungria, A.B.; Martinez-Arias, A.; Fuerte, A.; Fernandez-Garcia, M.; Anderson, J.A.; Conesa, J.C.; Soria, J. Nature and catalytic role of active silver species in the lean NO_x reduction with C_3H_6 in the presence of water. *J. Catal.* **2003**, *217*, 310–323. [CrossRef]

31. Kannisto, H.; Ingelsten, H.H.; Skoglundh, M. Ag-Al_2O_3 catalysts for lean NO_x reduction-influence of preparation method and reductant. *J. Mol. Catal. A Chem.* **2009**, *302*, 86–96. [CrossRef]

32. Shimizu, K.-I.; Satsuma, A. Reaction mechanism of H_2-promoted selective catalytic reduction of NO with NH_3 over Ag/Al_2O_3. *J. Phys. Chem. C* **2007**, *111*, 2259–2264. [CrossRef]

33. Shibata, J.; Shimizu, K.; Takada, Y.; Shichia, A.; Yoshida, H.; Satokawa, S.; Satsuma, A.; Hattori, T. Structure of active Ag clusters in Ag zeolites for SCR of NO by propane in the presence of hydrogen. *J. Catal.* **2004**, *227*, 367–374. [CrossRef]

34. Meunier, F.C.; Breen, J.P.; Zuzaniuk, V.; Olsson, M.; Ross, J.R.H. Mechanistic aspects of the selective reduction of NO by propene over alumina and silver-alumina catalysts. *J. Catal.* **1999**, *187*, 493–505. [CrossRef]

35. Tamm, S.; Fogel, S.; Gabrielsson, P.; Skoglundh, M.; Olsson, L. The effect of the gas composition on hydrogen-assisted NH_3-SCR over Ag/Al_2O_3. *Appl. Catal. B Environ.* **2013**, *136*, 168–176. [CrossRef]

36. Richter, M.; Fricke, R.; Eckelt, R. Unusual activity enhancement of NO conversion over Ag/Al_2O_3 by using a mixed NH_3/H_2 reductant under lean conditions. *Catal. Lett.* **2004**, *94*, 115–118. [CrossRef]

37. Kondratenko, V.A.; Bentrup, U.; Richter, M.; Hansen, T.W.; Kondratenko, E.V. Mechanistic aspects of N_2O and N_2 formation in no reduction by NH_3 over Ag/Al_2O_3: The effect of O_2 and H_2. *Appl. Catal. B Environ.* **2008**, *84*, 497–504. [CrossRef]

38. Poisson, R.; Brunelle, J.-P.; Nortier, P. *Catalyst Supports and Supported Catalysts. Theoretical and Applied Concepts*; Stiles, A.B., Ed.; Stoneham: Butterworth, Malasia, 1987; pp. 44–47.

39. Knozinger, H.; Ratnasamy, P. Catalytic aluminas—Surface models and characterization of surface sites. *Catal. Rev. Sci. Eng.* **1978**, *17*, 31–70. [CrossRef]

40. Digne, M.; Sautet, P.; Raybaud, P.; Euzen, P.; Toulhoat, H. Use of DFT to achieve a rational understanding of acid-basic properties of gamma-alumina surfaces. *J. Catal.* **2004**, *226*, 54–68. [CrossRef]

41. Sazama, P.; Capek, L.; Drobna, H.; Sobalik, Z.; Dedecek, J.; Arve, K.; Wichterlova, B. Enhancement of decane-SCR-NO_x over $Ag/alumina$ by hydrogen. Reaction kinetics and in situ FTIR and UV-vis study. *J. Catal.* **2005**, *232*, 302–317. [CrossRef]

42. Shi, C.; Cheng, M.J.; Qu, Z.P.; Bao, X.H. Investigation on the catalytic roles of silver species in the selective catalytic reduction of NO_x with methane. *Appl. Catal. B Environ.* **2004**, *51*, 171–181. [CrossRef]

43. Miao, S.J.; Wang, Y.; Ma, D.; Zhu, Q.J.; Zhou, S.T.; Su, L.L.; Tan, D.L.; Bao, X.H. Effect of $Ag+$ cations on nonoxidative activation of methane to C_2-hydrocarbons. *J. Phys. Chem. B* **2004**, *108*, 17866–17871. [CrossRef]

44. Musi, A.; Massiani, P.; Brouri, D.; Trichard, J.-M.; Da Costa, P. On the characterisation of silver species for SCR of NO_x with ethanol. *Catal. Lett.* **2009**, *128*, 25–30. [CrossRef]

45. Sato, K.; Yoshinari, T.; Kintaichi, Y.; Haneda, M.; Hamada, H. Remarkable promoting effect of rhodium on the catalytic performance of Ag/Al_2O_3 for the selective reduction of NO with decane. *Appl. Catal. B Environ.* **2003**, *44*, 67–78. [CrossRef]

46. Männikkö, M.; Skoglundh, M.; Ingelsten, H.H. Selective catalytic reduction of NO_x with methanol over supported silver catalysts. *Appl. Catal. B Environ.* **2012**, *119*, 256–266. [CrossRef]

47. Pestryakov, A.N.; Davydov, A.A. Study of supported silver states by the method of electron spectroscopy of diffuse reflectance. *J. Electron Spectrosc. Relat. Phenom.* **1995**, *74*, 195–199. [CrossRef] ·

48. She, X.; Flytzani-Stephanopoulos, M. The role of AgOAl species in silver-alumina catalysts for the selective catalytic reduction of NO_x with methane. *J. Catal.* **2006**, *237*, 79–93. [CrossRef]

49. Lv, J.; Kako, T.; Li, Z.; Zou, Z.; Ye, J. Synthesis and photocatalytic activities of $NaNbO_3$ rods modified by In_2O_3 nanoparticles. *J. Phys. Chem. C* **2010**, *114*, 6157–6162. [CrossRef]

50. Yang, X.; Xu, J.; Wong, T.; Yang, Q.; Lee, C.-S. Synthesis of In_2O_3-In_2S_3 core-shell nanorods with inverted type-I structure for photocatalytic H_2 generation. *Phys. Chem. Chem. Phys.* **2013**, *15*, 12688–12693. [CrossRef] [PubMed]

51. Zhu, G.; Guo, L.; Shen, X.; Ji, Z.; Chen, K.; Zhou, H. Monodispersed In_2O_3 mesoporous nanospheres: One-step facile synthesis and the improved gas-sensing performance. *Sens. Actuators B Chem.* **2015**, *220*, 977–985. [CrossRef]

52. Zhang, F.; Li, X.; Zhao, Q.; Zhang, Q.; Tade, M.; Liu, S. Fabrication of α-Fe_2O_3/In_2O_3 composite hollow microspheres: A novel hybrid photocatalyst for toluene degradation under visible light. *J. Colloid Interface Sci.* **2015**, *457*, 18–26. [CrossRef] [PubMed]

53. Yin, J.Z.; Huang, S.B.; Jian, Z.C.; Pan, M.L.; Zhang, Y.Q.; Fei, Z.B.; Xu, X.R. Enhancement of the visible light photocatalytic activity of heterojunction $In_2O_3/BiVO_4$ composites. *Appl. Phys. A Mater. Sci. Process.* **2015**, *120*, 1529–1535. [CrossRef]

54. Liu, Q.; Zhang, W.; Liu, R.; Mao, G. Controlled synthesis of monodispersed sub-50 nm nanoporous In_2O_3 spheres and their photoelectrochemical performance. *Eur. J. Inorg. Chem.* **2015**, *2015*, 845–851. [CrossRef]

55. Morterra, C.; Magnacca, G. A case study: Surface chemistry and surface structure of catalytic aluminas, as studied by vibrational spectroscopy of adsorbed species. *Catal. Today* **1996**, *27*, 497–532. [CrossRef]

56. Hadjiivanov, K.I. Identification of neutral and charged N_xO_y surface species by IR spectroscopy. *Catal. Rev. Sci. Eng.* **2000**, *42*, 71–144. [CrossRef]

57. Underwood, G.M.; Miller, T.M.; Grassian, V.H. Transmission FT-IR and knudsen cell study of the heterogeneous reactivity of gaseous nitrogen dioxide on mineral oxide particles. *J. Phys. Chem. A* **1999**, *103*, 6184–6190. [CrossRef]

58. Satsuma, A.; Shimizu, K. In situ FT/IR study of selective catalytic reduction of NO over alumina-based catalysts. *Prog. Energy Combust. Sci.* **2003**, *29*, 71–84. [CrossRef]

59. Adams, E.C.; Skoglundh, M.; Gabrielsson, P.; Carlsson, P.A. Passive SCR: The effect of H_2 to NO ratio on the formation of NH_3 over alumina supported platinum and palladium catalysts. *Top. Catal.* **2016**, *59*, 970–975. [CrossRef]

60. Adams, E.C.; Skoglundh, M.; Folic, M.; Bendixen, E.C.; Gabrielsson, P.; Carlsson, P.-A. Ammonia formation over supported platinum and palladium catalysts. *Appl. Catal. B Environ.* **2015**, *165*, 10–19. [CrossRef]

61. Acke, F.; Westerberg, B.; Skoglundh, M. Selective reduction of no by HNCO over PT promoted Al_2O_3. *J. Catal.* **1998**, *179*, 528–536. [CrossRef]

62. Centeno, M.A.; Carrizosa, I.; Odriozola, J.A. In situ drifts study of the SCR reaction of NO with NH_3 in the presence of O_2 over lanthanide doped V_2O_5/Al_2O_3 catalysts. *Appl. Catal. B Environ.* **1998**, *19*, 67–73. [CrossRef]

63. Wallin, M.; Grönbeck, H.; Spetz, A.L.; Skoglundh, M. Vibrational study of ammonia adsorption on Pt/SiO_2. *Appl. Surf. Sci.* **2004**, *235*, 487–500. [CrossRef]

64. Yu, L.; Zhong, Q.; Zhang, S. The enhancement for SCR of NO by NH_3 over the H_2 or CO pretreated $Ag/gamma-Al_2O_3$ catalyst. *Phys. Chem. Chem. Phys.* **2014**, *16*, 12560–12566. [CrossRef] [PubMed]

65. Tamm, S. The role of silver for the H_2-effect in H_2-assisted selective catalytic reduction of NO_x with NH_3 over Ag/Al_2O_3. *Catal. Lett.* **2013**, *143*, 957–965. [CrossRef]

66. Anderson, J.R.; Pratt, K.C. *Introduction to Characterization and Testing of Catalysts*; Academic Press Inc.: Melbourne, Australia, 1985.

67. Kannisto, H.; Karatzas, X.; Edvardsson, J.; Pettersson, L.J.; Ingelsten, H.H. Efficient low temperature lean NO_x reduction over Ag/Al_2O_3—A system approach. *Appl. Catal. B Environ.* **2011**, *104*, 74–83. [CrossRef]

68. Wang-Hansen, C.; Kamp, C.J.; Skoglundh, M.; Andersson, B.; Carlsson, P.-A. Experimental method for kinetic studies of gas-solid reactions: Oxidation of carbonaceous matter. *J. Phys. Chem. C* **2011**, *115*, 16098–16108. [CrossRef]

catalysts

MDPI

Article

Consideration of the Role of Plasma in a Plasma-Coupled Selective Catalytic Reduction of Nitrogen Oxides with a Hydrocarbon Reducing Agent

Byeong Ju Lee [†], Ho-Chul Kang [†], Jin Oh Jo and Young Sun Mok *

Department of Chemical and Biological Engineering, Jeju National University, Jeju 63243, Korea; qudwn211@jejunu.ac.kr (B.J.L.); khc0920@nate.com (H.-C.K.); zkfdh@jejunu.ac.kr (J.O.J.)
* Correspondence: smokie@jejunu.ac.kr; Tel.: +82-64-754-3682; Fax: +82-64-755-3670
† Two authors contributed equally to this work.

Received: 13 October 2017; Accepted: 28 October 2017; Published: 31 October 2017

Abstract: The purpose of this study is to explain how plasma improves the performance of selective catalytic reduction (SCR) of nitrogen oxides (NO_x) with a hydrocarbon reducing agent. In the plasma-coupled SCR process, NO_x reduction was performed with n-heptane as a reducing agent over Ag/γ-Al_2O_3 as a catalyst. We found that the plasma decomposes n-heptane into several oxygen-containing products such as acetaldehyde, propionaldehyde and butyraldehyde, which are more reactive than the parent molecule n-heptane in the SCR process. Separate sets of experiments using acetaldehyde, propionaldehyde and butyraldehyde, one by one, as a reductant in the absence of plasma, have clearly shown that the presence of these partially oxidized compounds greatly enhanced the NO_x conversion. The higher the discharge voltage, the more the amounts of such partially oxidized products. The oxidative species produced by the plasma easily converted NO into NO_2, but the increase of the NO_2 fraction was found to decrease the NO_x conversion. Consequently, it can be concluded that the main role of plasma in the SCR process is to produce partially oxidized compounds (aldehydes), having better reducing power. The catalyst-alone NO_x removal efficiency with n-heptane at 250 °C was measured to be less than 8%, but it increased to 99% in the presence of acetaldehyde at the same temperature. The NO_x removal efficiency with the aldehyde reducing agent was higher as the number of carbons in the aldehyde was more; for example, the NO_x removal efficiencies at 200 °C with butyraldehyde, propionaldehyde and acetaldehyde were measured to be 83.5%, 58.0% and 61.5%, respectively, which were far above the value (3%) obtained with n-heptane.

Keywords: selective catalytic reduction; plasma; nitrogen oxides; n-heptane; aldehydes

---

1. Introduction

Nitrogen oxides (NO_x), together with volatile organic compounds (VOCs), are the main contributors to the generation of particulate matters or photochemical smog [1]. Although there are several conventional technologies for NO_x removal, the ones based on catalysts are regarded as the most reliable and effective [2–5]. Generally, most of the catalysts used for the removal of NO_x from exhaust gases have excellent activity at high temperatures of 250–450 °C [6–8], but the catalytic activity sharply decreases at lower temperatures. Up to now, many researchers have made great efforts to improve catalytic NO_x reduction performances at low temperatures [9–12]. Particularly, when the temperature of exhaust gas fluctuates, it is very important to improve low-temperature catalytic activity to maintain a stable NO_x reduction efficiency.

In order to improve the NO_x reduction performance at low temperatures, a method of enhancing the reactivity in the catalyst by changing the NO/NO_2 ratio of the exhaust gas [11,12], or a method of improving the low-temperature catalytic activity by combining non-thermal plasma with catalysis [9], have been studied by many researchers. It has been reported that adjusting the NO/NO_2 ratio to around 1/1 in ammonia selective catalytic reduction (NH_3-SCR) results in the best catalytic activity, which is known as a fast selective catalytic reduction (SCR) reaction ($NO + NO_2 + 2NH_3 \rightarrow 2N_2 + 3H_2O$) [13,14]. It has also been reported that increasing the NO_2 concentration in the hydrocarbon SCR (HC-SCR) improves the reactivity [15]. In the case of plasma-coupled catalytic processes, NO_x can be more effectively removed by the oxidation of NO to NO_2 and partial oxidation of hydrocarbon [16,17].

A method of combining a plasma technique with a catalyst includes a one-stage method of generating plasma in a catalyst bed [17–20], and a two-stage configuration in which a catalytic reactor is installed downstream of the plasma reactor [16,21,22]. Jiang et al. [23] and Guan et al. [24] were able to effectively remove NO_x using a two-stage reactor configuration in NH_3-SCR. Besides, NO_x was successfully removed over a wide range of temperatures with a high efficiency using a hydrocarbon as a reducing agent in a two-stage or one-stage reactor configuration [16,17,25]. In any case, it has been proven by many researchers that plasma could enhance the catalytic activity for NO_x reduction, but it is still unclear how plasma promotes the catalytic reaction. It is understood that when the plasma is applied to the NH_3-SCR, the oxidation of NO to NO_2 by plasma plays a key role in the enhancement of NO_x removal, as already revealed in the fast SCR reaction [12,26]. However, in the case of HC-SCR, it is not clear why the plasma increases the catalytic reduction performance. Some researchers have interpreted that the formation of NO_2 by the oxidation of NO leads to an enhancement in NO_x removal [16,17,27], and others explain the improvement of NO_x reduction performance by the production of reactive intermediates from the decomposition of a hydrocarbon-reducing agent by plasma reforming [22,28].

In this study, systematic experiments have been conducted to understand the effect of plasma on the catalytic performance in HC-SCR and what mechanisms are involved in the enhancement of catalytic NO_x reduction. The catalyst used in this work was gamma alumina-supported silver (Ag/γ-alumina). A one-stage plasma-catalytic reactor configuration has been employed so that the plasma could directly affect the catalyst. The NO_x reducing agent used in this work was n-heptane (n-C_7H_{16}). First, in order to confirm the effect of NO_2 formation on the catalytic reaction, the NO/NO_2 ratio was artificially controlled, and its effect on the NO_x removal rate was carefully examined. Another experiment was about the effect of hydrocarbon decomposition products on the catalytic reactions. The kinds of decomposition products produced when n-heptane is decomposed by plasma were analyzed. The identified decomposition products were used as reducing agents, one by one, to see whether the decomposition products actually increase the NO_x removal performance, and if so, which decomposition product increases the NO_x removal the most. Through such systematic experiments, we have tried to identify the mechanisms by which plasma increases NO_x removal performance in the HC-SCR.

2. Results and Discussion

2.1. Plasma-Coupled SCR of NO_x

The discharge powers determined at different voltages and temperatures are given in Figure 1a. As shown, the discharge power was mainly affected by the discharge voltage. Figure 1b shows the measured NO_x conversions as a function of the applied voltage at several temperatures in the range of 100–250 °C, where the measurement at 0 kV indicates NO_x removed only by the catalyst. As can be seen, the NO_x conversion increased with increasing the voltage at all temperatures. Particularly, the higher the reactor temperature, the more rapid the increase of the NO_x conversion rate as the voltage increased. These results show that the plasma affects the catalytic reaction in some way, and similar results have been obtained in many previous studies [21–27]. In Figure 1c, n-heptane consumptions at different voltages and temperatures are presented. The higher the reaction temperature and voltage were, the higher the n-heptane consumption was, which agrees well with Figure 1b. Figure 1d compares the NO_x conversions of the plasma-coupled SCR with those of the SCR alone when the applied voltage was 25 kV. As shown in the figure, the NO_x conversion rate increased sharply by 30–80%, depending on whether plasma was generated in the catalyst bed or not. This result suggests that the plasma helps the catalyst to maintain its catalytic activity for NO_x reduction over a wide range of temperatures. Figure S1 shows a typical Fourier Transform Infrared (FTIR) spectrum of the effluent, where it can be observed that the main products from n-heptane were CO_2 and CO. During the plasma-catalytic reaction, some part of NO_x was converted into nitrous oxide (N_2O). The inset presents the concentration of nitrous oxide as a function of discharge voltage. The concentration of nitrous oxide increased from 16 to 29 ppm with increasing the voltage from 12 to 21 kV, and then decreased to 21 ppm at 25 kV.

In the case of NH_3-SCR, it is interpreted that the improvement of the catalytic activity by the plasma is attributed to the formation of NO_2 by oxidation of NO [23,24]. Although the fact that plasma improves the performance of HC-SCR has been experimentally proven in many studies [16,17,22,27,29], the role of the plasma has not yet been clearly elucidated.

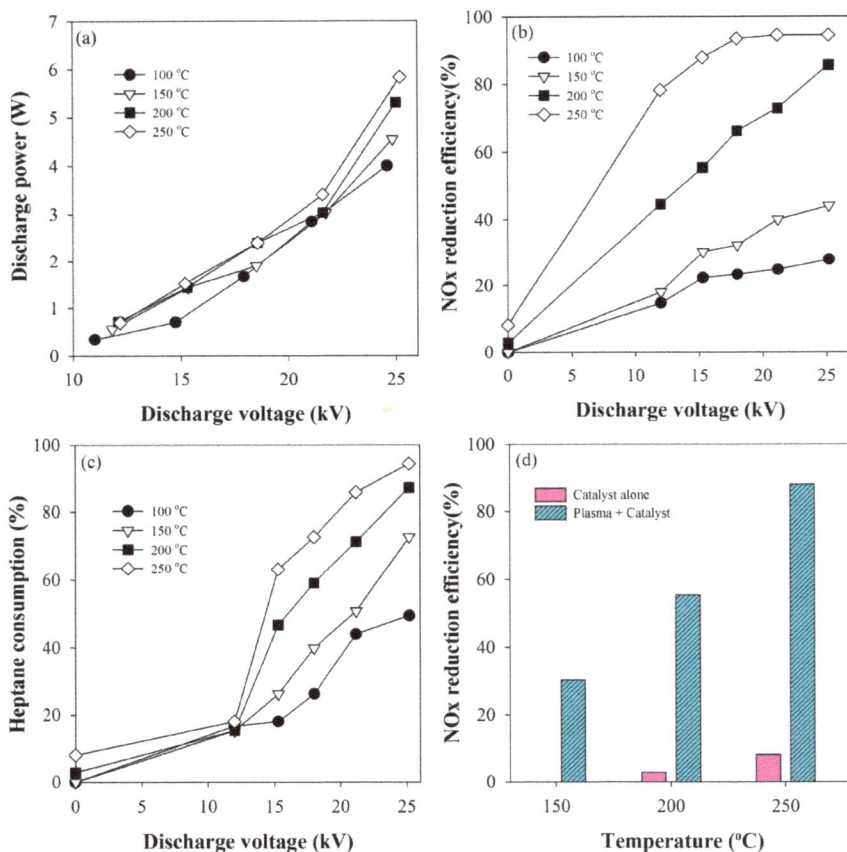

Figure 1. (**a**) Discharge powers at different voltages and temperatures; (**b**) NO_x conversions as a function of the applied voltage; (**c**) n-heptane consumption efficiencies at different voltages and temperatures; and (**d**) comparison of the NO_x conversions of the plasma-coupled HC-SCR with those of the HC-SCR alone.

2.2. Effect of NO_2 Fraction on the Catalytic Reduction of NO_x

It was a matter of concern whether the formation of NO_2 in the HC-SCR enhances NO_x reduction, as in the NH_3-SCR. First, in order to clarify the effect of plasma generation on the oxidation of NO to NO_2, several sets of experiments were carried out by filling bare γ-alumina in the quartz tube (i.e., without loading Ag). The feed consisted of NO 285 ppm and NO_2 15 ppm. For these NO oxidation experiments, n-heptane, a reducing agent, was not injected. Figure S2 in the Supplementary Materials shows the outlet NO and NO_2 concentrations as a function of discharge voltage. As expected, as the voltage increased, NO decreased and NO_2 increased, but the total NO_x (NO + NO_2) remained almost constant. The increase in the NO oxidation rate when the voltage increased is, of course, because more oxidative active species were produced. The main reaction scheme for NO oxidation can be found in the literature [25]. Anyhow, ozone and O radicals generated by the plasma can easily oxidize NO to NO_2 as shown in the figure. Ozone is one of key oxidants produced by the plasma that can easily oxidize NO to NO_2. Thus, tens to hundreds ppm of ozone generated by the plasma effectively increases the concentration of NO_2. Ozone reacts with NO in a one to one stoichiometry, and the rate of reaction for the oxidation of NO to NO_2 is very fast (almost completed in a few tens of milliseconds) [30]. Thus, unless the discharge power is unreasonably high, ozone slip is negligible. The concentration of ozone is proportional to the applied voltage (discharge power). At unreasonably high discharge power, NO_2 can be further oxidized to NO_3 and N_2O_5 [31,32]. Thus, the discharge power should be limited to avoid ozone slip and further oxidation of NO_2 to NO_3 and N_2O_5.

The following experiment was conducted to investigate the effect of NO_2 fraction in NO_x on the NO_x conversion. To investigate whether the NO_x conversion rate really increases as the NO_2 fraction increases, as reported in the literature [16,17,27], catalytic NO_x reduction was investigated by gradually increasing the NO_2 fraction. Figure 2a shows the NO_x conversion according to the NO_2 fraction in NO_x, and Figure 2b presents the concentrations of unreacted NO and NO_2. The NO_2 fraction in the feed gas was varied in the range of 5–85% with keeping the NO_x concentration at 300 ppm. The concentration of the reducing agent, n-heptane, was 257 ppm, which is equivalent to six times the NO_x concentration on a C_1 basis. As shown in Figure 2a, as the NO_2 fraction increased, the NO_x conversion tended to unexpectedly gradually decrease, implying that the increase in NO_x conversion by the plasma in the HC-SCR is not due to the oxidation of NO to NO_2. In the case of NH_3-SCR, the oxidation of NO to NO_2 by the active components produced by plasma has been reported to have the greatest effect on the conversion rate improvement [23,24]. However, in HC-SCR, the cause of NO_x conversion increase by plasma must be found elsewhere.

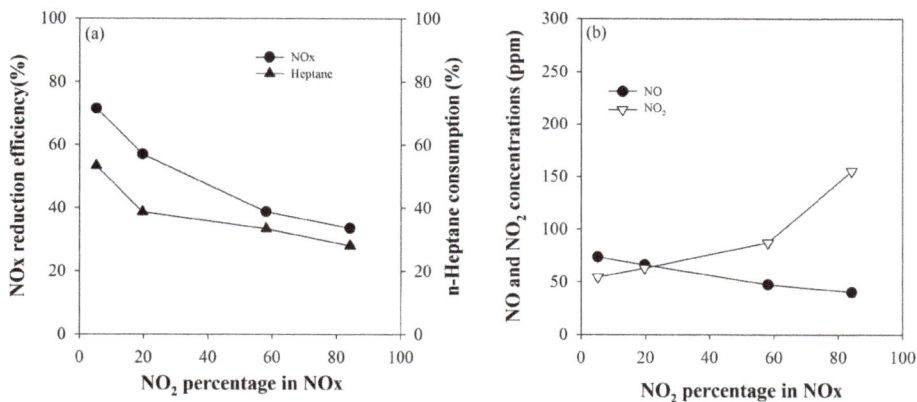

Figure 2. (a) Variations of the NO_x conversions according to the NO_2 fraction in NO_x and (b) concentrations of unreacted NO and NO_2.

2.3. Effect of n-Heptane Decomposition Products on the Catalytic Reduction of NO_x

When n-heptane is decomposed by plasma, decomposition products may show better performance as a reducing agent than n-heptane itself. According to the literature [28,33], when the diesel fuel was reformed from the outside of the reactor using plasma and injected into the reactor as the reducing agent, the catalytic NO_x conversion was substantially improved. This increase of NO_x conversion was interpreted by the production of oxygenated hydrocarbons, such as aldehydes and alcohols, which are reformed products of diesel fuel [28]. In a similar vein, n-heptane, a reducing agent used in this study, can decompose in the plasma-SCR reactor into oxygenated hydrocarbons and improve NO_x conversion. First, gas chromatography and FTIR spectroscopy were used to identify what kinds of decomposition products were formed by plasma. Figure 3a shows the gas chromatography (GC) chromatogram obtained by injecting only n-heptane (257 ppm) into the reactor without injecting NO_x to see what decomposition products are formed by the plasma from n-heptane. According to the GC chromatogram obtained at 25 kV, major decomposition products of n-heptane were found to be oxygenated hydrocarbons, like acetaldehyde, propionaldehyde and butyraldehyde. Even though many unidentified low molecular weight hydrocarbons other than aldehydes were also produced as shown, the most abundant n-heptane decomposition products were such aldehydes. Figure S3 in the Supplementary Materials shows the concentrations of acetaldehyde, propionaldehyde and butyraldehyde as a function of discharge voltage. As is well-known, plasma-induced reactive species such as energetic electrons, radicals (O, OH, N) and ozone are involved in the formation of oxygenated hydrocarbons. It has been shown above that the increased NO_x conversion by plasma is not due to the formation of NO_2, which suggests that the decomposition products of n-heptane promote NO_x reduction reactions. Figure 3b shows the GC chromatogram obtained by injecting both n-heptane (257 ppm) and NO_x (300 ppm) into the reactor when the temperature was 250 °C. As seen, the aldehyde concentrations largely decreased, indicating that they were consumed by quickly reacting with NO_x on the catalyst surface.

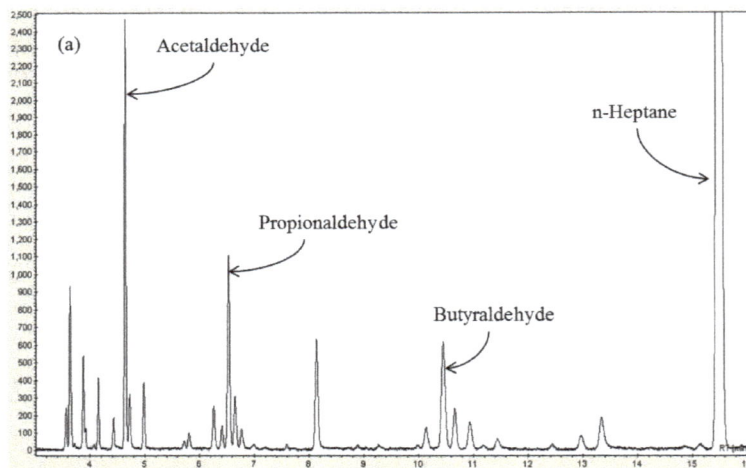

Figure 3. *Cont.*

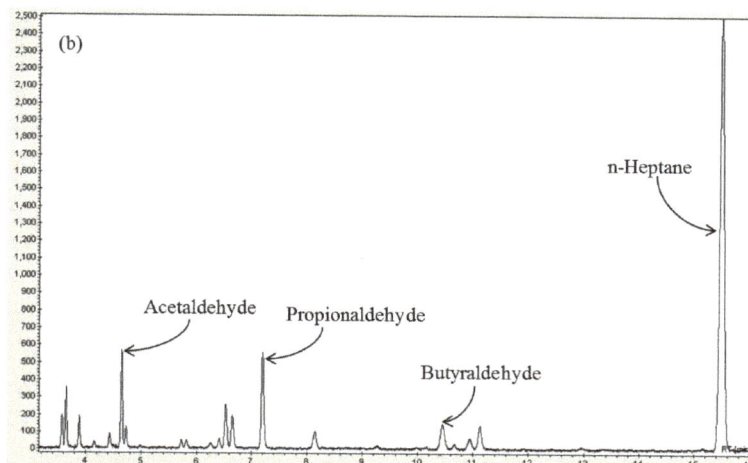

Figure 3. (**a**) GC chromatogram of the n-heptane decomposition products and (**b**) GC chromatogram of the reaction products of n-heptane and NO$_x$ (n-heptane: 257 ppm; temperature: 250 °C; applied voltage: 25 kV).

In order to investigate whether aldehyde is effective at NO$_x$ conversion, several sets of catalyst-alone NO$_x$ removal experiments were carried out by injecting acetaldehyde, propionaldehyde and butyraldehyde, one by one, as a reducing agent. The concentration of each reducing agent was six times the NO$_x$ concentration on a C$_1$ basis. The experimental results are shown in Figure 4. As seen in the figure, acetaldehyde, propionaldehyde and butyraldehyde exhibited better NO$_x$ removal performance than the parent reducing agent n-heptane; the higher the molecular weight among the aldehydes, the higher the NO$_x$ conversion. This result clearly shows that the main role of plasma in the plasma-coupled HC-SCR process is to decompose hydrocarbons to produce degradation products with better reducing power than the parent hydrocarbon itself. As shown in Figure 3a, n-heptane decomposition produces such aldehydes. It is thought that the HC-SCR reaction proceeds with the decomposition reactions of hydrocarbon [15,34]. In order for n-heptane to serve as a reducing agent for NO$_x$ removal, it must first be decomposed through a series of complicated reaction steps. Since aldehydes are already in such a decomposed state, it is natural that NO$_x$ reduction performance should increase when an aldehyde is used as a reducing agent. It has been reported that the reactions taking place in HC-SCR involve the production of ammonia (NH$_3$) via the formation of oxygenated hydrocarbons (C$_x$H$_y$O$_z$) [34–36]. For example, alkanes (RH) produce aldehydes like CH$_3$CHO through sequential oxidation reactions via alkoxy (RO) or alkyl (R) that are unstable intermediates, and aldehydes are eventually converted into NH$_3$ via isocyanate intermediate (NCO) on the catalyst surface [34]. For example, some of the pathways leading to NH$_3$ for the alkanes (C$_3$–C$_{10}$) are:

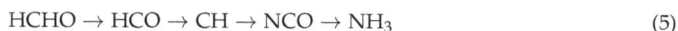

$$RH \rightarrow RO \rightarrow C_2H_5O \rightarrow CH_3CHO \tag{1}$$

$$RH \rightarrow R \rightarrow C_2H_5O \rightarrow CH_3CHO \tag{2}$$

$$CH_3CHO \rightarrow CH_3CO \overset{NO}{\rightarrow} CH_3NO_2 \rightarrow CH_2NO_2 \rightarrow NCO \rightarrow NH_3 \tag{3}$$

$$C_2H_5O \rightarrow C_2H_4 \rightarrow CHCH_3 \rightarrow CH_3O \rightarrow HCHO \tag{4}$$

$$HCHO \rightarrow HCO \rightarrow CH \rightarrow NCO \rightarrow NH_3 \tag{5}$$

Once NH_3 is formed, subsequent reactions are identical to those in NH_3-SCR:

$$4NO + 4NH_3 + O_2 \rightarrow 4N_2 + 6H_2O \tag{6}$$

$$6NO_2 + 8NH_3 \rightarrow 7N_2 + 6H_2O \tag{7}$$

When plasma is combined, aldehydes are formed by the decomposition of hydrocarbon, as discussed earlier. That is, since the step of converting the hydrocarbon to aldehydes is omitted, the above process can be performed more quickly in the presence of plasma. The concentration of NH_3 measured at the reactor outlet was below the detection limit. Based on this result, it is considered that ammonia immediately reacts with NO_x as soon as it is formed via the Reactions (1)–(5).

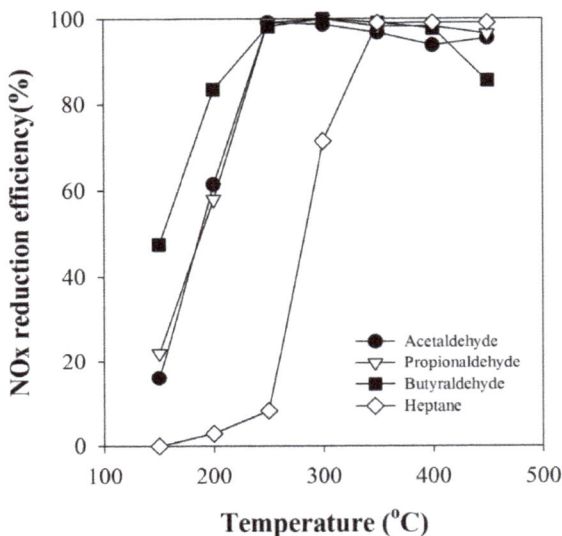

Figure 4. Catalyst-alone NO_x conversions obtained with n-heptane, acetaldehyde, propionaldehyde and butyraldehyde (C_1/NO_x: 6).

In relation to Figure 4, the concentrations of remaining aldehydes after the catalytic reactions without plasma are presented in Figure 5. As shown in the figure, the concentrations of the aldehydes decreased more rapidly than n-heptane. Among the aldehydes, the concentration of butyraldehyde decreased the fastest, followed by propionaldehyde and acetaldehyde, which shows a good agreement with the results in Figure 4.

As a further experiment, NO-rich feed gas (NO: 285 ppm; NO_2: 15 ppm) and NO_2-rich feed gas (NO: 45 ppm; NO_2: 255 ppm) were separately treated with acetaldehyde and n-heptane as a reducing agent, and the results are given in Figure 6. When n-heptane was used as a reducing agent, the NO_x conversion of the NO-rich feed gas was shown to be better than that of NO_2-rich feed gas, as in Figure 2. Similarly, it is clear in Figure 6 that even if acetaldehyde is used as a reducing agent, NO-rich feed gas is more easily reduced than NO_2-rich feed gas. With acetaldehyde, the NO_x conversions at 200 °C were 61.5% and 21% for the NO-rich and NO_2-rich feed gas, respectively, exhibiting about 40% difference in the NO_x conversion. At 250 °C, the difference in the NO_x conversion was also substantial; the NO_x conversions for the NO-rich and NO_2-rich feed gas were 99% and 68%, respectively. When the reaction temperature was further increased, the difference in the NO_x conversion between the NO-rich and NO_2-rich feed gas became negligible. In the case of NH_3-SCR, it has been reported that the best catalytic activity can be achieved by adjusting the NO/NO_2 ratio to around 1/1 [13,14]. Unlike NH_3-SCR, however, the formation of NO_2 in HC-SCR had a negative effect on the NO_x reduction, even with equal amounts of NO and NO_2 (see Figure 2). The results in Figure 6 show again that the improvement of HC-SCR performance by plasma is not due to the formation of NO_2.

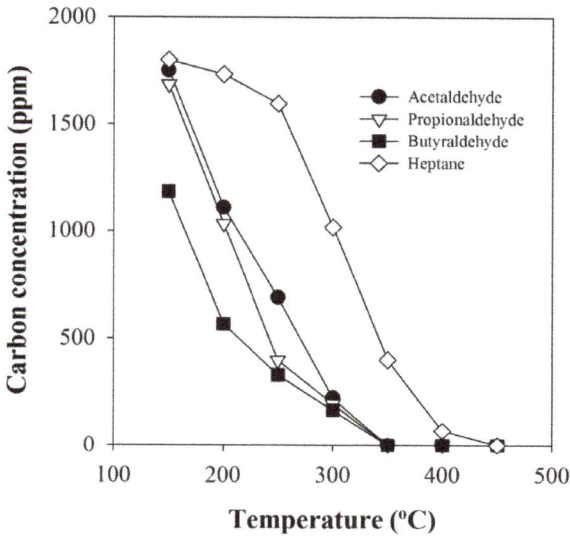

Figure 5. Concentrations of remaining aldehydes and n-heptane after the catalytic reactions without plasma (C_1/NO_x: 6).

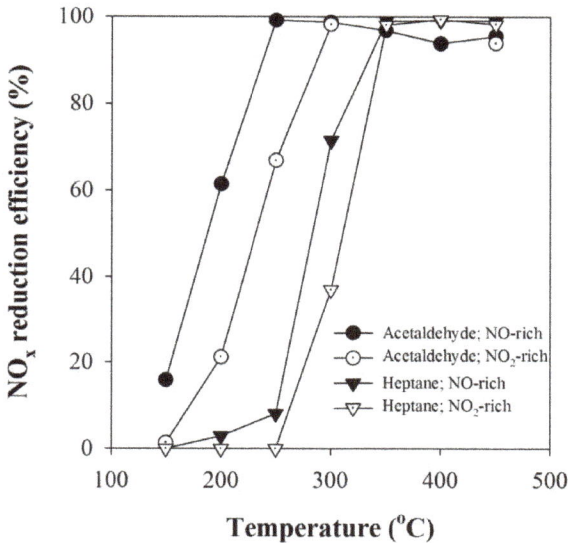

Figure 6. Comparison of the NO_x conversions between the NO-rich feed gas and NO_2-rich feed gas for the acetaldehyde and n-heptane reducing agents.

3. Experimental Section

3.1. Materials

The catalyst Ag/γ-Al$_2$O$_3$ was prepared by mixing 20-nm Ag nanoparticles (CNVISON Co., Ltd., Seoul, Korea) and γ-Al$_2$O$_3$ support (specific surface area: 175 m^2 g^{-1}; Alfa Aesar, Ward Hill, MA, USA). The mixture was well-blended so that the Ag nanoparticles were uniformly dispersed. The Ag loading in the prepared catalyst was 2 wt %. Previous studies reported that silver supported on γ-alumina exhibited high catalytic activity in the hydrocarbon SCR process [37,38]. The Ag-loaded γ-Al$_2$O$_3$ was dried overnight at 110 °C to drive off water, and then a thermal treatment was performed at 550 °C for 6 h to remove any possible impurities. After the thermal treatment, pelletizing, crushing and sieving were performed. Only the granules having a size of 1.18–3.35 mm were selected and used as the catalyst. Figure 7 shows transmission electron microscope (TEM, JEM-2100F, JEOL, Tokyo, Japan) images of the 20-nm Ag nanoparticles and the prepared catalyst. The analysis by means of a surface area and pore size analyzer (Autosorb-1-mp, Quantachrome Instruments, Boynton Beach, FL, USA) showed that the pore of the catalyst was in the form of a cylinder, a typical IV shape defined by IUPAC. Miessner et al. [29] reported that the pore size distribution is an important factor in the catalytic activity, and pore and pore size distributions below 5 nm are essential for the SCR process. The average pore size of the prepared catalyst was ~3 nm (pore volume: 0.45 cc g^{-1}), and confirmed to be suitable for the SCR process.

Figure 7. TEM images of (**a**) the 20-nm Ag nanoparticles and (**b**) the prepared Ag/γ-Al$_2$O$_3$ catalyst.

3.2. Methods

Figure 8 shows the schematic diagram of the plasma-coupled SCR reactor system. The plasma-SCR reactor is composed of a quartz tube having an inner diameter of 17.5 mm (thickness: 1.5 mm) and a discharging electrode (3.5 mm stainless steel) inserted coaxially in the center of the quartz tube. An aluminum foil (length: 5 cm) was wrapped around the quartz tube and used as the ground electrode. When an alternating current (AC) high voltage is applied between the discharging and ground electrode, plasma is generated inside the quartz tube. Plasma is created only in the area wrapped by the aluminum foil. The catalyst pellets were packed in the plasma generation region, and the filling length was 5 cm (8.8 g; 11.5 cm^3). Quartz wool was used to fix the catalyst bed at both ends. The plasma-SCR reactor was installed in a furnace (DTF-50300, Daeheung Science Co., Incheon, Korea) equipped with a proportional-integral controller (PID controller) to control the reaction temperature.

The total flow rate of the feed gas injected into the plasma-SCR reactor was 2 L min^{-1} (space velocity: 10,400 h^{-1}). The feed gas consisted of N_2 (87.3%), O_2 (9.7%), H_2O (3%) and NO (300 ppm). The flow rate of each gas was precisely controlled by a mass flow controller (MFC-500, Atovac Co., Yongin, Korea). The reducing agent, i.e., n-heptane (Sigma-Aldrich, St. Louis, MO, USA) was injected at the desired concentration by using its vapor pressure. As depicted in Figure 8, a porous gas diffuser was immersed in a bottle containing n-heptane and nitrogen gas was flowed through the diffuser to be saturated with n-heptane. The nitrogen saturated with n-heptane was mixed with N_2, O_2, H_2O and NO. The concentration of n-heptane is determined by the flow rate of nitrogen flowing through the diffuser. To keep the vapor pressure of n-heptane constant during the experiment, the bottle containing n-heptane was maintained at 10 °C using a water bath. The concentration of n-heptane was six times the NO_x concentration on a C_1 basis. In our previous work [39], it has been found that the optimum C1/NO_x ratio is 6. Bao et al. [33] also reported that the highest NO_x conversion efficiency of HC-SCR was achieved at a C1/NO_x ratio of 6–7. The method of injecting water vapor using the vapor pressure was the same as that of injecting n-heptane. The bottle containing distilled water was maintained at 25 °C.

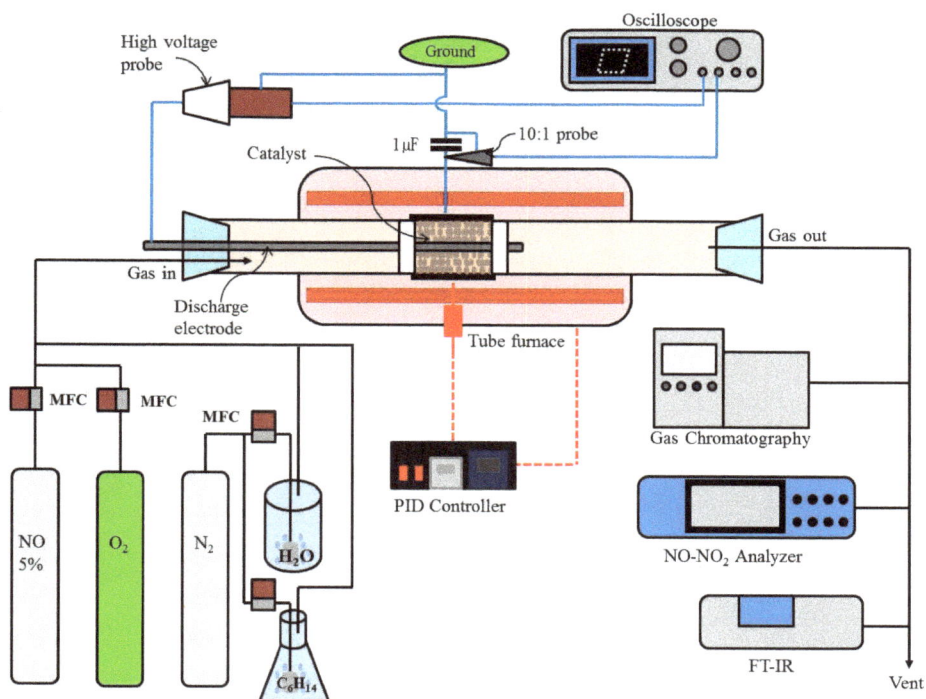

Figure 8. Schematic diagram of the plasma-coupled HC-SCR reactor system.

The electric power delivered to the plasma-SCR reactor was changed by varying the voltage applied between the discharging and ground electrode (operating frequency: 60 Hz). The power dissipated in the plasma-SCR reactor was determined by the Lissajous charge-voltage figure [19]. A capacitor was connected in series to the plasma-SCR reactor to measure the charge. The voltages across the electrodes of the plasma-SCR reactor and the capacitor were measured with a 1000:1 high-voltage probe (Probe P6015, Tekronix, Beaverton, OR, USA) and a 10:1 voltage probe (P69139, Tekronix, Beaverton, OR, USA), respectively. The measured voltages were recorded on a digital oscilloscope (TDS 3034, Tekronix, Beaverton, OR, USA). Figure 9 shows photos of the plasma discharge taken by increasing the applied voltage from 11 to 25 kV. At lower voltages, the plasma was created near the discharging electrode and inner wall of the quartz tube. At higher voltages, the plasma covered the whole catalyst bed, but the plasma created near the discharging electrode and inner wall of the quartz tube was still intense.

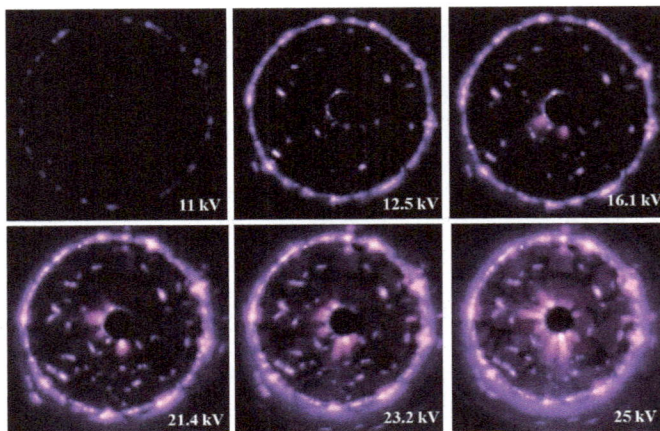

Figure 9. Photos of the plasma discharge at different voltages from 11 to 25 kV.

The energy consumed in the plasma–catalytic reactor depends on the discharge voltage. In order to estimate the energy consumption of the plasma–catalytic reactor, the specific input energy (SIE) was calculated based on the discharge power in Figure 1a as follows:

$$SIE \text{ (J/L)} = \frac{P \text{ (W)}}{Q \text{ (L/s)}} \tag{8}$$

where P is discharge power, and Q is gas flow rate. For information, the gas flow rate of 2 L min^{-1} is equivalent to 0.0333 L s^{-1}. The relation between the value of SIE and the discharge voltage is shown in Figure 10.

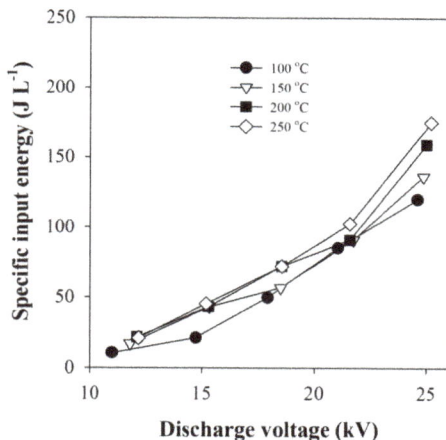

Figure 10. Effect of discharge voltage on the specific input energy.

The concentrations of NO and NO_2 were analyzed using a NO-NO_2 analyzer (rbr-Computertechnik GmbH, DE/dcom-KD, Iserlohn, Germany). A Fourier transform infrared spectrophotometer (FTIR-7600, Lambda Scientific, Edwardstown, Australia) was used to analyze the concentrations of n-heptane, CO and CO_2. The resolution of FTIR was set to 1 cm^{-1}, and the number of measurement iterations was set to 10. The path length of the infrared gas cell was 16 cm and the window material was CaF_2. The decomposition products of n-heptane were identified and quantified by a gas chromatograph (Bruker 450-GC, Fitchburg, WI, USA) equipped with a flame ionization detector (FID) and a 60-m long capillary column (BR-624ms, Bruker, Fitchburg, WI, USA). The concentration of ammonia was measured using a chemical detector tube (Product No. 3L; Measuring range 0.5–78 ppm; Gastec Co., Tokyo, Japan).

4. Conclusions

In the present catalytic NO_x reduction system with Ag/γ-Al_2O_3, the NO_x conversion was improved by 30–80%, depending on whether plasma was generated in the catalyst bed or not. This work focused on identifying the role of plasma and understanding the mechanisms involved in the plasma-enhanced HC-SCR. In the plasma-coupled catalytic system, plasma readily oxidized NO to NO_2 and decomposed the hydrocarbon reducing agent. Unlike NH_3-SCR, the formation of NO_2 in HC-SCR was found to negatively affect the conversion of NO_x, and the main cause of plasma-enhanced NO_x conversion was the formation of oxygenated hydrocarbons such as acetaldehyde, propionaldehyde and butyraldehyde. Through several sets of experiments carried out by injecting each aldehyde separately as a reducing agent, it was shown that the aldehydes had much higher NO_x reduction capability than n-heptane. Particularly, the more the number of carbons in the aldehyde molecule, the higher the NO_x conversion was. Thus, it can be concluded that the main role of plasma in the HC-SCR is to produce oxygenated compounds with better reducing capability.

Supplementary Materials: The following are available online at www.mdpi.com/2073-4344/7/11/325/s1, Figure S1: Typical FTIR spectrum of the effluent obtained at a voltage of 18 kV (the inset shows the N_2O concentration as a function of discharge voltage), Figure S2: NO and NO_2 concentrations at the reactor outlet as a function of discharge voltage, Figure S3: Concentrations of n-heptane decomposition products such as acetaldehyde, propionaldehyde and butyraldehyde as a function of discharge voltage.

Acknowledgments: This work was supported by the Basic Science Research Program through the National Research Foundation funded by the Ministry of Science, ICT and Future Planning, Korea (Grant No. 2016R1A2A2A05920703).

Author Contributions: Byeong Ju Lee, Ho-Chul Kang and Jin Oh Jo carried out the experimental work and analyzed the data; Young Sun Mok supervised all the study.

Conflicts of Interest: The authors declare no conflict of interest.

References

1. Jacobson, M.Z. *Air Pollution and Global Warming: History, Science, and Solutions*, 2nd ed.; Stanford University: Stanford, CA, USA, 2012; ISBN 9781107691155.
2. Schill, L.; Putluru, S.S.R.; Fehrmann, R.; Jensen, A.D. Low-temperature NH_3–SCR of NO on mesoporous $Mn_{0.6}Fe_{0.4}/TiO_2$ prepared by a hydrothermal method. *Catal. Lett.* **2014**, *144*, 395–402. [CrossRef]
3. Cheng, X.; Bi, X.T. A review of recent advances in selective catalytic NO_x reduction reactor technologies. *Particuology* **2014**, *16*, 1–18. [CrossRef]
4. Brandenberger, S.; Krocher, O.; Casapu, M.; Tissler, A.; Althoff, R. Hydrothermal deactivation of Fe-ZSM-5 catalysts for the selective catalytic reduction of NO with NH_3. *Appl. Catal. B Environ.* **2011**, *101*, 649–659. [CrossRef]
5. Metkar, P.S.; Harold, M.P.; Balakotaiah, V. Experimental and kinetic modeling study of NH_3-SCR of NO_x on Fe-ZSM-5, Cu-chabazite and combined Fe- and Cu-zeolite monolithic catalysts. *Chem. Eng. Sci.* **2013**, *87*, 51–66. [CrossRef]
6. Eranen, K.; Lindfors, L.E.; Klingstedt, F.; Murzin, D.Y. Continuous reduction of NO with octane over a silver/alumina catalyst in oxygen-rich exhaust gases: Combined heterogeneous and surface-mediated homogeneous reactions. *J. Catal.* **2003**, *219*, 25–40. [CrossRef]
7. Kim, Y.J.; Kwon, H.J.; Heo, I.; Nam, I.-S.; Cho, B.K.; Choung, J.W.; Cha, M.-S.; Yeo, G.K. Mn-Fe/ZSM5 as a low-temperature SCR catalyst to remove NO_x from diesel engine exhaust. *Appl. Catal. B Environ.* **2012**, *126*, 9–21. [CrossRef]
8. Yang, T.T.; Bi, H.T.; Cheng, X.X. Effects of O_2, CO_2 and H_2O on NO_x adsorption and selective catalytic reduction over Fe/ZSM-5. *Appl. Catal. B Environ.* **2011**, *102*, 163–171. [CrossRef]
9. Zhang, L.; Sha, X.-L.; Zhang, L.; He, H.; Ma, Z.; Wang, L.; Wang, Y.; She, L. Synergistic catalytic removal NO_x and the mechanism of plasma and hydrocarbon gas. *AIP Adv.* **2016**, *6*. [CrossRef]
10. Lee, T.Y.; Bai, H. Low temperature selective catalytic reduction of NO_x with NH_3 over Mn-based catalyst: A review. *AIMS Environ. Sci.* **2016**, *3*, 261–289. [CrossRef]
11. Ciardelli, C.; Nova, I.; Tronconi, E.; Chatterjee, D.; Bandl-Konrad, B. A "nitrate route" for the low temperature "fast SCR" reaction over a V_2O_5–WO_3/TiO_2 commercial catalyst. *Chem. Commun.* **2004**, *23*, 2718–2719. [CrossRef] [PubMed]
12. Iwasaki, M.; Shinjoh, H. A comparative study of "standard", "fast" and "NO_2" SCR reactions over Fe/zeolite catalyst. *Appl. Catal. A Gen.* **2010**, *390*, 71–77. [CrossRef]
13. Koebel, M.; Madia, G.; Elsener, M. Selective catalytic reduction of NO and NO_2 at low temperatures. *Catal. Today* **2002**, *73*, 239–247. [CrossRef]
14. Grossale, A.; Nova, I.; Tronconi, E.; Chatterjee, D.; Weibel, M. The chemistry of the NO/NO_2–NH_3 "fast" SCR reaction over Fe-ZSM5 investigated by transient reaction analysis. *J. Catal.* **2008**, *256*, 312–322. [CrossRef]
15. Piumetti, M.; Bensaid, S.; Fino, D.; Russo, N. Catalysis in diesel engine NO_x after treatment: A review. *Catal. Struct. React.* **2015**, *1*, 155–173. [CrossRef]
16. Tonkyn, R.G.; Barlowa, S.E.; Hoard, J.W. Reduction of NO_x in synthetic diesel exhaust via two-step plasma-catalysis treatment. *Appl. Catal. B Environ.* **2003**, *40*, 207–217. [CrossRef]
17. Pan, H.; Guo, Y.; Jian, Y.; He, C. Synergistic effect of non-thermal plasma on NO_x Reduction by CH_4 over an In/H-BEA catalyst at low temperatures. *Energy Fuels* **2015**, *29*, 5282–5289. [CrossRef]
18. Chen, H.L.; Lee, H.M.; Chen, S.H.; Chang, M.B.; Yu, S.J.; Li, S.N. Removal of volatile organic compounds by single-stage and two-stage plasma catalysis systems: A review of the performance enhancement mechanisms, current status, and suitable applications. *Environ. Sci. Technol.* **2009**, *43*, 2216–2227. [CrossRef] [PubMed]
19. Trinh, Q.H.; Mok, Y.S. Environmental plasma-catalysis for the energy-efficient treatment of volatile organic compounds. *Korean J. Chem. Eng.* **2016**, *33*, 735–748. [CrossRef]
20. Whitehead, J.C. Plasma–catalysis: The known knowns, the known unknowns and the unknown unknowns. *J. Phys. D Appl. Phys.* **2016**, *49*, 243001. [CrossRef]
21. Stere, C.E.; Adress, W.; Burch, R.; Chansai, S.; Goguet, A.; Graham, W.G.; De Rosa, F.; Palma, V.; Hardacre, C. Ambient temperature hydrocarbon selective catalytic reduction of NO_x using atmospheric pressure non-thermal plasma activation of a Ag/Al_2O_3 catalyst. *ACS Catal.* **2014**, *4*, 666–673. [CrossRef]

22. Yu, Q.; Liu, T.; Wang, H.; Xiao, L.; Chen, M.; Jiang, X.; Zheng, X. Cold plasma-assisted selective catalytic reduction of NO over B_2O_3/γ-Al_2O_3. *Chin. J. Catal.* **2012**, *33*, 783–789. [CrossRef]
23. Jiang, N.; Shang, K.-F.; Lu, N.; Li, H.; Li, J.; Wu, Y. High-efficiency removal of NO_x from flue gas by multitooth wheel-cylinder corona discharge plasma facilitated selective catalytic reduction process. *IEEE Trans. Plasma Sci.* **2016**, *44*, 2738–2744. [CrossRef]
24. Guan, B.; Lin, H.; Cheng, Q.; Huang, Z. Removal of NO_x with selective catalytic reduction based on nonthermal plasma preoxidation. *Ind. Eng. Chem. Res.* **2011**, *50*, 5401–5413. [CrossRef]
25. Talebizadeh, P.; Babaie, M.; Brown, R.; Rahimzadeh, H.; Ristovski, Z.; Arai, M. The role of non-thermal plasma technique in NO_x treatment: A review. *Renew. Sustain. Energy Rev.* **2014**, *40*, 886–901. [CrossRef]
26. Wang, J.; He, T.; Li, C. Majorization of working parameters for non-thermal plasma reactor and impact on no oxidation of diesel engine. *Int. J. Automot. Technol.* **2017**, *18*, 229–233. [CrossRef]
27. Pan, H.; Qiang, Y. Promotion of non-thermal plasma on catalytic reduction of NO_x by C_3H_8 over Co/BEA catalyst at low temperature. *Plasma Chem. Plasma Process.* **2014**, *34*, 811–824. [CrossRef]
28. Cho, B.K.; Lee, J.-H.; Crellin, C.C.; Olson, K.L.; Hilden, D.L.; Kim, M.K.; Kim, P.S.; Heo, I.; Oh, S.H.; Nam, I.-S. Selective catalytic reduction of NO_x by diesel fuel: Plasma-assisted HC/SCR system. *Catal. Today* **2012**, *191*, 20–24. [CrossRef]
29. Miessner, H.; Francke, K.; Rudolph, R. Plasma-enhanced HC-SCR of NO_x in the presence of excess oxygen. *Appl. Catal. B Environ.* **2002**, *36*, 53–62. [CrossRef]
30. Mok, Y.S.; Nam, I.-S. Reduction of nitrogen oxides by ozonization-catalysis hybrid process. *Korean J. Chem. Eng.* **2004**, *21*, 976–982. [CrossRef]
31. Jogi, I.; Stamate, E.; Irimiea, C.; Schmidt, M.; Brandenburg, R.; Hołub, M.; Bonisławski, M.; Jakubowski, T.; Kaariainen, M.-L.; Cameron, D.C. Comparison of direct and indirect plasma oxidation of NO combined with oxidation by catalyst. *Fuel* **2015**, *144*, 137–144. [CrossRef]
32. Jogi, I.; Levoll, E.; Raud, J. Plasma oxidation of NO in O_2:N_2 mixtures: The importance of back-reaction. *Chem. Eng. J.* **2016**, *301*, 149–157. [CrossRef]
33. Bao, X.Y.; Malik, M.A.; Norton, D.G.; Neculaes, V.B.; Schoenbach, K.H.; Heller, R.; Siclovan, O.P.; Corah, S.E.; Caiafa, A.; Inzinna, L.P.; et al. Shielded sliding discharge-assisted hydrocarbon selective catalytic reduction of NO_x over Ag/Al_2O_3 catalysts using diesel as a reductant. *Plasma Chem. Plasma Process.* **2014**, *34*, 825–836. [CrossRef]
34. Mhadeshwar, A.B.; Winkler, B.H.; Eiteneer, B.; Hancu, D. Microkinetic modeling for hydrocarbon (HC)-based selective catalytic reduction (SCR) of NO_x on a silver-based catalyst. *Appl. Catal. B Environ.* **2009**, *89*, 229–238. [CrossRef]
35. Gao, X.; Yu, Q.; Chen, L. Selective catalytic reduction of NO with methane. *J. Nat. Gas Chem.* **2003**, *12*, 264–270.
36. Yeom, Y.H.; Li, M.; Sachtler, W.M.H.; Weitz, E. A study of the mechanism for NO_x reduction with ethanol on γ-alumina supported silver. *J. Catal.* **2006**, *238*, 100–110. [CrossRef]
37. Furusawa, T.; Seshan, K.; Lercher, J.A.; Lefferts, L.; Aika, K. Selective reduction of NO to N_2 in the presence of oxygen over supported silver catalysts. *Appl. Catal. B Environ.* **2002**, *37*, 205–216. [CrossRef]
38. He, H.; Li, Y.; Zhang, X.; Yu, Y.; Zhang, C. Precipitable silver compound catalysts for the selective catalytic reduction of NO_x by ethanol. *Appl. Catal. A Gen.* **2010**, *375*, 258–264. [CrossRef]
39. Ihm, T.H.; Jo, J.O.; Hyun, Y.J.; Mok, Y.S. Removal of nitrogen oxides using hydrocarbon selective catalytic reduction coupled with plasma. *Appl. Chem. Eng.* **2016**, *27*, 92–100. [CrossRef]

MDPI

St. Alban-Anlage 66

4052 Basel

Switzerland

Tel. +41 61 683 77 34

Fax +41 61 302 89 18

www.mdpi.com

Catalysts Editorial Office

E-mail: catalysts@mdpi.com

www.mdpi.com/journal/catalysts